# Coastal Research Library

Volume 20

The aim of this book series is to disseminate information to the coastal research community. The Series covers all aspects of coastal research including but not limited to relevant aspects of geological sciences, biology (incl. ecology and coastal marine ecosystems), geomorphology (physical geography), climate, littoral oceanography, coastal hydraulics, environmental (resource) management, engineering, and remote sensing. Policy, coastal law, and relevant issues such as conflict resolution and risk management would also be covered by the Series. The scope of the Series is broad and with a unique cross-disciplinary nature. The Series would tend to focus on topics that are of current interest and which carry some import as opposed to traditional titles that are esoteric and non-controversial. Monographs as well as contributed volumes are welcomed.

More information about this series at http://www.springer.com/series/8795

Geoffrey N. Bailey • Jan Harff
Dimitris Sakellariou
Editors

# Under the Sea: Archaeology and Palaeolandscapes of the Continental Shelf

Springer

*Editors*
Geoffrey N. Bailey
Department of Archaeology
University of York
York, UK

Jan Harff
Institute of Marine and Coastal Sciences
University of Szczecin
Szczecin, Poland

Dimitris Sakellariou
Institute of Oceanography
Hellenic Centre for Marine Research
Anavyssos, Greece

ISSN 2211-0577 ISSN 2211-0585 (electronic)
Coastal Research Library
ISBN 978-3-319-53158-8 ISBN 978-3-319-53160-1 (eBook)
DOI 10.1007/978-3-319-53160-1

Library of Congress Control Number: 2017938016



Printed on acid-free paper

This Springer imprint is published by Springer Nature
The registered company is Springer International Publishing AG
The registered company address is: Gewerbestrasse 11, 6330 Cham, Switzerland

COST is supported by the EU
Framework Programme Horizon 2020

EUROPEAN COOPERATION
IN SCIENCE & TECHNOLOGY

COST (European Cooperation in Science and Technology) is a pan-European intergovernmental framework. Its mission is to enable break-through scientific and technological developments leading to new concepts and products and thereby contribute to strengthening Europe's research and innovation capacities. It allows researchers, engineers and scholars to jointly develop their own ideas and take new initiatives across all fields of science and technology, while promoting multi- and interdisciplinary approaches. COST aims at fostering a better integration of less research intensive countries to the knowledge hubs of the European Research Area. The COST Association, an International not-for-profit Association under Belgian Law, integrates all management, governing and administrative functions necessary for the operation of the framework. The COST Association has currently 36 Member Countries. www.cost.eu

# Preface

This volume is one of two volumes that are the outcome of a conference held at the University of Szczecin in September 2013 to mark the end of the SPLASHCOS Action. SPLASHCOS—Submerged Prehistoric Archaeology and Landscapes of the Continental Shelf—is a research network funded under the EU's COST (Cooperation in Science and Technology) programme as COST Action TD0902 (TD standing for Trans-Domain), which ran officially from November 2009 to November 2013. The conference itself served two purposes: first, to review progress in the work of the Action over the preceding 4 years and, second, to act as an open scientific meeting for anyone interested in the conference themes whether or not they were formal SPLASHCOS members. In the event, the meeting drew over 100 participants from across Europe and from further afield in Australia, China, South Africa and the USA.

The aim of this volume is to present a selection of the papers from the Szczecin meeting that cover the various themes addressed during the SPLASHCOS Action. It is aimed at all those with an interest in the sea floor of the continental shelf and the archaeological and social impact of sea-level change—especially archaeologists, marine scientists, geographers and cultural-heritage managers, but also commercial and governmental organisations, policymakers and interested members of the public.

The original aims of the SPLASHCOS Action, as set out in the COST Memorandum of Understanding, were to promote research on the archaeology, climate and palaeoenvironment of the drowned landscapes of the European continental shelf; to stimulate collaboration across national and disciplinary boundaries; to bring together interested parties from the worlds of academic science, commerce and government; to encourage participation and training of young and early-stage researchers; and to facilitate exchange of ideas, planning of research projects, application for research funds and dissemination through publications and other media. Target audiences included not only the disparate scientific disciplines concerned with researching the continental shelf—archaeology, marine geology, geophysics, biology, climatology, palaeoenvironment and oceanography—but also wider audiences including government officials responsible for the marine environment and its cultural heritage, industrial operators working on the seabed, funding agencies, school children and a wider public. In short, our aim was to promote an emerging new field, variously labelled as 'submerged landscape archaeology', 'submerged prehistoric archaeology', 'continental shelf archaeology' or 'continental shelf prehistoric research', as an integrated discipline in its own right, and to plant it more firmly on the international research agenda and in the wider public consciousness.

COST Actions are the longest-running form of European funding, intended to foster coordination of research and transfer of ideas and expertise across national boundaries within Europe and its neighbouring countries. Funds are provided for regular meetings, workshops, conferences, training programmes, research planning and coordination and publication, but not for the conduct of new research programmes (see http://www.cost.eu/about_cost). During its 4-year history, the SPLASHCOS Action grew to include 25 member states and an active membership of over 120 individuals from a wide range of institutions and disciplines. The Management Committee, comprising representatives from all member states, with Geoff Bailey and Dimitris Sakellariou as elected chair and vice-chair, respectively, held 8 major meetings in different European centres and a number of smaller workshops, organised 16 training schools or smaller short-term missions that provided experience and training to 65 early-stage researchers, stimulated a wide range of publications and successful applications to national and European funding agencies for new research and created a strongly international and interdisciplinary sense of common purpose. Details of SPLASHCOS activities and achievements can be found on the dedicated website at http://www.splashcos.org/.

In practice, activities were focussed around a core group of individuals who led the work through four formally constituted Working Groups (WGs): Archaeological Data and Interpretations (WG1) led by Anders Fischer; Environmental Data and Reconstructions (WG2) led by Jan Harff; Technology, Technical Resources and Training (WG3) led by Ole Grøn and Tine Missiaen; and Commercial Collaboration and Outreach (WG4) led by Julie Satchell. Some of this work is in preparation for publication elsewhere, notably the work of WG2 and WG1, dealing, respectively, with the Quaternary geology and palaeoenvironment of the European continental shelf and the underwater archaeological record. These volumes are intended to provide a comprehensive overview of the current state of knowledge around the European coastline and in all the major marine basins. A third volume already published is the outcome of a conference session organised at the 34th International Geological Congress in Brisbane in 2012, with papers made available online in 2014 and the final volume published in 2016 as a special publication of the Geological Society of London: *Geology and Archaeology: Submerged Landscapes of the Continental Shelf* (edited by Jan Harff, Geoff Bailey and Friedrich Lüth), which includes examples from across the world. Web-based guides to techniques and resources and to collaboration with marine industry are available online on the SPLASHCOS website, and a searchable website with maps and information about all known underwater prehistoric archaeological sites in European waters has recently been posted online at http://www.splashcos-viewer.eu.

The original conference was structured around the Working Groups, and we have adopted that structure as a basis for organising this volume, but with some modification in the light of the contributions finally delivered. All the chapters have been extensively rewritten, updated, comprehensively edited and independently reviewed. The geographical focus is primarily European, but we have not attempted to include a comprehensive range of examples from all the marine basins around the European coastline—that is the task of the other SPLASHCOS volumes. Our chapters focus on issues of method and interpretation and on wider issues of management and outreach. They also include examples from other parts of the world, and many of the discussions of method and interpretation presented here, though focussed on European case studies, have worldwide relevance.

The second volume arises from a parallel workshop incorporated in the Szczecin conference, representing the final meeting of the separately organised CoPaF Project *Coastline Changes of the Southern Baltic Sea—Past and Future Projection*, which ran from 2010 to 2014, in parallel with SPLASHCOS Working Group 2, led by Jan Harff and funded by the Polish Ministry of Science and Higher Education. The two projects thus have overlapping membership, but complement each other in their objectives and primary focus. In this volume, the emphasis is on examples from the marine basins of Western Europe and the Mediterranean, on archaeological investigations of submerged landscapes extending back into the Pleistocene or on the archaeological implications of environmental reconstruction. The CoPaF volume focusses on the geological and climatic conditions that have shaped changes in sea-level and coastline configuration during the past millennium in the southern Baltic and the likely trajectory of future changes and comprises contributions mainly from Germany, Poland, Lithuania and Estonia.

In producing these volumes, we thank, first and foremost, the COST Office, who provided administrative and financial support throughout the Action and funds for the Szczecin conference. We thank, in particular, the COST science officer, Luule Mizera; the COST administrative officer, Leo Guilfoyle;

the COST rapporteurs, Dr. Ipek Erzi (Scientific and Technological Research Council of Turkey, TUBITAK) and Prof. Daniela Koleva (Sofia University St. Kliment Ohridski); the COST external evaluators, Prof. Peter Veth (University of Western Australia) and Prof. Dr. Gerold Wefer (MARUM, University of Bremen); and the grant holder and administrative secretary of SPLASHCOS, Cynthianne DeBono Spiteri, all of whom gave invaluable advice and support.

We also thank the University of Szczecin for hosting the conference and generously providing facilities and hospitality and Helmholtz-Zentrum Geesthacht and Szczecińska Energetyka Ciepina (SEC) for additional financial support. We are also indebted to Prof. Andrzej Witkowski (University of Szczecin), who acted as chair of the Local Organising Committee and, together with the other Committee members, Prof. Marian Rębkowski, Dr. Przemyslaw Krajewski, Dr. Karolina Bloom, Marcin Wroniecki, Marta Chmiel and Michal Adamczyk, ensured the efficient organisation and smooth running of the whole enterprise.

We also express our gratitude to the many individual specialists who acted as peer reviewers and who have contributed significantly to the final outcome. Finally, we acknowledge the following institutions, who generously contributed towards the production of this volume: the European Research Council through ERC Advanced Grant 269586 DISPERSE, the German Archaeological Institute Berlin, the Hellenic Centre for Marine Research and the University of York.

York, UK
Szczecin, Poland
Anavyssos, Greece

Geoffrey N. Bailey
Jan Harff
Dimitris Sakellariou

# Contents

# Chapter 1
# Archaeology and Palaeolandscapes of the Continental Shelf: An Introduction

**Geoffrey N. Bailey, Jan Harff, and Dimitris Sakellariou**

**Abstract** The past decade has seen a rapidly widening interest in archaeological and scientific exploration of the submerged landscapes that were flooded by sea-level rise at the end of the Last Glacial period. That interest is shared by many different disciplines and constituencies, including archaeologists who recognise the potential significance of these underwater archives to improved understandings of world prehistory, palaeoclimatologists interested in modelling sea-level change, and government agencies charged by national and international legislation with managing the underwater cultural heritage in the light of expanding industrial exploitation of the seabed. This introductory chapter sets out the background to these developments and the role of the European network – SPLASHCOS – in promoting awareness of this new agenda to the many scientific disciplines involved in underwater research, government agencies, commercial and industrial interests, and a wider public. Here we set out the major themes that we have used to structure the chapters in this volume, ranging across techniques and strategies of underwater investigation, examples of underwater archaeological excavations, reconstructions of underwater landscapes, the role of the continental shelf in shaping patterns of early human dispersal and geographical expansion, and issues of training, outreach and management. Examples are drawn widely from Europe and other parts of the world. We summarise the individual chapters, identify their inter-relationships with each other and with the themes of which they form a part, and highlight their wider significance.

## 1.1 Background

It is by now a well-established fact that for most of human history on this planet over at least the past 1 million years, sea-levels have been substantially lower than the present, oscillating between short periods of extreme high sea level as today and low stands of −100 m or more at glacial maxima, with many lesser perturbations in between, and a long-term average some 40–50 m below the present level. When sea levels were low, extensive tracts of new territory, amounting to some 4 million km$^2$ around the coastlines of Europe and the Mediterranean, and at least 20 million km$^2$ at the global scale, became available for terrestrial plants and animals and, of course, for human populations. Much of that territory most likely represented some of the most fertile, well-watered and ecologically diverse land

G.N. Bailey (✉)
Department of Archaeology, University of York, the King's Manor, York YO1 7EP, UK
e-mail: geoff.bailey@york.ac.uk

J. Harff
Institute of Marine and Coastal Sciences, University of Szczecin, Mickiewicza 18, PL 70-383 Szczecin, Poland

D. Sakellariou
Institute of Oceanography, Hellenic Centre for Marine Research,
46.7 km Athinon-Souniou Av, Anavyssos, P.O. Box 712, 19013 Athens, Greece



G.N. Bailey et al. (eds.), *Under the Sea: Archaeology and Palaeolandscapes of the Continental Shelf*, Coastal Research Library 20, DOI 10.1007/978-3-319-53160-1_1

available, especially given the relatively arid climates that persisted during glacial periods. These coastal lowlands and their immediate environs presumably supported denser concentrations of human settlement than the more remote continental hinterlands where most of the currently-known surviving evidence of the Stone Age era has been recovered.

The last period of high sea level was about 130,000 years ago during the Last Interglacial period, commonly referred to Marine Isotope Stage (MIS) 5. As the continental ice caps grew again in the succeeding millennia, sea level dropped, reaching a maximum depth of −130 m at the Last Glacial Maximum 20,000 years ago. When the ice started melting, sea level rose again, progressively drowning these ancient landscapes, their coastlines and the material traces of their previous inhabitants until the present level was reached about 6000 years ago. This long period of lowered sea level spans some of the most important developments in human prehistory: the emergence and world-wide dispersal of anatomically modern humans with new technologies, skills and cognitive abilities; early experiments in sea travel including the earliest colonisation of New Guinea and Australia; exploitation of shorelines, marine resources and offshore islands; and the early development of agriculture and its dispersal around the coastlines and islands of the Mediterranean and NW Europe. It must follow from the history of sea-level change that a very significant part of the evidence for these developments lies hidden on the continental shelf.

In the past decade, the expanding industrial and commercial exploitation of the seabed, our growing knowledge of the inexorable and continuous effect of changes in sea level and climate on human affairs, and the development of national legislation and international conventions to protect the underwater cultural heritage, have all combined to focus attention on the potentially valuable archives of cultural and natural data locked up on the seabed of the continental shelf.

These underwater archives are of great interest and relevance to many different disciplines and constituencies—to archaeologists interested in the long-term record of human development and dispersal, to climatologists interested in obtaining well-dated index points for low sea-level stands to develop better predictive models of sea-level change, to geomorphologists interested in processes of coastal change and large-scale patterns of erosion and sedimentation, to geophysicists and geologists interested in the dynamics of plate motions and changes in the Earth's crust, to governmental agencies and heritage organisations charged with managing and protecting the underwater heritage, and to a wider public perennially fascinated by the idea of lost civilizations slipping beneath the waves.

Alongside this growing interest, industrial exploitation of the seabed has intensified. Beam-trawl fishing, drilling for oil and gas, sand and gravel extraction, laying of pipelines, building of wind farms, and engineering works associated with new harbours, bridges, tunnels and other infrastructure all pose a threat to the cultural and natural features of the pre-inundation landscape, to say nothing of natural changes in the underwater environment that lead to erosion or burial of ancient land surfaces. At the same time, as on land so underwater, these destructive processes can also have a potentially positive effect in exposing to discovery material that would otherwise remain deeply buried and hidden from view.

It was with all these factors in mind that the European SPLASHCOS research network came into being in November 2009 (COST Action TD0902 SPLASHCOS—Submerged Prehistoric Archaeology and Landscapes of the Continental Shelf), originating from an earlier series of meetings in 2008 and early 2009 to discuss ways of raising large-scale research funding for a pan-European research project entitled Project Deukalion (see Flemming et al. 2017). From the beginning, the SPLASHCOS network was organised around the need to address simultaneously not only the scientific evidence, both archaeological and palaeoenvironmental, but also issues of training, public outreach, management and industrial collaboration.

The chapters in this volume cover all of the above topics, and are organised around five principal themes that move progressively from method to interpretation, and from the study of individual sites to the wider landscape setting, beginning with a discussion of techniques and research strategies (Part I), progressing to examples of underwater archaeological investigation (Part II) and palaeolandscape reconstruction (Part III), moving on to issues of interpretation in relation to broader themes in prehistory (Part IV), and concluding with examples of wider public outreach and management (Part V). The organisation is thus primarily thematic, but with a subsidiary axis of organisation whereby chapters dealing with geographically neighbouring regions are put together, wherever possible and appropriate. The divisions between these major Parts or sections are not, of course, watertight, and some issues recur in different sections. In this introductory chapter, we highlight important issues that arise from the chapters in each section, cross-referencing wherever possible those topics that occur in more than one section.

## 1.2 Techniques and Strategies

The chapters in Part I address the three key questions that are usually the first to be raised by the interested outsider or the novice investigator. (1) How do we set about finding underwater traces of the submerged landscape and its archaeology, given the specialised skills and the costly technology required and the rarity of upstanding features that can be easily spotted on the seabed? (2) What are we likely to find preserved under water, given the likelihood of damage and destruction, or burial under marine sediments, during and after inundation by sea-level rise? (3) Having found a target, whether a landscape feature such as a palaeo-river channel or traces of an archaeological settlement, how do we extract the evidence for dating and other analytical investigations, and in the case of an archaeological site how do we conduct excavation to the same standards as on dry land?

Chapter 2 provides the first step in seeking answers to these questions, presenting a comprehensive review of the varied technologies available for underwater investigation. As Missiaen et al. emphasise, the challenge is considerable, with needs that range from survey and detection of submerged landscapes extending over thousands of square kilometres at one end of the scale, to discovery of tiny artefacts such as microlithic stone arrow heads at the other. Diving surveys, acoustic techniques and drilling or coring of sediments play a major role, augmented by underwater photography, airborne and satellite remote-sensing, and the development of underwater vehicles and robotics. New technologies and equipment are under continuous development, much of it driven by the needs of commercial exploitation of the seabed and the need to investigate deeply buried geological structures. Moreover, discovery of landscape features and especially archaeological remains often requires more sensitive instrumentation capable of producing high resolution images of shallow deposits and materials. Increasingly, the particular requirements of archaeological and geoarchaeological investigation are stimulating the development of new technological solutions. For example, during the course of the SPLASHCOS meetings, discussions between specialists in acoustics and archaeology led to pioneering experiments in the development of sensitive remote-sensing techniques for detecting buried flint artefacts (Hermand et al. 2011). These experiments are ongoing with underwater excavations currently in progress to test the validity of the acoustic signatures (Ole Grøn, pers. comm., 2016).

Turning from methods of detection and discovery to the question of what might survive inundation, the issue of sea-level change becomes paramount. There is now an extensive and growing body of knowledge about the rate and pattern of sea-level rise and fall during the Last Glacial (and earlier)—both absolute changes resulting from changes in the volume of sea water in the world's oceans, and relative changes resulting from vertical movements of the Earth's crust (e.g., Lambeck et al. 2002, 2014; Harff et al. 2017a, b). From the geoarchaeological point of view, it is not so much the rate of vertical movement that is important during sea-level rise as the rate of shoreline retreat in the spatial dimension. This in its turn is a complex function of the interaction between relative sea-level rise, the slope of the continental shelf and locally and regionally variable topography. Chiocci et al. address this issue in detail in Chap. 3 and develop an algorithm for measuring 'transgression velocity', the rate of shoreline retreat (or advance when sea level is dropping). This measure can be applied at different geographical scales depending on the detail and accuracy of the available information. The examples presented highlight significant variation at every geographical scale with important implications for evaluating the human impact of sea-level change and some predictive power in identifying areas most likely to have preserved material in the face of inundation.

The next challenge that has to be faced is the sheer complexity and cost of the technology needed, especially if it requires large research vessels and ship time. One solution is to enlist the cooperation of industrial companies engaged in exploitation of the seabed. Provided that archaeologists and government agencies engage with industry at an early stage in the planning process, the incorporation of underwater research can be achieved with minimal interruption to the timetable of the commercial operation. The scientific work gains enormously from free access to equipment and resources, and sometimes from direct financial contributions to research personnel and analytical work, the costs of which would otherwise far exceed even the most generous scientific grants, but which represent a relatively minor amount in relation to the total budget of the industrial operation. Commercial companies in their turn gain good publicity and can point to the wider educational and social benefits of the collaboration. Some excellent examples have taken place in NW Europe in recent years, for example with North Sea oil companies and with the Port of Rotterdam authority, producing quite spectacular and unexpected results generating wide interest and publicity (see also Glorstad et al. Chap. 19, Gaffney et al. Chap. 20, Momber and Peeters, Chap. 21, Satchell, Chap. 25, Sturt et al. Chap. 28).

Homlund et al. describe another type of commercial collaboration in Chap. 4, between an archaeological research institute and a company that provides commercial technological services in underwater investigation. The examples they present offer an illuminating illustration of the benefits of long-term collaboration, the technologies that can be deployed in underwater research, the results that can be achieved, and how such collaboration can be developed to mutual advantage at little extra cost, not only generating wider publicity for the commercial company involved, but stimulating new technological solutions in response to the demands of archaeological investigation that can have subsequent commercial application.

Not all underwater research requires big ships and big budgets. The last two chapters in this section (Chaps. 5 and 6) give good examples of what can be achieved in shallow water with divers and relatively simply equipment. Both projects were carried out as part of the SPLASHCOS training programme and provide good examples of the various constraints that limit the discovery and investigation of underwater archaeological sites, detailed guides as to how to go about conducting underwater excavation, the techniques and technologies required, and something of the detail and quality of material that can be preserved in underwater anaerobic sediments. They also demonstrate the importance of training and capacity building (see also Satchell, Chap. 25).

Denmark and Sweden are the pioneering countries for developing underwater research on submerged remains of prehistoric archaeology, as Uldum et al. emphasise in Chap. 5, with a long history of systematic investigation going back to the 1970s (Fischer 1996; Skaarup and Grøn 2004; Andersen 2013, see also Skriver et al. Chap. 8, Larsson, Chap. 11). In Denmark, professional archaeologists, with the support of the government agency responsible for the cultural heritage, currently known as the Danish Agency for Culture and Palaces, have over a long period developed protocols of investigation that include regular inspection of shorelines aided by reports from members of the public, methods of site monitoring and excavation, and predictive modelling of site discovery that takes account both of coastal locations attractive to prehistoric people and also of preservation conditions. The outcome is the discovery of many hundreds of underwater Stone Age finds, the largest concentration in the world, including well-known underwater settlements with unusual preservation of organic artefacts such as Tybring Vig and Møllegabet II.

The discovery and investigation of the Falden site described in Chap. 5 was the result of active exploration and test excavation made possible by the funding of a SPLASHCOS training exercise. The authors draw attention to the importance of public engagement and note that the majority of Denmark's underwater finds have been discovered by members of the public. Monitoring and reporting of chance finds, however, have their limitations. Lack of research funds except for rescue excavations of important finds under threat of destruction by new engineering developments, reporting only of those finds that are visible because of erosion near the modern shore, and lack of time and money for intrusive excavation or more extensive underwater survey, all mean that the many underwater sites in the Danish government records may represent a quite biased picture of the original site distribution.

Exploratory work at Falden included hand-fanning of sediments and test-pitting in an area deemed potentially productive of preserved prehistoric material. A 10-day test excavation revealed a site with undisturbed deposits rich in flint artefacts in an area that had not previously given any such indications. The training programme also made possible experimentation in methods of photogrammetry and computerised imagery, with great potential for future development and use elsewhere. Uldum et al. emphasise the importance of funding systematic investigation of sites such as Falden in correcting and advancing existing knowledge—and in protecting undisturbed sites from further loss and destruction—even if the resulting material is not as spectacular as the more famous sites. Above all, they emphasise the need to provide a sustained international training facility that can promote understanding and expertise. This is important not only for those intent on a research career but for those destined for employment in cultural-heritage organisations and government agencies responsible for informed evaluation and management of the underwater cultural heritage.

Underwater investigations have as long a history in Israel as in southern Scandinavia (Raban 1983), and have produced evidence of comparable importance. The 9000-year-old Pre-Pottery Neolithic site of Atlit-Yam, described by Galili et al. in Chap. 6 was first discovered in 1984 and has been subject to regular inspection and periodic excavation over a long period. It is arguably one of the best-known and best preserved underwater prehistoric sites in the world, with its remarkable evidence of a fishing-farming village with stone-structures and stone-lined wells. The details set out in Chap. 6 are an object lesson in the preparatory work, procedures and methods required for systematic excavation of an underwater settlement.

One of the interesting aspects of the Atlit-Yam site is the circumstances of its discovery. Originally buried under a protective layer of marine sand, the site is intermittently exposed by storms that temporarily remove the overlying cover. The time-window of opportunity for discovering the archaeological remains is clearly quite limited before they are either covered again by sand, or are destroyed and damaged by prolonged exposure to underwater currents. Discovery therefore depends to a very large degree on chance. But the chances can be improved by regular monitoring, whether by professional underwater archaeologists, sports divers, fishermen or other members of the public.

## 1.3 Underwater Archaeological Sites

Part II provides a range of examples of projects that amplify the results of the Falden and Atlit-Yam case studies, illustrating the various ways in which underwater archaeological sites are discovered and excavated, and the scientific impact of the resulting discoveries.

Examples of the ways in which material is exposed to discovery include: removal of sand cover by seasonal storms (Chap. 7, see also Chap. 6); commercial sand extraction (Chaps. 7 and 9); reporting of chance finds of artefacts or submerged tree trunks by sports divers (Chaps. 7, 9 and 10); expansion of harbour facilities, bridge building and other offshore constructions (Chaps. 7 and 11); dredging of marine channels (Chap. 11); natural erosion of overlying sediments by marine currents (Chap. 8); and scientific sampling of the sea bottom for zoological or other scientific investigations (Chap. 11). Additional examples of excavation and recovery methods include coring and auguring, the use of divers to excavate gridded trenches, and the application of induction dredges for removal and sieving of deposits.

These chapters also provide a range of examples of the types of evidence that can be recovered from submerged settlements, evidence of importance both to archaeological interpretation and to the development of more precise chronologies of sea-level change—and evidence that would rarely if ever be preserved on terrestrial sites.

In Chap. 7, Galili et al. present detailed findings from a group of eight Pottery Neolithic settlements discovered in the shallow offshore and intertidal zone of the Israeli coastline. As with the earlier Pre-Pottery Neolithic site of Atlit-Yam discussed in Chap. 6, the results present an extraordinarily diverse and well-preserved assemblage of artefacts, food remains and structures, including wooden implements and containers, basketry, stone- and timber-lined wells, house foundations, the earliest currently known evidence for olive-oil production, and a large group of human burials. The excellent conditions of preservation and the spatial mapping of site features over a considerable area also highlight one of the factors that may have led to the creation of separately demarcated burial grounds or cemeteries. In the relative confines of a narrow coastal strip, excavation in order to construct wells, house foundations and other structures, perhaps added to by the need for regular removal and renewal by a shifting shoreline, would have regularly disturbed the remains of earlier burials interred in domestic surroundings, leading to competition for sub-surface space and hence eventually to the need for separately defined sacred and domestic areas.

In Chap. 8, Skriver et al. present the results of recent investigations at Hjarnø Sund, another good example of a site first identified because of artefacts observed eroding out of intact sediment near the present shoreline. A small-scale excavation revealed a rich collection of flint, bone, antler and wooden artefacts, typical of other underwater Danish sites, and two other features more rarely recovered. The first is two wooden paddles with paintwork still preserved. Similar paddles have been recovered from two other Danish underwater sites, and the addition of the Hjarnø examples highlights regional differences in design. The authors suggest that these differences might indicate markers of different social or individual identities, especially since paddles would be highly visible artefacts and ones widely travelled during canoe journeys. The other feature is the presence of shell-midden deposits. These are common on shorelines above present sea level, often accumulating numerous thick mounds, most famously in the late Mesolithic Ertebølle examples of Denmark itself. However, underwater examples are rare, with only a few known—from the Danish site of Møllegabet II (Grøn 2004, pp. 41–45) and isolated examples in other parts of the world Faught 2014, pp. 43–44, Hayashida et al. 2014, pp. 283–284)—perhaps because unconsolidated shell deposits are especially vulnerable to erosion and destruction by wave action, or simply through lack of underwater survey. The Hjarnø Sund evidence demonstrates that shell deposits can survive if covered with a layer of protective sediment, but equally that they are vulnerable to rapid erosion by water currents once exposed.

Similar underwater conditions to Denmark are found in the German territorial waters of the western Baltic and in recent years these have begun to produce a comparable range and richness of underwater material (Harff and Lüth 2011). New discoveries from the Strande site LA 163 near Stohl Cliff in the Kiel Bay are presented by Goldhammer and Hartz in Chap. 9 and add to that growing body of information. The site has yielded a typically varied collection of artefacts made of stone, bone and wood, remains of a log boat, plant remains, shells of edible marine molluscs, animal bones including freshwater and marine fish, sea birds and land mammals, and fragments of human bone. Rapid survey over a larger area including the removal of the sand cover shows that more extensive terrestrial sediments with archaeological potential lie beneath the marine sand. Here, as at Hjarnø Sund, preliminary investigations have revealed the potential for the survival of more extensive and substantial archaeological deposits and the need to fund more detailed excavations if valuable material is not to be lost.

In addition, radiocarbon dating and high precision dendrochonological dating of the abundant wooden remains at Stohl Cliff have given a precise measure and date of sea level position. Similar evidence is present at the nearby location of Friedrichsort in the Kiel Fjord, and is discussed by Feulner in Chap. 10. This site is of interest in demonstrating the value of regular monitoring by an archaeological diver, and the preservation of intact sediments in a location subject to disturbance by regular ship movements. As at Stohl Cliff, the first indication of a submerged land surface was the discovery of remains of a submerged forest. Radiocarbon dating of a sample from a tree stump adds to the growing evidence of sea-level change in the area, and points to the possibility of discovering intact and well-preserved archaeological remains, given the frequent association of submerged forests and artefacts (e.g., Homlund et al., Chap. 4, Hansson et al. Chap. 13).

Finally, in Chap. 11, Larsson, who was involved in some of the earliest underwater work in southern Scandinavia in the 1970s and 1980s, brings together the results from a number of underwater discoveries in the Öresund Strait that separates Denmark and southern Sweden. None of these are individually spectacular or especially well-preserved, but in their collective totality they demonstrate how the cumulative growth of information from different underwater locations, combined with regional reconstructions of changing shorelines and palaeogeography, can lead to new interpretations at a regional scale of spatial and temporal patterns in the distribution of settlements, anticipating a theme that is developed in greater detail in Parts III and IV. Coastal settlements are notoriously absent from the earlier Holocene record, as in so many other regions, not least because the shorelines are mostly submerged and under-investigated, but perhaps also because earlier peoples showed less interest in coastal and marine resources. Hints of an earlier coastal interest are present from the terrestrial record in Sweden—coastal sites on uplifted shorelines further north, and hints of early contacts with the coast in the form of occasional marine shells and other marine indicators in inland sites. However, it is the underwater record that has provided fuller archaeological evidence of early coastal settlements, even if the evidence is not yet sufficient to show whether marine and coastal resources were exploited with the same intensity as demonstrated in the later coastal sites on shorelines above present sea level. In any case, interpreting the changing intensity of exploitation of coastal and hinterland resources, and their relative palaeoeconomic importance, is greatly complicated by the fact that the available environments and resources on the coast and in the hinterland in southern Scandinavia were continuously changing because of postglacial changes in climate, environment and forest cover (see also Hansson et al. Chap. 13, Glorstad et al. Chap. 19). Large samples of environmental and archaeological data from many points in time and space will be necessary to unravel these changing relationships, and further underwater research has a vital role to play.

One final point of interest in this Chapter is that one of the best known Mesolithic coastal settlements in southern Scandinavia on the present shoreline, including evidence of numerous human burials, the Swedish site of Skateholm II, was flooded by sea level rise and then exposed again because of land uplift, giving some insight into how well different sorts of materials survive inundation. In this case, interestingly, the human graves, perhaps because they were buried and covered over when first created, seem to have survived better than the remains from domestic activities.

Two final comments arise from a consideration of these chapters as a whole. The first is that they are dominated by evidence from just two major geographical centres: the coastlines of the western Baltic, especially Denmark; and the Mediterranean coastline of Israel. It could be argued that this results from uniquely favourable conditions for underwater archaeological preservation and discovery in these two regions. However, that is far from demonstrated. Rather, what the present concentration demonstrates is the importance of long-term engagement. Both regions have a long history, extending over many decades, of interest in and development of protocols for regular monitoring and recording of finds as they become exposed, in the refinement of methods of underwater survey and excavation, and in the continuity of expertise and interest sustained by the dedication of key individuals.

Conditions of survival and discovery are unlikely to be uniformly similar around all the coastlines of Europe—or indeed further afield. Nevertheless, increasing numbers of underwater finds are beginning to appear elsewhere (some are referred to in later chapters of this volume). There is no reason to doubt that similar settlements as substantial as the best-known from Israel and Denmark will eventually be forthcoming from other regions, and every reason for optimism given the number of new finds that are now appearing elsewhere—off almost every coastline in Europe (see the online data base at http://splashcos-viewer.eu/). Long-term scientific engagement and funding support, encouragement of public interest, regular offshore monitoring to check for the appearance of new finds, and training programmes for the next generation stand out as key ingredients for future success.

The second comment is that all the examples so far considered are from the early Holocene, and that is for the simple and obvious reason that the submerged shorelines of this period belong to the final stage of postglacial sea-level rise and are therefore relatively shallow and easily accessible to inspection on or close to the modern shoreline and by diving in shallow water. What about earlier and more deeply submerged shorelines? What if sites like Atlit-Yam and Tybrind Vig existed 25,000 or even 60,000 years ago, and now lie hidden on these earlier shorelines? Galili et al. (Chap. 6) address this question and suggest that the likelihood of encountering preserved prehistoric settlements decreases with depth: because of the technical challenges of working beyond the limit of SCUBA; because shorelines would have been exposed for shorter periods before inundation by sea-level rise; and because the remains left by hunter-gatherer activity might have been more ephemeral. The technical challenges of working at much greater depths are formidable and should not be played down. But there is no reason to suppose that periods of low sea-level at much earlier periods would not have stabilised long enough to create suitable conditions for the development of substantial and permanently occupied coastal settlements, assuming that environments and resources were available to support such a possibility, and that people were sufficiently numerous and sufficiently motivated to take advantage of it.

The belief that settlements with an elaborate material culture, permanent and substantial structures, and a sedentary way of life could not have existed 20,000 years ago (or earlier) is based on a circular argument, which assumes that the final rungs of ascent in a ladder-of-progress interpretation of human history—intensification of resources, especially marine resources, permanent village settlements, social stratification, population growth and ultimately agriculture—could not have happened before about 10,000 years ago at earliest. Why not, we might ask? Because there is no earlier evidence is the standard answer. And why is that? Because—as any informed reader must now be aware—the coastlines where such evidence would be mostly located are now deeply submerged and have scarcely begun to be investigated. We know from isolated artefacts and faunal remains recovered from the seabed that prehistoric material is preserved at great depth in the 60–>100 m depth range (e.g., Long et al. 1986; Fedje and Josenhans 2000; Stanford et al. 2014) . The discovery of settlements like Atlit-Yam or Tybrind Vig on much earlier and now deeply submerged shorelines would radically overturn current preconceptions about world prehistory, and for that reason alone should become a target for future investigation—even if we are still far from having the skills and resources to reach that target (but see chapters in Part IV for projects with that target in view).

## 1.4 Underwater Landscapes

The chapters in Part III move the focus of attention away from individual sites to the reconstruction of submerged palaeolandscapes at a larger geographical scale. The emphasis is mostly on geophysical and palaeoenvironmental techniques of reconstruction rather than underwater archaeological finds. Nevertheless, all the studies are motivated by archaeological questions and many involve multiple collaborations between archaeologists and geoscientists, underscoring the need to integrate many different types of expertise. They also illustrate a number of different archaeological uses of palaeogeographical and palaeoenvironmental reconstruction: as a means of identifying the distribution of significant concentrations of food resources and water supplies on the pre-inundation landscape (all Chapters); as a first step in assessing where underwater prehistoric sites and settlements are likely to be located and archaeological material is likely to be preserved (Chaps. 12, 13, 15, and 16); as a guide to the interpretation of underwater finds where they are already known (Chaps. 13 and 15); and as a means of refining the explanation and interpretation of archaeological sites and archaeological sequences on the present-day coastline (Chaps. 12 and 16).

Approaches presented in these chapters divide broadly into two categories: those that use bathymetry and other geophysical techniques to locate or infer the position of submerged shorelines and other physical features of the submerged landscape such as palaeo-river channels and physical topography (Chaps. 12 and 16); and those that use intrusive methods such as coring of sediments or sampling of speleothems, sometimes in combination with geophysical techniques, to recover dateable sequences of palaeobotanical and geochemical indicators of past environmental and climatic conditions (Chaps. 13, 14, 15, and 17). In an ideal world, the full range of geophysical and palaeoenvironmental techniques would be combined. However, the extent to which that is possible or necessary will depend on the nature of the offshore topography, the balance between sedimentation and erosion on the submerged shelf, the budget available to the research team, the availability of relevant equipment and expertise, and the nature of the archaeological problems to be addressed.

One important point about underwater mapping emphasised by Bicket et al. in Chap. 12, and one that recalls the approach developed by Chiocci et al. in Chap. 3, is that datasets of bathymetry or other underwater features are often already in existence and publically available and can be exploited to good effect (a point also made by Goldhammer and Karle in Chap. 15). In this Chapter, the authors show how far one can go in using existing data to map the changing configuration of coastlines and sea channels of relevance to routes of travel and communication by sea, to identify broad-scale changes in shoreline geomorphology and resource potential such as the varying width of the intertidal zone or the balance between rocky and sandy shores, to make sense of the location of existing settlements on the present shoreline, and to assess underwater areas promising for future exploration. Such publically available datasets, because they have usually been collected for other purposes, may not be ideal in their detail or resolution for archaeological purposes. Local shoreline positions taking account of glacio-isostatic movements may have to be estimated by reference to more generalised models of relative sea-level change (see also Harff et al. 2017a), so too the depth and character of the seabed in the absence of sub-bottom profiles. Nevertheless, such data provide usable information for many purposes, and a guide to where more detailed and localised investigation would be worth pursuing.

In Chap. 13, Hansson et al. provide a good example of how geophysical measurements and coring of sediments can be integrated to provide a detailed, high resolution reconstruction of environmental conditions on the submerged palaeolandscape. Detailed examination in this case was made possible through the collaboration of geoscientists, marine geologists and archaeologists with access to commercial ships and equipment and was focussed by the existence of already-known underwater archaeological remains (see also Holmlund et al. Chap. 4). One point that emerges very clearly from this presentation is the extraordinarily complex, dynamic and often subtle changes of shoreline configuration, environments and hydrogeological regime associated with the final stages of deglaciation and global sea level rise in the study area. The complexity arises from the interaction between sea-level rise, melting of ice barriers, regionally variable isostatic adjustments, and periodic damming back or release of Baltic waters. Such complexity places a high premium on the location, accuracy and detail of geophysical reconstructions and dateable palaeoenvironmental sequences.

Another point of interest, given the claims that have been made in other regions, notably the Black Sea, for catastrophic flood events and their human impact (Ryan et al. 1997; Lericolais et al. 2009; Yanko-Hombach et al. 2011), is the evidence in the western Baltic region for periodic, rapid and dramatic changes of water level through an amplitude of up to 25 m. As the waters of the Baltic were periodically dammed back to create a vast freshwater lake and then released again to connect with the world oceans, so they created large areas of newly exposed coastal lowland territory, only to flood them again, a sequence that occurred on at least two occasions over a period of less than 2000 years. Such a rapid complex of changes must have occasioned almost continuous disruption, re-adjustment and adaptation of settlement patterns and subsistence strategies for the human populations of the region, no doubt with more far-reaching social and demographic consequences.

This theme of dynamic and rapid change carries through to the work described in the following two chapters, both of which integrate a variety of geophysical and palaeoenvironmental techniques and reconstructions. Hepp et al. in Chap. 14, in another example of combined geophysical and palaeoenvironmental analysis, demonstrate what can be achieved by way of reconstruction over quite extensive areas, focussing on the submerged rivers systems of the 'Duck's Beak', which connected with the Palaeo-Elbe Valley, and the slightly later Palaeo-Ems. These are both part of a major system of river channels which drained northwards from the NW German land mass into the vast territory of 'Doggerland' exposed at lowered sea levels in the southern part of the North Sea Basin (see Glorstad et al. Chap. 19, Gaffney et al. Chap. 20, Momber and Peeters, Chap. 21). These valley systems would almost certainly have been a major focus of attraction for late Upper Palaeolithic populations already established in NW Europe during the closing stages of the Last Glacial, and most probably major arteries of communication, movement and animal migration (see Fischer 2004).

Karle and Goldhammer in Chap. 15 examine the southern rim of the North Sea, the Wadden Sea, one of the largest coastal and intertidal wetlands in the world. Here, because of long-term subsidence of the North Sea Basin, the sea has continued to encroach on dry land up to the present day, accompanied by rapid accumulation of marine, fluvial, estuarine and terrestrial sediments in a complex and ever-changing mosaic of drainage patterns and environmental conditions. The authors focus on the German sector of the Wadden Sea, between the estuaries of the Ems and Elbe Rivers, and thus complement the deeper offshore research of Hepp et al. discussed in Chap. 14. The approach adopted here is to exploit existing datasets including seismic profiles, borehole data representing some 3000 cores, geological and historical maps, nautical charts, and databases of archaeological finds. The primary output is environmental reconstruction with archaeological objectives: to audit the status of the information currently available with a view to ensuring effective management of the cultural heritage; to aid the interpretation of existing archaeological material; and to assess the potential for future discoveries and the most likely target locations.

Most of the known archaeological material, which is recorded from over 100 locations ranging from settlement data to individual finds, belongs to the past 2000 years, and that reflects the rapid rate of sedimentation and the fact that only the most recent millennia of the archaeological sequence have left a visible mark in the present-day offshore and intertidal mudflats. Much of that archaeological record is of intrinsic interest in demonstrating the various ways in which people attempted to protect themselves against encroaching flood waters, for example by building artificial mounds and constructing dykes. The corollary, of course, is that earlier material, particularly from the Stone Age, must be more deeply submerged beneath a thick overburden of later sediments, posing a considerable challenge to future discovery. Nevertheless, the authors point out that their mapping work has already led to the discovery of a previously unknown site, a 300-year-old farmstead destroyed by flooding, and to other locations with high archaeological potential where investigation should be carried out in advance of any offshore engineering or construction work.

Chapter 16 switches attention back to the Mediterranean and to the very different offshore conditions of the south Italian coastline, where steeper topography, predominance of erosion over sedimentation and tectonic activity pose very different challenges in assessing the effects of sea-level change on coastal topography. In this Chapter, Scicchitano et al. bring together a wide variety of geophysical and archaeological data to examine the relationship between coastline change and evidence of changing belief systems and social identities associated with a rich sequence of archaeological sites on the present-day coastline extending from the Neolithic through to the historical period. A combination of publically available mapping data, specially commissioned LiDAR and bathymetric survey, the use of underwater vehicles, isostatic modelling, and dating of uplifted sea-level indicators shows how a commanding promontory with a major settlement became progressively isolated and ultimately cut off from the mainland with corresponding changes from residential to ritual activity.

Finally, in Chap. 17, Radić Rossi and Cukrov highlight the significance of freshwater supplies located deep in coastal caves formed in karstic limestone country. The caves usually have a connection with the sea even if their entrances are above modern sea level. Accumulated freshwater floats on the surface of seawater and many of these freshwater sources are well below present sea level. Over 100 such caves are present in Croatia alone and they are well known to seafarers and the local inhabitants as a source of freshwater and in some cases for their medicinal benefits because of the presence of trace quantities of chemicals such as sulphur. They play an important role in supporting settlements in arid regions, and archaeological data within some caves show that they have been exploited since at least the Bronze Age. Many have been submerged since the time of formation and dating of speleothems provides evidence for the history of their formation and index points for refining sea-level curves. Little research has been devoted to their archaeological significance despite their potential importance. One intriguing question is how widespread they were during periods of low sea level. There is considerable interest in the significance of coastal springs in reinforcing the attractions of coastal regions during periods of low sea level when climate conditions were generally more arid (Faure et al. 2002, see also Bailey et al. Chap. 23). Anchialine caves in karst regions close to palaeo-shorelines could have further enhanced these attractions.

## 1.5 Landscapes of the Continental Shelf and Human Dispersals

In Part IV, discussion moves out onto a much broader geographical canvas, looking at whole sections of the continental shelf and deploying a range of archaeological, geophysical and palaeoenvironmental data and techniques, including more ambitious explorations of more deeply submerged landscapes. The central problem that motivates all these chapters is the grand theme of early human dispersals. This refers for the most part to human range-expansion into wholly new territory by our immediate ancestors, anatomically modern members of *Homo sapiens*, believed according to the current consensus to have originated in Africa and to have spread out from there at some time after about 200,000 years ago. This theme can also extend to include much earlier expansions from an African centre of origin by earlier members of the genus *Homo*. Range expansion may refer to territory never previously occupied, such as the Americas and Australia, territory made available again for human occupation after periods of abandonment, for example the recolonization of NW Europe following deglaciation, or territory already occupied by earlier human populations. The term may also refer, secondarily, to later colonising movements, for example the expansion of Neolithic farmers into European landscapes with already established hunter-gatherer populations. Since these processes of dispersal, for the most part, were in train when sea level was lower than present, they cannot be properly understood without systematic investigation of the submerged landscape, a point made by Flemming in Chap. 18 which provides a world-wide review of the issues relating to *Homo sapiens* dispersal.

As Flemming emphasises, an important source of new information lies in the rapidly expanding field of palaeogenetics, which has opened up many new ideas about the timing and pathways of population dispersal. However, new and sometimes contradictory interpretations are constantly emerging from this new research field along with conclusions that have not yet been tested against independent field data. Claims that genetic similarities between modern populations living in, for example, East Africa and the Indian subcontinent, can be used to derive not only a reliable date for the timing of the earliest dispersal between the two continents, but also the specific geographical pathway of movement—usually expressed as the shortest distance that can be drawn between them on a small-scale map—look over-optimistic and open to challenge. Uncertainties about the uniformity of mutation rates, methods of inferring dates that are subject to margins of error much greater than those typically associated with radiometric dates, the possibilities of multiple dispersals, overprinting of earlier genetic signatures by later dispersals, two-way movements and alternative geographical pathways, all need to be factored into the assessment. Ancient DNA, where preserved in human bone, offers an additional refinement and a check on palaeogenetic inferences.

At any rate, there is a need here for greater communication and collaborative research between the various disciplines involved, and especially more intensive research on the changing configuration of palaeoshorelines and palaeoenvironments on the submerged landscapes of the continental shelf, and a wider understanding of their significance. In his review, Flemming assesses what is currently known about all the relevant shelf areas in relation to these issues and the likelihood of further discoveries, drawing on the experience of submerged landscape reconstructions in Europe, and focussing on the key regions of particular significance for early human dispersal: the Mediterranean, the Red Sea, SE Asia, Australia-New Guinea, and Beringia.

Chapters 19, 20, and 21 form a group that deals with different aspects of 'Doggerland', the large area of the southern North Sea that united Britain, southern Scandinavia and the NW margins of continental Europe into a single territory during glacial periods, and which was progressively flooded by sea-level rise during the late Pleistocene and early Holocene. Since Coles (1998) first coined the term Doggerland and reviewed the existing evidence, new investigations have taken place, most notably the work of the Birmingham North Sea Palaeolandscapes Project (Gaffney et al. 2007, 2009). Palaeoenvironmental reconstructions and the recovery of artefacts and terrestrial fauna dredged up from the seabed by commercial operations all point to this region as a major focus of human settlement, and of movement and communication between the adjacent land masses.

Chapter 19 examines the timing and pathways of population expansion along the Norwegian coast following deglaciation. This issue is of particular significance because an ice-free corridor seemingly suitable for human occupation and accessible either from Doggerland or from southern Sweden and Denmark became available as early as 16,000 years ago, but there is no reliable archaeological evidence of settlement until about some 5000 years later (Bjerck 2008). One explanation for the time lag is that populations already established to the west and the south were primarily reindeer hunters, and that it would have taken time for them to adapt to the new conditions and develop the skills and motivation necessary to develop a way of life dependent on sea-mammal hunting and seafaring, both necessary for successful colonisation of the northern Norwegian coastline.

Glorstad et al. suggest a simpler explanation. Taking advantage of an evaluation project financed by the Norwegian Government in advance of offshore wind-farm construction, and exploiting already existing seismic profiles and bore-hole sediment data made available by oil companies working in this sector, they were able to reconstruct the sequence of landscape changes associated with the most likely pathway between Doggerland and Norway, using techniques for seismic profiles pioneered by the North Sea Palaeolandscapes Project referred to above. Their results demonstrate that this route would always have been blocked, initially by an ice-dammed lake and then by a marine channel in the vicinity of the present day Norwegian Trench, both too wide to be easily crossed by boat. To the south, the principal access route was blocked by an ice barrier, and it was only when that melted that easy passage northwards became available at about the time when archaeological evidence of settlement further north first appears.

In Chap. 20, Gaffney et al. outline the objectives of the newly funded Lost Frontiers Project to carry out more extensive reconstruction of the early Holocene Doggerland landscape, following on from their original work in the North Sea Palaeolandscapes Project. In the new project, they aim to bring together all the available seismic data to provide a comprehensive topographic mapping of the region, and to conduct a programme of coring in two target areas with evidence for well-established early Holocene river networks. The plan is to recover 100 cores and to conduct a range of dating and palaeoenvironmental analyses of their contained sediments.

One important innovation is the analysis of traces of ancient DNA incorporated in the sediments (sedaDNA) for clues to the plants and animals present on the now submerged palaeolandscape. Pilot research on sedaDNA has already been conducted at the submerged Mesolithic site of Bouldnor Cliff on the English south coast (Momber et al. 2011). The Bouldnor site, excavated over a period of many years by Garry Momber and his team, and referred to in the following chapter, is equivalent in importance to the Danish sites referred to earlier and elsewhere in this volume. Recent sedaDNA analysis there has established the presence of cultivated wheat in deposits sealed by later marine sediments, demonstrating the introduction of domestic cultigens from Europe at a much earlier date than previously suspected. This and the presence of Neolithic pottery recovered from the floor of the North Sea suggest that early farmers were already moving into this region before it was fully inundated by sea-level rise. Hence, another focus of this project will be to explore the significance of the Doggerland submerged landscape and the impact of its changing configuration on the socio-economic transitions associated with the spread of farming and the Mesolithic-Neolithic transition.

The theme of changing social and economic interactions and patterns of population movement and cultural diffusion is taken up by Momber and Peeters in Chap. 21. They examine the role of Doggerland in relation to the surrounding countries at a broad regional scale, looking at late Upper Palaeolithic and Mesolithic culture groupings and the mosaic of their changing distributions against the changing pattern of climate and sea-level change from 16,000 years ago onwards. They draw on recent archaeological discoveries both on existing coastlines and below modern sea level, particularly the underwater sites of Bouldnor Cliff on the English south coast and the Yangtze Harbour finds on the Dutch side of the English Channel (see Satchell, Chap. 25), to identify links and patterns of dispersal and communication as sea level rose. These include connections across the south of Doggerland from the earliest period after glacial retreat, early connections between Scotland and northern Europe perhaps involving seafaring around the northern coastlines and estuaries of Doggerland from as early as early as 12,000 years ago, and growing evidence in the Mesolithic of maritime connections around the coasts and archipelagos of northern and western Britain after about 10,000 years ago.

Also at about 10,000 years ago, semi-sedentary Mesolithic settlements with hut structures and a mixed terrestrial-marine palaeoeconomy appear on the modern coastline in northern Britain (where earlier shorelines are visible because of land uplift compared to southern Britain). The authors raise the interesting question of whether this phenomenon represents a new phase of intensification and adaptation to the loss of lowland territory with progressive sea-level rise, or is simply the first visible evidence of an adaptation that was already present at an earlier period on the now submerged coastlines of Doggerland—a question that cannot be resolved without further underwater investigations.

Dating from a little later, the Bouldnor Cliff site on the English south coast, with its well-preserved underwater finds, has evidence of precocious skills in axe manufacture, wood-working technology, including wooden structures and possible boat-construction, and DNA evidence for the presence of cultivated wheat (also discussed by Gaffney et al. Chap. 20) substantially earlier than elsewhere in Britain. These indicate close connections with France at a time when easy travel by boat between southern England and France was still possible around sheltered coastlines and estuaries before sea-level rise finally breached the land connection between the two regions.

The remaining chapters in this section move the geographical focus away from NW Europe, to the Aegean, the Red Sea, and Western Australia, all key regions in relation to early human dispersal patterns, and all with potential for new underwater research. These chapters have a common thread, which is the emphatic assertion in all three cases that the offshore environment, including the submerged landscape at lower sea levels and the marine environment created when sea level rose with its offshore islands, archipelagos and 'seascapes'—or 'aquapelagos', the term used by Ward and Veth in Chap. 24— should be treated as a coherent entity, forming a seamless whole with the adjacent mainlands, a focal centre of interest for the people who lived there as much as for the archaeological investigator.

The Aegean is of particular interest because it is on the major pathway for connections between Europe, the Near East and Africa, both for successive hominin expansions out of Africa during the Pleistocene and also for the later dispersal of farming communities, who left some of their earliest traces on Aegean islands as on islands elsewhere in the Mediterranean. The Aegean, along with Cyprus, has also provided some of the earliest evidence for seafaring—or at any rate sea crossings—in the northern hemisphere, with unequivocal evidence for a human presence on islands from at least

13,000 years ago, long before the spread of agriculture, and claims for even earlier dates on the island of Crete (Ammerman and Davis 2013–2014).

As Sakellariou and Galanidou emphasise in Chap. 22, the relative placement of land and sea, and the status of islands and their connections with—or separation from—their adjacent mainlands, have undergone dramatic changes on the Pleistocene time scale. This is not only because of sea-level change as elsewhere, but because of very active tectonics associated with the collision of the African Plate with Europe and subduction and volcanic activity along the Hellenic Trench, promoting long-term subsidence throughout the Aegean punctuated by more localised episodes of uplift. The authors tackle this puzzle by breaking the Aegean region down into nine sub-regions, and draw on available seismic, geological and archaeological data within each sub-region to reconstruct changing patterns of palaeogeography and the nature of connections and pathways of movement. As with similar mapping projects in other chapters, this exercise plays a double role, illuminating the interpretation of existing archaeological sites on land, and focussing attention on submerged areas worth closer investigation.

The southern Red Sea, discussed by Bailey et al. in Chap. 23, has acquired prominence in recent debates about the dispersal of *Homo sapiens* out of Africa as a key pathway for range expansion, following the so-called southern corridor from The East African Rift to SW Arabia and onward around the coastlines of the Indian Ocean into SE Asia and Australia. The project described here is part of a larger project designed to test these ideas, combining archaeological and landscape investigations on the mainland of SW Saudi Arabia, the Farasan Islands and the submerged shelf. Exploitation of existing core data and bathymetry combined with sea-level modelling shows that, at low sea levels, the Red Sea between Africa and Arabia would have been reduced to a narrow and easily crossable marine channel with the exposure of extensive coastal lowlands some 100 km wide on both sides. Targetted surveys with a research vessel and a full suite of acoustic, seismic and coring equipment and an ROV has demonstrated the presence of a complex topography on this submerged landscape with numerous fault-bounded basins, some of which are likely to have formed freshwater lakes, making this an important refugium during periods of climatic aridity. The results also demonstrate the very different character of the shoreline at lowest sea level. Offshore archipelagos were present that could have facilitated early experiments with seafaring and exploitation of marine resources, with further implications for possible patterns of dispersal via maritime corridors, a theme also discussed in the following chapter.

The Australian continent is the paradigm case of early sea travel as the primary means of colonising a new continent, since it would always have required sea crossings to travel from SE Asia even at lowest sea level stands. As Ward and Veth emphasise in Chap. 24, a maritime aspect to human activity was present from the very outset, with cave sites on steeply shelving or uplifted shorelines on the islands of Wallacea and the Bismarck Archipelago showing evidence of sea travel and intensive exploitation of marine resources from at least 40,000 years ago. Investigation of the offshore landscape and its submerged shorelines is therefore of more than usual importance in understanding the history of dispersal and settlement (see also Flemming, Chap. 18). In this chapter, the authors discuss the coastline of Western Australia. They are reliant mainly on bathymetry and models of sea-level change to chart the changing character of the submerged landscape. But they also have access to well-preserved archaeological sequences on a series of offshore islands that were progressively isolated as sea-level rose, as well as evidence from the adjacent mainland, giving a window into the changing character of the submerged landscape. The changing composition of food remains in these sites through time, together with evidence for the movement of raw materials for making stone artefacts, provide a measure of changing patterns of adaptation and movement in response to rising sea level, with evidence that coastlines and marine resources were important throughout the sequence. More detailed underwater work is planned to further test these conclusions.

## 1.6 Outreach and Management

The importance of engagement with a wider public, of training, and of collaboration with industrial companies and government agencies is a recurring theme in many of the earlier chapters. In this final section, four chapters address more fully these issues.

Satchell (Chap. 25) draws on the experience of the Maritime Archaeology Trust (MAT) in training, education and public outreach, and the initiatives developed within the SPLASHCOS Action, to identify examples of good practice. She identifies four different audiences that need to be addressed: scientists involved in coastal and underwater investigations of various sorts, who have often accumulated data archives that can be turned to advantage for archaeological purposes; commercial companies, who also often acquire data of archaeological and palaeoenvironmental value and who, in addition, need advice on how to implement national and international directives on the mitigation of damage to the underwater heritage; government agencies and planning authorities responsible for management of the onshore and offshore heritage, who can also benefit from archaeological and geological studies of coastline change; and members of the public, who ultimately pay for underwater research and help to make the wider case with local authorities and government agencies.

A key difficulty is the remoteness of underwater finds and difficulties of access especially for non-divers. However, MAT, which is also involved in the work on the submerged site of Bouldnor Cliff (see Momber and Peeters, Chap. 21) has developed a number of innovative practices for many different audiences ranging from 10-year-old school children to professional archaeologists. These include teaching packs for school teachers, writing of illustrated story books, hands-on experience with post-excavation work, and a travelling 'Maritime Bus' with videos, leaflets and examples of underwater artefacts, experience that has been extended more widely through collaboration with neighbouring European countries and through the SPLASHCOS network. She also gives detail on the training programmes developed during the SPLASHCOS Action (see also Uldum et al. Chap. 5, Galili et al. Chap. 6), and on a variety of initiatives in the UK setting for defining protocols to encourage commercial operators, fishermen and members of the public to record and report underwater artefacts and other features.

Tidbury et al. (Chap. 26) describe case studies that bring together evidence from underwater archaeology, palaeoenvironmental data, historical maps and archives, and old photographs and paintings to chart long-term changes in coastal geomorphology. Their approach has much in common with that described by Karle and Goldhammer (Chap. 15) with the difference that in this chapter the research is aimed primarily at providing evidence of coastal change to inform local authorities and policy makers responsible for coastal management and planning. This approach contributes to a wider research agenda in which the reconstruction of the geological past can be used for future projection of coastline change (see Harff et al. 2017b). For another example of the value of ancient maps for palaeogeographical reconstruction and the refinement of geomorphological models, see also Deng et al. (2016). The examples presented in this chapter give detail on the mapping of small-scale changes, for example in rates of erosion leading to the retreat of the coastline on a time scale of decades and centuries, as well longer-term changes in sea level, and evidence for their human impact and response.

Missiaen et al. (Chap. 27) take up the issue of the threats to the coastal and underwater heritage by intensifying commercial and industrial activity in the Belgian context, particularly in the shallow waters immediately offshore, which are especially vulnerable to damage by engineering and construction work. The authors provide interesting detail on the development of new and efficient methods of underwater survey in shallow water—a good example of how archaeological imperatives are leading to new technological solutions with practical and commercial benefits as well as archaeological ones. They also provide a useful review of the international conventions for protecting the underwater heritage and the issues involved in translating these into a workable legislative framework and in providing advice to commercial operators who have to implement it. The discussion highlights the benefits of three-way collaboration between government agencies, commercial companies and underwater archaeologists.

Finally, in Chap. 28, Sturt et al. reflect on the long-term history of developments in the study of submerged landscapes and archaeology in the English context over the past century, and what we can learn from this history in looking into the future. They bring out the very interesting history of thinking and practice extending back to the discussion of submerged forests at the beginning of the twentieth century, emphasising the changing but closely inter-twined and complex relationships between scientific and academic research, industrial exploitation of the seabed, government legislation, public interest and outreach, and the different historical trajectories of on-land and underwater archaeological research and cultural heritage management. They point to the great mutual benefits that have come about as a result of a co-evolution between these very different interests, particularly in recent decades, but also to the risk of complacency in persisting with a particular way of doing things, simply because it has worked in the past, or because of the particular technologies and datasets made available by current funding budgets or industrial activity. Deciding what scientific and archaeological questions are worth asking, and what techniques and data are needed to answer them, communication across the many boundaries between the different worlds of academic and scientific research, commercial activity and public policy, and above all the engagement of wider public interest, are all essential ingredients, as they see it, in shaping future research agendas. This final chapter usefully draws together the different strands of investigation discussed in the other chapters in this section, and indeed represents a fitting finale to the themes of the volume as a whole.

## References


Ammerman AJ, Davis TW (eds) (2013–2014) Island archaeology and the origins of seafaring in the eastern Mediterranean. Eurasian Prehistory 10–11

Andersen SH (2013) Tybrind Vig. Submerged Mesolithic settlements in Denmark. Jutland Archaeological Society Publications 77. Jutland Archaeological Society, Højbjerg

Bjerck HB (2008) Norwegian mesolithic trends: a review. In: Bailey G, Spikins P (eds) Mesolithic Europe. Cambridge University Press, Cambridge, pp 60–106

Coles B (1998) Doggerland, a speculative survey. Proc Prehist Soc 64:45–81

Deng J, Harff J, Giza A, Hartleib J, Dudzinska-Nowak J, Bobertz B, Furmanczyk K, Zölitz R (2017) Reconstructions of coastline changes by the comparisons of historical maps at the Pomeranian Bay, southern Baltic Sea. In: Harff J, Furmanczyk K, VonStorch H (eds) Coastline changes of the Baltic Sea from South to East – past and future projection. Coastal Research Library. Springer, Heidelberg

Faught MK (2014) Remote sensing, target identification and testing for submerged prehistoric sites in Florida: process and protocol in underwater CRM projects. In: Evans AM, Flatman JC, Flemming NC (eds) Prehistoric archaeology on the continental shelf: a global review. Springer, New York, pp 37–52

Faure H, Walter RC, Grant DR (2002) The coastal oasis: ice age springs on emerged continental shelves. Glob Planet Chang 33:47–56

Fedje DW, Josenhans H (2000) Drowned forests and archaeology on the continental shelf of British Columbia, Canada. Geology 28:99–102

Fischer A (1996) An entrance to the Mesolithic world below the ocean. Status of ten years' work on the Danish sea floor. In: Fischer A (ed) Man and sea in the Mesolithic. Coastal settlement above and below present sea level. Oxbow, Oxford, pp 371–384

Fischer A (2004) Submerged stone age – Danish examples and North Sea potential. In: Flemming NC (ed) Submarine prehistoric archaeology of the North Sea. Research priorities and collaboration with industry, CBA research report, vol 141. Council for British Archaeology, York, pp 23–36

Flemming N, Harff J, Moura D, Burgess A, Bailey G (eds) (2017) Submerged landscapes of the European continental shelf, Quaternary palaeoenvironments. Wiley-Blackwell, Chichester

Gaffney V, Thompson K, Fitch S (eds) (2007) Mapping Doggerland: the Mesolithic landscapes of the southern North Sea. Archaeopress, Oxford

Gaffney V, Fitch S, Smith D (2009) Europe's lost world, the rediscovery of Doggerland CBA research reports 160. Council for British Archaeology, York

Grøn O (2004) Description of the dwelling pit. In: Skaarup J, Grøn O (eds) Møllegabet II. A submerged Mesolithic settlement in southern Denmark, BAR international series, vol 1328. Archaeopress, Oxford, pp 41–74

Harff J, Lüth F (eds). (2011) Sinking coasts. Geosphere, ecosphere and anthroposphere of the Holocene southern Baltic Sea II. Bericht der Römisch-Germanischen Kommission 92. Römisch-Germanische Kommission des Deutschen Archäologischen Instituts, Frankfurt a.M. Philipp von Zabern, Mainz am Rhein

Harff J, Flemming N, Groh A, Hünicke B, Lericolais G, Meschede M, Rosentau A, Sakellariou S, Uscinowoicz S, Zorita E (2017a) Sea level and climate. In: Flemming N, Harff J, Moura D, Burgess A, Bailey GN (eds) Submerged landscapes of the European continental shelf, Quaternary palaeoenvironments. Wiley-Blackwell, Chichester

Harff J, Furmanczyk K, Von Storch H (eds) (2017b) Coastline changes of the Baltic Sea from South to East – past and future projection. Coastal Research Library, vol 19. Springer, Dordrecht

Hayashida K, Kimura J, Saski R (2014) State and perspectives of submerged sites in Japan. In: Evans AM, Flatman JC, Flemming NC (eds) Prehistoric archaeology on the continental shelf: a global review. Springer, New York, pp 275–290

Hermand, JP, Grøn O, Asch, M, Ren QY (2011) Modelling flint acoustics for detection of submerged Stone Age sites. Proceedings of OCEANS 2011 IEEE Conference, Santander, Spain. http://www.lamfa.u-picardie.fr/asch/f/p/ac/7816Ed01_final_opt.pdf. Accessed 20 June 2016

Lambeck K, Esat TM, Potter E-K (2002) Links between climate and sea levels for the past three million years. Nature 419:199–206. doi:10.1038/nature01089

Lambeck K, Rouby H, Purcell A, Sun Y, Sambridge M (2014) Sea level and global ice volumes from the last glacial maximum to the Holocene. Proc Natl Acad Sci U S A 111(43):15296–15303. doi:10.1073/pnas.1411762111

Lericolais G, Bulois C, Gillet H, Guichard F (2009) High frequency sea level fluctuations in the Black Sea since the LGM. Glob Planet Chang 66:65–75

Long D, Wickham-Jones CR, Ruckley NA (1986). A flint artefact from the northern North Sea. In: Roe DA (ed) Studies in the upper palaeolithic of Britain and Northwest Europe. British Archaeological Reports International Series 296, Oxford, pp 55–62

Momber G, Tomalin D, Scaife R, Satchell J, Gillespie J (eds) (2011) Bouldnor cliff and the submerged Mesolithic landscape of the Solent, CBA research report, vol 164. Council for British Archaeology, York

Raban A (1983) Submerged prehistoric sites off the Mediterranean coast of Israel. In: Masters PM, Flemming NC (eds) Quaternary coastlines and marine archaeology. Academic, London, pp 215–232

Ryan WBF, Pitman WC III, Major CO, Shimkus K, Moskalenko V, Jones GA, Dimitrov P, Gorür N, Sakinç M, Yüce H (1997) An abrupt drowning of the Black Sea shelf. Mar Geol 138:119–126

Skaarup J, Grøn O (2004) Møllegabet II. A submerged Mesolithic settlement in southern Denmark. BAR International Series 1328. Archaeopress, Oxford

Stanford D, Lowery D, Jodry M, Bradley BA, Kay M, Stafford TW, Speakman RJ (2014) New evidence for a possible Paleolithic occupation of the eastern north American continental shelf at the last glacial maximum. In: Evans AM, Flatman JC, Flemming NC (eds) Prehistoric archaeology on the continental shelf: a global review. Springer, New York, pp 73–93

Yanko-Hombach V, Mudie P, Gilbert AS (2011) Was the Black Sea catastrophically flooded during the Holocene? In: Benjamin J, Bonsall C, Pickard C, Fischer A (eds) Submerged prehistory. Oxbow, Oxford, pp 245–262

# Part I
# Techniques and Strategies

# Chapter 2
# Survey Strategies and Techniques in Underwater Geoarchaeological Research: An Overview with Emphasis on Prehistoric Sites

**Tine Missiaen, Dimitris Sakellariou, and Nicholas C. Flemming**

**Abstract** Underwater geoarchaeological studies typically involve case studies that vary widely in scale, environment and stage of application. As a result, the range of survey techniques and applied methods is very broad. This paper aims to present an overview of state-of-the-art techniques and survey strategies for submerged prehistoric site evaluation. We focus not only on conventional techniques but also on technologies that were designed and developed for other research applications but which can or could be effectively applied to submerged prehistoric studies. Different techniques discussed in this paper include remote sensing (acoustic seafloor and sub-seafloor imaging, Lidar, electric and (electro)magnetic techniques), direct investigations (coring, sampling and excavation), 2D and 3D photographic techniques, and the use of remotely operated vehicles. Notwithstanding the enormous technological progress that has been made in recent years, a large number of challenges remain not only regarding detection and excavation (especially in deeper water) but also with regard to the cost-effectiveness of submerged geoarchaeological surveys. The set-up of 'best practice' guidelines and close(r) collaboration with industry may provide some solutions.

## 2.1 Introduction

Systematic mapping of the continental shelf at various scales (regional, local, site) and by means of various techniques and methodologies is of crucial importance for the reconstruction of submerged landscapes and terrestrial palaeoenvironments and the evaluation of possible locations for prehistoric archaeological sites. At this moment a large spectrum of techniques is available, not only in the scientific world but also in the world of the military as well as offshore industry. However, because of

T. Missiaen (✉)
Renard Centre of Marine Geology, Department of Geology, Ghent University,
Krijgslaan 281 S8, 9000 Ghent, Belgium
e-mail: tine.missiaen@ugent.be

D. Sakellariou
Institute of Oceanography, Hellenic Centre for Marine Research,
46.7 km Athinon-Souniou Av, Anavyssos, P.O. Box 712, 19013 Athens, Greece
e-mail: sakell@hcmr.gr

N.C. Flemming
National Oceanography Centre, European Way, Southampton SO14 3ZH, UK

Sheets Heath, Benwell Rd, Brookwood, Woking GU24 0EN, UK
e-mail: n.flemming@sheetsheath.co.uk



G.N. Bailey et al. (eds.), *Under the Sea: Archaeology and Palaeolandscapes of the Continental Shelf*, Coastal Research Library 20, DOI 10.1007/978-3-319-53160-1_2

their cost and leading-edge technology, not everybody has easy access to these methods. Even in the more economically advanced countries, research institutes have difficulty in keeping up with the fast-evolving technology. Moreover, the techniques are not always used in the most efficient/effective way, and in many cases an optimum survey methodology is lacking.

New developments are needed to 'tune' the existing techniques to specific scientific targets, as they were often developed for other purposes—such as purely geological or terrestrial studies, or military research. This modification (or implementation) of existing technologies is probably at present the best possible approach for submerged archaeological and palaeolandscape studies, as there is still not enough industrial/economic/scientific pressure to develop specific, purpose-designed prospection tools. Once there is high probability or proof of a preserved site, further specialist techniques are needed for site survey, assessment, and preservation or excavation.

One of the main technological challenges in continental shelf prehistoric research is the wide variety encountered in scale and environment. Both strongly affect the applied methodology or technique (Fig. 2.1). For instance, the search for extensive palaeolandscapes across wide areas of the continental shelf requires a different approach from surveying or documenting small (buried) structures of a few square meters. Conversely, this also affects the resolution required: what we are trying to resolve in a record from a buried river valley is obviously different from that of a single flint. The need for increasingly higher resolution, in other words the detection/identification of increasingly smaller objects, is pushing the existing technology to the limit. As yet there is no software that can enhance target recognition or pattern recognition that would speed the identification of prehistoric underwater sites.

The local or regional natural environment also influences the technology that is used. Exposed or buried features, sandy or rocky seafloor, hard or soft sediments in most cases require entirely different technologies. In this respect, caves are a specific problem, not only because their detection remains very difficult but also the nature of the topography (inclined walls instead of a more or less horizontal surface) poses major challenges to imaging technology. Local environmental conditions such as currents, waves, sedimentation, and (importantly) visibility also significantly affect the techniques that are used.

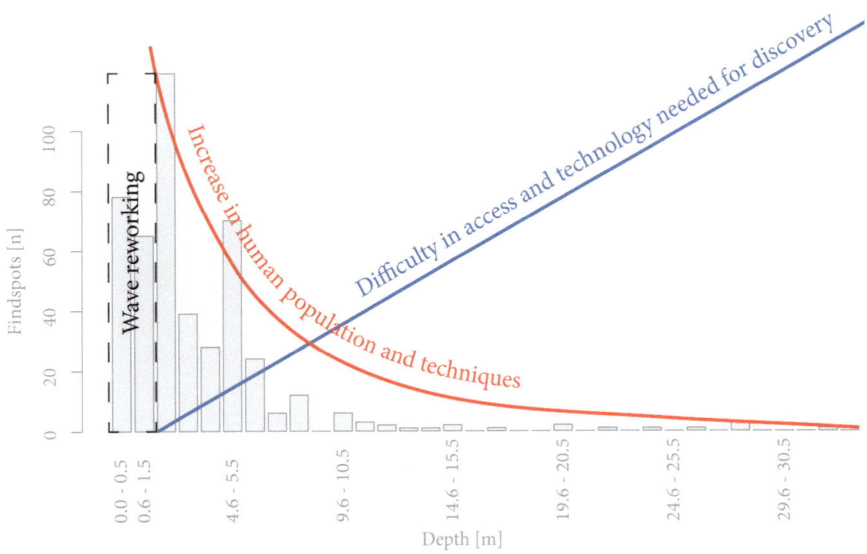


**Fig. 2.1** Depth of all submerged prehistoric sites registered by March 2014 in the SPLASHCOS data base. The number of known sites decreases rapidly with depth. The *red curve* uses arbitrary units to indicate that the population density and tool technology decreases as we go back in time and depth; while the *blue line* indicates the increasing cost and technical difficulty of working at depth (After Flemming et al. 2014)

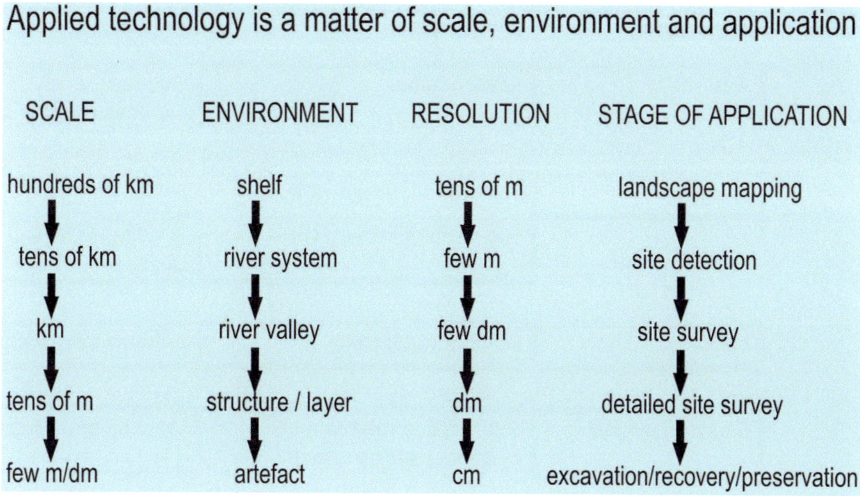


**Fig. 2.2** Simplified diagram showing the wide range of scale, environment and application stage encountered in submerged archaeological studies

Water depth plays a crucial role. Until now, the vast majority of submerged prehistoric landscapes and sites on the continental shelf have been found in relatively shallow waters (<10 m), often with good visibility and accessibility (Fig. 2.1, Flemming et al. 2014). Only a small (but significant) number of sites have been examined in the depth range 10–50 m. There is little doubt that a large number of well-preserved sites and landscapes are likely to exist in deeper waters but these are largely unexplored. To open this vast archive will require interdisciplinary cooperation between archaeologists, scientists and engineers.

Closely related to issues of scale is the stage of application, which will determine the technology to be used, going from large-scale, low-resolution landscape mapping through to high-resolution site imaging, sampling, excavation, artefact recovery, site stabilization, documentation, visualization, core interpretation, and site preservation or backfilling (see Fig. 2.2). Different technologies are applied in different phases of a survey e.g., (a) commercial seismic data for a rough reconstruction of the submerged landscapes—river systems, shorelines etc., (b) side scan sonar or multibeam echo sounder to register seafloor features such as preserved tree stumps and fallen trunks exposed on the bottom, (c) 2D or 3D high-resolution subbottom profilers to detect archaeological features or small terrain anomalies embedded in bottom sediments, (d) van Veen samplers and box corers to recover samples from the seafloor and sub-seafloor, (e) photogrammetry derived from imagery.

## 2.2 Current Survey Technology and its Shortcomings

Due to the large variety in scale, environment and stage of application, the range of survey techniques and methods applied in submerged prehistoric research is very wide (Fig. 2.3). It goes from remote sensing (acoustic seafloor and sub-seafloor imaging, Lidar, and (electro)magnetic techniques) through to direct investigations (coring, sampling and excavation) and 2D/3D photographic techniques and remotely operated vehicles.

In recent decades, enormous technical progress has been made in geophysical recording and documentation methods that are able to provide high resolution data about sediments and structures on and beneath the seafloor. Some of these remote sensing technologies were originally designed and

| METHOD | TECHNIQUE | TYPE OF DATA | TECHNOLOGY |
|---|---|---|---|
| REMOTE SENSING | Acoustic | Seafloor map | Side-scan sonar, multibeam echosounder |
| | Acoustic | Sub-seafloor image (2D) | Sub-bottom profilers |
| | | Sub-seafloor image (3D) | 3D Chirp, SES-2000 Quattro |
| | Lidar | Seafloor topography | Airborne Lidar Bathymetry |
| | (Electro-) magnetic | Seafloor and sub-seafloor magnetic/resistivity map | EM profilers, gradiometers |
| DIRECT INVESTIGATION | Coring and sampling | Sedimentological/environmental | Grabs (van Veen, Shipek) Boxcore, vibrocore, gravity core, piston core |
| | Dive surveys | Sedimentological/archaeological | Swim dive (corridor/jackstay/circular) |
| | | | Drift/contour dive |
| | | | Excavation |
| UNDERWATER PLATFORMS | Submersibles (manned/unmanned) | Wide spectre of data (acoustic maps, water/sediment samples, cores, video, ...) | HOV, ROV, AUV |
| PHOTOGRAPHIC | Photo, video, stereo | Exposed seafloor | Digital 2D/3D cameras, photo/video-mosaicing, video microsope |

**Fig. 2.3** Simplified diagram showing the wide range of technologies involved in submerged prehistoric research

developed for other research applications, such as mine detection and shipwreck studies, but they can be effectively applied (if needed with slight adaptation) to submerged prehistoric studies. Notwithstanding this technological progress, the identification of artefacts, or other physical evidence of a prehistoric site, on remote sensing data remains a huge challenge. This is even more so the case when the artefacts are buried below the sea floor.

Geophysical recording therefore always needs to be complemented with genuine archaeological investigations. In recent years direct observation methods with regard to documentation, sampling, excavation and preservation of cultural deposits have become increasingly efficient, especially in relatively shallow water environments. However the major challenge lies in the development of specialized techniques for archaeological investigations in deeper water (10–150 m), and/or environments with poor or non-existent visibility.

## 2.3 Acoustic Mapping Techniques

### *2.3.1 Acoustic Seafloor Mapping*

Where palaeolandscape areas are exposed or shallowly buried, *side-scan sonar* and *multibeam echo sounders* are very powerful tools since they provide detailed 'acoustic maps' of the sea bed. The main advantage of these techniques is that a large area can be scanned relatively fast (on average a few hours for a few square km) and with a high precision (decimeter to meter-range resolution). Over the years side-scan and multibeam systems have been improving towards increasingly higher resolutions, in some cases up to cm-range. This has resulted in an ever finer image of the seafloor morphology, its texture and sediments, and any objects or structures on the seafloor. The continuous advance in materials has moreover resulted in highly compact systems that can be deployed on very small boats and in increasingly shallow water, down to a few metres, e.g., Fig. 2.4 (Sakellariou et al. 2011). Though much improvement has recently been made in the identification of artefacts by a smart combination of backscatter and bathymetry (e.g., Bates et al. 2011), being able to distinguish anthropogenic features from natural phenomena remains very difficult.

Recently a new generation of compact, high resolution imaging sonars also known as 'acoustic cameras' or '3D sonars' has been developed for engineering purposes such as pipeline inspection. These sonar cameras (also called imaging sonars), often mounted on AUVs and ROVs, offer a 3D field-of-view (up to 130°) of the sea bed and are especially useful in low or zero visibility conditions

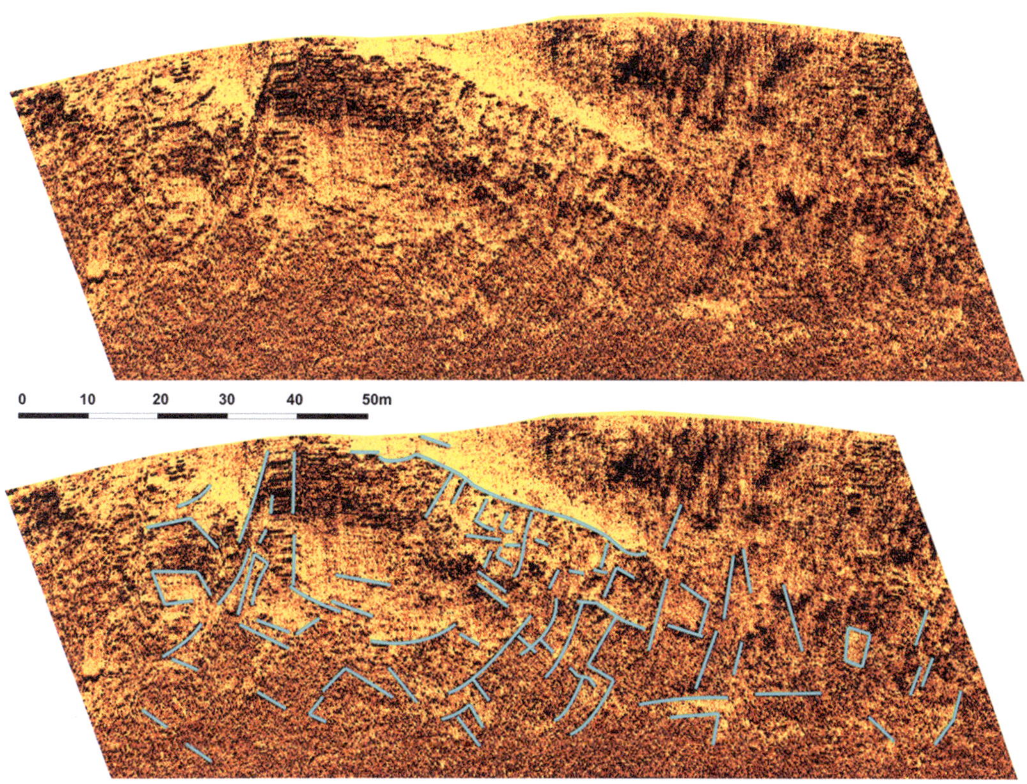


**Fig. 2.4** Side-scan sonar image (400 kHz, 2–4 m water depth) of the submerged prehistoric city of Pavlopetri in Greece. *Blue lines* highlight the remains of the walls of the houses (After Sakellariou et al. 2011)

(Johnson-Roberson et al. 2010). Another noteworthy development concerns multibeam echo sounder systems with vertical angular adjustment (by the use of a tilt motor), which can provide accurate images of inclined (even vertical) surfaces. This technique offers a huge potential for cave research, where conventional multibeam systems with a wide swath are not sufficient (Mallios 2014).

### 2.3.2 Acoustic Mapping of the Seabed Sub-Surface

In the case of buried features *seismic remote sensing* (also called sub-bottom profiling) is needed to image the palaeolandscape and possible archaeological remnants therein. The seismic technique involves a wide range of sources (e.g., boomer, sparker, chirp, echosounder) and receivers (single and multichannel streamers). These have been available for decades and they yield increasingly detailed images of the sub-bottom with a resolution ranging between 20 cm and 1–2 m. Yet their full potential for submerged prehistory research is often not well exploited. Ongoing studies in Belgium show that

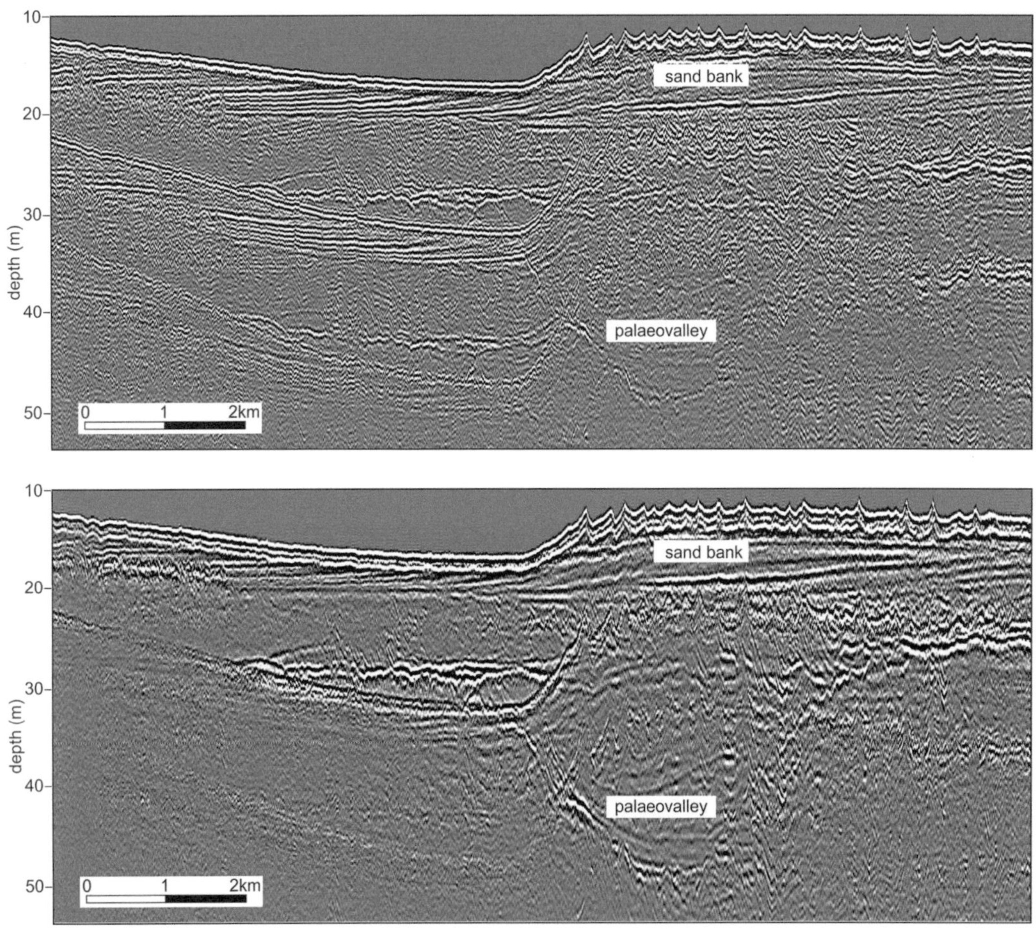


**Fig. 2.5** Acoustic sparker profile from the southern North Sea showing the present sea floor (marked by a *sand bank* superimposed with large sand waves) and the underlying buried landscape (including a palaeovalley). *Top image*: single channel data. *Bottom image*: multichannel (24 ch) data. Applying multichannel recording proved advantageous, even for the relatively shallow depths encountered here (After Zurita Hurtado et al. 2014)

with a smart combination of sources and receivers and resourceful data processing the efficiency and performance of sub-bottom profiling for archaeological studies can be significantly increased (Fig. 2.5, Zurita Hurtado et al. 2014).

Prime targets for seismic studies include relief features such as buried palaeochannel systems and shell midden accumulations, but also wooden remains and organic deposits (e.g., peat layers), the latter being good indicators of past coastlines and showing a high preservation potential (e.g., Wunderlich et al. 2005; Grøn et al. 2007; Plets et al. 2007; Missiaen 2010). Unfortunately these organic-rich deposits often give rise to biogenic gas which can completely obscure the seismic image (Missiaen et al. 2002). Moreover the detection of subsurface layers containing archaeological material remains problematic since different features, for instance hard layers or fine-grained deposits, may sometimes yield a similar reflectivity and can therefore be easily mistaken. In the southern North Sea, furthermore, the widespread occurrence of sand banks poses an additional challenge as the short wavelength (i.e., high frequency) sound waves are quickly absorbed in the heterogeneous sandy sediments, resulting in a decrease in penetration depth (e.g., Van Lancker et al. 2009).

A major drawback of conventional seismic profiling is that small buried features or artefacts are extremely difficult to identify. Indeed this technique provides vertical 2D profiles, and (if line spacing is sufficiently close) through data interpolation a (pseudo) 3D image of the sub-seafloor can be constructed that allows adequate imaging of large-scale topographical features (e.g., tens to hundreds of metres in size); but it will easily miss small features of metre and sub-metre scale. The latter require *true 3D imaging techniques*. Yet 3D imaging of the sub-seafloor with high spatial resolution is a complex matter due to the physical constraints placed on sampling and positioning accuracy (Missiaen 2005; Plets et al. 2009).

In recent years, however, two unique acoustic systems have been developed in the UK (3D Chirp) and Germany (SES-2000 Quattro) that allow true 3D imaging of small buried structures with dm-scale horizontal and vertical resolution. The 3D Chirp system, which can be surface-towed from a small survey vessel, incorporates a rigid frame (2.75 × 2.3 m) containing 4 source transducers and 60 receiver elements, and integrated real-time-kinematic GPS (Bull et al. 2005; Gutowski et al. 2008). The SES-2000 Quattro is a multi-transducer parametric echosounder system that consists of a line array of four transducers that are fix mounted onto the survey platform (often a vertical pole) of a small vessel; the distance between two transducers can be varied depending on the investigated target (Lowag 2010; Innomar 2014). Promising results were obtained for buried engineering structures and archaeological wooden artefacts (Fig. 2.6). Most recently, tests in Belgium (2015) have achieved images of peat and salt excavation features in the intertidal zone in the highest detail, thereby setting new standards for shallow water geoarchaeological research (SeArch 2015).

## 2.4 Non-acoustic Mapping Techniques

Although less frequently used than acoustic methods, *magnetic (or gradiometric)* and *electromagnetic (EM)* techniques may also be valuable tools to image the (sub-)sea floor. This is especially useful when the sediments or archaeological remnants have a magnetic imprint which cannot be detected using acoustic methods (e.g., small ferrous objects, pit hearth/oven). Since the level of detail obtained with these techniques decreases rapidly with the distance to the sensor, towing the equipment close to the seafloor is crucial.

Magnetic surveys at sea often involve the use of two (sometimes more) spatially separated sensors (gradiometric method), which provides a better resolution and is able to detect smaller phenomena. Surveys can be carried out from relatively small vessels. For shallow depths (less than a few metres) the magnetometers can be mounted on a fixed structure; for greater depths they are often mounted on a depth-controlled towfish. Increasingly high-performance magnetometers (Fig. 2.7) and smart data

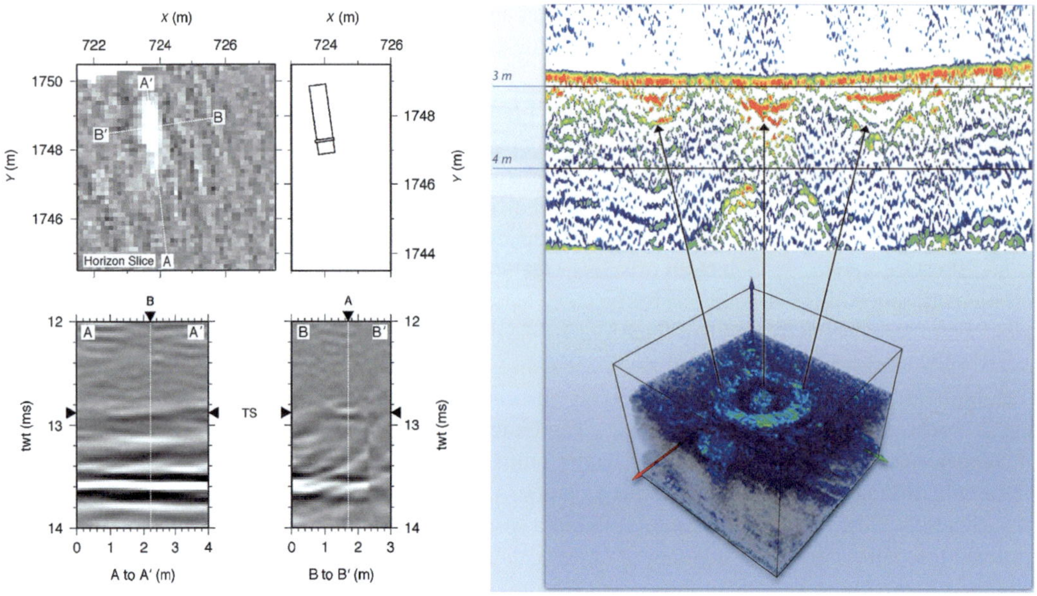


**Fig. 2.6** *Left*: **a** Horizontal slice and **b**, **c** vertical slices through a buried acoustic target (location map **d**) in a tidal basin off the SE coast of England (after Vardy et al. 2008). *Right*: Subbottom image in 2D (*top*) and 3D (*bottom*) of a circular buried structure in Wismar Bay, Germany. Dimension of the 3D volume is 40×40×3 m. Data obtained with the new 3D parametric echosounder system SES-2000 Quattro (© Innomar)

processing have resulted in increasingly high resolutions (e.g., Missiaen and Feller 2009). This opens perspectives for the identification of submerged archaeological sites (e.g., Camidge et al. 2010; Boyce and Reinhardt 2004).

Marine EM techniques conventionally involve the use of heavy and bulky equipment which implies the use of relatively large boats. The resolution of the obtained data is generally much lower compared to acoustic techniques. In recent years, marine EM techniques have evolved considerably through the use of sensitive, and smaller, sources and sensors (e.g., Klein et al. 2005). However until now their main use seemed to lie in shipwreck studies, and effective application in submerged prehistoric research lacking. The recent development of a marine EM profiler ('GEM-Shark') in Germany, however, opens up new perspectives for archaeological studies (Fig. 2.7, Müller 2009).

Electric resistivity imaging (ERI) is a well-known geophysical method applied on land, but at sea this technique is much less used. Most of the current marine electrical resistivity work concerns groundwater studies or rock/sediment identification, e.g., for dredging operations or harbour works, (Henderson et al. 2010; Rucker et al. 2011; Tarits et al. 2012). A recent case study in Italy applied the method to shipwreck detection, but the results were not very convincing (Passaro 2010). For the moment the applicability of ERI for underwater archaeological studies seems to be limited to lithological/structural information on palaeolandscapes, complementary to acoustic data.

*Airborne LiDAR* (Light Detection and Ranging) bathymetry techniques (aka ALB) may be a valuable tool in shallow water areas. This remote sensing technology for 3D mapping of the seafloor has undergone very rapid development over recent years (Pe'eri et al. 2011) and the quality of the data is nowadays often comparable to multibeam (Pastol 2011). Important benefits of this technology are the ability to survey seamlessly across the land-sea boundary (Fig. 2.8), and to map extremely shallow areas with complex and irregular topography that are off-limits and sometimes dangerous to conventional bathymetric acquisition techniques. On retreating coasts where major storms may result in short exposure of lagoon or backshore environments before they are buried again or wiped out by waves, the availability of shallow water remote sensing tools will be especially crucial.

Fig. 2.7 *Left*: Marine magnetometer array in vertical gradient acquisition mode (© G-Tec). *Right*: Electromagnetic profiler array "GEM-Shark" being prepared at sea (© University of Bremen)

Fig. 2.8 Seamless integration of offshore multibeam data and terrestrial/intertidal LiDAR bathymetry data in Bantry Bay, Ireland, superimposed in Google Earth (© INSS/ INFOMAR)

## 2.5 Sampling, Coring and Excavation

In order to ground-truth the remote sensing data, samples of the seabed and sub-seabed are necessary. This can be done from boats using either *grab samplers* (e.g., Hamon, Day, Shipek, Van Veen grabs) or *coring devices* (e.g., box, gravity, piston, vibro-corer), or under water by *divers*. By studying how the sediments, and the fossils within them, change over time, a picture emerges of how the landscape, climate and sea-level have fluctuated in the past. In general it can be said that increasing the amount of sampled, cored or excavated sediment will also increase the likelihood of site discovery.

**Grab Samplers and Corers** Grabs are easy to deploy and can give a very large sample, but they only provide information about the seabed surface. Corers can theoretically reach several tens of metres beneath the seabed, but in reality rarely exceed 10 m. Box corers can recover large undisturbed samples of the seafloor and the uppermost sedimentary deposits up to a few tens of decimetres below the seafloor. Most corers are limited in diameter (roughly 10 cm) and require powerful winches for their deployment and recovery. They are therefore mostly used by purpose-built oceanographic or geotechnical vessels, not easily accessible to the scientific community. Moreover cores or grabs cannot be operate in rocky substrates.

**Diver-Controlled Excavation** This involves the careful removal of sediment in order to expose archaeological layers and artefacts so they can be carefully recorded. Excavation often involves the use of hydraulic and airlift dredges. Whereas the latter performs better at greater depths, the former is very effective in shallow water (Faught 2014). The dredged material can be sieved to recover any artefacts. When dealing with very small items (e.g., prehistoric flints) it can be appropriate to recover material in sample boxes. In the case of loose deposits, coffer structures are necessary for deeper excavations to prevent slumping of the sides. Rigid frames may also be used to provide fixed points from which to measure, or as support for divers to help them keep clear of delicate material (Fig. 2.9).

So far, surprisingly little research has been done to identify the sedimentary signatures of the deposits that make up submerged archaeological sites. Yet Gagliano et al. (1982) showed already several decades ago that detailed recording of colour, bedding, and contact descriptions

**Fig. 2.9** Divers excavating the Mesolithic pit dwelling at Møllegabet II, Denmark, from a wooden platform positioned above the cultural layer under excavation (Photo M. Gull)

complemented with point counting, grain size analysis and multiple geochemical analysis make it possible to distinguish probable archaeological sites from 'natural' sites. This is an important technique for further research and calibration of proxy indicators.

## 2.6 Submersibles and Underwater Vehicles

Manned underwater submersibles (or Human operated vehicles, HOVs) and remotely operated vehicles (ROVs) have been used by marine scientistists, military and offshore industry for over 40 years. During recent years the use of AUV[1] (autonomous underwater vehicle) technology for the survey of the seafloor has become increasingly available and economical. Due to the relatively limited bottom time (5–6 h on average), slow speeds and human pilots, HOVs are best suited for direct-observation mapping and sampling rather than fine-resolution surveys. ROV surveys do not have the constraint on bottom time, but often require a dynamically positioned support ship. However, the tethered configuration limits the efficiency and effectiveness, as robot and surface ship have to move in concert, and strong currents may also be a problem. AUVs have proven their utility as a stable, controlled near-bottom survey platform. They are capable of flying precisely controlled fixed-altitude survey lines, making full use of the sonar resolution. They can operate from modest support ships (or from shore) and can survey large areas of the seafloor for 24–72 h without returning to the surface (Bingham et al. 2010).

The most commonly used sensors mounted on ROVs, HOVs and AUVs include navigation sensors for positioning, optical sensors (video, photographic, stereoscopic still cameras), sonar sensors for mapping the seafloor and its features (multibeam, side scan sonar, subbottom profiler) and chemical/environmental sensors for quantifying the oceanographic environment. Modern digital image recording, combined with accurate position fixing, facilitates the merging of hundreds or thousands of digital images into continuous optical maps (Mahon et al. 2011; Mallios 2014).

Though considerable progress has been made in underwater robotics, and a variety of underwater vehicles has been used in shallow and deep water for different applications, from mapping to sampling and excavating deep water wrecks and carrying different payloads and sensors, so far their application to seabed prehistory and submerged shoreline studies has been astonishingly limited. One of the few exceptions is the Pavlopetri project where an AUV was used for the optical identification of small surface artefacts (Fig. 2.10, Henderson et al. 2011). This approach could be developed for deeper work.

## 2.7 Photogrammetry

The aim of photographic techniques is to produce a precise, three-dimensional map and image of the archaeological site. In recent years rapid progress has been made, as increasingly high precision navigation and vehicle control permit high precision positioning of the acquired photographic data (McCarthy and Benjamin 2014; Uldum et al., Chap. 5). Furthermore, photo- and video-mosaic techniques can combine photographic images with precise positioning.

Most common techniques for automated creation of mosaics make use of simultaneous localization and mapping (SLAM), augmented with techniques from computer vision and photogrammetry, to create a consistent set of image transformations (e.g., Bingham et al. 2010). These techniques enable automated generation of strip mosaics, using data association between sequential images to

[1] AUVs operate independently from the ship whereas ROVs are connected by a cable to an operator on the ship.

**Fig. 2.10** Launching of an Autonomous Underwater Vehicle (AUV, *left*) and Remotely Operated Vehicle (ROV, *right*) for underwater archaeological surveys (© MMRG and Archivo ARQUA)

produce a mosaic representing a single pass over the site. Currently, active research is ongoing with regard to multiple omnidirectional transects (Eustice et al. 2008).

One of the latest advance in underwater photographic techniques is the creation of 3D reconstructions in parallel with generation of 2D photomosaics (Pizarro et al. 2009) and the development of visually augmented navigation (VAN) (Fig. 2.11, Eustice et al. 2008). These techniques extract three-dimensional bathymetry estimates based on the collected photographic images Recent software developments allow three-dimensional models of outcrops reconstructed from video and photographic data, facilitating quantitative measurements and photo mosaics within a georeferenced framework. The compiled data enable post-cruise 'virtual fieldwork' on the seafloor, providing the possibility for field data interpretations at a variety of scales from hand specimens to outcrops based on the quantified optical data (Kwasnitschka et al. 2013). Integration of photographic and sampling technology is perfectly illustrated with the 'Sandcam' microscope camera, a so-called contact video microscope, which provides high magnification CCTV images of seafloor sediments that are in contact with the viewing port of the camera (Puleo and Pearre 2008).

Most of the discussed improvements in 2D and 3D optical imagery are related to the use of autonomous underwater vehicles, since AUVs often prove more stable platforms compared to ROVs and HOVs and are capable of travelling at constant altitude and speed above the seafloor (Bingham et al. 2010). Given that acoustical detection of small surface artefacts is still problematic, optical identification of artefacts (in reasonably clear water), for instance using an AUV patrolling close to the seabed, would greatly help to identify sites.

One of the drawbacks of photography and videography is that these techniques do not provide accurate dimensional information of underwater archaeological sites. The latter is possible with underwater laser scanning (USL), which allows extremely high-detail (mm or sub-mm level) 3D images of exposed targets (Gillham 2011). USL can be deployed by ROV, AUV or divers, and has a maximum range of 5 m. The technique was applied successfully in 2014 on the Monohansett shipwreck site, where the 3D scans enabled detailed understanding of the current condition of the wreck (2GRobotics 2015).

**Fig. 2.11** 3D coloured relief (*top left*) and 2D photo-mosaic (*top right*) showing a chamber tomb (approx. 4 × 5 m, cut 3 m into the surrounding bedrock) at the submerged archaeological site of Pavlopetri, Greece. The image was produced by geo-referenced stereo imagery from a diver-propelled platform (shown below) combined with mapping techniques (After Henderson et al. 2013; Mahon et al. 2011

## 2.8 Challenges for the Future

Notwithstanding the current availability of high-performance technology and the technological advances that are being made, a large number of challenges for the future remain. Perhaps the most important one is to develop methods that can increase the ratio of archaeological finds discovered by purposeful exploration to those discovered by chance. This will not only require more reliable remote sensing data (i.e., the confirmation of a prehistoric site without actual seabed examination or large-scale physical sampling), but also better cost-effectiveness (i.e., improved search efficiency and integration of existing data). A second important overall challenge lies in direct observations—once the site has been found—i.e., efficient sampling, documentation and excavation, especially in deeper water.

In addition to these general challenges a number of specific technological challenges exist. A few of these are highlighted here.

**Detection of Small Buried Artefacts and Bones** Identifying small objects buried beneath the seabed is still a major challenge. This is due not only to the resolution of the current acoustic imaging systems (limited to a few dm at best), but also to the likelihood that the (acoustic) contrast between the object and the surrounding sediments may not be sufficient. Ongoing Danish-Belgian research focuses on the acoustic features of worked flint, specifically differential resonance patterns (Hermand et al. 2011; Ren et al. 2011). Although the first results from laboratory studies are promising, extensive further research is needed.

**Effective Image Recognition of the Seabed** Notwithstanding the recent advances in sonar and video technology, there still seems to be no system available that can recognize prehistoric relics on the seabed in an efficient way. Accurate object identification on sonar data is a difficult task, and automated classification methods are often not able to differentiate between seafloor areas that contain features of archaeological significance and areas that are barren (Bates et al. 2007). Future research is needed into combining the information of full backscatter maps, high resolution bathymetry and photographic data in a smart and efficient way. An important recent development is an efficient and easy to deploy Underwater Laser Scanning system (Gillham 2011).

**Excavating in Deep Water** Underwater expeditions with HOVs, ROVs and AUVs have yielded spectacular findings on the seafloor and produced high quality results particularly in deep and shallow water shipwreck archaeology. So far the main use of HOV, ROV and AUV is in documenting, remote sensing and sampling (and at its best digging trial trenches). An important challenge for the future is to advance their use from visual surveys and incidental salvage to real excavations. This will require highly delicate (and remotely controlled) robotic manoeuvring which poses a high engineering challenge.

**Detection and Excavation of Cave Sites** Conventional seafloor mapping is based on a plan view projection, which is not suitable for exploration of nearly vertical limestone walls where cave entrances may be found. Very-large-swath multibeam systems, currently used for shallow water inspection of harbour structures, can be adapted to ROV for deeper water use. DEM-rendering software used for multibeam and side scan sonar data interpretation could be modified to manage 3D data (indeed some software already support this). Examination of the inside of a cave is still a major problem, and so far no efficient techniques for this seem to exist, as the risk of failure has deterred trials. This area needs intensive study to find a way forward.

**Survey of Intertidal Areas** Intertidal areas are problematic due to the extreme shallow water, tidal effects and wave disturbance, and the often widespread presence of shallow gas. Archaeological exploration is very difficult (especially in high-energy areas) and seldom attempted, which is unfortunate since these areas are often known to be rich in archaeological remains. Recent studies have indicated that a smart combination of complementary techniques, both acoustic and non-acoustic, and preferably combining marine (during high tide) and terrestrial (during low tide) investigations, is crucial (Missiaen et al. 2008; Evangelinos and Missiaen 2012; Delefortrie et al. 2013). Furthermore, ongoing studies in Belgium in the framework of the SeArch project focus on the potential of acoustic shear waves and surface waves for archaeological assessment of intertidal and nearshore areas (Kruiver et al. 2013; Missiaen et al., Chap. 27).

## 2.9 Conclusions

Due to the wide variety of observational scales and environmental conditions encountered in submerged prehistory, research involves a large spectrum of technologies. This ranges from acoustic, (electro)magnetic and Lidar remote sensing to underwater coring and sampling, diving surveys and the use of underwater vehicles. Many of the techniques were developed for other purposes (e.g., geological or terrestrial studies, military research), and they are often aimed at shallow water environments. In order to adapt the technology to the specific needs of submerged prehistory research and to advance to deeper water investigations, improvements in survey techniques are needed. This also involves new developments in software to allow the integration of large data volumes from a wide range of sources. All these developments require sufficient resources, which can only be achieved through increased scientific budgets, both on a national and on an international scale. Close collaboration with marine industry may facilitate technologically demanding research and also reduce costs.

Industrial projects in underwater settings make elaborate use of seismic surveying, dredging/extraction and coring, and these data can be of great interest for submerged prehistory research. Yet they are an under-utilized source. Petroleum industry (3D seismic) data, for instance, cover large areas of the continental shelf but with relatively low spatial resolution (line spacing at a few hundreds of metres). The Doggerland project has shown that the use of 3D industrial seismic data, combined with novel data analysis techniques developed by the petroleum industry, provides an efficient way to map the Late Pleistocene and Holocene palaeomorphology (Fitch et al. 2005; Gaffney et al., Chap. 20; Holmlund et al., Chap. 4). Industrial cores are originally carried out for engineering purposes, but the resulting information about the substrate is extremely useful for the geoarchaeologist, and collaboration with industry can seriously reduce the cost of sampling.

At the same time the expansion of industrial activities on the continental shelf forms a threat for the archaeological material (either by destruction or by removal from its context). This is especially the case with the aggregate industry, where the key resource zones often coincide with zones of high archaeological potential, at least with regard to Palaeolithic material (e.g., relict fluvial terraces). A close collaboration with the aggregate industry is therefore needed, the benefits of which are well demonstrated by the success of the work supported by the Aggregates Levy Sustainability Fund (ALSF) in English waters between 2002 and 2011 (Gaffney et al., Chap. 20; Sturt, Chap. 28). This will not only help to avoid unnecessary loss of archaeological information, but will also increase the number of finds.

With regard to industrial projects at sea the set-up of a 'best practice' approach is crucial, going from general reconnaissance surveys to the localisation and investigation of small sites or targets, keeping in mind the different settings and environments in different sea areas. This will allow critical decisions to be taken at every step of the research regarding more detailed survey, assessment, test excavation, preservation in situ, etc. Furthermore such a best practice approach can become a strong tool in Environmental Impact Assessments (before research is carried out).

A final, but important, aspect is the availability of new technology. As shown above there is a large technological potential but not everyone has access to the technology and related infrastructure. Collaboration between national institutions to share expertise and techniques is therefore crucial. High-technology data acquisition and processing requires experienced people and trained operators. This collaboration is largely interdisciplinary in nature, given the wide range of techniques and data that need to be integrated, and increasingly international.

**Acknowledgements** This review study was carried out in the framework of COST Action TD0902 'Submerged Prehistoric Archaeology and Landscapes of the Continental Shelf (SPLASHCOS)' and the Flemish IWT-SBO project 'Archaeological Heritage in the North Sea (SeArch)'.

## References

2GRobotics (2015) www.2grobotics.com/industries/marine-archaeology/

Bates M, Dean M, Lawrence M, Robertson P et al. (2007) Innovative approaches to Rapid Archaeological Site Surveying and Evaluation (RASSE). Final Report. English Heritage, report nr. 3837. http://www.st-andrews.ac.uk/rasse/library/index.html

Bates R, Lawrence M, Dean M, Robertson P (2011) Geophysical methods for wreck-site monitoring: the rapid archaeological site surveying and evaluation Programme. Naut Archaeol 40(2):404–416

Bingham B, Foley B, Singh H, Camilli R, Delaporta K, Eustice R, Mallios A, Mindell D, Roman C, Sakellariou D (2010) Robotic tools for deep water archaeology: surveying an ancient shipwreck with an autonomous underwater vehicle. J Field Robot 1–16. doi:10.1002/rob.20350, Wiley Periodicals, Inc

Boyce JI, Reinhardt EG (2004) Marine magnetic survey of a submerged roman harbor, Caesarea Maritima, Israel. Int J Naut Archaeol 33(1):122–136

Bull J, Gutowski M, Dix J, Henstock T, Hogarth P et al (2005) Design of a 3D chirp sub-bottom imaging system. Mar Geophys Res 2(4):157–169

Camidge K, Holt P, Johns C, Randall L, Schmidt A (2010) Developing magnetometer techniques to identify submerged archaeological sites. Theoretical Study Report. Cornwall Council Historic Environments Projects, report nr. 2010R012. http://archaeologydataservice.ac.uk/archsearch/

Delefortrie S, Saey T, Van De Vijver E, De Smedt P, Missiaen T, Demerre I, Van Meirvenne M (2013) Frequency domain electromagnetic induction survey in the intertidal zone: limitations of low-induction-number and depth of exploration. J Appl Geophys 100:14–22

Eustice R, Pizarro O, Singh H (2008) Visually augmented navigation for autonomous underwater vehicles. IEEE J Ocean Eng 33(2):103–122

Evangelinos D, Missiaen T (2012) Acoustic mapping of the intertidal area in Raversijde and comparison with electromagnetic data. Arch-Manche unpublished report, Renard Centre of Marine Geology, University of Ghent

Faught M (2014) Remote sensing, target identification and testing for submerged prehistoric sites in Florida: process and protocol in underwater CRM projects. In: Evans A, Flatman J, Flemming NC (eds) Prehistoric archaeology on the continental shelf: a global review. Springer, New York

Fitch S, Thomson K, Gaffney V (2005) Late Pleistocene and Holocene depositional systems and the palaeogeography of the Dogger Bank, North Sea. Quat Res 64:185–196

Flemming NC, Cagatay MN, Chiocci FL, Galandiou N, Lericolais G, Joens H, Missiaen T, Moore F, Rosentau A, Sakellariou D, Skar B, Stevenson A, Weerts H (2014) Land beneath the waves: submerged landscapes and sea level change. A joint geoscience-humanities strategy for European continental shelf prehistoric research. In: Chu NC, McDonough N (eds) Position paper 21 of the European marine board. European Marine Board, Ostend

Gagliano SM, Pearson CE, Weinstein RA, Wiseman DE, McClendon CM (1982) Sedimentary studies of prehistoric archaeological sites: criteria for the identification of submerged archaeological sites of the northern Gulf of Mexico continental shelf. Prepared for the U.S. Department of the Interior National Park Service, Division of State Plans and Grants, Contract No. C35003(79)

Gillham J (2011) Development of a field deployable underwater laser scanning system. PhD thesis, University of Waterloo, Canada

Grøn O, Nørgård Jørgensen A, Hoffmann G (2007) Marine archaeological survey by high-resolution sub-bottom profilers. Norsk Sjøfartsmuseums Årbok. pp 115–144

Gutowski M, Bull J, Dix J, Henstock T et al (2008) Three-dimensional high-resolution acoustic imaging of the sub-seabed. Appl Acoust 69:412–421

Henderson R, Day-Lewis F, Abarca E, Harvey C et al (2010) Marine electrical resistivity imaging of submarine groundwater discharge: sensitivity analysis and application in Waquoit Bay, Massachusetts, USA. Hydrogeol J 18(1):173–185

Henderson J, Gallou C, Flemming NC, Spondylis E (2011) The Pavlopetri underwater archaeology project: investigating an ancient submerged town. In: Benjamin J, Bonsall C, Pickard C, Fischer A (eds) Submerged prehistory. Oxbow Books, Oxford, pp 207–218

Henderson J, Pizarro O, Johnson-Roberson M, Mahon I (2013) Mapping submerged archaeological sites using stereo-vision photogrammetry. Int J Naut Archaeol 42(2):243–256

Hermand, JP, Grøn O, Asch, M, Ren QY (2011) Modelling flint acoustics for detection of submerged Stone Age sites. In: Proceedings of OCEANS 2011 IEEE conference, Santander, Spain. http://www.lamfa.u-picardie.fr/asch/f/p/ac/7816Ed01_final_opt.pdf. Accessed 20 June 2016

Innomar (2014) SES-2000 Quattro parametric subbottom profiler. www.innomar.com

Johnson-Roberson M, Pizarro O, Williams S, Mahon I (2010) Generation and visualization of large-scale three-dimensional reconstructions from underwater robotic surveys. J Field Robot 27(1):21–51

Klein G, Bohlen T, Forbriger T (2005) Acquisition and inversion of dispersive seismic waves in shallow marine environments. Mar Geophys Res 26:287–315

Kruiver P, Zurita-Hurtado O, Diafera G, Missiaen T (2013) Non-conventional techniques for marine archaeological investigations. SeArch report, Renard Centre of Marine Geology, University of Gent. http://www.sea-arch.be/nl/resultaten

Kwasnitschka T, Hansteen TH, Devey CW, Kutterolf S (2013) Doing fieldwork on the seafloor: photogrammetric techniques to yield 3D visual models from ROV video. Comput Geosci 52:218–226

Lowag J (2010) Application of multi-transducer parametric sub-bottom profiler for three-dimensional archaeological investigations in shallow waters. www.innomar.com

Mahon I, Pizarro O, Johnson-Roberson M, Friedman A, Williams SB, Henderson JC (2011) Reconstructing Pavlopetri: mapping the world's oldest submerged town using stereo-vision. In: 2011 IEEE international conference on robotics and automation (ICRA), pp 2315–2321. doi:10.1109/ICRA.2011.5980536

Mallios A (2014) Sonar scan matching for simultaneous localization and mapping in confined underwater environments. PhD thesis, University of Gerona
McCarthy J, Benjamin J (2014) Multi-image photogrammetry for underwater archaeological site recording: an accessible, diver-based approach. J Marit Archaeol 9:95–114
Missiaen T (2005) VHR marine 3D seismics for shallow water investigations: some practical guidelines. Mar Geophys Res 26:145–155
Missiaen T (2010) The potential of seismic imaging in marine archaeological site investigations. Relicta 6:219–236
Missiaen T, Feller P (2009) Detection of chemical munition buried below the seabed using seismic and magnetic techniques. First Break 27:69–74
Missiaen T, Murphy S, Loncke L, Henriet JP (2002) Very high resolution seismic mapping of shallow gas in the Belgian coastal zone. Cont Shelf Res 22(16):2291–2301
Missiaen T, Slob E, Donselaar ME (2008) Comparing different shallow geophysical methods in a tidal estuary, Verdronken land van Saeftinge, western Scheldt, the Netherlands. Neth J Geosci 87(2):151–164
Müller H (2009) Characterization of marine near-surface sediments by electromagnetic profiling. PhD thesis, University of Bremen. http://elib.suub.uni-bremen.de/diss/docs/00011853.pdf
Passaro S (2010) Marine electrical resistivity tomography for shipwreck detection in very shallow water: a case study from Agropoli (Salerno, southern Italy). J Archaeol Sci 37:1989–1998
Pastol Y (2011) Use of airborne LiDAR bathymetry for coastal hydrographic surveying: the French experience. J Coast Res 62:6–18
Pe'eri S, Morgan L, Philpot W, Armstrong A (2011) Land-water interface resolved from airborne LiDAR bathymetry (ALB) waveforms. J Coast Res 62:75–85
Pizarro O, Eustice RM, Singh H (2009) Large area 3-D reconstructions from underwater optical surveys. IEEE J Ocean Eng 34(2):150–169
Plets R, Dix J, Bastos A, Best A (2007) Characterization of buried inundated peat on seismic (chirp) data, inferred from core information. Archaeol Prospect 14:261–272
Plets R, Dix J, Adams J, Bull J, Henstock T, Gutowski M, Best A (2009) The use of a high-resolution 3D chirp sub-bottom profiler for the reconstruction of the shallow water archaeological site of grace Dieu (1439), river Hamble, UK. J Archaeol Sci 36(2):408–418
Puleo JA, Pearre N (2008) Coastal currents. Monitoring coastal Delaware – the SANDCam System. University of Delaware's Center for Applied Coastal Research. www.deseagrant.org/products/coastal-currents-monitoring-coastal-delaware-sandcam-system. Accessed Nov 2015
Ren QY, Grøn O, Hermand, JP (2011) On the in-situ detection of flint for underwater Stone Age archaeology. In: Proceedings of OCEANS 2011 IEEE conference, Santander, Spain. doi:10.1109/Oceans-Spain.2011.6003529
Rucker DF, Noonan GE, Greenwood WJ (2011) Electrical resistivity in support of geological mapping along the Panama Canal. Eng Geol 117:121–133
Sakellariou D, Rousakis G, Maroulakis S, Georgiou P et al. (2011) The submerged city of Pavlopetri. In: Poseidons Reich XVI, (DEGUWA 2011), Heidelberg, 18–20 Feb 2011
SeArch (2015) Archaeological heritage in the North Sea. http://www.sea-arch.be/
Tarits P, D'Eu JF, Balem K, Hautot S, Prevot J, Gaspari F. (2012). Mapping seismically masked seabed structures with a new DC resistivity streamer. In: 18th European meeting of environmental and engineering geophysics EAGE, Paris, France. doi:10.3997/2214-4609.20143374
Van Lancker V, Dufour I, Mathys M, Versteeg W, De Batist M (2009) Towards a high-resolution 3D–analysis of sandbank architecture on the Belgian Continental Shelf. Final Report. Belgian Science Policy, Brussels. https://www.belspo.be/belspo/fedra/proj.asp?l=en&COD=AP/20
Vardy M, Dix J, Henstock T, Bull J, Gutowski M (2008) Decimeter-resolution 3D seismic volume in shallow water: a case study in small-object detection. Geophysics 73(2):33–40
Wunderlich J, Wendt G, Müller S (2005) High-resolution echo-sounding and detection of embedded archaeological objects with nonlinear sub-bottom profilers. Mar Geophys Res 2–4(2005):123–133
Zurita Hurtado OJ, Missiaen T, De Clercq M, Versteeg W, Kruiver P, De Kleine M (2014) Comparative seismic source study for buried palaeolandscape investigations in the southern North Sea. In: Near surface geoscience conference abstracts, Mo Myce 06. http://lib.ugent.be/nl/catalog/pug01:5786258

# Chapter 3
# Relative Sea Level Rise, Palaeotopography and Transgression Velocity on the Continental Shelf

**Francesco Latino Chiocci, Daniele Casalbore, Francesca Marra, Fabrizio Antonioli, and Claudia Romagnoli**

**Abstract** After the Last Glacial Maximum, some 21,000 years BP, the sea level rose from −130 m to its present-day position. This process of marine transgression inundated or eroded palaeolandscapes to varying degrees, resulting in the landward movement of the shoreline. The transgression velocity (TV), i.e., the velocity at which the shoreline migrated landwards, depends on evaluating the balance between the rate of relative sea level rise and the slope of the transgressed palaeotopography. It has a key role in determining the possibilities for reconstructing palaeoenvironments, the potential preservation of archaeological sites and the socio-economic and psychological impact of sea-level rise on past human populations. In this chapter we present a simple conceptual and computational approach to reconstructing the transgression velocity on shelf areas, making use of Digital Terrain Models (DTMs) of seafloor topography coupled with relative sea level curves, and discuss the different outcomes and limitations at different spatial scales, ranging from the continental (European seas) to the ultra-local scale.

## 3.1 Introduction

The interest for prehistoric archaeologists in continental shelves is steadily increasing as it becomes clear that present-day submerged shelf areas afforded palaeolandscapes for human exploitation with a range of coastal, marine and terrestrial resources as well as access to transportation and migration

F.L. Chiocci • D. Casalbore (✉)
Department of Earth Science, University of Rome La Sapienza, Piazzale Aldo Moro 5, 00185 Rome, Italy

Institute of Environmental Geology and Geoengineering, National Research Council (CNR IGAG), Area della Ricerca Roma 1, Via Salaria km 29,300, 00015 Rome, Italy
e-mail: francesco.chiocci@uniroma1.it; daniele.casalbore@uniroma1.it

F. Marra
Department of Earth Science, University of Rome La Sapienza, Piazzale Aldo Moro 5, 00185 Rome, Italy
e-mail: fmarra19@gmail.com

F. Antonioli
Italian National Agency for New Technologies, Energy and Sustainable Economic Development (ENEA), Via Anguillarese 301, 00060 S. Maria di Galeria, Rome, Italy
e-mail: fabrizio.antonioli@enea.it

C. Romagnoli
Department of Biological, Geological and Environmental Sciences, University of Bologna, Piazza di Porta San Donato 1, 40127 Bologna, Italy
e-mail: claudia.romagnoli@unibo.it


G.N. Bailey et al. (eds.), *Under the Sea: Archaeology and Palaeolandscapes of the Continental Shelf*, Coastal Research Library 20, DOI 10.1007/978-3-319-53160-1_3

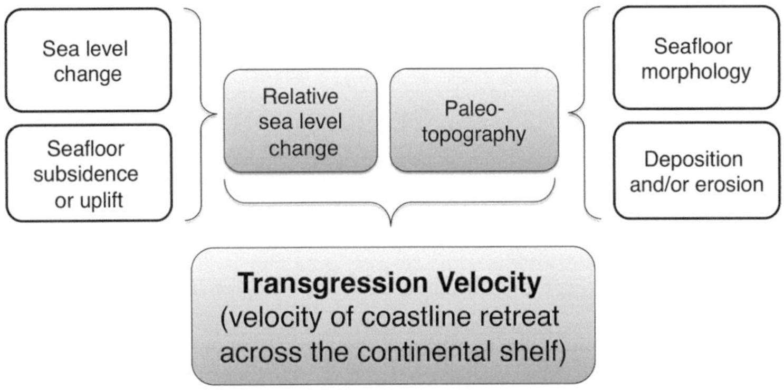


**Fig. 3.1** The velocity of coastline retreat (transgression) across the continental shelf emphasising the interplay between relative sea level change, the pre-existing seabed topography and the present-day seabed topography as modified by erosional and depositional processes

routes during the glacial and early postglacial periods (Bailey and Flemming 2008; Flemming et al. 2014; Evans et al. 2014). Many archaeological findings have been discovered on the seabed and the environmental features of the submerged landscape have been reconstructed with increasing detail (Anzidei et al. 2014; Evans et al. 2014). The expectation of future discoveries is very high, due to the growing interest in this new research field and to technological advances in high-resolution underwater investigation (Flemming et al. 2014). Underwater prehistoric archaeology crosses disciplinary boundaries and a variety of data is needed for a precise definition of the position of palaeoshorelines and the reconstruction of palaeolandscapes, both for palaeoenvironmental reconstruction, for the identification of possible sites where prehistoric artefacts and remains can be found, and for an evaluation of the social, economic and psychological impact of past sea-level rise (Evans et al. 2014; Flemming et al. 2014).

A commonly underestimated but crucial factor is the velocity at which the high-energy shoreline environment migrated across the continental shelf in the last 21 ka BP, from more than −100 m to its present position (Flemming et al. 2014). The velocity of coastline retreat, or transgression velocity (TV), on a continental shelf depends on the often-complex interplay between (1) sea-level change, (2) seafloor subsidence or uplift, (3) seafloor morphology and (4) erosional or depositional processes that may have altered the pre-existing land surface and created the present-day bathymetry (Fig. 3.1).

Conceptual and computational models reproducing the effects of this interplay of factors are useful for prehistoric archaeology but they may also give useful information on the morphogenesis of specific features of the shelf, rates and amounts of eroded sediment available for past littoral dynamics, formation of specific sediments or vegetational assemblages, and so on (e.g., Trenhaile 2001, 2002, 2011; Zecchin et al. 2015).

Here, we propose a simple computational approach taking into account the combination of relative sea-level change and present-day bathymetry and discuss its application, potential and significance at different scale ranges.

## 3.2 Relative Sea-Level Change

Sea-level during the last glacial cycle rose from about −130 m at the Last Glacial Maximum (LGM), 21,000 years BP, to the present-day level (Caruso et al. 2011). For instance, in the Mediterranean, at about 6800 cal years BP, the sea-level curves show a clear inflexion with a slowing down of the sea level rise from 6 to about 1 mm/year (Lambeck et al. 2011).

The overall rate of sea-level rise since the LGM was about 1 cm/year but it did not occur at a steady rate. It was very slow at the beginning and at the end of the rise; for example, the sea level in the last 2000 years has been rising at an average rate of 0.1 mm/year (Lambeck et al. 2014). In contrast, periods of very fast sea level rise, probably linked to abrupt input of freshwater produced by melting ice caps, are recognised worldwide and reached rates of up to 3–4 cm/year (Fairbanks 1989; Bard et al. 1996; Liu and Milliman 2004).

However, the absolute (eustatic) sea-level curve can be significantly modified regionally and even locally by uplift or subsidence of the continental margin due either to (1) tectonics (including volcanism) or (2) isostatic readjustment of the Earth's crust due to the loading and unloading of ice caps thousands of meters thick, and of water masses on the continental shelf with depths of up to 100 m or more (for a detailed analysis, see Lambeck and Purcell 2005).

In recent years, geophysical modelling of the rheology of the Earth crust's in response to these glacio-isostatic and hydro-isostatic processes has been proposed and verified in tectonically stable areas (Lambeck et al. 2006, 2011; Bailey et al. 2011). As far as tectonics is concerned, the behaviour of a given coastal area is usually derived from studies of coastal terraces (Ferranti et al. 2006 and Anzidei et al. 2014 for the Mediterranean Sea, Harff and Meyer 2011 for the Baltic Sea) suggesting that non-linear or alternating vertical movements often occur, and that these can differ even in closely neighbouring coastal regions. Recently, insular shelves and submarine depositional terraces have been also used to constrain vertical movements (Quartau et al. 2014; Casalbore et al. 2017 and references therein).

For this reason each sector of the shelf should refer to a specific curve that takes into account the different components contributing to the relative sea-level changes in that sector. In enclosed seas the situation may be even more complex due to the opening or damming of seaways that resulted in rapid flooding or delayed onset of transgression in specific basins, as for instance in the Black Sea and Baltic Sea (Ryan et al. 2003; Uścinowicz 2006, respectively, see also Hansson et al. Chap. 13).

## 3.3 Palaeotopography

At a broad scale, topography of the shelves may be derived from nautical charts or from bathymetric compilations such as EMODnet (European Marine Observation and Data Network) for European seas and GEBCO (General Bathymetric Chart of the Oceans) worldwide, having a spatial resolution of 500 m or more. At a higher resolution, the reconstruction of the present-day topography of the continental shelf is best based on acoustic swath-bathymetry techniques providing Digital Terrain Models (DTMs) with a grid-size of metre to sub-metre scale.

In order to reconstruct the palaeotopography when the coastline migrated over a given area, depositional or erosional processes also have to be taken into account (Flemming 1972). Deposition of highstand sediments and, to a lesser extent, of transgressive parasequences (*sensu* Van Wagoner et al. 1988) made up of submerged beach facies, may have modified the seafloor morphology, blanketing the underlying transgressive surface and often infilling topographic depressions. In this case, the reconstruction of the original topography has to rely on high-resolution (hereafter HR) reflection seismic profiles that can see through the overlying sediments and map the underlying surface, although only in a two-dimensional section and therefore with a limited spatial resolution (see Sect. 3.3.4.2 on sources of error).

Much more complex and uncertain is the reconstruction of erosional processes that altered the pre-existing topography. For instance, the unconformity that was created during the last glacial cycle by the emergence of the continental shelf with consequent erosion by subaerial processes (river, wind, glaciers, etc.) has usually been re-worked during subsequent sea-level rise, creating a transgressive erosional surface known as a *ravinement surface* (Nummedal and Swift 1987). This surface usually shows a flat or smooth morphology, showing no evidence of unevenness due to subaerial erosion, nor the presence of continental deposits (Fig. 3.2). The main exception to this is where the original land surface comprised hard bedrock and where eroded remnants may be preserved (e.g. Zecchin et al.

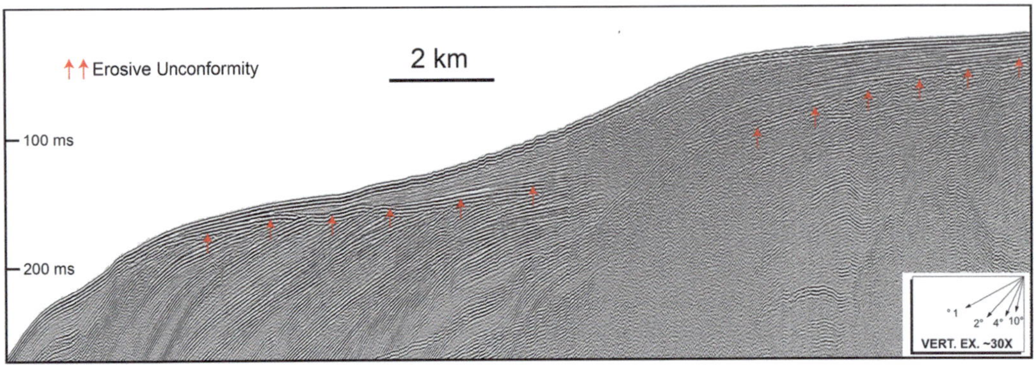


**Fig. 3.2** High resolution seismic profile off the mouth of the Tiber River. The presence of postglacial marine sediments means that the present-day bathymetry is very different with respect to the palaeosurface transgressed by the rising sea level. Note the flatness of the erosional unconformity (ravinement surface) at the base of the postglacial depositional sequence. For location, see Fig. 3.6

2015 and references therein). Hence the amount of seafloor removed or modified by erosion due to shoreline transgression strongly depends on the nature of the transgressed substratum, wave regime and tidal range (Belknap and Kraft 1981; Bernè et al. 1998; Cattaneo and Steel 2003).

## 3.4 Transgression Velocity on the Continental Shelf

Transgression velocity is a combined function of relative sea level rise and the slope of the transgressed topography (Fig. 3.3). For example, slow sea-level rise and high-gradient topography result in a low TV, while high velocity is the result of fast sea-level rise coupled with low-gradient topography. Analytically, this velocity can be expressed by the ratio between the rate of relative sea-level rise and the slope of the transgressed surface. For the most complete results, a three-dimensional reconstruction of seafloor topography is desirable, and for this purpose we use Digital Terrain Models (DTMs). If the relative sea-level curve between specified bathymetric intervals is taken into account and the rate of sea-level rise is considered, by estimating the seafloor slope at each point of the DTM over that bathymetric interval, one may apply the following:

$$TV = \frac{RSLR}{SFS} = \frac{H / TH}{V / O} = \frac{H \times O}{V \times TH} \tag{3.1}$$

Where: *TV* transgression velocity, *RSLR* rate of sea-level rise, *SFS* seafloor slope, *H* relative sea-level rise, *TH* time interval over which sea-level rise occurred, *O* horizontal distance, *V* vertical distance.

If the sea level rise is computed as corresponding to an increase of sea level of 1 m and the slope is computed as the horizontal distance corresponding to 1 m on the vertical scale, then:

$$TV = \frac{\cancel{H} x O}{\cancel{V} x TH} = \frac{horizontal\,distance\,for\,1m\,vertical\,offset}{time\,interval\,over\,which\,1m\,sea\,level\,rise\,occurred} \tag{3.2}$$

In this way, for any given depth at any given point, the TV can be computed from information on sea floor topography and the relative sea level curve.

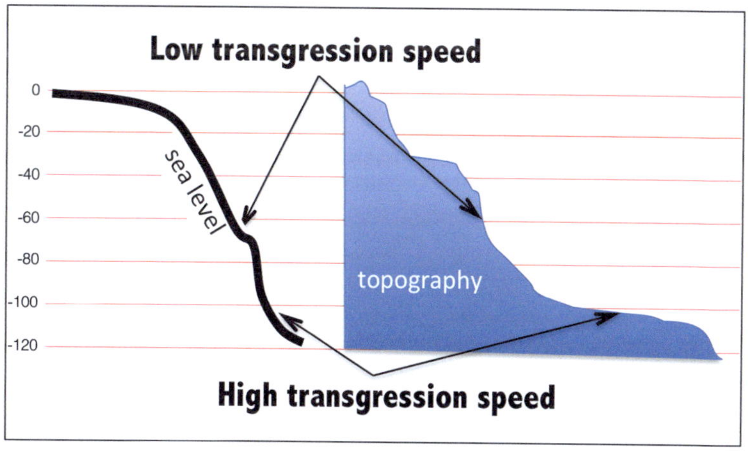


**Fig. 3.3** The transgression velocity, i.e. the velocity at which the shoreline moved landwards, is the combination of relative sea level rise and the slope of the transgressed topography

### *3.4.1 Computational Procedure*

We have created a MATLAB routine (named TRASPEED) to combine the grid data and the relative sea-level curve(s) in order to obtain the TV. The TRASPEED routine allows us to handle large data-sets, to compute the values needed for the TV algorithm, and to generate ascii output files that can drape successive surfaces onto the available DTM (Fig. 3.4). In detail:

1. ***trasdec.m*** deletes possible null depth values or depth errors present in the DTM, and then transforms slope gradient units from degrees to radians
2. ***RSLR.m*** defines, for reference sites, the vertical rate of sea-level rise at each depth, using linear or spline interpolation of sea-level curves
3. ***DWM.m*** associates the matrices produced by the previous two subroutines and is the core of the MATLAB routine. Using an inverse distance weight (Dodonov and Dodonova 2011), it averages two neighbouring curves to compute the values for each grid node
4. ***Trasp.m*** computes horizontal rates of movement as described in Sect. 3.3.4 and Eq. 3.2

### *3.4.2 Sources of Error*

When reconstructing the velocity of the shoreline transgression, it is important to be aware that errors and inaccuracies are inherently present in the computation. These include the following:

**Eustatic Curve** The sea level rise is reconstructed as a continuous record from variations in the ratio (Δ) $^{18}$O and $^{16}$O isotopes as a proxy for ice volume and therefore for water subtracted from the oceans. The composite curves so obtained are refined by additional field observations such as dates of sea-level-related corals or marine deposits (Bard et al. 1996; Waelbroeck et al. 2002; Siddall et al. 2003). Different authorities, however, offer different curves so that margins of uncertainty involving at least some metres have to be taken into account.

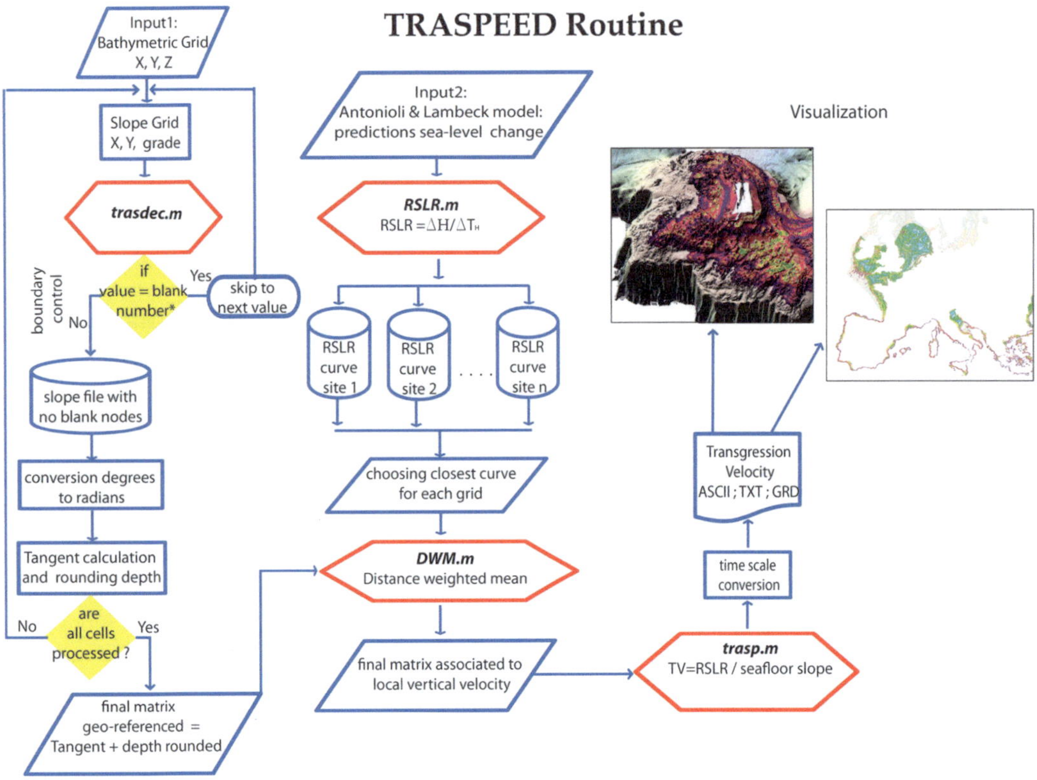


**Fig. 3.4** Flow chart of the TRASPEED MATLAB routine

**Tectonic/Isostatic Component of Sea Level Rise** Vertical behaviour of a coastal area is usually derived from the height/depth of the MIS5.5 shoreline. The reconstructed eustatic height of this feature is between +3 m and +7 m in the northern hemisphere (Ferranti et al. 2006; Pedoja et al. 2014). The offset of the local MIS5.5 shoreline with respect to this reconstructed height is used to compute the average subsidence or uplift over about the last 124–130,000 years. This raises two possible sources of error: (a) averaging vertical movements on this time period may be not correct, as tectonics may not behave linearly, especially in geologically active areas; and (b) continental margins are commonly affected by subsidence (resulting from thermal, loading or compaction factors) that increases towards the outer edge of the shelf, with a hinge-line located near the coast. Therefore extrapolation of vertical movements reconstructed on the coast to the submerged shelf area needs to take into account an additional contribution for shelf subsidence. Moreover, vertical movements may also vary over very short distances because of local fault activity and volcanic dynamics (e.g. De Guidi and Monaco 2009) as well as complex patterns of ice melt in periglacial regions (Hetherington et al. 2004).

**Depositional Processes** As already explained, the deposition of marine sediments over the transgressed surface may alter the bathymetry especially in front of major deltas or relevant sedimentary sources (Fig. 3.2). HR seismics allow reconstruction of the surface underlying the post-transgression sediments, but the resulting DTM is at low-resolution. This is due to the two-dimensional nature of seismic profiles and the spacing between them, and results in a smoothed three-dimensional surface interpolated from the available data points.

**Erosional Processes** Errors can also arise from ravinement processes, as discussed above (Sect. 3.3.3).

## 3.5 Differences of Scale and Resolution

The nature of the results obtainable from computation of transgression velocities will vary depending on the geographical scale of observation and the resolution of the available datasets, and this can be illustrated by contrasting analysis of continental margins with regional and local analyses.

At the continental scale, estimation usually relies on low-resolution bathymetric data (grid cells of hundreds of metres). Moreover, high-resolution relative sea-level curves appropriate to the regional or local scale are only available for some parts of the European coastline, notably the coastlines of Italy (Lambeck et al. 2011). In consequence, there are some large uncertainties. However, these uncertainties are gradually being reduced as new sea-level curves become available for more locations and HR swath bathymetry is extended over larger areas of the shelf. On the other hand, inherent and unavoidable errors or uncertainties due to the presence of postglacial deposits of unknown thickness or to erosional processes of uncertain extent are less relevant at this scale. The computation of transgression velocities at the continental scale is most useful for highlighting differences between sectors that were exposed to very different processes of coastline migration due to major differences in slope gradients, such as for instance between the North Sea and the Mediterranean, between the northern and the southern sectors of the Black Sea, and between the Adriatic and the Tyrrhenian Seas (Fig. 3.5).

At the regional or local scale, in contrast, full-coverage swath bathymetry and densely spaced and local relative sea level curves may be available. In these cases estimation of the TV is much more reliable, but the results are more vulnerable to possible sources of error. Where the postglacial sediment thickness is negligible, e.g., shelves around islands, in arid or karst regions, and offshore of major lagoons and marshland areas, we can assume that multibeam bathymetry will give a reliable indication of palaeotopography. Elsewhere the method has to be applied to isochrones (i.e. isobaths of the transgression surface reconstructed from seismic profiles), and this often implies a lower resolution.

The use of regional and local transgression velocities is useful in depicting specific sectors where fast transgression in combination with particular coastline morphological configurations may have resulted in the preservation of past terrestrial features and archaeological materials. In this case, it is also important to consider tectonic data if available.

### 3.5.1 *Continental Scale*

The lack of a complete dataset of relative sea level curves for all the European coasts does not allow us to produce a reliable and detailed map of TV at a continental scale. Nevertheless, we can compute approximate transgression velocities using EMODnet bathymetric data and taking into account the different combination of eustatic changes and isostatic crustal movements due to the unloading of the ice caps and collapse of the glacial fore-bulge (Fig. 3.5). Thus in areas far from the ice caps (Mediterranean, Black Sea, Atlantic Iberia and Bay of Biscay) a relative sea level curve derived from the stable Apulia coast has been applied (Lambeck et al. 2011). On the shelf surrounding the British Isles, a relative sea level curve originally created for southern England (Houghton et al. 2001) has been applied. In areas near the Scandinavian and Scottish ice sheets, the TV was not computed because of the occurrence of continuous uplift, or a combination of uplift and subsidence that caused minor fluctuations of relative sea level close to its present position (Smith et al. 2007).

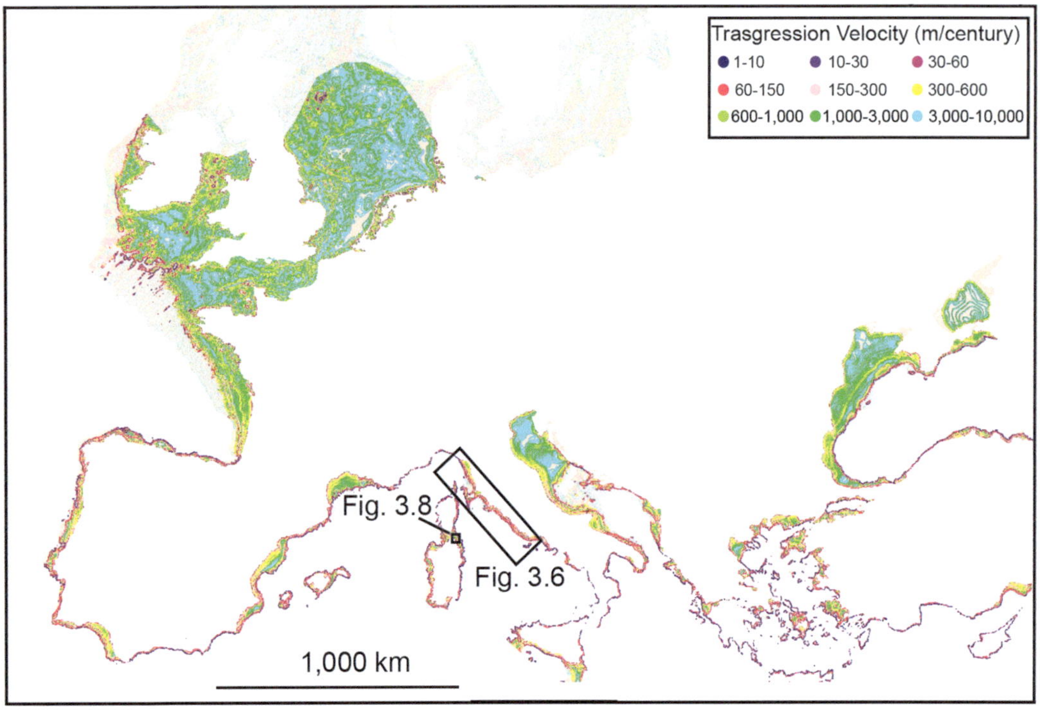


**Fig. 3.5** Distribution of transgression velocity on European shelves computed on the basis of EMODnet bathymetric data. In areas close to the Scandinavian and Scottish ice caps the velocity has not been computed because relative sea level was falling or nearly stable during the past 21,000 years B.P. (See text for details). *Black squares* locate Figs. 3.6 and 3.8

We are aware that this procedure is an oversimplification of the very complex pattern of crustal behaviour of European margins during deglaciation; moreover we have not taken account of tectonics and hydro-isostasy at this scale. However, it can be seen that the map highlights striking differences between continental shelves that experienced very dissimilar transgression rates, with values ranging from more than 2 km per century in the North Sea and Adriatic, to less than 1 m per century in the Tyrrhenian and Iberian margins (i.e. from 20 m/year to 1 cm/year, respectively).

### 3.5.2 *Regional Scale*

The availability of closely-spaced relative sea-level curves for the post-glacial sea-level rise in the northern Tyrrhenian Sea (Lambeck et al. 2011) allows the weighted computation of the TV in different sectors of the shelf (Fig. 3.6). In the case of regional-scale computations, multibeam data are not always available everywhere so that one sometimes has to rely on single-beam bathymetric data. In any case, the presence of a thick sedimentary cover – up to a maximum thickness of 80 m of postglacial sediments offshore of the Tiber River mouth (Fig. 3.2, Chiocci and Milli 1995) means that the use of bathymetric data may be misleading in shelf sectors fed by major river mouths (white arrows in Fig. 3.6). In contrast, the map is much more reliable where the postglacial sedimentary cover is thin (dashed black circles) or absent (thick black circles) such as on the Tuscan archipelago or in southern Latium.

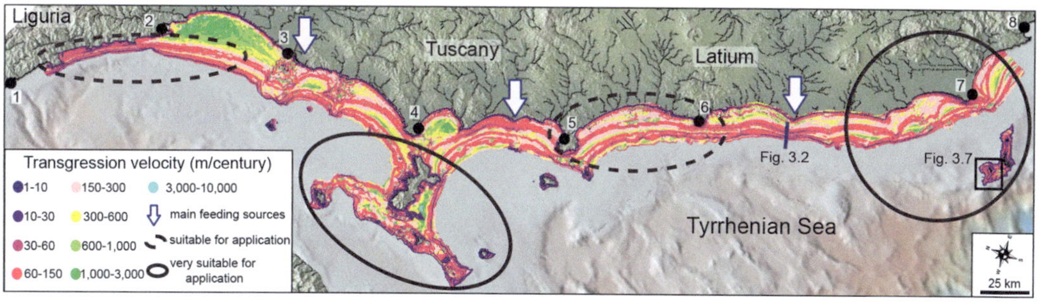


**Fig. 3.6** Transgression velocity in the northern Tyrrhenian Sea continental shelf (see Fig. 3.5 for location). *Dots* with numbers indicate reference sites for relative sea level curve according to Lambeck et al. (2011). *Black square* locates Fig. 3.7. *Blue line* locates Fig. 3.2

### 3.5.3 *Local Scale*

The shelf surrounding Palmarola Island (southern Latium) is starved of sediments and dominated by bioclastic postglacial sedimentation a few metres thick at most (Chiocci and Martorelli 2014). Therefore, the site is suitable for the local reconstruction of palaeotopography using multibeam bathymetry. Here, we have computed TVs using the relative sea-level curve derived from the neighbouring Circeo site of Lambeck et al. (2011, see dot no. 7 in Fig. 3.6). We have then created a 3-D reconstruction based on the bathymetry of the area (Fig. 3.7a), and draped TVs onto this reconstruction (Fig. 3.7b).

The distribution of TVs appears to be mainly controlled by the uneven topography, i.e., low-gradient = high transgression speed. However, the influence of the sea-level curve, though minor is not irrelevant. If we apply different relative sea level curves from Lambeck et al. (2011), derived respectively from the isostatically subsiding Trieste site in north-eastern Italy (Fig. 3.7c) and from the uplifting eastern coast of Sardinia (Fig. 3.7d), we obtain some interesting contrasts. Despite a generally similar pattern, the position of areas characterized by high and low TVs differs between the two cases, indicating a not insignificant contribution from the rate of sea-level rise when analysing the data at a local scale. This effect is of course even more apparent when considering localised sites in detail (see Sect. 3.5.4 below).

### 3.5.4 *Ultra-Local Scale*

A further advance in interpretation can be achieved when combining the TV distribution with a very detailed morphological analysis based on multibeam bathymetry in an area with little or no postglacial sedimentation (Fig. 3.8).

In this case, bedrock outcrops are present in the middle-outer shelf around Palmarola Island that would have acted as sheltering headlands or islets during sea level rise. The presence of areas characterized by high TVs close to these outcrops (Fig. 3.8) indicates specific locations where a very fast inundation of the land surface occurred during sea level rise with little wave reworking, thus creating a higher possibility for the preservation of archaeological sites. Note how the visualization of TV by colour dots helps to detail very local values at the maximum possible resolution.

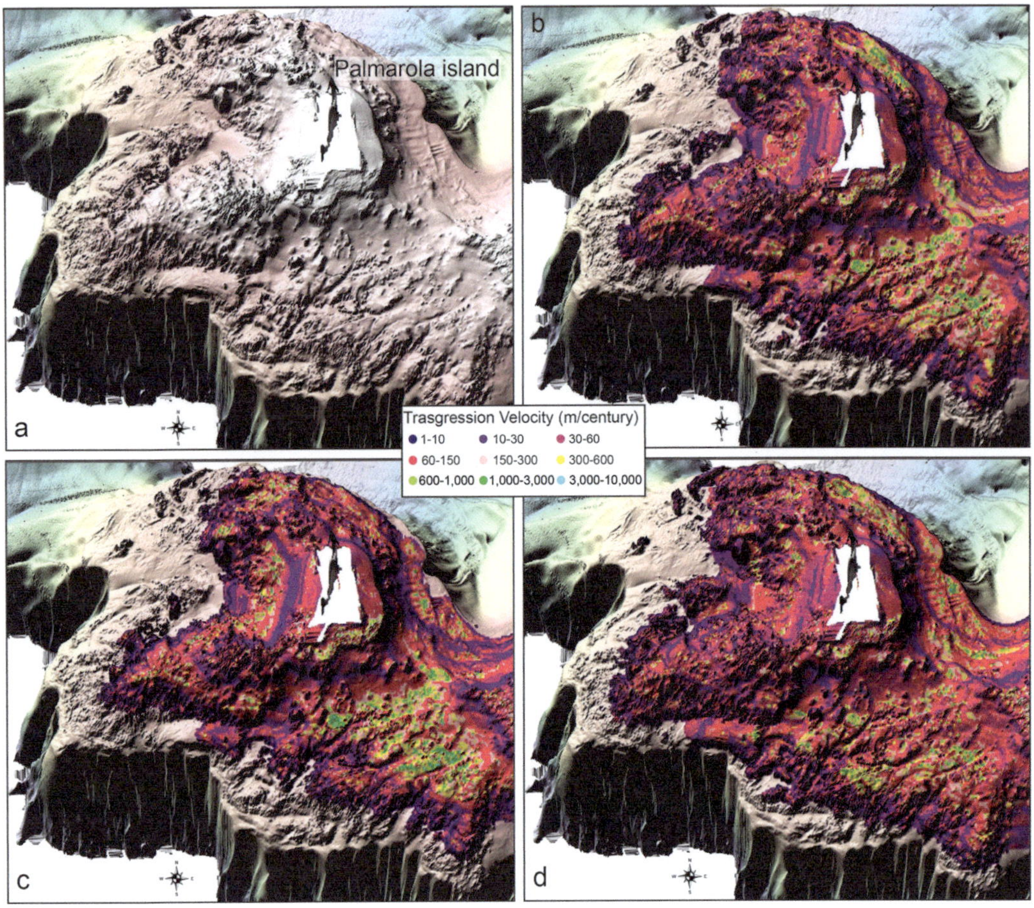


**Fig. 3.7** Estimated transgression velocity draped on a 3D view of the shelf offshore of Palmarola Island (see Fig. 3.6 for location). (**a**) shaded relief from multibeam bathymetry; (**b**) transgression velocity computed by using an RSL curve from the neighbouring Circeo site (Fig. 3.6); (**c**) transgression velocity computed by using an RSL curve from an isostatically subsiding site; (**d**) transgression velocity computed by using an RSL curve from an isostatically uplifting site (See text for details)

## 3.6 Concluding Remarks

From 21,000 to 7000 cal years BP, the sea level rose from −130 m to its present position at variable rates, with peak rates of 4 cm/year. This process directly controlled the velocity at which the shoreline migrated landwards (i.e., the transgression velocity) and, consequently, the amount of time during which each point on the coastline was exposed to wave action, the type of environment that developed there and the likelihood that prehistoric sites would have been protected from wave erosion during the transgression.

A conceptual method and a computational MATLAB routine (TRASPEED) have been described in order to calculate the velocity of transgression (TV) during the last sea level rise on shelf areas at different spatial scales. TVs result from the interplay between relative sea-level rise and the palaeotopography of the submerged landscape. The impact of errors associated with the measurement of these processes varies, depending on the scale of reconstruction – continental-margin compared to regional and local coastlines. In the latter case, high-resolution bathymetry and local sea-level curves are often available, but due to the detailed scale, other factors such as the thickness of postglacial marine

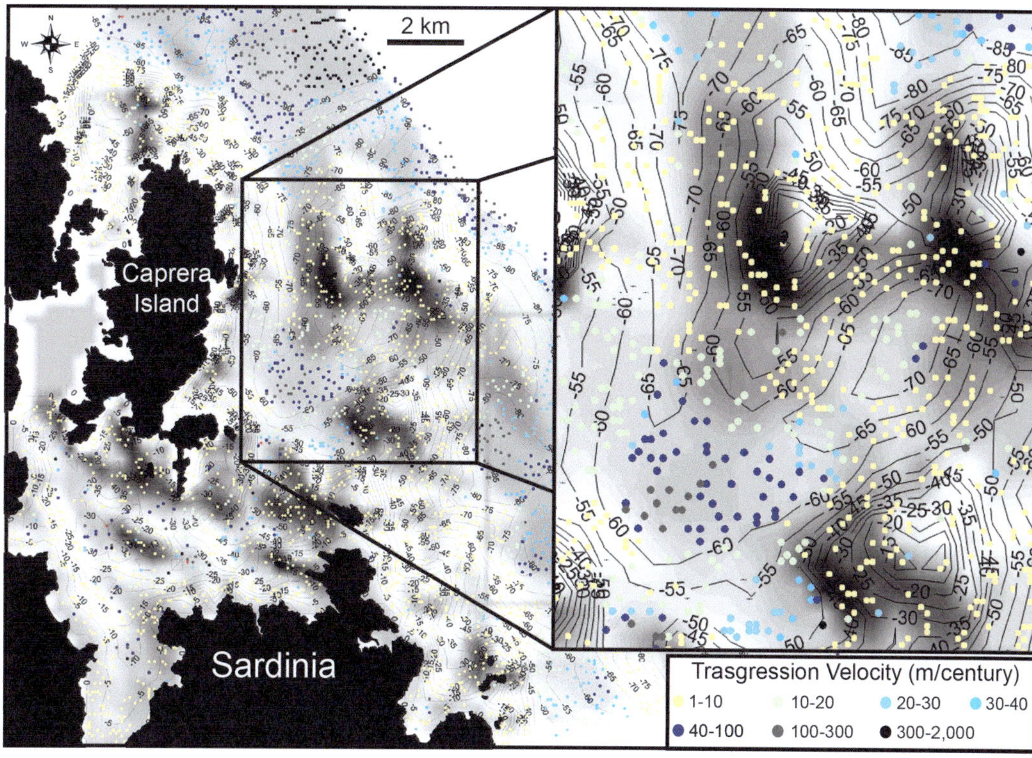


**Fig. 3.8** Distribution of the estimated transgression velocity offshore of Caprera Island (location in Fig. 3.6). The presence of irregularities on the seafloor (most likely corresponding to headlands or islets during shoreline transgression) created sheltered areas of high-transgression velocity such as those depicted in the *boxes* in the zoom. This setting may have enhanced the possibility of preserving prehistoric sites. The numbers in the inset are the transgression velocities in m/century

sediments or the amount of seafloor eroded by ravinement have to be taken into account. In addition, the relative sea-level rise (i.e. absolute or eustatic sea level rise combined with vertical movements of the margin due to subsidence or uplift) has to be computed at a local scale, but these data are not always available. Where they are, the estimate may be affected by errors due to the long time span over which vertical movements are commonly averaged (125,000 years or more using uplifted marine terraces) and the fact that they are reconstructed at the coastline and do not incorporate the possibility of additional subsidence on the outer shelf.

At the continental scale, the computation of the transgression velocity becomes less accurate because of the low resolution of bathymetric data, but the errors have less impact.

These differences of scale have different archaeological implications in relation to the sorts of social and cultural processes that they may illuminate. For example, comparison of TVs in different European sea basins may be useful in relation to issues of human dispersal and migration, and the potential impact of more or less rapid inundation and transformation of human landscapes. At the local scale, identification of very high TVs may help to identify locations where archaeological material has the best chance of preservation because of rapid inundation and localised topographic conditions that protected material from the full effects of wave erosion.

The method we propose might be applied to many different shelves (the TRASPEED routine is available and we are open to collaboration) and its usefulness will be enhanced in future as more detailed bathymetry and refinements in crustal modelling become available, allowing the reconstruction of relative sea level changes at many different locations along the coast and across the shelf.

**Acknowledgments** The ideas expressed in this paper have benefited from discussion with Nic Flemming and other SPLASHCOS participants. We also thank Francesca Argiolas for developing the preliminary MATLAB routine, and Francesco Falese for providing material for the figures. For Figs. 3.5 and 3.6, the Digital Terrain Model data products have been derived from the EMODnet Bathymetry portal at http://www.emodnet-bathymetry.eu. We also gratefully acknowledge anonymous reviewers and Geoff Bailey for their helpful suggestions.

# References

Anzidei, M, Lambeck K, Antonioli F, Furlani S, Mastronuzzi G, Serpelloni E, Vannucci G (2014) Coastal structure, sea-level changes and vertical motion of the land in the Mediterranean. In: Martini IP, Wanless HR (eds) Sedimentary coastal zones from high to low latitudes: similarities and differences. Geological Society of London, Special Publication 388: 453–479. http://dx.doi.org/10.1144/SP388.20

Bailey GN, Flemming NC (2008) Archaeology of the continental shelf: marine resources, submerged landscapes and underwater archaeology. Quat Sci Rev 27:2153–2165

Bailey GN, Reynolds SC, King GCP (2011) Landscapes of human evolution: models and methods of tectonic geomorphology and the reconstruction of hominin landscapes. J Hum Evol 60(3):257–280

Bard E, Hamelin B, Arnold M, Montaggioni L, Cabioch G, Faure G, Rougerie F (1996) Deglacial sea-level record from Tahiti corals and the timing of global meltwater discharge. Nature 382(6588):241–244

Belknap DF, Kraft JC (1981) Preservation potential of transgressive coastal lithosomes on the U.S. Atlantic shelf. Mar Geol 42:429–442

Bernè S, Lericolais G, Marsset T, Bourillet JF, de Batist M (1998) Erosional shelf sand ridges and lowstand shorefaces: examples from tide and wave dominated environments of France. J Sediment Res 68:540–555

Caruso A, Cosentino C, Pierre C, Sulli A (2011) Sea-level changes during the last 41,000 years in the outer shelf of the southern Tyrrhenian Sea: evidence from benthic foraminifera and seismostratigraphic analysis. Quat Int 232:122–131

Casalbore D, Falese F, Martorelli E, Romagnoli C, Chiocci FL (2017) Submarine depositional terraces in the Tyrrhenian Sea as a proxy for paleo-sea level reconstruction: problems and perspective. Quat Int. http://dx.doi.org/10.1016/j.quaint.2016.02.027

Cattaneo A, Steel RJ (2003) Transgressive deposits: a review of their variability. Earth Sci Rev 62(3–4):187–228

Chiocci FL, Martorelli E (2014) Note illustrative della carta geologica d'Italia alla scala 1:50.000 foglio 413 – Aree Marine. ISPRA-Servizio Geologico d'Italia

Chiocci FL, Milli S (1995) Construction of a chronostratigraphic diagram for a high-frequency sequence: the 20 ky B.P. to present Tiber depositional sequence. Il Quaternario 8:339–348

De Guidi G, Monaco C 2009. Late Holocene vertical deformation along the coast of Pantelleria Island (Sicily Channel, Italy). In: Catto NR, Yim, WW-S, Antonioli F, Yokoyama Y (eds) Continental shelves: sea levels and environments. Contributions from 17th INQUA Congress, Cairns. INQUA Commission on Coastal and Marine Processes, and International Geoscience Programme Project 526. Quat Int 206: 158–165

Dodonov YS, Dodonova YA (2011) Robust measures of central tendency: weighting as a possible alternative to trimming in response time data analysis. Psikhologicheskie Issledovaniya 5(19):1–11

Evans AM, Flatman JC, Flemming NC (eds) (2014) Prehistoric archaeology on the continental shelf: a global review. Springer, New York

Fairbanks RG (1989) A 17,000-year glacio-eustatic sea level record: influence of glacial melting rates on the Younger Dryas event and deep-ocean circulation. Nature 342(6250):637–642

Ferranti L, Antonioli F, Mauz B, Amorosi A, Dai Pra G, Mastronuzzi G, Monaco C, Orrù P, Pappalardo M, Radtke U, Renda P, Romano P, Sansò P, Verrubbi V (2006) Markers of the last interglacial sea-level high stand along the coast of Italy: tectonic implications. Quat Int 145:30–54

Flemming NC (1972) Relative chronology of submerged pleistocene marine erosion features in the Western Mediterranean. J Geol 80:633–662

Flemming NC, Cağatay MN, Chiocci FL, Galanidou N, Jöns H, Lericolais G, Missiaen T, Moore F, Rosentau A, Sakellariou D, Skar B, Stevenson A, Weerts H (2014) Land beneath the waves: submerged landscapes and sea level change. A joint geoscience-humanities strategy for European continental shelf prehistoric research. In: Chu NC, McDonough N (eds) Position paper 21 of the European Marine Board. Ostend, Belgium

Harff J, Meyer M (2011) Coastlines of the Baltic sea – zones of competition between geological processes and a changing climate: examples from the southern Baltic. In: Harff J et al (eds) The Baltic sea basin, Central and Eastern European 149 Development Studies (CEEDES). Springer, Heidelberg, pp 149–164

Hetherington R, Barrier JV, Reid RGB, MacLeod IL, Smith DT (2004) Paleogeography, glacially induced crustal displacement, and late quaternary coastlines on the continental shelf of British Columbia, Canada. Quat Sci Rev 23:295–318

Houghton JT, Ding Y, Griggs DJ, Noguer M, van der Linden PJ, Dai X, Maskell K, Johnson CA (eds) (2001) IPCC Climate change 2001: the scientific basis. Contribution of working group 1 to the third assessment report of the intergovernmental panel on climate change. Cambridge University Press, Cambridge

Lambeck K, Purcell A (2005) Sea-level change in the Mediterranean sea since the LGM: model predictions for tectonically stable areas. Quat Sci Rev 24:1969–1988

Lambeck K, Purcell A, Funder S, Kjær KH, Larsen E, Moller P (2006) Constraints on the late Saalian to early Middle Weichselian ice sheet of Eurasia from field data and rebound modelling. Boreas 35(3):539–575

Lambeck K, Antonioli F, Anzidei M, Ferranti L, Leoni G, Scicchitano G, Silenzi S (2011) Sea level change along the Italian coast during the Holocene and projections for the future. Quat Int 232:250–257

Lambeck K, Rouby H, Purcell A, Sun Y, Sambridge M (2014) Sea level and global ice volumes from the last glacial maximum to the Holocene. Proc Natl Acad Sci 111(43):15296–15303

Liu JP, Milliman JD (2004) Reconsidering melt-water pulses 1A and 1B: global impacts of rapid sea-level rise. J Ocean Univ China 3:183–190. doi:10.1007/s11802-004-0033-8

Nummedal D, Swift DJP (1987) Transgressive stratigraphy at sequence-bounding unconformities: some principles derived from Holocene and Cretaceous examples. In: Nummedal D, Pilkey OHJ, Howard D (eds) Sea level fluctuation and coastal evolution, Society of Economic Paleontologists and Mineralogists, Special Publication 41: 241–260

Pedoja K, Husson L, Johnson ME, Melnick D, Witt C, Pochat S, Nexer M, Delcaillau B, Poprawsku Y, Authemayou C, Elliot M, Regard V, Garestier F (2014) Coastal staircase sequences reflecting sea-level oscillations and tectonic uplift during the quaternary and neogene. Earth Sci Rev 132:13–38

Quartau R, Hipólito A, Romagnoli C, Casalbore D, Madeira J, Tempera F, Roque C, Chiocci FL (2014) The morphology of insular shelves as a key for understanding the geological evolution of volcanic islands: insights from Terceira Island (Azores). Geochem Geophys Geosyst 15:1801–1826

Ryan WBF, Major C, Lericolais G, Goldstein SL (2003) Catastrophic flooding of the Black sea. Annu Rev Earth Planet Sci 31:525–554

Siddall M, Rohling EJ, Almogi-Labin A, Hemleben C, Meischner D, Schmelzer I (2003) Sea level_fluctuations during the last glacial cycle. Nature 423:853–858

Smith DE, Cullingford RA, Mighall TM, Jordan JT, Fretwell PT (2007) Holocene relative sea level changes in a glacio-isostatic area: new data from south-west Scotland, United Kingdom. Mar Geol 242(1–3):5–26

Trenhaile AS (2001) Modeling the effect of late quaternary interglacial sea levels on wave-cut shore platforms. Mar Geol 172:205–223

Trenhaile AS (2002) Modeling the development of marine terraces on tectonically mobile rock coasts. Mar Geol 185:341–361

Trenhaile AS (2011) The effect of Holocene changes in relative sea level on the morphology of rocky coasts. Geomorphology 114:30–41

Uścinowicz S (2006) A relative sea-level curve for the Polish southern Baltic sea. Quat Int 145–146:86–105

Van Wagoner JC, Posamentier HW, Mitchum RM, Vail PR, Sarg JF, Loutit TS, Hardenbol J (1988) An overview of the fundamentals of sequence stratigraphy and key definitions. In: Wilgus CK, Hastings BS, Kendall CGStC, Posamentier HW, Ross CA, Van Wagoner JC (eds) Sea-level changes: an integrated approach. Society of Economic Paleontologists and Mineralogists, Special Publication No. 42: 39–45

Waelbroeck C, Labeyrie L, Michel E, Duplessy JC, Lambeck K, McManus JF, Balbon E, Labracherie M (2002) Sea-level and deep water temperature changes derived from benthic foraminifera isotopic records. Quat Sci Rev 21:295–305

Yokoyama Y, Lambeck K, De Deckker P, Johnston P, Fifield LK (2000) Timing of the last glacial maximum from observed sea-level minima. Nature 406(6797):713–716

Zecchin M, Ceramicola S, Lodolo E, Casalbore D, Chiocci FL (2015) Episodic, rapid sea-level rises on the central Mediterranean shelves after the last glacial maximum: a review. Mar Geol 369:212–223

# Chapter 4
# Joint Explorations of the Sunken Past: Examples of Maritime Archaeological Collaboration Between Industry and Academia in the Baltic

**Joakim Holmlund, Björn Nilsson, and Johan Rönnby**

**Abstract** This chapter examines the benefits and constraints of collaboration between an archaeological research unit and a commercial company, using as examples joint research conducted by MARIS (Maritime Archaeological Research Institute at Södertörn University) and the Swedish commercial marine survey company MMT. The examples presented here included the detailed reconstruction by remote sensing of deeply submerged shipwrecks and the mapping and discovery of submerged archaeological landscapes and associated artefacts such as fish traps, which can then be examined more closely by archaeological divers. The benefits to archaeologists of collaborating with well-equipped commercial companies are obvious, but the benefits are mutual. The demands of archaeological research can generate new technological solutions that have commercial application, as well as producing results with wider educational and social benefits. Provided that archaeological investigations are embedded in the normal commercial operations of the company, such collaboration can be cost-effective for both parties, and is further enhanced by collaboration with film companies, which generates wider public interest and publicity for all concerned.

## 4.1 Introduction

The Baltic Sea is one of the best places in the world for underwater archaeology thanks to the unique conditions for preserving wood and other organic material in the brackish and cold water. The Maritime Archaeological Research Institute, MARIS, based at Södertörn University in Stockholm, is focused on investigations and interpretations of this special 'Baltic sunken past'. However, archaeological studies of shipwrecks, maritime defence systems and submerged landscapes require special techniques, skills and equipment. This is particularly evident for work carried out at great depths, where underwater robotics, multibeam bathymetric systems and other sub-sea survey technologies are necessary. Developing and improving different methods for underwater survey, mapping and excavation has therefore also been a part of MARIS activities. Since 2008 this has been done in close cooperation with the marine survey company MMT and with Deep Sea Productions, a film production company. In 2011 the tech diving company Ocean Discovery also became a part of this collaboration.

J. Holmlund (✉)
MMT, Sven Källfelts Gata 11, SE-426 71 Västra Frölunda, Sweden
e-mail: joakim.holmlund@mmt.se

B. Nilsson • J. Rönnby
MARIS (Maritime Archaeological Research Institute), Södertörn University, 141 89 Huddinge, Sweden
e-mail: bjorn.nilsson@ark.lu.se; johan.ronnby@sh.se



G.N. Bailey et al. (eds.), *Under the Sea: Archaeology and Palaeolandscapes of the Continental Shelf*, Coastal Research Library 20, DOI 10.1007/978-3-319-53160-1_4

The marine survey company MMT is based in Gothenburg but the company works worldwide. MMT's clients are primarily found in the oil, gas, renewable energy and cable industries, and in governmental organizations. MMT operates a fleet of fully equipped Survey and ROV vessels and smaller workboats (Fig. 4.1). All ships are equipped with precision navigation and work on seabed mapping assignments. The company employs 200 individuals with various specialties in the field of marine survey work. MMT combines experience in marine survey operations with new techniques, the latest sub-sea equipment, and different types of survey vessels in order to meet the demands of the commercial market. It is important for the company to be flexible, innovative and reliable in order to deliver cost-efficient marine survey solutions.

The differences in perspectives and aims between a small research-based archaeological institute such as MARIS and a commercial survey company such as MMT are obvious. But there are, nevertheless, interests in common, and our aim in this chapter is to describe our ability to work together, assess the benefits of industrial collaboration, and outline some of the ways in which the relationship can be optimized for archaeological purposes.

The collaboration between MARIS and MMT has focused on the improvement and adaptation of underwater technologies and methods for archaeological use (Holmlund 2014a, b). A particular aim has been to develop new standards for documentation and evaluation of source material on the sea floor that is out of range of ordinary diving or where diving is too time consuming or risky. The work includes both the documentation of minute details on the exterior and inside of shipwrecks and the mapping of large areas of underwater landscape.

**Fig. 4.1** MMT research vessels *Franklin* (*left*) and *Triad* (*right*) are two of the vessels which have been used for the fieldwork described in this chapter

Although the emphasis of this volume is on submerged landscapes, we will first describe some joint work on shipwreck projects, in order to demonstrate the benefits of our collaboration, and then proceed to present new technologies and techniques for deep-investigation that also have potential application to studies of prehistoric landscapes.

## 4.2 Shipwrecks

In 2003 Deep Sea Productions (DSP) and MMT discovered a shipwreck in the middle of the Baltic Sea, which even for the Baltic was exceptionally well-preserved (Figs. 4.2 and 4.3). In complete darkness and 130 m below the surface lay an almost intact Dutch *fluite* (a sailing cargo vessel) from the mid-seventh century. Since 2009, the wreck, called 'The Ghost Ship' by the team, has been studied as part of an international scientific project coordinated by MARIS and with MMT providing the key logistical platform. The work has also been documented for TV by Deep Sea Productions (Dixelius et al. 2011; Eriksson and Rönnby 2012a,b).

Sampling, salvaging and archaeological documentation have been achieved using Remotely Operated Vehicles (ROVs). The robots have in essence replaced the work of the diving archaeologists. The total darkness at this depth in the Baltic demands considerable artificial light. For detailed inspection the limited view from standard ROV video is sufficient, but to achieve an overview, the entire vessel needed illumination. This was provided by four LED lights mounted above the ROV as well as a 50,000-lumen light ramp lowered from the aft A-frame of the survey vessel. The lights had to be

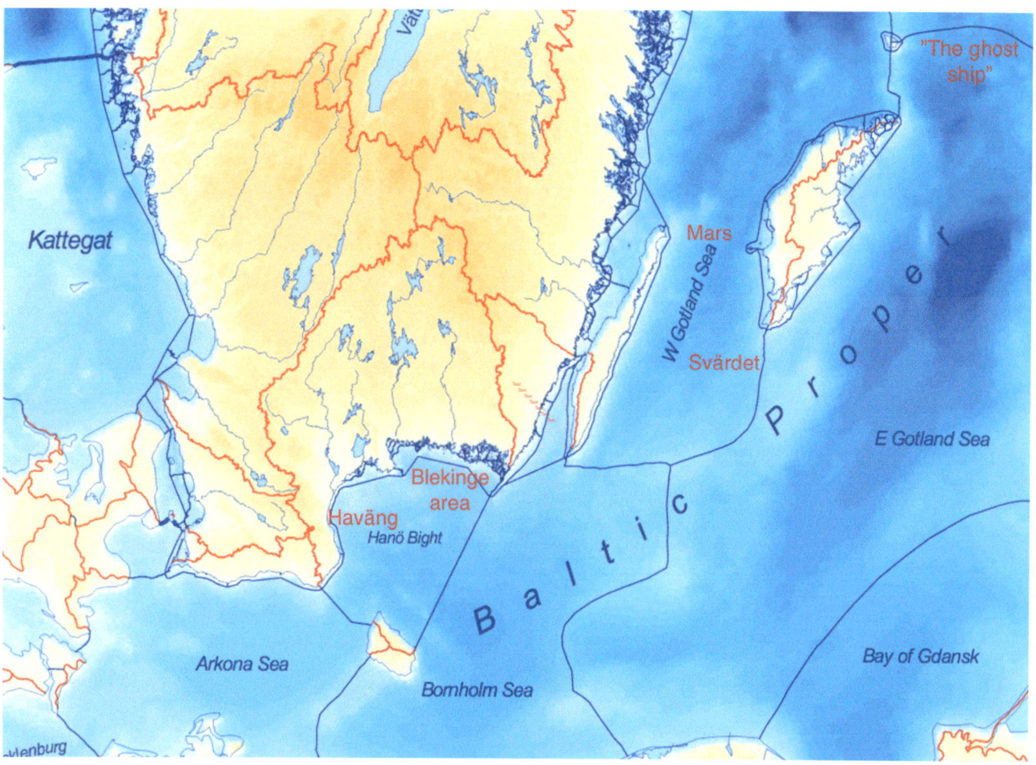


**Fig. 4.2** The Baltic Sea in the region of southern Sweden showing the location of sites and other features mentioned in the text

lowered between the wreck's mast tops. This required very precise position-holding by the vessel, with a capacity to ensure that ship movements never exceeded 0.2 m (Dixelius et al. 2011).

An almost intact three-dimensional ship is an archaeological challenge. Thorough video documentation was recorded for the archaeological site plans and sketches of the ship. Thanks to accurate measurements by laser technology, these plans could be drawn with precision and are correct in scale. This work combines archaeological drawing skills with the use of advanced techniques (Eriksson 2012; Adams 2013, pp. 85–95). Mini-robots and cameras mounted on an extension arm have also allowed the research team to 'board the ship', and see actual details on the inside of the hull.

During the 2010 expedition, MMT's vessel *Ice Beam* was equipped with a single transducer Reson 7125 multibeam echo sounder mounted under a SubAtlantic Mohican ROV. It recorded reference points for the entire wreck site (Fig. 4.3). The beams of the echo sounder penetrated the upper deck and the holds, so that very accurate measurements of the inside of the hull, the quarters, the holds and the forecastle were taken and presented in detail. The final 3-D model of the Ghost Ship makes it possible to look inside the ship, study its inner construction and the location of bulkheads and deck levels, which allows for interpretation of the various functions performed in different areas of the ship. The model, which collates over six million depth soundings, can also create cross-sections of the ship, both lengthwise and across the beam from bow to stern. This can be turned into a construction design for a small seventeenth-century ship more than a 100 years before such design drawings were made.

In the summer of 2011, two new spectacular shipwrecks were discovered in the central Baltic Sea. After years of searching, the well-preserved remains of both *Mars* (1564) and *Svärdet (The Sword)* (1676) were found. Both were large Swedish royal naval ships that went down after tough and lengthy sea battles (Rönnby 2012; Rönnby 2013a, b, pp. 12–14; Eriksson and Rönnby In press).

MARIS and MMT worked together in the exploration of these new ships together with Deep Sea Productions and Ocean Discovery. The ongoing documentation of the two wrecks highlights the violent nature of the conflict and the chaotic environment on board during the course of the battles. The sites with all their heavy bronze guns and the ships themselves are well-preserved 'maritime battlefields'.

*Mars* is situated at a depth of 70 m and *Svärdet* at 90 m. Some of our work at the sites was achieved through technical diving, but most of the documentation has been completed using ROV and multibeam echo sounder. In this respect, the experience from the Ghost Ship was invaluable.

**Fig. 4.3** 'The Ghost Ship', a Dutch fluitship from around AD 1650, found at 120 m depth. The archaeological and hydroacoustic information was acquired with ROVs (Photo MMT/Deep Sea Production)

As part of the ongoing effort to find new solutions, MARIS and MMT have initiated a bottom based Blueview scan of *Mars* (Fig. 4.4) which is stitched together with the ROV MBES data from above the wreck. The equipment used is a 3D BV5000 1.35 MHz Blueview and Subatlantic Mohican ROV equipped with a R2Sonic 700kHz UHR MBES. The equipment was operated by MMT staff and the results were processed jointly by the surveyor and the archaeologist working together. The scanner sends out 256 beams with a range resolution of 1.5 cm in a 45° swath; this swath is then rotated with a mechanical motor through 360° with different tilt angles, creating a spherical image of the surroundings. Maximum range with 1.35 MHz is around 30 m.

The preliminary results are very promising, and the final data outcome will form a three-dimensional point cloud of the wreck that the investigating archaeologist can view and turn on all axes, while taking measurements and obtaining curvatures of the hull. It is also possible to place the scanner inside the wreck to obtain a three-dimensional image of the inside.

Initial work on some other sites, for example *Gribshunden* (1495) (Rönnby 2015) and also on a Viking age pole blockage (defensive structure) in Gamlebyviken indicates that the method is suitable for various types of archaeological remains. The BV50000 3D is further rated to 4000 m, hence it can be placed on a deep-water rated ROV and sent down to investigate wrecks at depths that are unsuitable for divers.

A spectacular photo-mosaic of the total wreck-site of *Mars* was completed in 2012 (Fig. 4.4). The mosaic was made by combining more than 600 separate high-resolution photos. Recent fieldwork included documentation with photogrammetry. Our ongoing work combining the Blueview results with the photogrammetry is very promising. This can be used for archaeological interpretation and can also provide 3D virtual reconstructions for museum exhibitions.

**Fig. 4.4** Blueview scan over the stern of *Mars* (Rendering: Joakim Holmlund, MMT)

## 4.3 The 'Landscapes Lost' Project

A third joint project that MARIS and MMT have been working on is the investigation of the submerged Stone Age landscapes in the southern Baltic (Nilsson and Sjöström 2012; Törnqvist 2013; Nilsson et al. 2016; Nilsson et al. 2017).

During the Early Holocene in southern Scandinavia sea-levels changed dramatically (cf. Björck 2008; Hansson et al., Chap. 13). It has long been known that there exist large areas in the Atlantic and the Baltic which for long periods of time were dry land. The scientific knowledge of such inundated landscapes has gained new prominence in recent decades, not least in archaeology and Quaternary geology (Benjamin et al. 2011). This is partly due to increased commercial interest in the continental shelf particularly from oil, mining and aggregate industries, wind farms, and the growing need to create infrastructure for communication and energy transfer over marine areas. In addition, and partly an effect of the above, improved availability of high-resolution hydro-acoustic mapping (at sub-metre scale) and cost-effective photogrammetric 3D modelling (sub-centimetre scale) has made surveys, fieldwork and visualization of the submerged landscapes easier.

Consequently, industrial companies, researchers, policy makers concerned with the natural and cultural heritage, and other stakeholders have converged in their need for a better understanding of the sea-bed environment. It is in this context that the Landscape Lost project was developed. The aim is to bring together archaeology and other landscape disciplines from the fields of geology, geography and biology. An overarching question is how a better understanding of the inundated landscapes of the Early Mesolithic affects the general understanding of this period, and how, at a national level, we can map, monitor and preserve these submerged cultural heritage sites (Nilsson 2012; Nilsson et al. 2016).

One goal is archaeological survey, examination and interpretation of the submerged landscapes of the early Holocene. Two main study areas have been chosen (Fig. 4.2): The Haväng area, located at the Early Holocene river mouth of Verkeån in south-eastern Scania, and the Blekinge archipelago with several now-submerged estuaries. So far, hydrographic survey and archaeological exploration have been completed from the MMT research vessel *Triad* using MBES and Topas parametric sub bottom sonar. Nearly 15 km$^2$ have been measured so far.

The main focus is on the period that comprises the Early Mesolithic Yoldia Sea, the Ancylus Lake and the initial Littorina Sea (c. 11,500–8,500 years BP). The Geology Department at Lund University has played a significant role, including since 2014 a PhD student involved in the underwater exploration (Hansson et al., Chap. 13). In Blekinge, the project is a collaboration with Martin Jakobsson, Professor of Marine Geology and Geophysics at the Department of Geological Sciences, Stockholm University.

Here we summarise work in the Haväng area, which has turned out to be the most rewarding so far, as well as the most complex area. The work here builds on a 30-year history of geological research (Björck and Dennegård 1988; Gaillard and Lemdahl 1994) and involves new surveys and sampling which have revealed fish traps dated to 9000 years BP (Nilsson and Sjöström 2012). These constructions have been found in Mesolithic lagoonal sediments, and consist of baskets, weirs and different kinds of wattle made of selected hazel withies (Fig. 4.5). The river sediments are still visible on the sea bed, as are preserved stumps and logs from the forests which were growing there 11,000 years ago.

As in the two earlier collaborations, the archaeological questions have determined the mapping methods. Thus, the main goal of the survey was to find eroded sediments and sections, remains of old submerged beach ridges, river mouth palaeochannels and even standing tree stumps. We were in great need of high-resolution 'diving-maps' to facilitate georeferenced exploration and sampling (Fig. 4.6).

Diving on the submerged sites also revealed possible threats (Fig. 4.7). Hydro-acoustic measurements can provide a future monitoring system to monitor such threats, but this is still under development and evaluation since it requires continuous surveys over a longer time.

**Fig. 4.5** At the Haväng site several remains of different types of stationary fish traps and fish weirs have been found. The fish weir in the lower right image measures 0.75 m. The oldest have been dated to 9000 cal BC (Photo: Arne Sjöström, LU/MARIS)

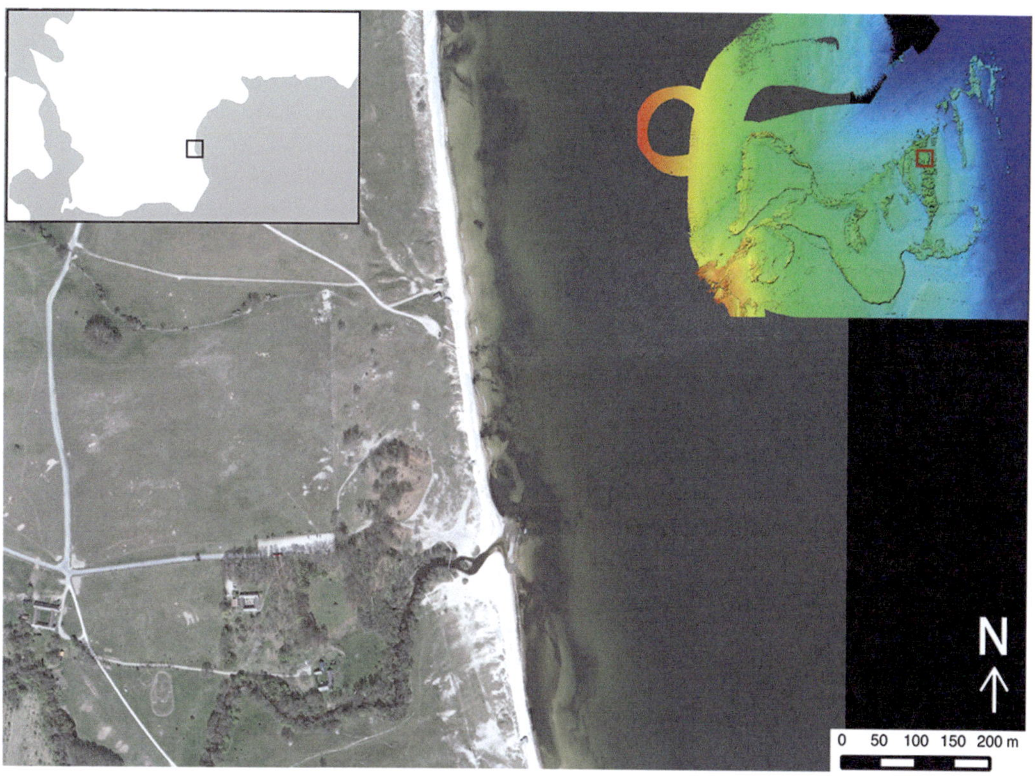


**Fig. 4.6** Multibeam measurements showing the submerged lagoon landscape of the Early Holocene Verkeån River mouth, SE Scania, Sweden. For further detail see Hansson et al., Chap. 13, Fig. 4.1

**Fig. 4.7** A threatened heritage. Wires tied to nets or anchors cut through the sediments. The actual photo lies in the vicinity of the site pictured in Fig. 4.5 (Photo Arne Sjöström, LU/MARIS)

In addition to the acquisition of archaeological and geographical data, the project was also an attempt to optimize the way different specialists work together on board. The skipper, the surveyor and the archaeologist together contribute to a dynamic process of investigation in which it is possible to alter a previously established route plan. In this way it is possible to cover large areas, but also to concentrate on places with the highest archaeological potential (Fig. 4.8). The surveyors and the archaeologists have a mutual interest in learning how each discipline operates and gathers sea floor information. The archaeological demand for 'high-resolution' and 'unfiltered' data generated considerable discussion and new technical solutions, and the integration of different skills and disciplines created new and valuable scientific knowledge and optimal use of research funds.

## 4.4 General Constraints and Benefits of Collaboration

The high daily cost of a vessel equipped with ROV/survey systems, including the ship and survey personnel, makes it important to have a detailed plan for the vessel at all times. Any interruption in commercial production, or dead time at the pier, will result in extra costs for the survey company. In all projects, one has to consider these costs when it comes to mobilization and transit time. Hence in order to undertake non-profit projects, such as those involving collaboration with MARIS, it is necessary to have a good plan and a lot of patience. Ideally, one tries to make use of the vessel while it is passing by, or when it is in the proximity of the research area. A model so far for our joint work has also been to develop new techniques or evaluate new systems for industrial applications, testing them on archaeological sites, an activity of mutual benefit to both partners. We have also found it helpful to establish simple means of communication involving personal contact. The more information and understanding that the commercial operator has about the needs of archaeological research—and the more archaeologists understand about the constraints of commercial operations—the more

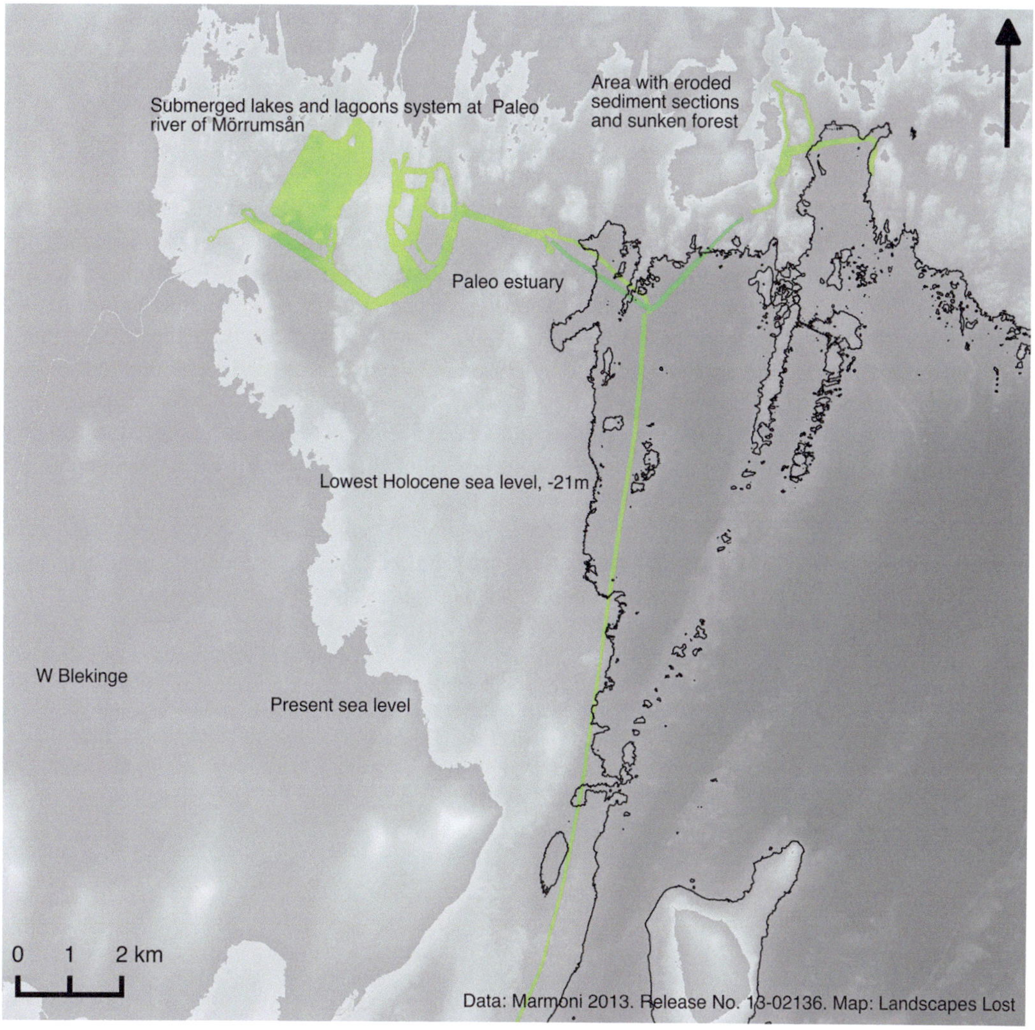


**Fig. 4.8** The irregular trajectory of an archaeologically effective survey, showing 48 h of continuous survey in the Hanö Bight. LU/MMT

efficient the use of the vessels is. This also helps to ensure the optimal methods for a specific archaeological project.

Our intention is to continue to develop the ongoing collaboration between MMT and MARIS (and our other partners) regarding investigation, documentation and interpretation of archaeological material on the seabed. For MARIS the benefit of this is obvious; together with MMT we have been able to explore places and sites which we would never have been able to access otherwise. The fact that several of our mutual projects have been carried out as part of filmmaking projects with Deep Sea Productions and Ocean Discovery has also stimulated the archaeological research—the need 'to tell a story' has had a creative effect on the projects.

For MMT, challenging new sub-sea projects need new creative solutions, and finding novel ways to solve archaeological problems can be a way for the company to develop skills and technologies that attract other jobs and commercial contracts. The archaeological knowledge and experience can also assist in the development of faster and more efficient interpretation of survey results of benefit to commercial contracts, for example mapping for new pipelines or wind farms. A series of lectures on basic principles of maritime archaeology has been held at the MMT office for project and offshore managers and further courses and lectures are also planned. PhD students from Södertörn University and MARIS have been able to join expeditions with MMT, and have used the resulting data in their work (see Eriksson 2014a, b).

A further important step in this collaboration is our new joint near-shore survey and diving vessel. MMT and MARIS have together obtained a new small research catamaran, *Svanen* (the *Swan)* equipped during 2015 for archaeological survey and sampling. This will further strengthen our collaboration. The vessel will be able to carry out detailed and effective near-shore investigations and surveys, important not least in finding shipwrecks and inundated prehistoric settlements in shallow water. An important challenge for us to tackle regarding this in the future is to find old archaeological material which is located beneath a covering layer of sediment. New methods for acoustic penetration of sediments combined with core sampling will be tried in selected areas with high potential for prehistoric human activity.

Our experience convinces us that this kind of cooperation between academia and industry can give new insights and perspectives, generate new scientific results and also contribute to strengthening a company's abilities and competitiveness.

Finally, regardless of current political agendas, economic outcomes and the possible commercial benefits of collaboration, we want to stress that the archaeological aspect is the real core of our joint work. What really links us together is our mutual interest in exploring the sunken past.

**Acknowledgements** Many dedicated friends and colleagues have been (and continue to be) involved in the projects described above, and we want to thank them all. We are very pleased that they share our interest in 'solving mysteries'. We especially wish to thank three people who have been instrumental in making this unique collaboration between MARIS and MMT possible: Malcolm Dixelius, Carl Douglas and Ola Oskarsson. Besides the institutions and companies already mentioned in the text, we also acknowledge support from The Knowledge Foundation, The Foundation for Baltic and East European Studies, National Geographic and Stiftelsen Olle Engkvist Byggmästare.

## References

Adams J (2013) Experiencing shipwrecks and the primacy of vision. In: Rönnby J, Adams J (eds) Interpreting shipwrecks. Maritime archaeological approaches. Southampton Monographs in Archaeology New Series 4 and Södertörn Academic Studies, vol. 56. Highfield Press, Southampton, pp 85–95

Benjamin J, Bonsall C, Pickard C, Fischer A (eds) (2011) Submerged prehistory. Oxbow Books, Oxford

Björck S (2008) The late quaternary development of the Baltic Sea basin. In: The BACC Author Team (ed) Assessment of climate change for the Baltic Sea Basin. Springer, Heidelberg, pp 398–407

Björck S, Dennegård B (1988) Preliminary stratigraphic studies on the late Weichselian and Holocene development of the Hanö Bay, Southeastern Sweden. Geogr Pol 55:51–62

Dixelius M, Oskarsson O, Nilsson O, Rönnby J (2011) The ghost ship expedition: Frontline deepwater archaeology in the Baltic Sea. Hydro Int 15(1):14

Eriksson N (2012) Recording a large three-dimensional ship-structure – thoughts rendered from the Dalarö-wreck project. In: Henderson JC (ed) IKUWA 3: beyond boundaries. Römisch-Germanische Kommission, Frankfurt a.M., Eurasien-Anteilung, Berlin des Deutschen Archaeologischen Instituts. Kolloquien zur Vor- und Fruhgeschichte, Band, vol 17. Dr. Rudolph Habelt GmbH Habelt, Bonn, pp 193–198

Eriksson N (2014a) Urbanity at sail. Urbanism under sail: an archaeology of fluitships in early modern everyday life. Södertörn Doctoral Dissertation 95, Södertörn University

Eriksson J (ed) (2014b) Skeppet Mars (1564). Fältrapport, etapp 3. Södertörn arkeologiska rapporter och studier. Södertörns högskola

Eriksson N, Rönnby J (2012a) The ghost wreck. An intact fluit ship from around 1650 in the middle of the Baltic Sea. Int J Naut Archaeol 41(2):350–361

Eriksson N, Rönnby J (2012b) Stormaktstidens krigsskeppsvrak: del av nationell berättelse. Populär arkeologi 1:16–18

Eriksson N, Rönnby J (In press) 'Smoking guns': new research on two early-modern maritime battlefields in the Baltic, Mars (1564) and Svärdet (1676). In: Proceedings of the 13th International Symposium on Boat and Ship Archaeology, Amsterdam, 2012

Gaillard M-J, Lemdahl (1994) Early-Holocene coastal environments and climate in southeast Sweden: a reconstruction based on macrofossils from submarine deposits. The Holocene 4(1):53–68
Holmlund J (2014a) Ny verktygslåda för marinarkeologer. Marinarkeologisk tidskrift 2:4–9
Holmlund J (2014b) Developing tools for non-intrusive electronic marine archaeological wreck excavations: the Mars project. Quest 15 (1). https://www.globalunderwaterexplorers.org/quest
Nilsson B (2012) Flooded stone age – towards an overview of submerged settlements and landscapes on the continental shelf. Session report. Eur Archaeol 38:84–86
Nilsson B, Sjöström A (2012) Forskningsprojekt undersöker världens äldsta fasta fiske. Marinarkeologisk tidskrift 2:9–10
Nilsson B, Sjöström A, Persson P (2016) The submerged seascapes of the early mesolithic in southern Scandinavia, In Persson, P (ed)
Nilsson B, Sjöström A, Persson P (2017) Seascapes of stability and change: the archaeological and ecological potential of the early mesolithic seascapes with examples from Haväng in SE Baltic, Sweden. In: Persson P, Riede F, Skar B, Breivik HM, Jonsson L (eds) The ecology of early settlement in Northern Europe: conditions for subsistence and survival, Volume 1. Equinox, Sheffield
Rönnby J (ed) (2012) Skeppet Mars (1564). Fältrapport, etapp 1. Södertörn arkeologiska rapporter och studier. Södertörns högskola
Rönnby J (ed) (2013a) Skeppet Mars (1564). Fältrapport, etapp 2. Södertörn arkeologiska rapporter och studier. Södertörns högskola
Rönnby J (2013b) The archaeological interpretation of shipwrecks. In: Rönnby J, Adams J (eds) Interpreting shipwrecks. Maritime archaeological approaches. Monographs in Archaeology NS nr 4 and Södertörn Academic Studies, vol. 56. The Highfield Press, Southampton, pp 9–24
Rönnby J (ed) (2015) Gribshunden (1495) Skeppsvrak vid Stora Ekön, Ronneby, Blekinge. Marinarkeologisk rapport 2013–15. MARIS Södertörns högskola/Bleking läns museum. Karlskrona
Törnqvist O (2013) Jakten på Blekinges tidigaste stenålder. Marinarkeologisk tidskr nr 1:14–22

# Chapter 5
# The Late Mesolithic Site of Falden, Denmark: Results from Underwater Archaeological Fieldwork and a Strategy for Capacity-Building Based on the SPLASHCOS Mission

**Otto Uldum, Jonathan Benjamin, John McCarthy, Frederick Feulner, and Harald Lübke**

**Abstract** In 2010, a submerged Late Mesolithic settlement site was discovered near the hamlet of Falden during a week-long archaeological dive survey of Helnæs Bay, Denmark. The survey was carried out as part of a training exercise, funded by the SPLASHCOS network. The discovery was a rare occurrence in recent years because priority has been given to rescue investigation of sites affected by the threat of erosion or otherwise potentially destructive modern development. This was not the case at Falden. Two short field seasons subsequent to the site's discovery (2011–2012) yielded a large inventory of worked flint, faunal remains and various other archaeological materials. The site is presented here within a larger discussion surrounding issues in training and capacity building, as well as management and research strategies concerned with prehistoric underwater cultural heritage. The fieldwork was based on methods used for many years by Langelands Museum as part of a combined survey and public outreach programme, with the additional integration of SPLASHCOS participants, mainly Early Stage Researchers. The lessons learned during this integrated dual-purpose capacity-building and archaeological research mission serve as a valuable experience for a proposed training centre with the aim of providing the opportunity for researchers and practitioners of underwater archaeology to gain the necessary experience to properly undertake research and advise (or themselves become) competent authorities working in underwater heritage management.

O. Uldum (✉)
Langelands Museum, Jens Winthersvej 12, 5900 Rudkøbing, Denmark
e-mail: otul@langelandkommune.dk

J. Benjamin • J. McCarthy
Department of Archaeology, Flinders University, PO Box 2100, Adelaide 5001, Australia
e-mail: jonathan.benjamin@flinders.edu.au; john.mccarthy@flinders.edu.au

F. Feulner
Department of Archaeology, University of York, The King's Manor, York Y01 7EP, UK
e-mail: freder@palaeome.org

H. Lübke
Centre for Baltic and Scandinavian Archaeology, Schleswig-Holstein State Museums, Schloss Gottorf, 24837 Schleswig, Germany
e-mail: harald.luebke@schloss-gottorf.de



G.N. Bailey et al. (eds.), *Under the Sea: Archaeology and Palaeolandscapes of the Continental Shelf*, Coastal Research Library 20, DOI 10.1007/978-3-319-53160-1_5

## 5.1 Introduction

The importance of the coast to Stone Age hunter-gatherer populations has been addressed by many prehistorians (e.g., Rowley-Conwy 1983; Van Andel 1989, 1990; Larsson 1990; Bailey 2004) and should not be understated. Advantages of coastal living are (simplified from Bailey 2004) as follows: (1) Ease of travel and communication for people and cultural ideas, including trade, social interaction, movement of people and material culture, and travel by sea; (2) Access to food resources, specifically the high availability and variety of marine and terrestrial plants and animals; (3) Access to other (non-food) resources, including fresh water in high water-table environments and at coastal river-mouths, and raw materials including pebbles, river rocks, driftwood, and other organic materials for structures, tools and fuel. The coastal prehistoric archaeological resource in the Baltic offers some of the clearest evidence of this relationship and more prehistoric sites have been discovered and excavated from submarine environments in southern Scandinavia than anywhere else in the world.

In this chapter, we summarise the development of underwater prehistoric archaeology in southern Scandinavia and then present the results of a SPLASHCOS training mission that resulted in the discovery of a previously unknown late Mesolithic underwater settlement, the site of Falden, in Helnæs Bay, Funen. We set out the methods used in the survey and excavation of the Falden site, including experimentation with novel photogrammetry techniques, present the results and discuss their wider significance. We use this case study to reflect on the tensions between cultural heritage management and research, and highlight the need to provide opportunities for gaining experience and training in underwater prehistoric archaeology both for researchers and cultural heritage managers.

## 5.2 Underwater Prehistoric Archaeology in Southern Scandinavia

Underwater archaeological methodology has been practised on prehistoric sites in southern Scandinavia since the 1970s (Larsson 1983, Chap. 11, Andersen 1987), and survey methods for submerged site discovery have proved, in certain cases, to be adaptable to other coastal locations (Fischer 1993a, b, 1995; Benjamin 2010). As a result of the richness and preservation of archaeological material and a sustained programme of research and investigation, archaeologists in southern Scandinavia have set a standard for underwater methodology and its application to Stone Age archaeology. Material from such sites has been used to address a variety of research questions within the wider field of prehistory including subsistence (Pedersen 1995, 1997), trade and social stratification through prestige goods (Klassen 2000; Fischer and Kristiansen 2002), Mesolithic dwellings (Grøn 2003), settlement patterns (Schilling 1997) and early prehistoric seafaring (Andersen 1987, 2011).

The Western Baltic is shallow, sandy and almost brackish with a salinity of less than 0.1%. Its shores are predominantly sandy beaches, and many bays and coves offer sheltered waters. Late glacial and postglacial climate and landscape have been a focus of Quaternary geologists and archaeologists in the Baltic region for over a century (Björck 1995; Fischer and Kristiansen 2002). Even prior to the 1920s, attempts were made to reconstruct the late glacial and postglacial palaeoenvironments of the Baltic region (Antevs 1922, Fig. 5.1), illustrating the rich history of scientific interest in the region. Discussions related to the late Quaternary climate and environment of southern Scandinavia include geology, topographic change, temperature fluctuations, eustatic and regional sea-level change, isostasy and hydrology, with particular emphasis on the development of the Baltic Sea, which went through multiple transitions (Björck 1995; Christensen et al. 1997). All of these variables have had an effect on the landscape of southern Scandinavia, and therefore impacted the human populations inhabiting the region during times of rapid change (Larsson 1990; Price 1991; Fischer 1995). Climatic indicators used to study these physical changes include seismic records, pollen data, macrofossil

**Fig. 5.1** Map of Denmark with the location of the Falden site indicated. West of Kattegat is the Swedish coast, south of the Jutland peninsula is Germany (Graphics: Langelands Museum)

analysis and the presence of microorganisms, the latter with particular reference to distinguishing freshwater deposits from brackish deposits (Bennike et al. 1998, 2004).

The landscape of Denmark and the Baltic coastal regions of Germany and Poland is generally characterised by low moraine hills on land, and a sea that rarely exceeds 30 m depth. The shallow topography means that changes in sea level have a major impact on the position of shorelines (see Chiocci et al. Chap. 3), and the Jutland Peninsula and the Danish Isles looked markedly different in the Boreal and Atlantic periods. This expanded terrain was densely forested with predominantly deciduous species, offering habitats for a wide spectrum of fauna. The Mesolithic populations in the low-lying areas that make up present-day Denmark are thought to have been relatively dense during

**Fig. 5.2** Historic vessel 'Mjølner' with the 2011crew. Standing: Arne Jensen, Kieran Westley, Freder Feulner, Ehud Galili, Radek Szmelak, Hans Toft Nielsen (skipper), sitting: Anne Margrethe Walldén, Jonathan Benjamin, Otto Uldum, Flemming Sørensen (Photo: Langelands Museum)

the Atlantic period (Brinch-Petersen et al. 1998). A large proportion of this land was inundated during the Atlantic period, and extensive hunting grounds and coastal habitats were submerged (cf. Mathiassen 1997, Fig. 2).

Since coastal Mesolithic peoples in the Atlantic period lived during a period of dramatically changing environments, they had to contend, physically and mentally, with constant adaptation (Larsson 1990). Settlement directly on and adjacent to the shore was frequent, due to a subsistence strategy that relied heavily on marine resources and stationary fishing structures (Pedersen 1997). The relationship between inland and coastal settlements is still not very well established, because the inland dwellings have suffered poor conditions of preservation, while the coastal settlements have been relatively well preserved, often under sediments in shallow water. Equally, sea-level change would have inundated coastal settlements, and affected soils and vegetation while creating new environments, including those valued for their fishing potential. This process submerged Early and Middle Mesolithic sites, but also affected later Mesolithic (Ertebølle) sites (Malm 1995).

Pioneering investigations on submerged Mesolithic sites took place in Denmark in the early 1970s. Among the first to be professionally excavated were Møllegabet on the Island of Ærø, undertaken by Langelands Museum (Skaarup 1995; Skaarup and Grøn 2004) and Tybrind Vig on Funen, led by Moesgaard Museum (Andersen 1987, 2013, see also Uldum 2011). The waters surrounding Funen have proven to be among the richest in Denmark—and the world—with regards to the amount of preserved settlements and inundated Mesolithic landscapes. Although relative sea-level change is a complex and regional process, dropping the water level by 1–4 m gives a rough estimate of the past shoreline, which corresponds with the location of coastal settlements of the late Ertebølle culture.

The Ertebølle culture (c. 5600–3950 cal BC) is known for its use of flint blades, its simple flake industry and pursuit of intense coastal-resource subsistence and near-shore habitation sites (Price

1991). Ertebølle people carved bone and antler, lived in houses and buried their dead close to their everyday living space (Larsson 1990; Grøn 2003). Though there may have been knowledge of food production and agricultural peoples of neighbouring regions, the Ertebølle maintained a highly productive use of marine resources (Pedersen 1995). This shows a diversity of food resources, which, in terms of quantity of species, was much greater than in the earlier Kongemose and Maglemose cultures (Price 1991) and was perhaps sufficient to deter the introduction of animal husbandry (Fischer and Kristiansen 2002). Importantly, the Ertebølle represents a transitional Mesolithic culture, which adopted and used ceramics from neighbouring Neolithic groups around 4600 cal BC and lived in permanent or semi-permanent sedentary communities. However, the Eretebølle did not adopt agricultural subsistence and continued as a hunter-fisher society for several hundred years despite the presence of agriculture in neighbouring regions to the south and east.

The inventory of submerged sites from this period is now approaching 500 entries in the National Finds & Monuments database for the Fyn area, managed by the Langelands Museum. This consists of finds and sites ranging from a few pieces of worked flint (sometimes out of context), to large areas of in situ settlement or refuse layers. In particular the Møllegabet and Tybrind Vig sites both show excellent conditions of preservation resulting from the compact layers of peat and gyttja that have encapsulated every detail. Small, fragile organic specimens and a range of hitherto unknown artefacts have been excavated in both sites, such as paddles, shafted axes and textiles (Andersen 2013, see also Skriver et al. Chap. 8).

## 5.3 The Helnæs Bay Survey and Discovery of the Falden Site

In 2010 a site was discovered in the Helnæs Bay on the South-western part of the island of Funen (Fig. 5.1). The Bay had been visited previously by Langelands Museum with the aim of locating previously unknown Mesolithic sites, and also to search for remnants of an early Medieval pole blockage, a military defensive construction, which had been investigated by the Danish National Museum in the early 1970s. The Bay, both in its present form, and as part of the Atlantic period archipelago, forms a distinct sheltered marine environment, and up to modern times has been the source of livelihood and port of departure for the local population. For the most part, the underwater prehistoric sites found in the area were those that were clearly undergoing erosion, and seemingly without the presence of any peat or gyttja layers, but the general potential of the inner Danish waters for locating submerged Mesolithic sites is illustrated by the fact that a handful of sites were found in the Bay during a few days of diving in 1995.

On a reconnaissance project in 2003 the museum located a site near the south-eastern shore of the bay which contained layers of brown peat. Embedded in the organic material were preserved tree trunks from the transgressed Mesolithic forest, and considerable amounts of worked flint including many unpatinated and sharp pieces, which suggested a primary deposit. The new site discovered in 2010 during the SPLASHCOS training mission off the hamlet of Falden, south of the village of Faldsled, also contained peat layers, and worked flint in very good condition in quantities previously not seen in Helnæs Bay.

### 5.3.1 *Rationale: Strategic Research and Active Heritage Management*

Although Danish waters have yielded some of the world's most spectacular submerged Mesolithic sites, the number of investigations carried out by archaeologists is quite limited. Apart from the few extensively excavated sites such as Møllegabet I–II, and Tybrind Vig, the many hundreds of dots on the Sites & Monuments Inventory (compiled into a digital database/GIS) are mainly the result of only single-day reconnaissance dives. Only very limited information can be gathered from a few hours of diving, typically spent by swimming over the patches of eroded settlement or collecting worked flint. Often, very little can be done to gather information about sedimentation, and even the spatial extent of the site in question is often not systematically confirmed. In the absence of a remit and budget to secure material for absolute dating, the possibility of dating such a site depends on chance finds of artefacts with diagnostic characteristics suitable for typologically-based sequence dating. Although it is significantly better than nothing, this method of limited reconnaissance diving also carries a number of inherent biases or even faults. It is a superficial approach to archaeological survey that establishes and perpetuates an inventory with biased data. Because heritage management seeks not only to quantify its resource, but rather qualify the significance for public benefit, such a superficial approach is insufficient and does too little to promote scientific and archaeological advancement given the demonstrated international importance of this resource.

Adding to the further bias of survey results, only sites that are actively eroding will be discovered by reconnaissance missions alone. There is no correlation between this near-shore erosion and the preserved Mesolithic shoreline, and as such the settlements that are surveyed are dictated by modern conditions; they are not necessarily representative of prehistoric activity centres themselves or, as a dataset, an accurate distribution of prehistoric sites. The Falden site, on the other hand, is rare in that it was only found because it was actively sought through test excavation justified by and based on existing information and local knowledge.

The difference between a single day and a couple of weeks of field investigation at a site is dramatic. After a site is discovered and the initial information is gathered, a research plan must be devised. All submerged prehistoric sites in Denmark are protected by law. Therefore it is a prerequisite to obtain permission to carry out any intrusive methods (such as test pitting) from the Agency of Culture, the central government agency responsible for cultural heritage. Often in cases of newly discovered sites, preliminary research goals will be to establish the basic elements of the site: age/date(s), extent/size, state of preservation, characterisation of the site activities and, where possible, site function. In most cases, while a brief visual survey does help to establish the most basic of records, it is insufficient to document with confidence the core archaeological information needed to establish the significance of the site. For an inventory of underwater cultural heritage to be useful, the data on which it is built must be reliable and substantial. Strategic research must go beyond the minimum prerequisites needed to 'put a dot on the map'. Of course, similar challenges are faced in terrestrial archaeology; however, the challenges of working in the underwater environment magnify this problem, resulting in large databases of poorly understood sites, as discussed by Adams (2014) with specific reference to shipwreck studies.

Although the discussion above highlights the low percentage of in-depth site investigations undertaken in recent years, the long history of submerged prehistoric site investigation in the Baltic has resulted in the evolution of an efficient and practical methodology. Wherever possible, a firm date should be obtained, ideally by radiometric techniques, to secure an age-range of an artefact, or from clearly cultural material provenanced from an intact (settlement or refuse) layer, which is sequentially related to other preserved layers. To achieve this, some intrusive methods are necessarily involved. Similarly the spatial extent of the site needs to be based on thorough documentation of the stratigraphy; samples may be obtained by hand-coring and test trenches, and at spatial intervals to demonstrate appropriate coverage within the site. Trenches should reveal the state of preservation, and where they yield artefacts, evidence of past activity. Remote sensing, which can be very useful in understanding a landscape on the regional scale, still cannot yet provide the resolution obtained by observation during traditional, diver-based underwater archaeological field practices (but see Bailey et al. Chap. 1, Missiaen et al. Chap. 2). This is especially the case where sites are known or considered likely to exist at a local scale, and certainly there is no way to replicate excavation remotely. Remote sensing and sampling techniques (such as geotechnical coring and grab-sampling) have, however, proved useful in surveying and establishing potential for the discovery of new sites, especially at the medium or large (regional) scale where sites are not necessarily expected, or where limited data have demonstrated some possibility of discovery (e.g., Plets et al. 2007, Tizzard et al. 2011, Westley et al. 2011, Missiaen et al. Chap. 2).

## 5.4 A Programme of Community Engagement: Langelands Museum

Local knowledge has been cited as being the primary resource for archaeological discovery. More submerged sites in Denmark were discovered by members of the public than by any other means during the heyday of discovery in the 1970s and 1980s (Smed 1987).

Langelands Museum has taken a proactive approach to community engagement. Since the start of its underwater activities in 1972, when surveying for submerged Mesolithic sites was initiated, the museum has acted in cooperation with amateur divers interested in archaeology (Skaarup and Grøn 2004). The investigations of the Møllegabet I site began in 1976, and the majority of excavators were not qualified archaeologists, though they were directed by professionals.

In more recent times, an annual 'cruise' (in the humblest sense) with the aim to survey for submerged Mesolithic sites has taken place since 2000, with the museum's (active) historic vessel 'Mjølner' used as a research platform (Fig. 5.2). The vessel is suitable for up to eight people as a live-aboard, but can accommodate another six on board during a day's work. The participants are required to demonstrate adequate dive qualification and skills, but need not have any formal qualification in archaeology. A short introduction addressing ethics, protocols and diving safety sends the new amateur archaeologist off on a first non-disturbance, visual survey. The sites are chosen in advance, mainly based on Fischer's (1993a, 1995) predictive model for stationary fishing sites, and are marked with buoys enclosing the area in a square formation.

During the time Langelands Museum has hosted these tours, more than 100 one-week slots have been offered, but since a core group of participants has taken part in more than one tour, this roughly equates to 60 different individuals. More than a dozen have been undergraduate archaeology students who have taken this opportunity to gain their first hands-on experience with underwater archaeology as an entry-level underwater archaeological field school. Some of those have pursued careers in the field of underwater and maritime archaeology, while others have stayed in terrestrial archaeology but with an insight into the practical realities and first-hand experience of submerged cultural landscapes.

About 60 individual locations have been surveyed by visual reconnaissance, and this has added roughly 30 new entries into the Danish national finds and monuments database, ranging from a few displaced pieces of worked flint to well-preserved and intact settlement layers.

## 5.5 Field Results at Falden

The discovery of the Falden site was made in 2010 during a 'Short Term Scientific Mission'—a field course under the SPLASHCOS banner funded as part of the European Cooperation in Science and Technology (COST) Trans-Domain Action 0902. Helnæs Bay was chosen because earlier reconnaissance surveys using a small team and rapid survey methods, as described above, had identified finds worth further evaluation. Also, because the waters in the area are reasonably well sheltered, the risk of having to cancel diving due to bad weather was reduced. The need for potential 'call-off-days' for underwater fieldwork and training missions is an especially important real-world consideration and imposes a need to manage expectations for everyone involved, from students to staff to funding agencies.

During the initial days of the 5-day training course, a number of dive locations along the northern shore of the bay were subjected to the usual visual survey, with the addition that the divers were allowed to hand-fan the surface sediments. During the first 3 days, the limitations of visual survey were apparent and archaeological results were thin. A spot on the Eastern shore had been chosen in advance, but initially nothing but sandy seabed with *Zostera* (sea grass) and boulders was to be seen through non-disturbance methods. At this point, at the end of the survey, one of the most experienced members of the SPLASHCOS survey team (and indeed one of the most experienced underwater archaeologists specialising in prehistoric material), Dr. Ehud Galili, decided to investigate the seabed further. His efforts hand-fanning a small test pit resulted in the recovery of worked flint with sharp edges and unpatinated surfaces, characteristics of a primary, undisturbed deposit. A basic grid was laid on the seabed, with a N–S/E–W cross, measuring 10 m, and more hand-fanned test pits were dug. All of the test pits contained unusually high concentrations of worked flint, even by Danish standards.

### 5.5.1 Test Excavations

During the relatively few, though often long, dives undertaken in 2010, the minimally-intrusive field work had proved that the site contained intact layers of peat with substantial quantities of worked flint. The site's spatial extent was measured to be at least a 10 × 10 m square. The museum had not investigated a submerged site in Helnæs Bay or its vicinity, though Ertebølle-sites had previously been studied on land on Western Funen (Andersen 1978). The new site had the potential to add new data to the general inventory, with the potential to provide insights into the history of transgression in the region. Also, the fact that the site would not have been discovered but for active investigative methods (hand-fanning and hand-dug test pits) on the seabed gave it a unique status, with the potential to reveal new knowledge about southern Danish submerged sites, their preservation and the possibilities for future significant scientific and archaeological discovery.

In 2011 another field course in the SPLASHCOS-network was organised, with the aim of investigating the Falden site further and training early career researchers. The historic vessel Mjølner once again served as a working platform, and was moored just off the site, allowing divers to enter the water directly from the ship. The basic grid from the previous year was extended (with the intersection between the axes labelled 500/500, to avoid negative coordinates). The x axis was marked with a red plastic baseline, with a tape measure pinned alongside. During the course of a week, five test pits, each 1 m × 1 m, were excavated south of the baseline intersection, and two east of the intersection (Fig. 5.3). In total, the team managed seven test pits in the course of 4 days of diving, which was satisfactory. The last day had to be spent securing the site and backfilling with sandbags and geotextile, since the protective layer of *Zostera* roots and surface sand had been removed.

The 2012 season was a straightforward continuation, with the grid re-established, consistent with the previous year. Further test pits were excavated (Figs. 5.3 and 5.4, bringing the total to 15). The observations previously made were substantiated, namely that the conditions of preservation for flint are good, but only fair regarding organic materials (when compared with the best-preserved submerged sites in the south-west Baltic).

### 5.5.2 Material Culture, and Faunal Remains and Chronology

The sediments differed considerably between two different test pits. The time needed to excavate the shallow pits amounted to a morning's dive, while the southernmost pit required 2 days. The stratigraphy below the *Zostera* roots on the site was a layer of coarse sand in all pits, which contained the bulk of

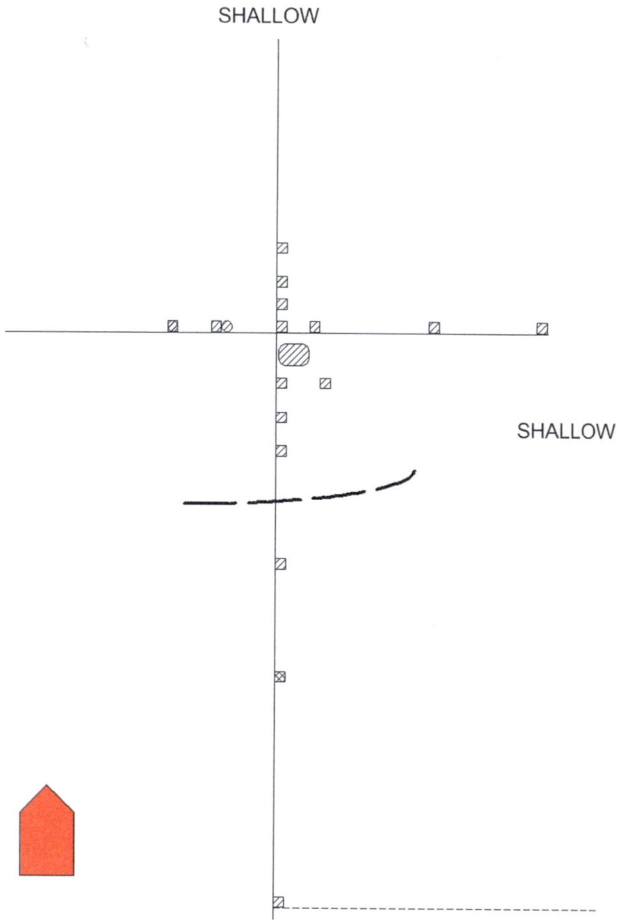


**Fig. 5.3** Grid with trenches indicated. The shore is 200 m to the east of the grid centre, towards the north is a now transgressed small spit of land. The *hatched line* shows the approximate location of the prehistoric shoreline (Graphics: Langelands Museum)

the worked flint. Under this, a layer of fibrous peat appeared around y = 490, but not extending to y = 470, and was also seen in patches north of y = 490. At y = 470, the layer of coarse sand was more than 50 cm thick. This could be explained by the presence of the prehistoric shoreline somewhere between these two points. The thick layer of coarse sand containing large amounts of worked flint would thus represent a refuse layer deposited in the shallows just off the beach. The dwelling area would have been to the north, on a small spit of land projecting to the west, still visible today as a shallow mound on the seabed.

The majority of the finds consisted of worked flint. At the time of writing, cataloguing is not complete, but many of the test pits contained several hundred pieces, most of course being flint waste, but including many tools and diagnostic pieces. Notable diagnostic artefacts are the transverse arrowheads, as they date the inventory to the Ertebølle culture, most likely its middle phases (Fig. 5.5). These are made from blades, and of a very uniform markedly splayed type, known as Stationsvej type in Danish archaeology (Vang Petersen 1993). Other evidence of blade production is documented in numerous 'raw' (unretouched) blades, retouched blades in the form of flint knives and scrapers, and cores in their final discarded stage. Larger tools are common in the form of flake axes. They are made from one large flake, often resembling a large transverse arrowhead and ranging in length from 4 cm to the more common 12–15 cm. Many large flakes, presumably blanks intended for being shaped into axes, were also found. No core axes have been found on the site, which is quite unusual and likely to be indicative of its function and character. Scrapers are predominantly circular flake scrapers, made

**Fig. 5.4** John McCarthy dredging a 1 × 1 m trench marked by stainless 25 × 25 cm frames. *Top layer* was sand, in places covering peat and/or gyttja. The sediment ejects into a bag made of fine-mesh fishing net (Photo: Jonathan Benjamin)

by a direct steep retouch of approximately one third of the circumference from the ventral side. Most of the worked flint is well preserved, with varying degrees of patina (Figs. 5.6 and 5.7). Patinated lithics in the region generally demonstrate whitening when exposed to the seawater, whereas those which are well protected in the anaerobic gyttja are usually dark grey or black in colour. Based on the patina encountered, only very little worked flint had been exposed by erosion. The excavated finds show no mechanical damage, such as rolling or fracturing, as would be expected if they had been transported and redistributed by water action. The site and its material were deemed to have been in situ.

In contrast to the numerous flint artefacts, only a few faunal remains have been recovered so far. During the 2011 campaign, twelve bones were found (Table 5.1). They include a skull fragment and a piece of antler of roe deer, a left humerus and two molars of wild boar and five unidentifiable pieces. Furthermore, two fish bones, from cod and flatfish, are present. The small number of finds limits further statements, but in general roe deer and wild boar are typical prey, and fishing of marine species also is not unusual during this period. The faunal results of the 2012 field season have yet to be fully analysed.

Thus far, two radiocarbon samples have been dated from the Falden site. These are a molar (M3) and a left humerus, both from wild boar (Table 5.2). They come from the same test pit 500/489, but from different strata. The tooth was found on top of the peat or detritus mud just under the coarse sand, the bone roughly 15 cm lower in the same organic sediment. Both samples produced ample collagen yields and the AMS $\delta^{13}C$ values are in the normal range for collagen. Therefore the results are regarded as reliable and because the samples are from terrestrial mammals a marine environmental reservoir correction is not required. The bone from the lower level yielded a date of 4991–4798 cal BC and the tooth from the upper level a date of 4520–4363 cal BC. The dates are significantly different and in the

**Fig. 5.5** Typical examples of worked flint from the site. A flake axe at *top left* is slightly damaged, and has a white patina, as has the transverse *arrowhead* at *top right*. Two flakes at *bottom left* are presumably by-products from rough shaping of a flake axe. At *bottom right* is a blade-yielding core (Photo: Langelands Museum)

**Fig. 5.6** The amount of worked flint dredged during one single dive in one trench. Blades, flakes and various chips show varying patination (Photo: Langelands Museum)

**Fig. 5.7** John McCarthy emptying a net taken off the dredge, and brought topside to be sorted. Sediment is blown through the mesh, and all objects kept inside. Un-worked flint, pebbles and marine life is thrown back in the sea (Photo: Jonathan Benjamin)

**Table 5.1** Animal bones from the 2011 survey

| Provenance | Number | Quadrat | Depth | |
|---|---|---|---|---|
| ØHM 15264 | X2 | 499/490 | 0–10 | 1 *Capreolus capreolus*; Os frontale with antler tine, left |
| ØHM 15264 | X12 | 500/489 | 30–40 | 1 *Sus scrofa*, Humerus, left |
| ØHM 15264 | X13 | 500/489 | 20–30 | 1 *Sus scrofa*, tooth M3 |
| | | | | 1 indet. |
| ØHM 15264 | X14 | 500/492 | 0–15 | 1 indet. |
| ØHM 15264 | X15 | 500/479 | 0–40 | 1 Pleuronectidae (Vertebra caudalis), Flounder? |
| | | | | 1 *Gadus morhua* (Vertebra precaudalis) |
| ØHM 15264 | X16 | 500/489 | 0–20 | 1 *Capreolus capreolus*; Antler |
| | | | | 1 *Sus scrofa*, tooth M2(?) |
| | | | | 3 indet. |

Determinations by Dirk Heinrich and Wolfgang Lage, Schleswig-Holstein Archaeological State Museum, Schleswig, Germany

**Table 5.2** Radiocarbon results from the 2011 survey, test trench 500/489, calibrated using the IntCal09 calibration (Reimer et al. 2009) and OxCal v4 (Bronk Ramsey 2009)

| Inventory number | Layer depth | Laboratory number | Identification | δ13C (‰)[a] | Conventional 14C age (BP) | Calibrated date (95% confidence) |
|---|---|---|---|---|---|---|
| OHM15264-X13 | Peat 20–30 cm | KIA-45623 | Wild boar (*Sus scrofa*), molar (M3) | −21,34 | 5620 ± 30 BP | 4520–4363 cal BC |
| OHM15264-X12 | Peat 30–40 cm | KIA-45622 | Wild boar (*Sus scrofa*), L humerus | −20,27 | 6005 ± 35 BP | 4991–4798 cal BC |

[a]Note that the δ13C includes the fractionation occurring in the sample preparation as well as in the AMS measurement and therefore cannot be compared to a mass-spectrometer measurement

right stratigraphic order; however, it remains uncertain if these dates hint at two different occupations or one continuous occupation of the site. More data, including radiometric dates, will be needed to resolve this question.

If the dates are compared with the age of other submerged Ertebølle sites south of Funen, the older Falden date corresponds with Tybrind Vig B, horizon 1, and the younger Falden date with Tybrind Vig B, horizon 2 (Andersen 2013, 51) and with Ronæs Skov (Andersen 2009, 38). Both sites are situated north-west of Falden in the Little Belt while the other well-known sites of Møllegabet I and II are in the south-east off the north-eastern coast of Ærø Island. In this case it seems that Møllegabet II is older than Falden, in contrast to Møllegabet I, which was occupied more recently than the Falden site (Skaarup and Grøn 2004, 103). Further to the south on the German coast of the Western Baltic, sites like Timmendorf-Nordmole II, Rosenfelde, and Rosenhof are more or less contemporaneous with the Falden site (Hartz and Lübke 2006; Hartz et al. 2014).

### 5.5.3 Photogrammetry

The nature of the investigations at Falden as a training exercise allowed scope for experimentation with novel techniques. During the 2012 season, the authors conducted some preliminary trials of underwater photogrammetry on some of the dives using a ruggedized consumer-grade compact Lumix DMC-TS3 without a housing, waterproofed to a depth of 12 m. It was hoped that this would demonstrate methods for achieving a richer and more accurate record of submerged prehistoric sites than had previously been possible. Conditions on the site, such as depth, general light level and visibility were found to be suitable to take photos of sufficient quality. However, the presence of *Zostera* did present a major challenge. Photogrammetric recording requires a largely static subject and the *Zostera* were in constant motion due to the swell and current. After some disappointing results, we concentrated on limited areas which had been manually cleared (for more detail see McCarthy and Benjamin 2014).

We also made successful surveys of individual metre-square trenches using large numbers of oblique images taken in a roughly circular pattern. We also surveyed in a similar manner selected trenches at different stages of excavation (Fig. 5.8) and small individual in-situ archaeological finds including an in-situ flake axe (Fig. 5.9). Trench recording was also possible once *Zostera* was manually removed. Large numbers of images (up to 500 exposures) were captured in several trenches. Despite the large numbers of photographs taken, survey was rapid, in each case taking less than 10 min per dataset. We also found that the use of large numbers of photographs had the effect of filtering out sediment in the water column, making for clearer imagers.

Trench models were output with approximately five million polygons. Processing was undertaken using dedicated geomatics workstations with Agisoft Photoscan software. Small finds were also recorded, including a 15 cm flint flake axe, output as a model of 19.8 million polygons (Fig. 5.9). Although these datasets took over 24 hours of computer time to process, the resulting models of the

**Fig. 5.8** Orthographic photogrammetric plan views of Trench 4 before and after removal of archaeological deposits. The trench is one metre square and images are north-orientated (Graphics: John McCarthy)

**Fig. 5.9** Perspective view of photogrammetric model of an in-situ flake axe on the seabed. The axe is ca. 15 cm from edge to butt (Graphics: John McCarthy)

trenches and the axe proved to be accurate representations of the features encountered and after colour and contrast correction offered a better record than was possible through photography alone, possibly as a result of the filtering described above; for example, compare one of the original photographs (Fig. 5.10) to the photogrammetric trench plans (Fig. 5.8). The 3D trench models also facilitated the creation of virtual reconstructions of the hand dredging excavation, and enabled us to create an animated video showing one of the trenches before and after excavation, and the removal of the overlying

**Fig. 5.10** An example of one of the photographs of Trench 4 used in the photogrammetric recording. The trench is 1 m square (Graphics: John McCarthy)

**Fig. 5.11** A screen grab from an animated reconstruction of the dredging process (Graphics: John McCarthy)

sediment by a virtual dredger to reveal the underlying surface (Fig. 5.11). This will be used as a teaching tool in future work. The application of these experimental survey techniques proved valuable not only as a record of the site but also in demonstrating the value of interdisciplinary approaches and international collaboration.

## 5.6 Discussion

### *5.6.1 Archaeological Significance*

Apart from its initial discovery, the Falden site has been investigated by two, 1-week mini-seasons, totalling less than ten full days of work. The results show that much can be gained even by this admittedly limited approach, and certainly much more than by a single-day rapid reconnaissance survey. The ideal would be a comprehensive full season (or multiple seasons) of archaeological excavation. However, given real-world considerations, resources and time, we suggest that the type of activity we conducted at Falden does indeed provide good value for money where full-scale excavation is not justifiable or the resource is simply unavailable. It could be considered on more sites as part of a broader research strategy, and this would both enhance our knowledge of the past and also inform good decision making in the future. It would also facilitate collection of additional absolute dates from the Danish submerged settlement sites, which is necessary if we are to fully incorporate the submerged Mesolithic with what is known generally from terrestrial deposits and with submerged sites further south in the German Baltic (see Goldhammer and Hartz, Chap. 9; Hartz et al. 2014).

The material encountered at Falden may seem a quite unremarkable and typical example of a Late Ertebølle settlement. The work carried out does not proclaim the discovery of the world's (or Europe's) 'oldest evidence'—for example of boating, art, or major cognitive or cultural advancement. However, there is significance in making incremental contributions to the dataset of the past: a record that slowly and steadily fills in the picture of the human experience during prehistoric periods where modern keyhole sampling exercises provide the only information available, information that may be biased because of the limited sample of available data. In the Falden case, the results are of particular interest in referring to peoples who were on the cusp of the transition from foraging to farming. The archaeology of the average site may be just that, but it should not be under-appreciated. Care should be taken to ensure that regional and national inventories are based on methodical data gathering and recording techniques. In this vein, it is important that funding agencies do not expect every archaeological site to produce sensational results. Indeed this will only serve to produce sensationalised reporting or, put differently, unrealistic expectations and interpretations. Substantive archaeological research takes time, and the analysis of the 'mundane' in the Mesolithic may require hundreds of sites before real breakthroughs and interpretive advancements are made. The site at Falden and work carried out there is a good example of a solid contribution to the incremental approach required to build a reliable record of the past and a reliable basis for future interpretation and decision-making.

Because the Falden site was not discovered by visual inspection, but by active search in the form of hand-fanned pits, the result is that a site has been found where almost no erosion had taken place. This was achieved through the use of the fishing-site-location model and the availability of many diving archaeologists at one time. This approach can surely be refined by further experience. It is the first site in Helnæs Bay (a substantial body of water) that shows more than just displaced artefacts in secondary position. Judging from the seabed characteristics, many more sites like this must be present in the Bay and in the waters of the South Funen Archipelago.

### *5.6.2 Training and Capacity-Building*

This site and the archaeological project associated with it also presented an excellent opportunity for capacity-building and training of Early Stage Researchers, many of whom would not have had the opportunity to work, first hand, on such a submerged prehistoric site.

In modern underwater archaeology, SCUBA diving and the ability to breathe and thus work under water are taken for granted. There are, however, numerous considerations which must be acknowledged when working under water:

> It is almost impossible to describe to someone who has never dived what it is like to work underwater; for various technical reasons, even the most realistic of underwater films give a partially false impression. Much of the peculiarity of the experience can be traced to the fact that the diver is effectively weightless under water... (Muckelroy 1978, p. 24).

Coupled with gravity, pressure and mobility issues, limited visibility and communication make underwater archaeology physically different from archaeology on land. In addition to the reality of working underwater, there are numerous basic principles of dive theory which must be considered. It is within this context that we identify a need for training of qualified underwater archaeologists.

Even more specific to this discussion is the opportunity for young practitioners and early career researchers to train on submerged prehistoric sites. Since the vast majority of prehistoric sites are investigated in terrestrial contexts, and most underwater cultural heritage (the broad legal term encompassing archaeological sites of all ages) is concerned with sites of Historic or Classical periods (Bass 1966; Muckelroy 1978; Delgado 1997), there are few opportunities for early stage researchers destined for employment in the heritage profession to gain suitable training and qualifications, and especially to the standard required by competent authorities responsible for the management and, importantly, the authorization of research. Indeed, if such competent authorities are entrusted with checking and weighing the considerations involved (see Maarleveld et al. 2011), it is imperative that they are qualified and experienced beyond mere theoretical familiarity or minimal competence. There is an urgent need for practical training in the applied methodology of this branch of archaeology and heritage management, especially beyond the Danish sphere.

It is also important that when archaeological mitigation is required in a pre-development context, trained professionals are available who are capable of responding to the threats to the cultural heritage of all site-types which may be encountered under water. A lack of training and experience in the discipline of submerged landscape archaeology will limit both the legal framework within which decisions are taken and also the capacity of the profession to respond appropriately to real-world conditions. In the worst-case, with managers and practitioners who lack the necessary knowledge and experience, protection of prehistoric sites may not be considered at all. Also of grave concern under such conditions would be the tendency for heritage authorities to refuse intrusive activity, such as test excavation, even when legitimate reasons of archaeological research and capacity-building justify it. Such a stance would have prohibited the discovery and investigation of the Falden site, and many other sites investigated in the past which we now know to have made both incremental and step-change contributions to our understanding and appreciation of early prehistory. In a world of increased scrutiny and accountability, and with seemingly fewer resources for archaeology, it may perhaps seem safest to do nothing. However, this is a direct threat to scientific progress, to the development of submerged landscape archaeology as an emerging discipline and to archaeology in general.

It is fortunate that presently such a scenario does not exist in Denmark, with the balance of protection and intrusive investigation weighed along with the appropriate needs for scientific and archaeological research as well as an understanding and appreciation of the need for capacity building. However, with an increased focus on in situ preservation and more jurisdiction and responsibility placed on the shoulders of already-stretched heritage managers around the world, the described scenario may not be far-fetched, especially if action is not taken to train future generations in the theoretical and applied aspects of all types of underwater archaeology.

### 5.6.3 *Proposal for an International Centre of Excellence*

Through network-building between EU-based participants, the Danish field training undertaken in collaboration with SPLASHCOS participants of various countries and the Langelands Museum can be considered as a highly important first step—a foundation for education and international capacity-building in submerged prehistoric archaeology. Many years of collaborative research with local community members and professional archaeologists in Denmark have provided increased training in the practical aspects of underwater site recording. It is to be hoped that the work undertaken by the SPLASHCOS network will continue; the momentum and good will, as well as awareness in public life, must be harnessed in order to ensure a future in the field of study. Only by continuing research and training—on actual sites, not only theoretical 'landscapes'—can the next generation of archaeologists and heritage professionals be prepared to encounter real-world scenarios. The Danish Baltic, with its regional experts and practitioners (and international friends) is uniquely positioned to make this globally important contribution to underwater archaeology and cultural heritage management.

The Danish coast and nearshore waters provide an internationally recognized opportunity for a centre for excellence in underwater research, training and capacity building. Building on its history within the field of study, its leadership representation within SPLASHCOS (and other international consortia), Danish marine archaeology has an opportunity to offer a safe, accessible and scientifically significant heritage resource in its shallow coastal waters, a resource that can be considered relatively abundant in its concentration when compared with what is known on a global scale. Denmark has also been at the forefront of developing what is now considered an internationally important field, both in terms of scientific and archaeological contribution as well as heritage management. It now has a unique opportunity to demonstrate its successful development of the field of study since the 1970s. In this vein, but perhaps from a different perspective, it could equally be argued that through the investment in and success of Danish marine archaeology, Danish practitioners have a professional responsibility to share their experience and to assist a growing number of interested members of the professional community of researchers and heritage managers in Europe and worldwide who need training. A recent workshop held near Washington DC, USA, has demonstrated an international demand for such expertise. This should be viewed as an opportunity for Danish practitioners to make a considerable and lasting contribution to World Prehistory.

Without deliberate succession planning for long-term capacity building (following on from SPLASHCOS), a sustained field of research of submerged prehistoric landscape archaeology may not develop to its full potential. A case can be made for the establishment of a training centre, based in Denmark and with international participants, on an annual basis, to ensure the future of suitably trained specialists and practitioners in submerged landscape archaeology worldwide. This would be both a contribution to capacity building and to submerged archaeology and can also be developed to the mutual benefit of all participants and institutions.

## 5.7 Conclusion

The SPLASHCOS Short Term Scientific Mission in 2010 funded the training of eight international Early Stage Researchers. The Langelands Museum provided training as part of their existing community archaeology programme for the public benefit. This proved to be not only a fruitful exercise, but an example of what is truly necessary to train future generations of researchers and managers both in Denmark and worldwide. The archaeological results presented here are significant in their own right as an original contribution to the European Mesolithic and differ in character from the sites previously known in Helnæs Bay. The Falden site is therefore important in both the incremental sense on the large scale and for its uniqueness at a local scale. It would not have been encountered without some minimally intrusive investigation which was legitimately carried out by trained professionals and inexperienced Early Stage Researchers under appropriate supervision and direction. It is under these conditions that the SPLASHCOS training missions represent an important step forward in capacity building and preparation for the future through appropriate training and real-world experience that involves underwater archaeological survey and excavation of prehistoric sites.

**Acknowledgements** The authors wish to thank skipper Hans Toft Nielsen and first mate Ann Katrine Toft Lauritzen who safely manoeuvred 'Mjølner'. We thank Dr. John Meadows for the $^{14}$C-dates, Wolfgang Lage for the analysis of the faunal remains, and the crew of archaeologists who took part: Kieran Westley, Jörg Ewersen, Anne Margrethe Walldén, Ehud Galili, Radek Szemelak, Jesper Frederiksen and Claus Skriver. The mainstay of amateur archaeologists during many years, Flemming Sørensen and Arne Jensen are warmly thanked for their unwavering enthusiasm.

## References

Adams J (2014) A maritime archaeology of ships. Oxbow Books, Oxford

Andersen SH (1978) Ertebøllekultur på Vestfyn. En oversigt. Fynske Minder 1977. Fyns Stiftsmuseum, Odense, pp 7–36

Andersen SH (1987) Mesolithic dug-outs and paddles from Tybrind Vig, Denmark. Acta Archaeol 57:87–106

Andersen SH (2009) Ronæs Skov. Marinarkæologiske undersøgelser af kystboplads fra Ertebølletid. Jutland Archaeological Society Publications 64. Jutland Archaeological Society, Højbjerg

Andersen SH (2011) Ertebølle canoes and paddles from the submerged habitation site of Tybrind Vig, Denmark. In: Benjamin J, Bonsall C, Pickard C, Fischer A (eds) Submerged prehistory. Oxbow Books, Oxford, pp 1–14

Andersen SH (2013) Tybrind Vig: submerged Mesolithic settlements in Denmark, Jutland Archaeological Society Publications, vol 77. Jutland Archaeological Society, Højbjerg

Antevs E (1922) On the late-glacial and post-glacial history of the Baltic. Geogr Rev 12(4):602–612

Bailey G (2004) The wider significance of submerged archaeological sites and their relevance to world prehistory. In: Flemming NC (ed) Submarine prehistoric archaeology of the North Sea. Research priorities and collaboration with industry, CBA Research Report, vol 141. Council for British Archaeology, York, pp 3–10

Bass GF (1966) Archaeology under water. Thames and Hudson, London

Benjamin J (2010) Submerged prehistoric landscapes and underwater site discovery: reevaluating the 'Danish model' for international practice. J Island Coast Archaeol 5(2):253–273

Bennike O, Lemk W, Jensen JB (1998) Fauna and flora in submarine early-Holocene lake-marl deposits from the southwestern Baltic Sea. The Holocene 8(3):353–358

Bennike O, Jensen JB, Lemke W, Kuijpers A, Lomholt S (2004) Late-and postglacial history of the Great Belt, Denmark. Boreas 33:18–33

Björck S (1995) A review of the history of the Baltic Sea 13.0–8.0 ka BP. Quat Int 27:19–40

Brinch-Petersen E, Meiklejohn C, Alexandersen V (1998) The later Mesolithic population of Sjælland, Denmark, and the Neolithic transition. In: Zvelebil M, Dennell RW, Domanska L (eds) Harvesting the sea, farming the forest. Sheffield Academic Press, Sheffield, pp 203–212

Bronk Ramsey C (2009) Dealing with offsets and outliers in radiocarbon dating. Radiocarbon 51(3):1023–1045

Christensen C, Fischer A, Mathiassen DR (1997) The great sea rise in the Storebælt. In: Pedersen L, Fischer A, Aby B (eds) The Danish Storebælt since the ice age – man, sea and forest. A/S Storebælt Fixed Link, Copenhagen, pp 45–54

Delgado JP (ed) (1997) Encyclopaedia of underwater and maritime archaeology. British Museum Press, London

Fischer A (1993a) Stenalderbopladser i Smålandsfarvandet: en teori afprøvet ved dykkerbesigtigelse (Stone Age settlements in the Småland Bight). Miljøministeriet, Skov- og Naturstyrelsen, Copenhagen

Fischer A (1993b) Stenalderbopladser på bunden af Øresund. Afprøvning af en model. Miljøministeriet, Skov- og Naturstyrelsen, Copenhagen

Fischer A (1995) An entrance to the Mesolithic world below the ocean. Status of ten years' work on the Danish sea floor. In: Fischer A (ed) Man and sea in the Mesolithic. Oxbow Books, Oxford, pp 371–384

Fischer A, Kristiansen K (eds) (2002) The neolithisation of Denmark: 150 years of debate, Sheffield Archaeological Monographs, vol 12. J.R. Collis, Sheffield

Grøn O (2003) Mesolithic dwelling places in south Scandinavia: their definition and social interpretation. Antiquity 77:685–708

Hartz S, Lübke H (2006) New evidence for a chronostratigraphic division of the Ertebølle culture and the earliest Funnel Beaker culture on the southern Mecklenburg Bay. In: Kind CJ (ed) After the ice age. settlements, subsistence and social development in the Mesolithic of Central Europe, Materialhefte zur Archäologie in Baden-Württemberg, vol 78. Theiss, Stuttgart, pp 59–74

Hartz S, Jöns H, Lübke H, Schmölcke U, Von Carnap-Bornheim C, Heinrich D, Klooß Lüth F, Wolters S (2014) Prehistoric settlements in the southwestern Baltic Sea area and development of the regional Age economy. In: Harff J, Lüth F (eds) SINCOS II – sinking coasts: geosphere, ecosphere and anthroposphere of the Holocene southern Baltic Sea. Bericht der Römisch-Germanischen Kommission 92 (2011). Philipp von Zabern, Mainz am Rhein, pp 77–210

Klassen L (2000) Waterborne exchange and Late Ertebølle social structure. In: Von Schmettow H, Koldeweij AR, Ter Molen JR (eds) Schutz des Kulturerbes unter Wasser-Veränderungen europäischer Lebenskultur durch Fluß-und Seehandel. Beiträge zum Internationalen Kongreß für Unterwasserarchäologie (IKUWA "99) 18. – 21. Februar 1999 in Sassnitz auf Rügen. Beiträge zur Ur- und Frühgeschichte Mecklenburg-Vorpommerns, Band 35, Lübstorf, pp 43–51

Larsson L (1983) Mesolithic settlement on the sea floor in the Strait of Öresund. In: Masters P, Flemming NC (eds) Quaternary coastlines and marine archaeology. Academic, New York, pp 283–301

Larsson L (1990) The Mesolithic of Southern Scandinavia. J World Prehistory 4(3):257–309

Maarleveld T, Guerin U, Egger B (2011) Underwater cultural heritage and the rules of the UNESCO 2001 convention: a manual. UNESCO, Paris

Malm T (1995) Excavating submerged Stone Age sites in Denmark – the Tybrind Vig example. In: Fischer A (ed) Man and sea in the Mesolithic. Oxbow Books, Oxford, pp 385–396

Mathiassen D (1997) The changing landscapes of the Storebælt from the retreat of the ice to the sea flood. In: Pedersen L, Fischer A, Aby B (eds) The Danish Storebælt since the ice age – man, sea and forest. A/S Storebælt Fixed Link, Copenhagen, pp 22–28

McCarthy J, Benjamin J (2014) Multi-image photogrammetry for underwater archaeological site recording: an accessible, diver-based approach. J Marit Archaeol 9:95–114

Muckelroy K (1978) Maritime archaeology. Cambridge University Press, Cambridge

Pedersen L (1995) 7000 years of fishing: stationary fishing structures in the Mesolithic and afterwards. In: Fischer A (ed) Man and sea in the Mesolithic. Oxbow Books, Oxford, pp 75–85

Pedersen L (1997) They put fences in the sea. In: Pedersen L, Fischer A, Aby B (eds) The Danish Storebælt since the ice age – man, sea and forest. A/S Storebælt Fixed Link, Copenhagen, pp 124–143

Plets R, Dix J, Bastos A, Best A (2007) Characterization of buried inundated peat on seismic (chirp) data, inferred from core information. Archaeol Prospect 14:261–272

Price TD (1991) The Mesolithic of northern Europe. Annu Rev Anthropol 20:211–233

Reimer PJ, Baillie MGL, Bard E, Bayliss A, Beck JW, Blackwell PG, Bronk Ramsey C, Buck CE, Burr GS, Edwards RL, Friedrich M, Grootes PM, Guilderson TP, Hajdas I, Heaton TJ, Hogg AG, Hughen KA, Kaiser KF, Kromer B, McCormac FG, Manning SW, Reimer RW, Richards DA, Southon JR, Talamo S, Turney CSM, Van der Plicht J, Weyhenmeyer CE (2009) IntCal09 and Marine09 radiocarbon age calibration curves, 0–50,000 years cal BP. Radiocarbon 51(4):1111–1150

Rowley-Conwy P (1983) Sedentary hunters: the Ertebølle example. In: Bailey G (ed) Hunter-Gatherer economy in prehistory. Cambridge University Press, Cambridge, pp 111–126

Schilling H (1997) The Korsør Nor site. The permanent dwelling place of a hunting and fishing people in life and death. In: Pedersen L, Fischer A, Aby B (eds) The Danish Storebælt since the ice age – man, sea and forest. A/S Storebælt Fixed Link, Copenhagen, pp 93–98

Skaarup J (1995) Hunting the hunters and fishers of the Mesolithic – twenty years of research on the sea floor south of Funen, Denmark. In: Fischer A (ed) Man and sea in the Mesolithic. Oxbow Books, Oxford, pp 397–401

Skaarup J, Grøn O (2004) Møllegabet II: a submerged Mesolithic settlement in southern Denmark, BAR International Series, vol 1328. Archaeopress, Oxford

Smed P (1987) Argus Grund – feltundersøgelser (1984). Fortidsminder og Kulturhistorie, Antikvariske Studier 8:125–133

Tizzard L, Baggaley P, Firth A (2011) Seabed prehistory: investigating palaeo-landsurfaces associated with a Palaeolithic tool find, North Sea. In: Benjamin J, Bonsall C, Pickard C, Fischer A (eds) Submerged prehistory. Oxbow Books, Oxford, pp 65–73

Uldum O (2011) The examination of a Mesolithic double burial from Tybrind Vig, Denmark. In: Benjamin J, Bonsall C, Pickard C, Fischer A (eds) Submerged prehistory. Oxbow Books, Oxford, pp 15–20

Van Andel TH (1989) Late quaternary sea-level change and archaeology. Antiquity 63:733–745

Van Andel TH (1990) Addendum to 'Late Quaternary sea-level changes and archaeology'. Antiquity 64:151–152

Vang Petersen P (1993) Flint fra Danmarks oldtid. National Museum of Denmark, Copenhagen

Westley K, Bell T, Plets R, Quinn R (2011) Investigating submerged archaeological landscapes: a research strategy illustrated with case studies from Ireland and Newfoundland, Canada. In: Benjamin J, Bonsall C, Pickard C, Fischer A (eds) Submerged prehistory. Oxbow Books, Oxford, pp 129–144

# Chapter 6
# Atlit-Yam: A Unique 9000 Year Old Prehistoric Village Submerged off the Carmel Coast, Israel – The SPLASHCOS Field School (2011)

**Ehud Galili, Jonathan Benjamin, Israel Hershkovitz, Mina Weinstein-Evron, Irit Zohar, Vered Eshed, Deborah Cvikel, Jehuda Melamed, Yaacov Kahanov, Jean Bergeron, Clive Ruggles, Avraham Ronen, and Liora Kolska Horwitz**

**Abstract** The site of Atlit-Yam is one of the best preserved and most thoroughly investigated submerged prehistoric settlements in the world, with a wealth of finds of material culture and organic remains characteristic of a Pre-Pottery Neolithic village based on a mixed economy of farming and fishing 9000 years ago. Stone-lined water wells were also found, providing a precise measure of sea-

E. Galili (✉)
Zinman Institute of Archaeology, University of Haifa, Haifa, POB 180 Atlit, 30300, Israel

Israel Antiquities Authority, POB 180, Atlit 30300, Israel
e-mail: galilish@netvision.net.il

J. Benjamin
Department of Archaeology, Flinders University, PO Box 2100, Adelaide 5001, Australia
e-mail: jonathan.benjamin@flinders.edu.au

I. Hershkovitz
Department of Anatomy and Anthropology, Faculty of Medicine, Tel-Aviv University, Tel Aviv, Israel
e-mail: anatom2@post.tau.ac.il

V. Eshed
Israel Antiquities Authority, POB 586, Jerusalem 91994, Israel
e-mail: veshed@gmail.com

D. Cvikel • Y. Kahanov
The Leon Recanati Institute for Maritime Studies, University of Haifa, Haifa 3498838, Israel
e-mail: dcvikel@research.haifa.ac.il; yak@research.haifa.ac.il

J. Melamed
The Hyperbaric Treatment Institute, Elisha Medical Center, The Technion, Haifa, Israel
e-mail: hbomed@barak.net.il

J. Bergeron
Alpha-Zoulou films, Montreal, Canada
e-mail: bergeron@gmail.com

C. Ruggles
School of Archaeology and Ancient History, University of Leicester, Leicester, UK
e-mail: cliveruggles@btinternet.com

A. Ronen • M. Weinstein-Evron • I. Zohar
Zinman Institute of Archaeology, University of Haifa, Haifa, 180 Atlit, 30300, Israel
e-mail: aronen@research.haifa.ac.il

L.K. Horwitz
National Natural History Collections, Faculty of Life Science, The Hebrew University, Berman Building, E. Safra Campus, Jerusalem 91904, Israel
e-mail: lix1000@gmail.com



G.N. Bailey et al. (eds.), *Under the Sea: Archaeology and Palaeolandscapes of the Continental Shelf*, Coastal Research Library 20, DOI 10.1007/978-3-319-53160-1_6

level position when the site was in use, as well as a megalithic structure and human burials. Eventually the site was abandoned in the face of progressive sea-level rise, and later Neolithic settlements, were occupied at a higher level, and are now submerged closer to the shore. SPLASHCOS funding to support a Training School, allowed renewed investigations in 2011, providing an unusual opportunity for early stage researchers to gain experience and training on a submerged prehistoric settlement which also resulted in the discovery of some new features. This chapter provides a summary of the finds recovered from Atlit-Yam, the evidence for sea-level change, and a detailed description of the methods used in underwater survey and excavation.

## 6.1 Introduction

Today the remains of submerged settlements, and indeed, most of the inundated prehistoric landscape, are, generally speaking, buried beneath post-inundation marine and fluvial sediments or biogenic rock and are mostly invisible and inaccessible. The likelihood of encountering such remains depends on exposure of the material by natural processes of coastal or underwater erosion or by human intervention. In principle, submerged prehistoric remains may occur at any depth on the continental shelf. In practice, however, the more deeply submerged the original landscape is, the greater the difficulty of locating archaeological material. This is partly because of the technical and technological difficulties of surveying the seabed and locating material beyond the range of normal SCUBA diving. Additional factors are that deeper landscapes would have been exposed as dry land for shorter periods of the sea-level cycle, and settlements from earlier periods are likely to have been more ephemeral, lacking the substantial structures and other features that make sites of later periods more easily visible on the seabed (see Galili et al. 2013, in press; Galili 2016, for further detail).

On the Carmel coastline of northern Israel, the most promising areas to find inundated human settlements, in terms of preservation, concentration of material and accessibility, are those which were located on dry land close to the ancient coastline during the Mesolithic, Neolithic and Chalcolithic periods (c. 12,000–6,500 BP). Today these sites are at 0–40 m below present sea level, with the optimal depth for finding traces of human settlement at 0–15 m. These submerged settlement sites are covered most of the year by shifting surface-sands. Storms occasionally remove the sand cover and expose wide areas on the sea bottom including archaeological sites. However, exposure, in its turn, makes the archaeological material vulnerable to fairly rapid and massive erosion and loss, unless survey and excavation are carried out immediately, sometimes requiring a rapid or rescue-style approach (Hershkovitz and Galili 1990; Galili et al. 1993, 2005a, 2010).

The 2011 SPLASHCOS field school, held at the Atlit-Yam site on the northern Carmel coast of Israel, aimed at providing practical experience for early career researchers by enabling them to engage in underwater excavation of in situ archaeological features. The Atlit-Yam submerged Pre-Pottery Neolithic village was chosen as an ideal location since it is one of the best preserved submerged prehistoric sites and represents a well-known and archaeologically important site for the Neolithic period as a whole (Fig. 6.1A). Other documented submerged sites of later date are also present along this coastline at shallower depths (see Galili et al. Chap. 7, Fig. 6.1B). Excavating at Atlit-Yam provided an opportunity for students to become familiar with the environmental conditions, site taphonomy and practical and scholarly working procedures of exploring a submerged prehistoric site. In addition, the participants were instructed in local and regional material culture, as well as faunal, floral and bioanthropological assemblages. Special emphasis was placed on methods of working with materials preserved in a submerged environment.

In this chapter, we summarize the finds recovered from excavations at Atlit-Yam including evidence of progressive sea-level rise and its impact on prehistoric settlements along this coastline, and set out the details of the training school and the methods employed to undertake renewed excavations at the site in 2011.

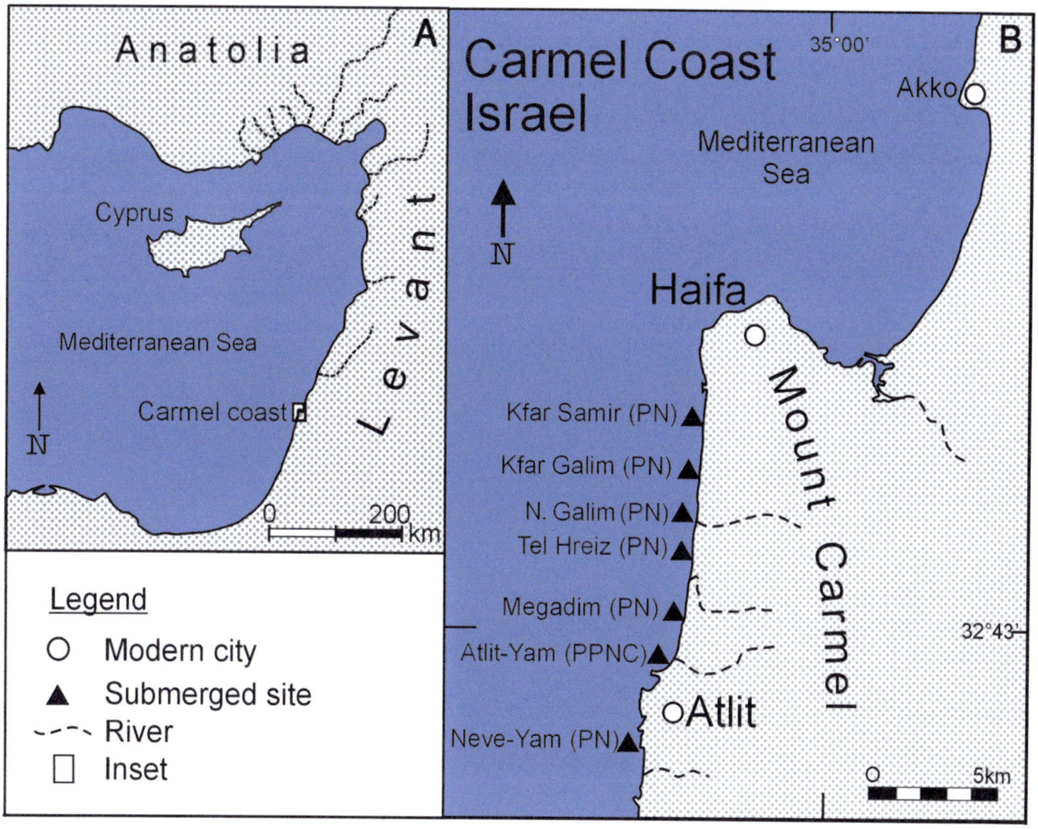


**Fig. 6.1** Location maps: **A** showing position of Carmel Coast within the Eastern Mediterranean, and **B** showing detail of archaeological sites mentioned within the text

## 6.2 The Atlit-Yam Site

The site is now located approximately 200–400 m offshore on the north bay of Atlit. The remains of the Neolithic village are dispersed on the seabed at a depth of 8–11 m below modern sea level (Galili et al. 1988, 2005a; Galili and Rosen 2011a, b). The site covers an area of about 40,000 m$^2$. It is exceptionally well preserved and is unique in its scale, composition and richness of finds relative to other known prehistoric sites on continental shelves the world over. At Atlit-Yam the earliest known constructed (stone-walled) fresh-water wells have been excavated together with other stone-built constructions such as rectangular dwellings (Figs. 6.2 and 6.3) and megaliths, which have been associated with ritual behaviour (Fig. 6.4). The site also contains installations and facilities for the production and storage of food and abundant remains of tools made of flint, stone, wood and bone, numerous remains of animals and plants that were consumed at the site, and tens of human burials (Fig. 6.5).

Galili et al. (2002) proposed that Atlit-Yam is the oldest example of a Mediterranean fishing village, based on the simultaneous utilization of marine and terrestrial resources. Remains of about 100 different plants, cultivated or collected from the wild, were recovered as well as bones of fish, domestic and wild animals, indicating that the village's subsistence was based on a mixed economy of agriculture with animal husbandry supplemented by hunting, gathering and fishing. Such foods form the basis of what is recognized today as the traditional Mediterranean diet (Galili et al. 2002, 2004). Analysis of human remains demonstrates that the population had to cope with diseases such as

**Fig. 6.2** Model of a rectangular dwelling from Atlit-Yam (Photo by J Galili 2002)

**Fig. 6.3** Divers covering rectangular dwelling, using sand bags to prevent erosion (Photo by J Galili 1995)

tuberculosis and malaria, the latter associated with the local marshes (Hershkovitz et al. 1991; Galili and Rosen 2011a). An ear pathology found in some of the human skeletons is symptomatic of diving in cold water, most probably associated with fishing activities (Galili et al. 2005a). In spite of these diseases, a substantial part of the population reached the age of 50 years, an exceptional age relative to the Neolithic inhabitants of the Levant (Eshed and Galili 2011). It is possible that the balanced diet based on a broad spectrum of terrestrial and marine resources contributed to the relatively good health and longevity of the inhabitants.

**Sea-Level Change** Atlit Yam also provides direct evidence of sea-level change and insight into how the prehistoric inhabitants adapted to a continuously rising sea level. This is highly relevant to the

**Fig. 6.4** *Top*: Megalithic structure under excavation (Photo by I Greenberg 1996); *Lower*: Reconstruction of the structure in use (Drawing by S Ben Yehuda)

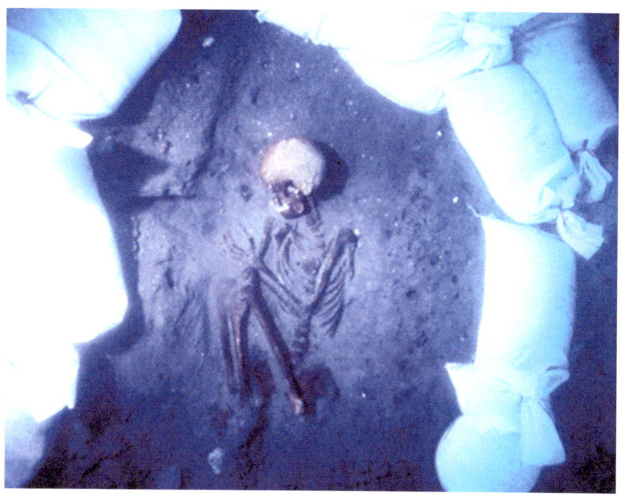

**Fig. 6.5** Human burial protected with sand bags (Photo by A Zaid 1988)

modern situation, given evidence of sustained global sea-level rise during the twentieth century, and predictions of a continued rise during the twenty-first century (Vermeer and Rahmstorf 2009; Gehrels 2010, p. 31; Church and White 2011). Even the modest sea-level rise in the past century has resulted in flooding, coastal erosion and damage to, or destruction of, modern coastal facilities, as well as

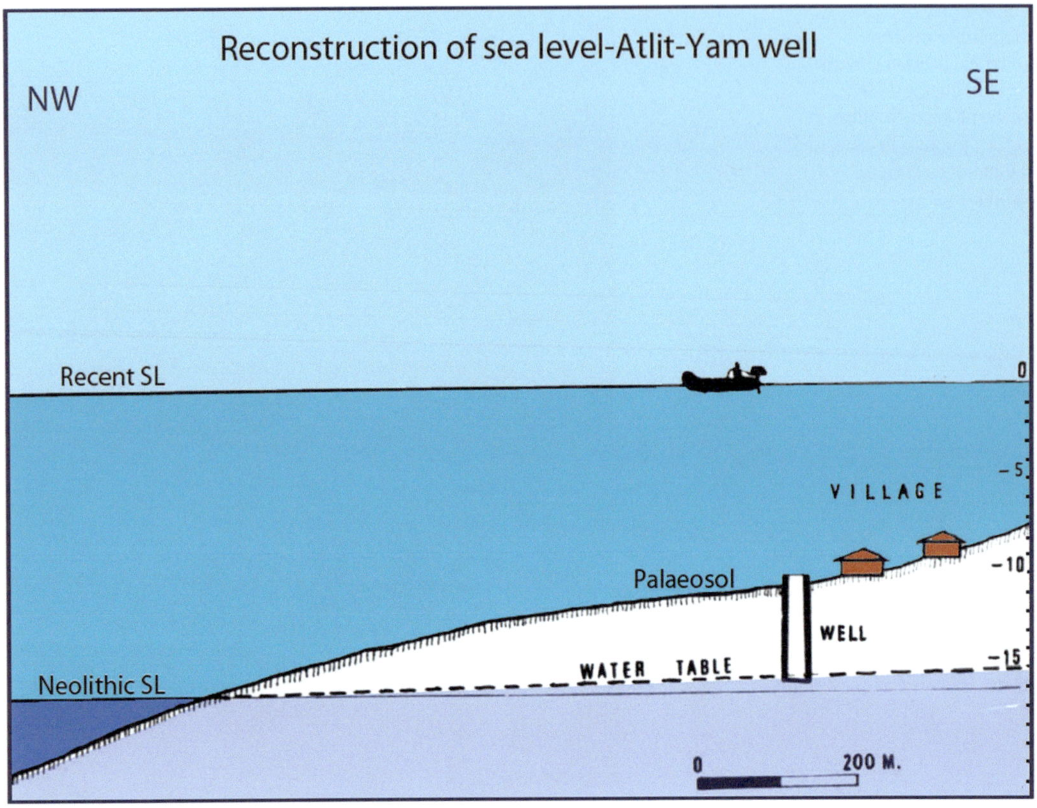


**Fig. 6.6** Reconstruction of sea level at time of occupation at Atlit-Yam using water wells (E Galili)

ancient archaeological sites located on the modern coastline. These destructive effects are likely to intensify in the coming century, with ongoing threats to the offshore and onshore cultural heritage.

At the Last Glacial Maximum 20,000 years ago, the coastline in the Carmel region was some 10 km west of its present position. By the time Atlit-Yam was established, sea-level rise had brought the shoreline to within ca. 500 m of the site, at which time the sea was some 16 m below the present level. This is indicated by the most fully excavated water well (Feature 11), with a base at 15.5 m below present sea level indicating the position of the freshwater table at that time (Fig. 6.6). Subsequently, sea-level continued to rise at a mean annual rate of 5–6 mm between 9200 and 7000 BP and 1–4 mm between 7000 and 4000 BP. That led to progressive salinization, flooding and ultimately abandonment of the village, and the establishment of later settlements of the Pottery Neolithic period at a higher elevation. These sites are presently submerged closer to the modern shore. The combination of archaeological evidence from water wells and other features provides an insight into the human response to progressive sea-level rise and narrowing of the coastal plain, as well as an unusually detailed and precise chronology for the pattern of sea-level change on this coastline over the past 9000 years (Fig. 6.7).

## 6.3 The 2011 Field School

Altogether there were 50 participants, investigators and students, most of them divers, including 12 international SPLASHCOS members. A total of 150 dives were conducted. During the excavation, the participants exposed structures previously identified and studied at the site. These included a water-well (Feature 11) (Fig. 6.8) and an upstanding megalithic monument (Feature 56, Fig. 6.4). As part of

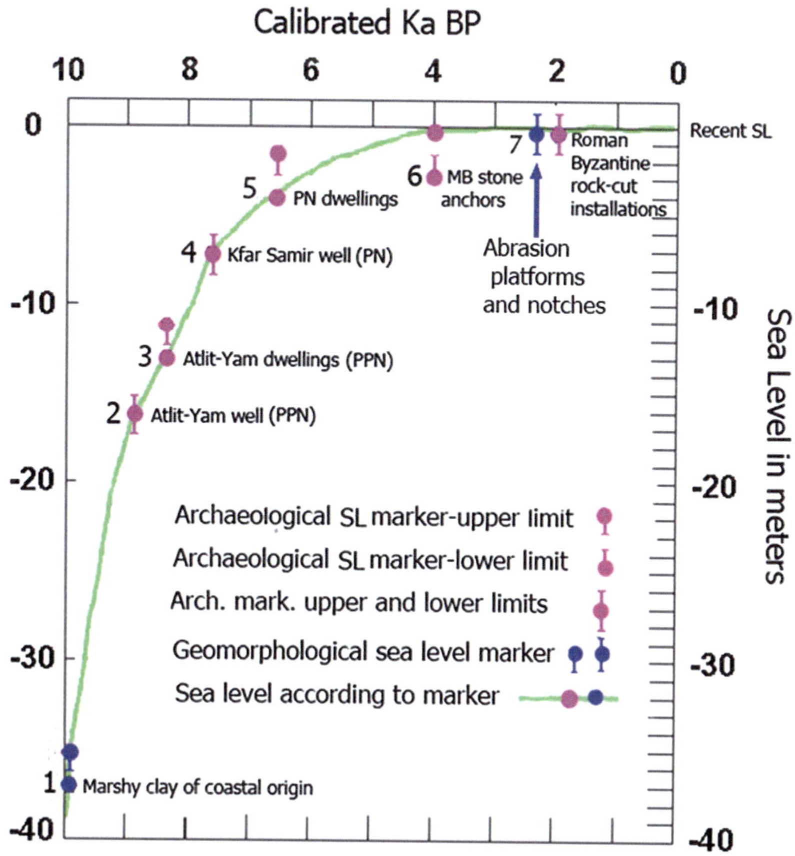


**Fig. 6.7** Curve showing the sea-level changes in the Carmel coast based on archaeological evidence (E Galili)

the terrestrial work, Clive Ruggles and students carried out a Total Station survey from the shore in order to connect this megalithic structure to the visible horizon and to determine its possible astronomical significance. An additional circular structure, appearing to be another water-well, was also discovered and partially excavated. This is an exciting and significant new contribution from the 2011 season as the wells tend to yield a wealth of well-stratified material and information. The new feature was excavated by the SPLASHCOS divers as part of their training. Only the top layers of the well, down to 20 cm below the present seabed, were removed, and the material from this structure was bagged and wet-sieved on-shore. It contained animal bones, flint and bone tools, as well as waterlogged plant remains.

In addition to activities in the field, the training program also included lectures by experts, excursions to archaeological sites, visits to museums and demonstrations in laboratories. These included the Hecht Museum at the University of Haifa; the Prehistoric Museum of Haifa (where the exhibition presenting the submerged settlements off the Carmel coast is displayed), and the National Maritime Museum in Haifa and its underwater archaeology section. A visit to the UNESCO Heritage prehistoric site of Nahal Me'arot, where an unparalleled time-depth of prehistoric cultures ranging from the Lower Paleolithic to the Neolithic was discovered, was guided by M. Weinstein-Evron, the excavator of el-Wad Cave and Terrace. The program also included visits to the archaeozoological laboratory of the University of Haifa, guided by I. Zohar and a workshop on human bones from the excavations at Atlit-Yam conducted by I. Hershkovitz in the Tel Aviv University laboratories where these bones are curated. In addition, a dive excursion was undertaken to the Underwater Archaeological Park located in an inundated canyon, at 20 m depth, west of Haifa.

**Fig. 6.8** Divers excavate well no 11 in Atlit-Yam (Photo by I Greenberg 2011)

## 6.4 Excavation Methods

Underwater archaeological excavation methods developed by the maritime archaeology community have largely focused on shipwreck sites. Excavating a submerged settlement requires coping with issues similar to terrestrial archaeology, such as multi-period stratigraphy, duration of site occupation and the function of structures. To these are added additional complications associated with the marine environment and sea conditions when working underwater.

Participants were exposed to many aspects of fieldwork and post-excavation treatment of materials on a submerged site. However, excavation of such sites entails much pre-excavation work and other elements that could not be demonstrated to all of the participants of the field school. The following section details the steps taken prior to, during and following the excavations at Atlit-Yam since initiation of research at the site in 1984. This may serve as an outline to guide researchers planning to work on similar inundated sites.

### *6.4.1 Pre-fieldwork Collection of Information*

Any long-term plan for locating and studying inundated archaeological sites should take advantage of information received from members of the public associated with maritime activity (e.g. divers, fisherman and amateur underwater archaeologists). Furthermore, raising public awareness of the importance of archaeological investigation and recruiting the public as active partners is essential. This was a feature of the Atlit-Yam research from the beginning.

### 6.4.2 *The Use of Remotely Operated Devices*

Sub-bottom profiler (SBP) surveys at the site were carried out to map the sub-surface, assess the thickness of the overlying sand and the configuration of the paleo-landscape. The IOLR (Israel Oceanographic and Limnological Research Institute) conducted an initial survey using 3.5 kHz equipment (Adler 1985). Another SBP survey aimed at locating prehistoric structures overlain by sand was conducted in 1984 by Harold Edgerton, with no conclusive results (Galili 1985). A multi-beam echo sounder was used to map bathymetry. Experiments for locating concentrations of flint implements are currently being conducted at the site by Ole Grøn and others (see Grøn and Boldreel 2014).

### 6.4.3 *Shallow Water Surveying by Divers and Locating Submerged Sites and Features*

The following surveying, mapping and documentation activities were conducted all year round and are ongoing:

- Walking survey on the beach, retreaving prehistoric artefacts washed ashore
- Searching for areas where underwater changes caused by coastal erosion have occurred
- Snorkelling surveys
- Underwater surveys by SCUBA divers
- Sub sea-bed probing using water-jet
- Applying a variety of techniques in documentation, photography and mapping
- Tagging of structures and site features (using iron poles and buoys)

It is important to consider how one can practically work in shallow water on a coast that is exposed to wave action. Investigating submerged prehistoric sites in the inter-tidal and the surf zones on an open coast requires special techniques and skills. The shallow sectors of the sites are too shallow to apply the underwater excavation methods developed and described below for Atlit-Yam (Galili et al. 2005a). Traditional terrestrial excavation methods are also inadequate since the sites and finds are submerged. Waves interfere with the excavation and the visibility is poor. Usually in these conditions in Israel such sites are covered by 1–2 m of sand and their exposure is accidental, unpredictable and cannot be pre-planned. The exposure may last a few hours to a few days and then the site is again covered by mobile sands. During exposure, the site erodes and finds may shift or be damaged. Removing the sand for excavation is complicated and extremely expensive. It requires the use of heavy dredging and the building of protective caissons of sheet piles which are costly and time consuming.

Along the Carmel coast an alternative excavation method was developed: the archaeologists simply adopted the strategy of allowing the sea to do the job of removing the overlying sediments. After every storm, sites were surveyed to locate newly exposed areas. When an exposure occurred, a rapid rescue and conservation operation including excavation and documentation was carried out by archaeologists, either by diving or, more often, by snorkelling. The loose sand that covered the delicate finds was removed by manual hand-fanning. As a result, after several decades, a rather large portion of a jigsaw puzzle of randomly documented sections of sites has become available.

## *6.4.4 Excavation and Documentation*

### 6.4.4.1 Pre-excavation

Before the actual underwater archaeological excavation of a submerged prehistoric settlement can begin, operational considerations and preparation must be undertaken.

These activities include:

- Choosing the season for excavation: the preferred seasons for underwater excavations along the Israeli coast are spring and autumn. During these seasons the sea is relatively calm and the visibility is much better underwater
- Organizing all aspects of health and safety and carrying out risk assessments for both onshore activity and diving work
- Setting up the excavation base camp (Fig. 6.9)
- Organizing and managing diving and boating operations including transportation of divers to and from the site according to a planned timetable
- Arranging a floating inspection and supervision platform (on a boat or a pontoon) for operating the dredging system and monitoring divers (Fig. 6.10)
- Preparing and running a registration/documentation system including: (1) coastal base-camp diary containing the reports of the dive teams, (2) graphics diary containing the drawings produced by the dive teams, and (3) a white board with the daily schedule of activities and dives
- Preparing excavation equipment for the diving teams
- Instructing and guiding the excavating teams before diving (Fig. 6.11)

**Fig. 6.9** Setting the coastal excavation camp for the Atlit-Yam 2011 field school (Photo by E Galili 2011)

**Fig. 6.10** Setting the dredging system on a boat (Photo by E Galili 2011)

**Fig. 6.11** Instructing the excavating teams before diving (Photo by E Galili 2011)

- 'Dry demonstration' on land of how to operate the equipment
- 'Dry demonstration' on land of the dredging systems operation
- Tagging of structures and site features (using iron poles and buoys)

The following descriptions refer mainly to excavations which took place prior to 2011, but many of these activities were also undertaken during the SPLASHCOS field school.

#### 6.4.4.2 General Excavation

The upper layer of loose sand and gravel covering the palaeosol in which the site is embedded was removed manually by fanning or assisted by the dredging system until the clay containing the archaeological material was reached. The archaeological deposit was excavated in 10 cm spits in gridded squares of 0.5 by 0.5 m. The excavated material was collected in tagged plastic find-bags, marked by square and layer, and transported to the shore laboratory for sieving (Fig. 6.12). Small or fragile artefacts were collected in plastic jars and core samples of in situ clay were taken for pollen and sediment analysis.

**Fig. 6.12** Floating and transferring bags containing the excavated material to shore for sieving (Photo by I Greenberg 1998)

#### 6.4.4.3 Documentation and Excavation of Individual Structures

At Atlit-Yam, each individual structure, installation or burial that was identified was marked with a galvanized iron rod bearing an identifying plastic tag. The rod protruded 0.7–1.0 m above the seabed to enable relocation in case the feature was covered by sediments. From the top of each rod, a numbered buoy was floated that could be located from the coast by a laser distance meter. Additionally, a marked and weighted rope that served as a baseline was laid in a straight line on the sea floor and the divers mapped each feature relative to the baseline. The two ends of the baseline were mapped from the coast and the site plan was prepared by coordinating the buoy measurements taken from the coast and the measurements taken by the divers.

Underwater excavation of individual structures was carried out using standard SCUBA gear and an induction dredge operated by a water pump set on a boat (Fig. 6.13) as is standard practice in maritime and underwater archaeology. Each diver also carried equipment for documentation during the excavation. A grid made of synthetic fibre or metal rods was set over the structure to be excavated (Fig. 6.14). The excavation team included three divers. One diver worked in the suction area holding the plastic hose to remove the suspended particles disturbed by the working process; this substantially improves visibility. A second diver monitored the exhaust end of the dredging system, where a collecting box was set. A third diver excavated using only a spatula or a trowel. The material dredged or excavated from each square was collected and placed in a separately tagged plastic bag. Delicate artefacts were stored in tagged plastic jars filled with sea water.

#### 6.4.4.4 Excavating Vertical Shafts and Water-Wells

In order to excavate a shaft that penetrates deep into the seabed, special procedures and techniques were used at Atlit-Yam. The excavating diver wore a safety helmet and knelt down inside the pit without diving fins, where the shaft diameter allowed. If the shaft was too narrow, the diver was positioned upright with the head down. A second diver was responsible for the excavator's safety and was positioned at the opening of the shaft and also lifted stones from the shaft bottom using a lifting box tied to a rope. A third diver monitored the dredger's exhaust, ensuring that the excavated fill entered the collecting box. The clayey fill within the water-wells was excavated manually in 10 cm spits, and the finds—including stone, bone and wooden artefacts, and faunal and floral remains—were stored in tagged bags and jars. After every 10 cm of deposit, the third diver transferred the excavated material from the box into tagged bags and prepared them to be lifted ashore. The murky water and some of the fill was dredged. Stones were removed manually into the lifting box and were taken out of the pit by the second diver assisting the excavator. Core samples of undisturbed clay were taken every 10 cm for pollen and sediment analysis (Fig. 6.15). In the case of the deep wells, the walls were supported by metal rings arranged at 0.7 m intervals, to prevent collapse (Fig. 6.16).

#### 6.4.4.5 Documenting and Excavating In-Situ Human Burials

Before excavating burials, observations and documentation procedures were undertaken of the grave structure or grave type, the location of the grave in the site relative to adjacent structures and installations, traces of ceremonial activities and other activities associated with the burial (fireplaces, remains of offerings and ceremonial meals, grave goods around the burial). The burials were classified according to the following categories: (1) Primary burials (including a subcategory of disturbed burial) relating to skeletons with bones in complete or partial articulation, showing no signs of removal from their original burial site; this category may also include complete, articulated skeletons that had been disturbed in antiquity, or were disturbed by post-depositional processes; (2) Secondary burial is a burial

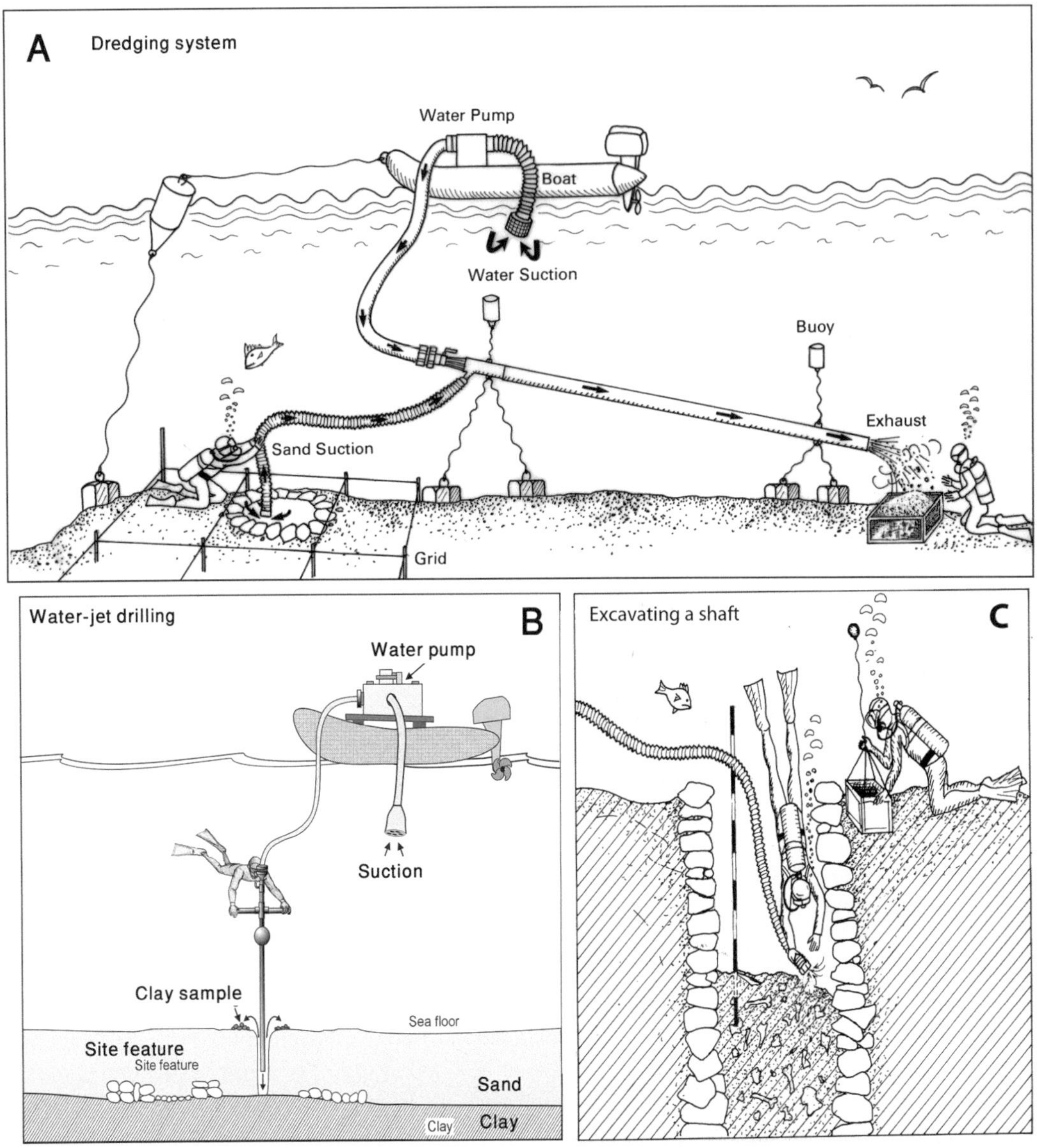


**Fig. 6.13** Excavations methods: **A** the dredging system, **B** water jet drilling and sampling, **C** excavating water wells (Modified after Galili and Rosen 2011a)

of an individual that was deliberately removed from its original grave site to a new location in antiquity. In these cases, the bones are not in articulation (Galili et al. 2005a); (3) Skull or skull fragments found (usually on living floors) without post-cranial bones; (4) Isolated bones/teeth scattered randomly on-site, which are often found during surface surveys.

After the completion of the spatial documentation of the grave site and its surroundings, the upper layer of loose sand and gravel covering the skeleton was removed manually. The human skeleton was then carefully excavated, using spatula and trowel to remove the clay from the bones. Following partial exposure of the skeleton to enable recording of sufficient details of the position, the skeleton and adjacent artefacts within the grave were measured, drawn and photographed. Undisturbed core samples of sediments were taken in the pelvic area (to obtain possible traces of food), and in other locations to search for plant pollen that might have been associated with the burial (Fig. 6.15).

**Fig. 6.14** Excavating an archaeological feature by squares using a metal grid (Photo by I Greenberg 1998)

**Fig. 6.15** Taking undisturbed cores for pollen and sediment analysis (Photo by J Galili 1989)

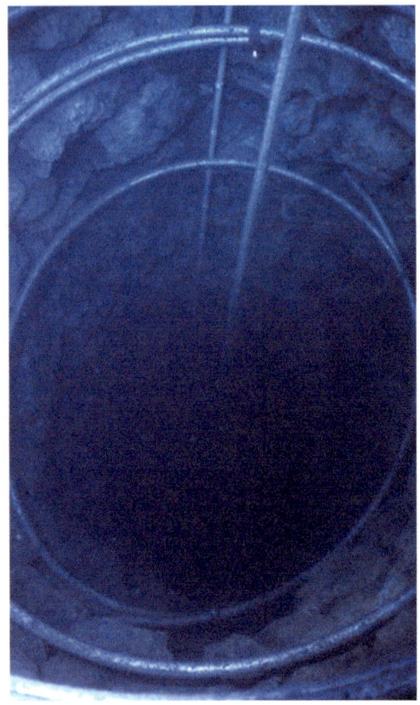

**Fig. 6.16** Supporting a Pre-Pottery Neolithic well in Atlit-Yam with iron rings during the excavation to prevent collapse (Photo by I Greenberg 1995)

The burial position of each skeleton was described as flexed, semi-flexed or straight and was recorded using a graphic scheme based on the angles of the elbow, shoulder, hip and knee joints (Galili et al. 2005a). In addition, various observations related to the orientation of the inhumations were made, including direction of the head, lying position (on the left or right side, on the back or belly) and direction of the face. The possibility of group burials, articulated bones indicating a primary burial, and non-articulated bones indicating secondary burial was noted. Given the limitations of time and sea conditions, no attempt was made to remove an entire skeleton within its context. Removing a whole block of clay that size would not have been feasible, but such an operation may be considered in areas where there is no danger of immediate erosion or where sea conditions are favourable. Instead, the bones of each skeleton were removed separately one by one. Where possible, the bones of each section of the skeleton (hands, legs, pelvic area, ribs spine and skull) were removed together to a solid container to prevent breakage and post-excavation exposure damage. If possible, skulls were removed whole by placing them in a container and covering them with fine sand to avoid movement and damage while transferring to the land base. When the work had to be stopped overnight or for several days due to limitations dictated by sea conditions, the exposed skeletons were temporarily protected from wave erosion. In such instances, the skeletons were covered by a woven plastic sheet that was, in turn, covered by bags partly filled with sand (Fig. 6.5).

#### 6.4.4.6 Processing Archaeological Material at the Coastal Base

All excavated materials from the site underwent a series of post-excavation processes which varied depending on the material and related requirements for conservation and analysis. Sediments containing the archaeological material retrieved from the site were soaked in freshwater tanks and were separated and sorted into different elements by material and size. The coarse materials (artefacts and stones larger than 2 cm) were removed manually and were dried in wooden trays. Waterlogged plant material was then separated from the rest of the sediments in the water by manually spinning the water in the tank and pouring it through a 0.5 mm plastic sieve. This plant material was kept in sealed jars in a freshwater and alcohol solution and preserved in a refrigerator for further study. Some of the plant remains were kept in sea water for future radiocarbon dating to prevent contamination. The coarser material remaining in the water was sieved using a 1 mm mesh and dried in wooden trays. After drying, the material underwent coarse separation and fine picking. The material was separated into categories: artefacts, including stone, bone, and flint artefacts (tools, flakes, blades, cores, debris, and waste elements); bones; waterlogged wood and plant remains; and charred plant remains. All bones and tools recovered from the site were later soaked in running freshwater for a few days, to dissolve the salts.

## 6.5 Summary

Decades of surveys and exploration of the submerged settlements off the Carmel coast have revealed remarkably well-preserved Neolithic villages. These submerged archaeological resources require frequent monitoring in order to preserve and document the unique archaeological remains of Levantine prehistoric coastal communities, so important to world prehistory.

Research carried out to date indicates that the majority of such sites are covered most of the time, hidden under mobile sands, and therefore remain undetected. Most known sites were exposed only briefly—sometimes for a few days—every few decades, and were covered soon after their exposure. Therefore, raising public awareness of the importance of reporting the exposure of such sites is of high priority.

The 2011 field school at Atlit-Yam offered an intensive and extensive training program in excavation of a submerged prehistoric site for new and experienced researchers. The methods and techniques used are of value for investigations of any inundated archaeological site. It is expected that the international participation will have benefited underwater and prehistoric archaeology in the Mediterranean, Europe and further afield. The field school gave a unique opportunity for participation of numerous early career researchers at this underwater prehistoric village. The support of COST and SPLASHCOS was instrumental in providing basic and advanced training in excavating and studying submerged prehistoric archaeological sites.

The fieldwork was supplemented by lectures and visits to relevant archaeological sites, laboratories and a hyperbaric medical centre. This enabled the participants to broaden their understanding of the types of activities that would be essential for long-term research on submerged settlements and landscapes.

Finally, the 2011 field school was a valuable capacity-building exercise, with the deliberate aim of ensuring long-term succession planning and enhancing the continuation of submerged prehistoric archaeology. Also, it yielded new data and discoveries that benefit the ongoing research program at Atlit-Yam.

**Acknowledgements** The 2011 excavation at Atlit-Yam was undertaken within the framework of a field school organized by the Israel Antiquities Authority, the Leon Recanati Institute for Maritime Studies and the Zinman Institute of Archaeology, both at the University of Haifa. Other associated institutions included The Israel Prehistoric Society, The Department of Anatomy and Anthropology, Faculty of Medicine at the Tel-Aviv University and the Eco-Ocean organisation. The entire activity was supported and funded by COST (European Cooperation in Science and Technology) Trans-Domain Action TD0902 SPLASHCOS, with a contribution also coming from Alpha-Zoulou Films (a Canadian production company), who filmed the process and interviewed the team as part of a scientific documentary. Other supporters of the Atlit-Yam project included: The National Geographic Foundation, The Irene Levi Care Archaeological Foundation, The Dan David Foundation, The Margolis Foundation, the Honor Frost Foundation and the Leplestat family. We are grateful to them all. Last but not least, we thank all the divers and the volunteers and the underwater photographers J. Galili, I. Greenberg and A. Zaid.

## References

Adler E (1985) The submerged Kurkar ridges off the northern Carmel coast. Unpublished MA Thesis, Department of Maritime Civilizations, University of Haifa (In Hebrew)

Church J, White N (2011) Sea-level rise from the late 19th to the early 21st century. Surv Geophys 32:585–602

Eshed V, Galili E (2011) Palaeodemography of Southern Levantine pre-pottery neolithic populations: regional and temporal perspectives. In: Pinhasi R, Stock J (eds) Human bioarchaeology of the transition to agriculture. Wiley-Blackwell, Chichester, pp 403–428

Galili E (1985) Clay exposures and archaeological finds on the sea bottom, between Haifa and Atlit. Unpublished MA Thesis, Department of Maritime Civilizations, University of Haifa (In Hebrew)

Galili E (2016) Book review. Prehistoric archaeology on the continental shelf: a global review. J Island Coast Archaeol 00:1–3. doi:10.1080/15564894.2016.1225323

Galili E, Rosen B (2011a) Submerged Neolithic settlements off the Mediterranean coast of Israel. In: Benjamin J, Bonsall C, Pickard C, Fischer A (eds) Submerged prehistory. Oxbow Books, Oxford, pp 272–286

Galili E, Rosen B (2011b) Submerged neolithic settlements off the Mediterranean Carmel coast of Israel and water mining in the Southern Levant. Neo-Lithics 2(10):47–52

Galili E, Weinstein-Evron M, Ronen A (1988) Holocene sea-level changes based on submerged archaeological sites off the Northern Carmel coast in Israel. Quat Res 29:36–42

Galili E, Weinstein-Evron M, Hershkovitz I, Gopher A, Kislev M, Lernau O, Horowitz LK, Lernau H (1993) Atlit-Yam: a prehistoric site on the sea floor off the Israeli coast. J Field Archaeol 20:133–156

Galili E, Rosen B, Gopher A, Horwitz LK (2002) The emergence and dispersion of the Eastern Mediterranean fishing village: evidence from submerged Neolithic settlements off the Carmel coast, Israel. J Mediterr Archaeol 15:167–198

Galili E, Lernau O, Zohar I (2004) Fishing and marine adaptations at Atlit-Yam, a submerged Neolithic village off the Carmel coast, Israel. Atiqot 48:1–34

Galili E, Eshed V, Gopher A, Hershkovitz I (2005a) Burial practices at the submerged PPNC site of Atlit-Yam, northern coast of Israel: what do they tell us about the final phase of the Pre-Pottery Neolithic culture? BASOR 339:1–19

Galili E, Zviely D, Weinstein-Evron M (2005b) Holocene sea-level changes and landscape evolution on the northern Carmel coast (Israel). Méditerranée 12:1–8

Galili E, Eshed V, Rosen B, Kislev M, Simchoni O, Hershkovitz I, Gopher A (2010) Evidence for a separate burial ground at the submerged Pottery Neolithic site of Neve-Yam, Israel. Paléorient 351:31–46

Galili E, Horwitz LK, Eshed V, Rosen B, Hershkovitz I (2013) Submerged prehistoric settlements off the Mediterranean coast of Israel. Skyllis 13(2):181–204

Galili E, Nir Y, Vachtman D, Mart Y (in press) Physical characteristics of the continental shelves of the East Mediterranean basin, submerged settlements and landscapes – actual finds and potential discoveries. In: Flemming N, Harff J, Moura D, Burgess A, Bailey G (eds) Submerged landscapes of the European continental shelf: vol. 1, Quaternary Palaeoenvironments. Wiley-Blackwell, Chichester

Gehrels R (2010) Sea-level changes since the last glacial maximum: an appraisal of the IPCC fourth assessment report. J Quat Sci 25:26–38

Grøn O, Boldreel LO (2014) Chirping for large-scale maritime archaeological survey: a strategy developed from a practical experience-based approach. J Archaeol. 10.1155/2014/147390

Hershkovitz I, Galili E (1990) 8000 years-old human remains on the sea floor near Atlit, Israel. Hum Evol 5:319–358

Hershkovitz I, Ring B, Spears M, Galili E, Kislev M, Edelson G, Hershkovitz A (1991) Possible congenital hemolytic anemia in prehistoric coastal inhabitants of Israel. Am J Phys Anthropol 85:7–13

Vermeer M, Rahmstorf S (2009) Global sea level linked to global temperature. Proc Natl Acad Sci U S A 106(51):21527–21532

# Part II
# Underwater Archaeological Sites

# Chapter 7
# Submerged Pottery Neolithic Settlements off the Coast of Israel: Subsistence, Material Culture and the Development of Separate Burial Grounds

**Ehud Galili, Liora Kolska Horwitz, Vered Eshed, and Baruch Rosen**

**Abstract** Eight inundated archaeological sites dating to the Pottery Neolithic period (Wadi Rabah culture), 8000–6500 cal. BP, have been exposed under water off the Carmel coast of Israel. The sites represent in situ settlements with architectural remains comprising domestic stone-built structures and water wells built of wood and stone. Rich assemblages of flint tools, ground stone artefacts and pottery were recovered in addition to organic remains (wooden bowls, baskets etc.). Faunal and botanical remains demonstrate that the subsistence economy consisted of animal husbandry, hunting and fishing complemented by cultivation of domestic crops and gathering of wild plants. Special features include the beginning of olive oil extraction, a major component of the Mediterranean subsistence economy, demonstrated at the site of Kfar Samir, while at the Neve-Yam site, the earliest separate burial ground in the region was found with a concentration of stone-built cist graves.

## 7.1 Introduction

The Mediterranean coast of Israel is about 200 km long. It is slightly curved with mostly sandy beaches in the south, eroding coastal cliffs in the centre and some rocky areas in the north along the Carmel and Galilee coasts (Fig. 7.1). Eight submerged Pottery Neolithic (PN) settlements inundated by Holocene sea-level rise have been discovered along a 15 km strip of the northern Carmel coast (Fig. 7.1b).

E. Galili (✉)
Zinman Institute of Archaeology, University of Haifa, Haifa, POB 180 Atlit, 30300, Israel

Israel Antiquities Authority, POB 180, Atlit 30300, Israel
e-mail: galilish@netvision.net.il

L.K. Horwitz
National Natural History Collections, Faculty of Life Science, The Hebrew University, Berman Building, E. Safra Campus, Jerusalem 91904, Israel
e-mail: lix1000@gmail.com

V. Eshed
Israel Antiquities Authority, POB 586, Jerusalem 91994, Israel
e-mail: veshed@gmail.com

B. Rosen
Israel Antiquities Authority, POB 180, Atlit 30300, Israel
e-mail: rosenbar@netvision.net.il



G.N. Bailey et al. (eds.), *Under the Sea: Archaeology and Palaeolandscapes of the Continental Shelf*, Coastal Research Library 20, DOI 10.1007/978-3-319-53160-1_7

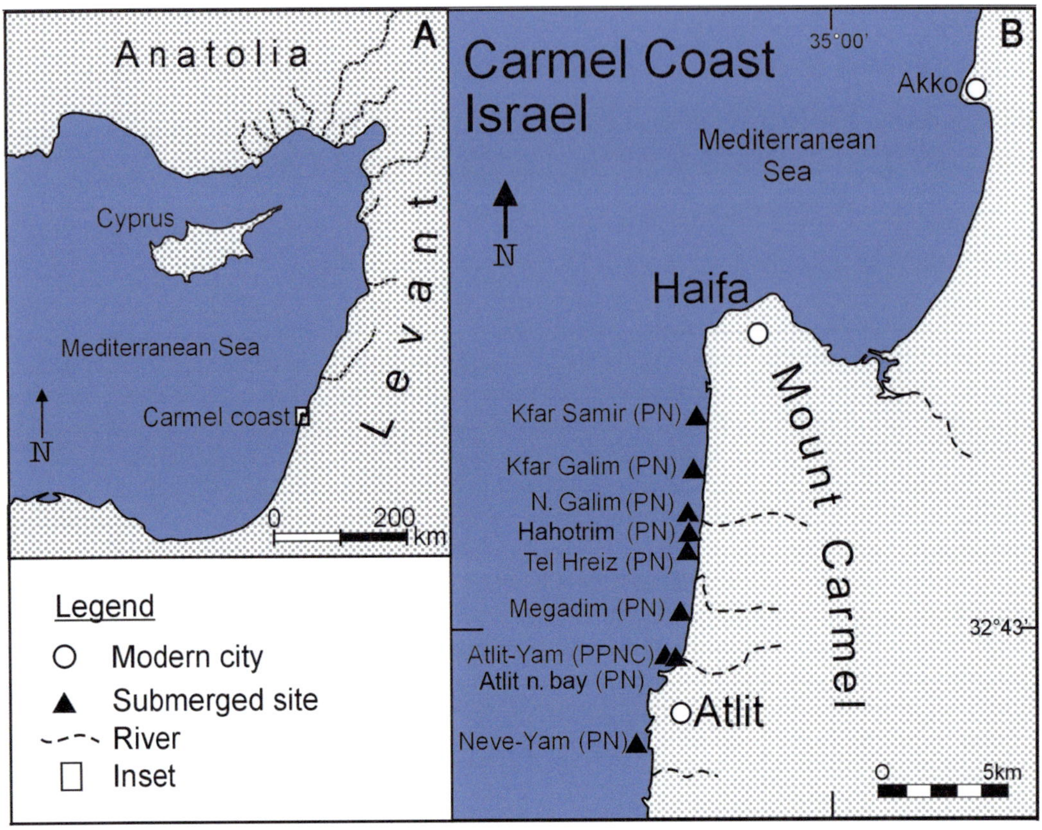


**Fig. 7.1** Location map: **a** The Israeli coast; **b** Submerged settlements off the Carmel coast (E. Galili)

These settlements are embedded in the upper part of a hard clay palaeosol (Galili 2004). Due to continuous sea-level rise (Galili et al. 1988, 2005; Galili and Rosen 2011b), the duration of each site was relatively short and there were no earlier or later occupations. Thus, mixing of layers and post-depositional disturbances were minimal. Following inundation, the sites were covered by a protective layer of sand 1–2 m thick that prevented disturbance and intrusions. In recent years these submerged settlements have been exposed by human activities – mainly sand quarrying and construction of marine structures – together with seasonal sea storms that have removed the sand layer and exposed parts of the sites (Wreschner 1977a, b; Galili and Weinstein-Evron 1985; Galili and Inbar 1987; Galili et al. 1988).

Until recently, information on Levantine coastal Neolithic communities was relatively scarce and their role in the Neolithic revolution was barely understood. The unique physical conditions of the submerged sites and the circumstances of their preservation and exposure, together with intensive underwater rescue surveys and excavations over the past four decades, have yielded a mosaic of randomly uncovered portions of the sites. Collating the accumulated data has enabled us for the first time to draw a comprehensive picture of the material culture, economy, and socio-cultural practices of these coastal societies, and to reconstruct the palaeoenvironment of the Mediterreanean shore during two critical transitional periods: the Pre-Pottery Neolithic C (PPNC) c. 9250–8000 cal. BP (Hershkovitz and Galili 1990; Galili et al. 1993; Galili et al. Chap. 6) and the late Pottery Neolithic (PN), or Wadi Rabah Culture[1] c. 8000–6500 cal. BP (Fig. 7.2, Table 7.1). In this chapter we describe the sites of the

[1] The Wadi Rabah (WR) culture is attributed by some to the late Pottery Neolithic period (Gopher and Gophna 1993) and by others to the Early Chalcolithic period (Garfinkel and Dag 2008).

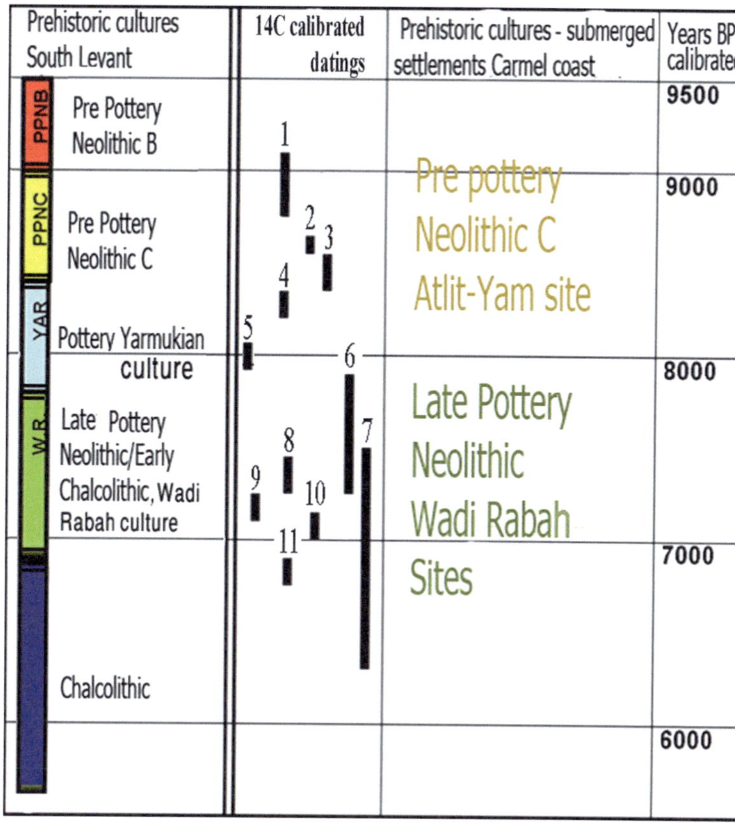


**Fig. 7.2** Cultures of the 7th–10th millennia BP in the Levant (B. Galili) and radiocarbon dates obtained from the submerged sites: (1) Atlit-Yam, charcoal taken near structure 13; (2) Atlit-Yam, charcoal from the fill of well 66; (3) Atlit-Yam, charcoal taken near structure 10A; (4) Atlit-Yam, charcoal from the fill of well 11; (5) Kfar Samir, wooden bowl; (6) Kfar Samir, wood taken from the construction of the wells; (7) Kfar Samir, olive pits, woven mat and wooden branches; (8) Neve-Yam south, charcoal; (9) Megadim, jaw of a mammal; (10) Tel Hreiz, wooden poles; (11) Kfar Galim, wooden branches from a well

Pottery Neolithic Wadi Rabah phase and discuss the main changes in subsistence and material culture that occurred from the PPNC to the later Wadi Rabah PN.

## 7.2 Submerged Pottery Neolithic Sites

Eight submerged PN settlements were discovered between Haifa and Atlit on the Northern Carmel coast: Kfar Samir (north, centre and south), Kfar Galim North, Kfar Galim South, Nahal Galim, Hahotrim, Tel Hreiz, Megadim, Atlit northern bay and Neve-Yam (north and south) (Fig. 7.1b). Recently another PN site was located off Habonim, ca. 2.4 km south of Neve-Yam are located close to the present coastline at depths of 0–5 m below sea level (Ronen and Olami 1978; Olami 1984; Galili and Weinstein-Evron 1985; Galili and Inbar 1987; Galili et al. 1988, 1989, 1997, 1998, 2002; Galili and Schick 1990; Galili 2004). Most of the archaeological material was collected in the course of underwater surveys and sometimes on the coastline after storms, while in a few instances limited excavations were carried out. Radiocarbon dates, and ceramic and lithic typologies, place the sites in the late Pottery Neolithic period i.e., the Wadi Rabah culture (Table 7.1).

**Table 7.1** List of radiocarbon dates from sites mentioned in the text

| Site and lab no | C14 age – uncalibrated BP | C14 age – calibrated BP | Sample material and feature | Location at site |
|---|---|---|---|---|
| Kfar Samir | | | | |
| BETA –82851 | 5860± 140 | 7005–6385 | Wood from well 113 | Centre |
| RT–282B | 6470 ± 130 | 7453–7257 | Wood from well 5 | Centre |
| RT–682A | 6670 ± 160 | 7666–7625 | Wood from well 3 | Centre |
| PTA–3820 | 6830 ± 80 | 7698–7582 | Wood from well 5 | Centre |
| PTA–3821 | 6830 ± 160 | 7730–7530 | Wood from well 3 | Centre |
| BETA–82850 | 6940 ± 60 | 7890–7615 | Wood from pit 10 | Centre |
| BETA–82845 | 6080 ± 70 | 7165–6765 | Olive from pit 6 | Centre |
| BETA–82846 | 6210 ± 150 | 7385–6740 | Olive from pit 6 | Centre |
| BETA–82847 | 6210 ± 80 | 7240–6885 | Olive from pit 6 | Centre |
| BETA–82848 | 6230 ± 80 | 7255–6900 | Olive from pit 6 | Centre |
| BETA–82715 | 6500 ± 70 | 7480–7230 | Olive from pit 6 | Centre |
| RT–1898 | 5790 ± 55 | 6669–6519 | Olive from pit 6 | Centre |
| RT–1930 | 5870 ± 70 | 6785–6575 | Olive from pit 6 | Centre |
| BETA–82843 | 6100 ± 60 | 7165–6805 | Olive from pit 7 | Centre |
| BETA–82844 | 6290 ± 60 | 7270–7020 | Olive from pit 7 | Centre |
| RT–1929A | 5630 ± 55 | 6669–6519 | Olive from pit 7 | Centre |
| RT–1929 | 5870 ± 70 | 6466–6317 | Olive from pit 7 | Centre |
| BETA–82849 | 6350 ± 90 | 7390–7020 | Branch from it 9 | Centre |
| RT–855 | 6420 ± 120 | 7517–7038 | Mat from pit 8 | Centre |
| RT–1360 | 7230 ± 80 | 8115–7949 | Wooden Bowl | South |
| Kfar Galim | | | | |
| RT–1748 | 5985 ± 70 | 6910–6670 | Wooden structure | South |
| RT–1749 | 5985 ± 55 | 6890–6740 | Wooden structure | South |
| RT–1750 | 6890 ± 50 | 7790–7670 | Wooden branch | South |
| Tel Hreiz | | | | |
| RT–779A | 7330 ± 120 | 8330–7970 | Wooden pole | North |
| PTA–3460 | 6310 ± 70 | 7210–6980 | Wooden branch | North |
| RT–779B | 6269 ± 150 | 7160–7000 | Wooden pole | North |
| Megadim | | | | |
| PTA–3652 | 7960 ± 70 | 8990–8650 | Clay outside pit | South |
| PTA–3648A | 6310 ± 70 | 7310–7020 | Canine bone (jaw) | South |
| PTA–4339A | 6270 ± 50 | 7270–7100 | Canine bone (jaw) | South |
| Neve-Yam | | | | |
| HV–4256 | 6510 ±395 | 7600–6750 | Charcoal | Centre |
| RT–1723 | 6390± 70 | 7480–7270 | Charcoal | South |
| RT–1724 | 6565 ± 70 | 7580–7320 | Charcoal | South |

### 7.2.1 *Kfar Samir*

Kfar Samir lies at a depth of 0.3–5 m, some 10–200 m offshore (Fig. 7.1b). The prehistoric remains are scattered along a 1200 m strip on the sea bottom, parallel to the coastline. The site may be divided into northern, central and southern sectors.

#### 7.2.1.1 Northern Sector

This part of the site is located at 34° 57′ 20″ E, 32° 47′ 58.5″ N Fig. 7.1b). In the north sector of the site, four vertical stone slabs (1.3 × 0.9 × 0.5 m) and some similar sized tilted slabs (angled at ~45° angle) were recovered at a water depth of 1.5–2.5 m, some 30–60 m offshore. One end of the slabs was partly inserted in the clayey bottom and they may represent symbolic or ritual features, burial markers or other locations of importance.

#### 7.2.1.2 Central Sector

This part of the site is located at 34° 57′ 8.65″ E, 32° 47′ 6.72″ N (Fig. 7.1b), 1–200 m from the water line, at a depth of 0–5.5 m.

**Architectural Remains** In the central sector, floors made of unworked local stones (8–10 cm in diameter) were found partially embedded in the clay palaeosol, as well as floors constructed of flat stone slabs. A few hearths (c. 0.5 m in diameter) were located. In one, the bottom of the hearth was lined with stones and contained fragments of burnt bones and charcoal. Several dozen metres to the south of this hearth, a pit contained crushed olive pits and pulp, evidently waste from olive oil extraction (Galili and Sharvit 1994–1995; Galili et al. 1997, 1989; Galili and Rosen 2007). The radiocarbon dates of the olive pits range from 7480–7230 to 6669–6519 cal. BP (Table 7.1).

**Water Wells** Three water wells were identified on the north-west edge of the central sector submerged at a depth of 5 m, some 200 m offshore. They were constructed of alternating courses of wooden branches and undressed kurkar and limestone pebbles (Fig. 7.3). The southern well (No. 3) had a rectangular opening 1 × 0.8 m. It was excavated to a depth of 2 m, but the bottom was not reached. With increasing depth it widens and becomes more circular in plan. In its lower part, two courses of stones were laid between the wooden beams. The well's fill consisted of soft clay, small pieces of stone as well as several bird bones, olive pits, some pot sherds (one with an incised design), flint flakes and waterlogged plant remains such as tree branches and straw. Wood samples from the well construction were radiocarbon-dated with ages ranging between 7890–7615 and 7005–6385 cal BP.

**Olive Oil Extraction** A feature associated with producing olive oil was recovered at a water depth of 1 m. It comprised a pit dug in the clay (Fig. 7.4a). Its base was paved with stones and it contained thousands of crushed and whole stones and traces of pulp, evidently olive oil extraction waste (Fig. 7.4b, Galili et al. 1997, 1989; Galili and Rosen 2007). Several stone basins recovered from the site (e.g., Fig. 7.5a) could have been used for crushing the olives. These finds suggest that about 8000 years ago the Neolithic inhabitants of the Carmel coast started to produce olive oil and they represent the earliest known evidence for its production.

**Flint Implements** Two groups of flint tools can be identified; those which represent the period of site occupation (flint axe, burin) and a few Middle Palaeolithic implements that had been found by the Neolithic inhabitants and apparently reused.

**Special Finds** At a depth of 0.5 m and 15 m from the shoreline, on a section of stone paving, a fragment of a pot-shaped wooden bowl was found. The fragment included a part of the flat base, a straight wall that is slightly inverted and a section of the rim. An elongated knob handle on the upper part of the wall is perpendicular to the rim and has a narrow lateral perforation.

At a depth of 2.3 m, a pit 0.9 m in diameter and 0.55 m deep had been dug into the clay palaeosol. It contained water-logged pieces of a braided basket (Fig. 7.5a), tree branches and dozens of olive pits. The basket was made of braided pieces 3–5 mm in diameter. One of the braided pieces was round and

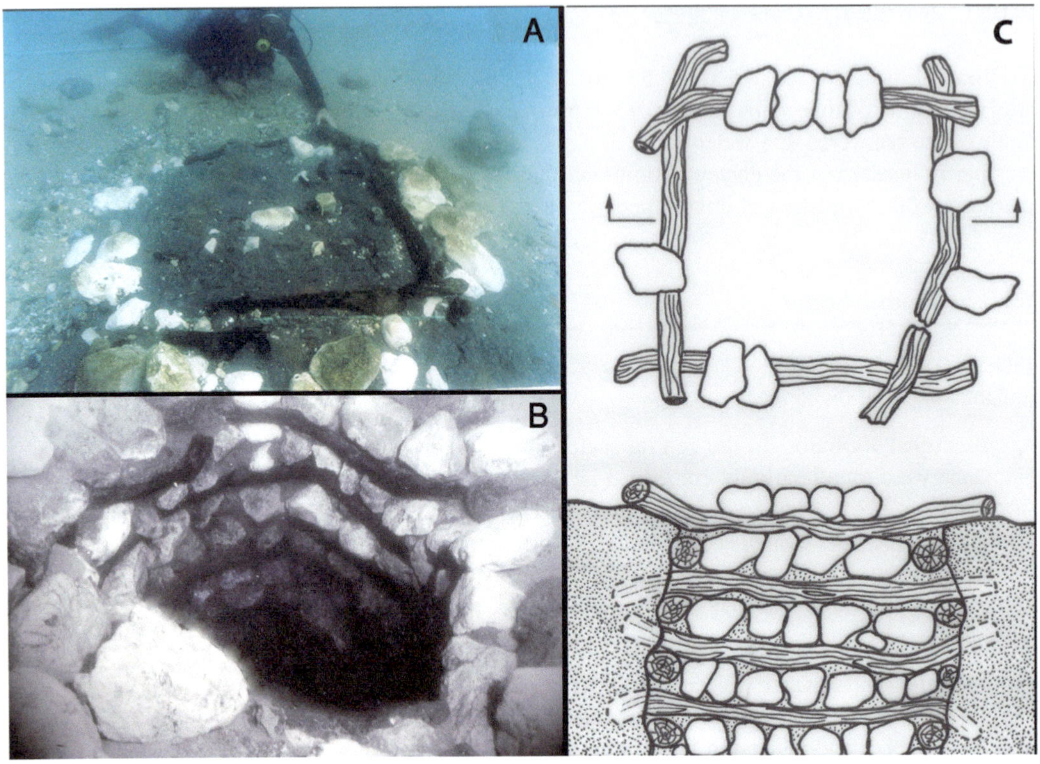


**Fig. 7.3** Kfar Samir Central Sector: water-well 3 constructed of alternating courses of wooden branches and stones: **A** before excavation; **B**, **C** after excavation (E. Galili)

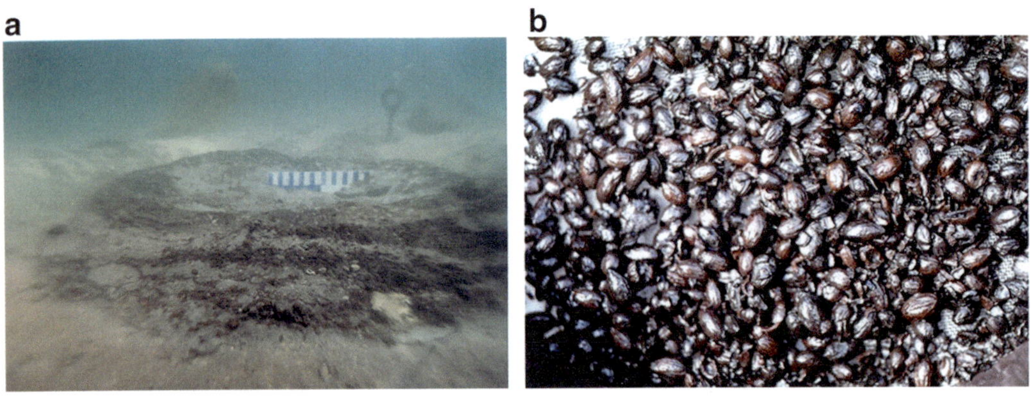


**Fig. 7.4** (**a**) Kfar Samir Central Sector: Pit filled with olive oil extraction waste of crushed olive pits and pulp (scale = 20 cm), (**b**) Crushed olive pits from the pit (J. Galili)

may have formed the base. The braiding method used was alternate pair twining: the warps emerge from the centre of the base, perpendicular to its rim and parallel to each other while the wefts are alternately twined around two warps and cross over after each pass. In several places the wefts were twined around three warps or only around a single warp. Usually the warps are made of one branch, but in a few places they consisted of a pair of thin branches (Galili et al. 2007). We interpret this artifact as a strainer, similar to modern ones known as *aqal* (Fig. 7.5b), which are used to squeeze the oil from the olive pulp after the olives have been crushed. Several stone basins recovered nearby could

**Fig. 7.5** Kfar Samir Central Sector: finds associated with olive oil extraction: **a** Basketry item made of woven twigs, interpreted as a strainer for olive oil extraction (Israel Antiquities Authority); **b** Example of modern strainer for olive oil extraction still in use today; **c** Stone basin from the Kfar Samir site, possibly used for crushing olives (E. Galili)

have been used for crushing the olives (Fig. 7.5c). Another mat fragment 70 × 160 mm was found at a depth of 1.5 m in an unlined pit no. 8 dug in the upper clay level. It is made of bundles of unidentified material, perhaps rushes or straw. Technologically, it belongs to a type known as 'coiled basketry with intricate stitch'. The mat fragment was dated by radiocarbon to 7517–7038 cal BP (Table 7.1) (Galili and Schick 1990).

**Faunal Remains** In the central sector of Kfar Samir, faunal remains were scarce. A few bird bones were recovered from well No. 3 and included seven bones of mallard duck (*Anas platyrynchos*) and one unidentified mammalian bone fragment (Galili and Weinstein-Evron 1985, p. 40, Horwitz et al. 2002).

**Fig. 7.6** Kfar Samir Southern Sector: elliptical building constructed of upright stone slabs (E. Galili)

#### 7.2.1.3 Southern Sector

This part of the site is located at 34° 57′ 8.6″ E, 32° 47′ 6.72″ N (Fig. 7.1b).

**Architecture** An elliptical structure of upright stone slabs was exposed near the shore in the southern part of the sector (Fig. 7.6). Hundreds of olive pits were discovered in the vicinity of this structure (Galili and Rosen 2013).

**Special Finds** A complete wooden bowl (Fig. 7.7) was found in a pit in the clay c. 300 m south of the Kfar Samir central sector (Galili and Schick 1990). It was made of *Ceratonia siliqua*, the carob tree or St John's bread. Pieces of waterlogged tree branches and straw, perhaps remains of a mat or a basket, were found next to it. Tool marks of an adze or a chisel were visible on the outer face, as well as traces of an imperfection left by a fracture in the flint tool used to form the bowl. The inner surface of the bowl was made using a different tool and it is rougher compared to the outer surface – perhaps because of difficulties in forming such a surface using an adze or a chisel-like tool. The base and the rim of the bowl are very smooth, either polished intentionally or abraded by lengthy use. The bowl was radiocarbon-dated to 8115–7949 cal. BP (Table 7.1).

### *7.2.2 Kfar Galim North*

This sector of the site is located at 34° 57′ 64″ E, 32° 46′ 14″ N (Fig. 7.1b) some 50–120 m off the coast at a depth of 1.5–4 m. Surveys revealed several round structures 1–1.5 m in diameter made of undressed stones, representing either refuse pits or water wells. Several rows of undressed stones, some forming right angles, imply the existence of rectangular structures, possibly dwellings. Other structures investigated were pits, probably water wells, walled with water-logged tree branches and stone pebbles. A test excavation carried out in one of these round structures yielded a few potsherds

**Fig. 7.7** Kfar Samir Southern Sector: complete wooden bowl made from a carob tree (J. Galili)

and flint flakes and fragments of plant material in the fill. A cylindrical 'tower' 1.2 m high consisting of six courses of undressed stones was found above one of these round stone structures. Its function is unknown, but it may have served as a superstructure over a well.

Finds of material culture from the site include an abundance of ground-stone artefacts made of limestone, sandstone and basalt, including mortars, grinding stones, basalt goblets on high cylindrical bases ('chalices'), rectangular troughs or mangers, and a few potsherds. Flint artefacts include several flake tools and four bifacials (adzes and an axe). The fauna comprise two remains, a horncore of an adult male mountain gazelle (*Gazella gazella*) and a mandible of Palestine molerat (*Spalax ehrenbergi*) (Horwitz et al. 2002). A single radiocarbon date of 7790–7670 cal BP taken from a wooden branch places this site well within the range of the Wadi Rabah phase (Table 7.1).

### 7.2.3 *Kfar Galim South*

This sector of the site is located at 34° 57′ 8″ E, 32° 46′ 5″ N (Fig. 7.1b), some 30–100 m off shore, at a depth of c. 2–4 m.

**Architecture** Architectural remains consist of ten round stone-built pits including water wells lined with stones and tree branches, some branches up to 0.2 m thick (Fig. 7.8). Four water wells c. 1 m in diameter were at a depth of 1.5–3.0 m below sea level. In one of these, there were three pieces of waterlogged wood about 25 cm long and 15 cm in diameter that had been worked at both ends (Fig. 7.9). Their dimensions and shape suggest that they were pre-forms intended for the production of wooden bowls as found at Kfar Samir. Test excavations, in two of the round structures revealed fills of sherds, flint artefacts, water-logged plant remains, animal bones, and wooden branches radiocarbon-dated to the PN at 6910–6670, 6890–6740 cal BP.

**Fig. 7.8** Kfar Galim South: water-well built of tree branches and stones (I. Greenberg)

**Artefacts** Pottery finds include 15 fragments of bases and handles, all typical of the Wadi Rabah culture, as well as a few non-diagnostic body sherds. Flint implements include several undiagnostic flakes.

**Faunal Remains** Seven animal species were identified (NISP = 31 bones). Remains of domestic pig (*Sus scrofa*) dominated at 55% and these were mainly immature animals, while domestic cattle (*Bos taurus*) at 22% were the next most common taxon, all adults. Present in far lower frequencies were remains of domestic sheep/goat (*Ovis aries/Capra hircus*) at 3.5%, dog (*Canis familiaris*) at 3.5%, grey heron (*Ardea cinerea*) at 3.5%, and vertebrae of an unidentified species of snake (Ophidia) at 6%. Two fish bones belonging to the family Sparidae (sea bream) were also identified (Horwitz et al. 2002).

### 7.2.4 Nahal Galim

The site is located at 34° 56′ 55″ E, 32° 45′ 23″ N, opposite the outlet of the river Nahal Galim (Fig. 7.1b). At a depth of 3–5 m, there were two round structures constructed of undressed stones 1–1.5 m in diameter, probably water wells. One of these structures, located some 150 m offshore at 4 m depth (Fig. 7.10), was marked by three courses of stones projecting above the sea bottom, apparently a superstructure forming the well mouth (Galili and Rosen 2013).

### 7.2.5 Hahoterim

The site is located at 34° 56′ 59″ E, 32° 44′ 59″ N (Fig. 7.1b). Several round structures constructed of undressed stones c. 1 m in diameter, probably water wells or storage pits, were identified at a water depth of 2–5 m. Hundreds of waterlogged olive pits and wooden branches were embedded in the clay palaeosol, indicating olive exploitation, probably for oil.

**Fig. 7.9** Kfar Galim South: waterlogged piece of wood cut at both ends, perhaps intended for the production of wooden bowls (E. Galili)

**Fig. 7.10** Nahal Galim: water-well constructed of undressed stones (I. Greenberg)

## *7.2.6 Tel Hreiz*

This site is located at 34° 56′ 55″ E, 32° 44′ 45″ N, at a depth of 0–5 m adjacent to the terrestrial Tell of the same name (Fig. 7.1b). The submerged site was first identified in the 1960s and initially identified as Chalcolithic or Early Bronze Age. Further surveys since 1984, as well as radiocarbon dates of 8330–7970, 7210–6980 and 7160–7000 cal BP, indicate that it represents a late PN Wadi Rabah settlement.

**Architecture** The 1965 survey revealed stone paving, hearths containing charred remains of wood, animal bones including a jaw of Persian fallow deer (*Dama mesopotamica*), flint artefacts, potsherds and a basalt mortar. Likewise the 1984–1985 and 1993 surveys noted the presence of stone structures and a row of upright wooden poles. During surveys in 2012, a large section of the southern part of the site was exposed from the coastline to a depth of 4.5 m (Galili and Rosen 2013). A megalithic structure c. 60 m long and 1 m high built of boulders up to 1 m in length was discovered at a water depth of 3–4 m. The structure comprises an elongated concentration of boulders broadly parallel to the coastline at some 80–100 m offshore and 2.5–3.5 m depth. A tail-like row of boulders protrudes from the southern end of the concentration and two additional rows, parallel to each other, protrude from its northern end. This structure could have been a coastal defence representing an effort to minimize damage from the rising sea. Also recorded at the site were several upright wooden poles (12–17 cm in diameter) located some 15 m seward from the row of boulders. These may have been the foundations of a hut. Other architectural finds include the remains of a square building 3 × 4 m constructed of undressed fieldstones, a hearth with charcoal and two round structures built of undressed stones.

**Artefacts** Flint artefacts include bifacial axes and adzes, sickle blades and other blade tools, but no arrowheads. The pottery comprised numerous fragments of bowls and scores of handles typical of the Wadi Rabah culture, and a few handles of ceramic churns. Ground stone remains included bowls and grinding stones made of sandstone or limestone, and mortars and chalices made of basalt.

**Botanical Remains** These included hundreds of waterlogged olive pits, numerous tree branches and a circular ring woven from twigs. The olive remains suggest that oil extraction took place on-site.

**Faunal Remains** Eight animal species were identified (NISP = 106 bones), the majority representing domestic animals. Remains of cattle (*Bos taurus*) were the most common (53%), while sheep/goat (*Ovis aries/Capra hircus*) amounted to 16% and pigs (*Sus scrofa*) to 14%. Based on size, cattle were identified as domestic animals. The majority of pig bones came from immature individuals but included domestic pigs as well as wild boar. Dogs (*C. familiaris*) comprised a significant proportion of the remains at 14%. The several canine skulls belong to adult domestic dogs, resembling those at Kfar Galim. Game species were limited in number and comprise Persian fallow deer (*Dama mesopotamica*), mountain gazelle (*Gazella gazella*) as well as two fish Families (Serranidae and Tilapia sp.), sea water and freshwater taxa respectively (Horwitz et al. 2002).

**Human Remains** During 2015 a winter storm exposed two stone-built cist graves, two disturbed skeletons buried in the clay with no grave structure, comprising a lower jaw and a few fragmentary ribs and calvaria.

### 7.2.7 *Megadim*

Megadim site is located north of Atlit (Fig 7.1B) at 34° 56′ 40″ E, 32° 43′ 39 N″, some 80–120 m off shore, at a depth of 2.5–3.5 m (Galili and Weinstein-Evron 1985). Finds included three round pits lined with undressed stones, possibly water wells, found some 50 m apart. The southern pit was partly excavated to a depth of 0.4 m. Finds included a few flint flakes, ground tools including three basalt chalices and two limestone bowls, some waterlogged botanical remains of small tree branches and straw and a canid mandible, probably a dog, radiocarbon dated to 7310–7020 cal BP.

### 7.2.8 *Neve-Yam*

The Neve-Yam site actually comprises two separate areas (Figs 7.11a,b and Fig. 7.11c). The north area (C) has very few remains, including a round stone structure 1 m in diameter, probably a well, and

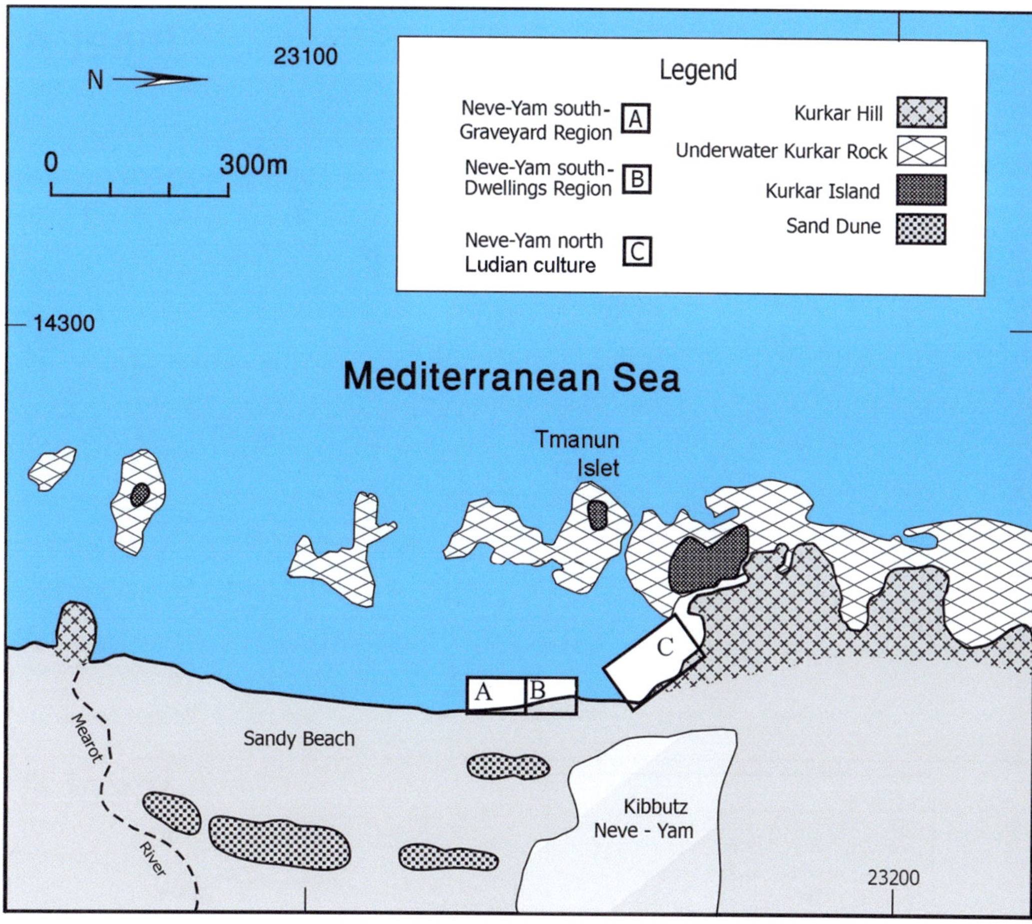


**Fig. 7.11** Location map of Neve-Yam sites: (A) Neve-Yam south, Wadi Rabah site, the graveyard area, (B) the dwelling area, (C) Neve-Yam north, site of the Lodian culture (For details see inset in Fig. 7.12) (E. Galili)

pottery fragments—rims, handles and bases of bowls and containers—attributed to the Lodian culture as defined by Gopher and Gophna (1993), which predates the Wadi Rabah culture.

A rather important area is Neve-Yam south (Fig. 7.11a,b) which is located at 34° 55′ 40″ E, 32° 40′ 30″ N. It is one of the southern submerged Neolithic sites on the Carmel coast and is located between the coast line and a submerged ridge of Kurkar (aeolian sandstone) at a water depth of 1–5 m. Some 5000 $m^2$ have been investigated to date. The site can be divided into northern and southern sectors (Figs. 7.11A, B and 7.12). A portion of the coastal section of Neve-Yam was first exposed during the late 1960s following a sea storm (Wreschner 1977a). During 1968, a small rescue excavation was carried out on the shore in the northern sector of the site (B). Over the period 1983–1995, parts of the southern sector were exposed and underwater rescue surveys carried out (Galili et al. 1998, 2009; Galili and Rosen 2011b). The southern sector mainly comprises stone-built cist graves (Fig. 7.12a), the northern sector mostly rectangular dwellings (Fig. 7.12b). Three radiocarbon dates ranging from 7580 to 7270 cal BP place both sectors in the Wadi Rabah culture.

**Architectural Finds** In the north sector (Fig. 7.12b), there were foundations of rectangular structures, probably dwellings, and straight sections of walls c. 0.5 m wide built of two rows of undressed pieces of sandstone (Fig. 7.13). Also found were unpaved pits, paved surfaces made of small undressed stones, stone slabs, or postholes, probably of huts.

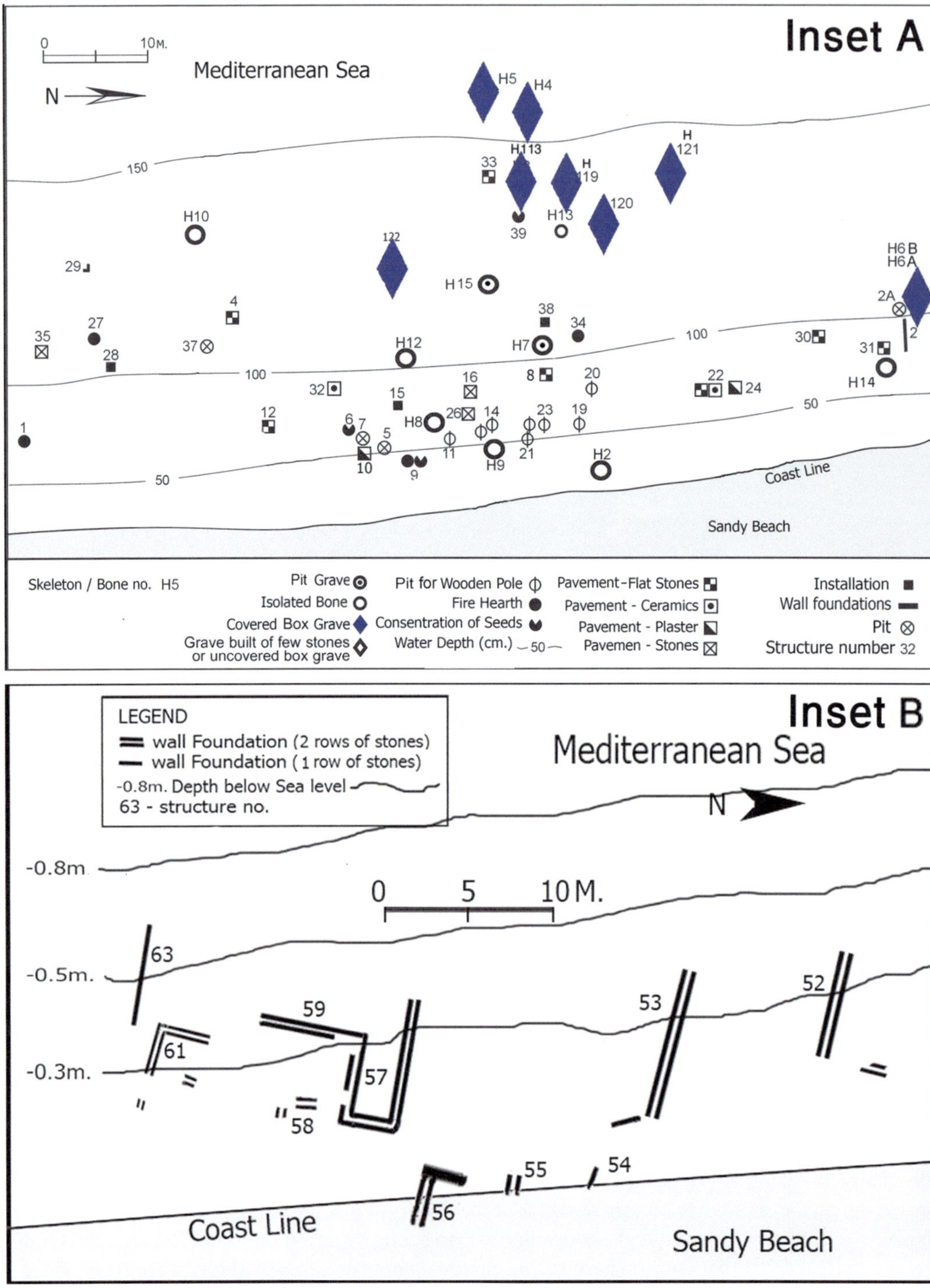


**Fig. 7.12** Plan of the Neve-Yam south site: (A) Graveyard area, (B) Dwelling area (E. Galili)

**Fig. 7.13** Rectangular dwellings at Neve-Yam (J. Galili)

The southern sector contained graves, hearths, paved surfaces and remains of small structures, pottery vessels (Fig. 7.14), flint and ground stone artefacts and plant remains. Of particular interest is the fact that this south sector was used primarily for burial in stone-built cist graves arranged in an organized pattern (Figs. 7.12.A, 7.15a, b). Thus, a clear division is discernible between the northern sector of the site, used as a residential area for the living, and the southern 'suburb of the dead', containing the graves.

**Flint Tools** The flint assemblage includes a high percentage of narrow chisels of plano-convex section, adzes, wide but short rectangular sickle blades which are backed and truncated, polished bifacial tools with polished working edges, including adzes, chisels and a low percentage of axes. Noteworthy is the absence of arrowheads.

**Ground Stone Assemblage** This includes basalt chalices, basalt and limestone grinding tools, limestone bowls and large basins made of Kurkar. About 6% of the ground stones in the site and also in the earlier Atlit-Yam site are perforated stone weights. These could have been used for fishing nets (Galili et al. 2002, 2004).

**Ceramics** The ceramic assemblage of Neve-Yam South is very similar to that initially collected by Wreschner in the 1960s, and to other Wadi Rabah sites in the Southern Levant. It consists mainly of bowls, spouted vessels, hole-mouth jars, bow-rim jars and pithoi, all decorated by painted, incised and applied elements. Of special interest is a sherd fragment with an incised decoration of fishes and a herringbone design (Fig. 7.14a).

**Figurines** A schematic, anthropomorphic stele made of Kurkar was recovered near the burial area. Also recovered from the burial area was an anthropomorphic figurine engraved on bone (Fig. 7.16). A third anthropomorphic figurine made of greenstone was found on the shore (Fig. 7.17) (Galili et al. 2016).

**Fig. 7.14** Neve-Yam: **a** Fragment of a pottery vessel with incised fishes and herringbone design, **b** Pottery vessel with two openings (E. Galili)

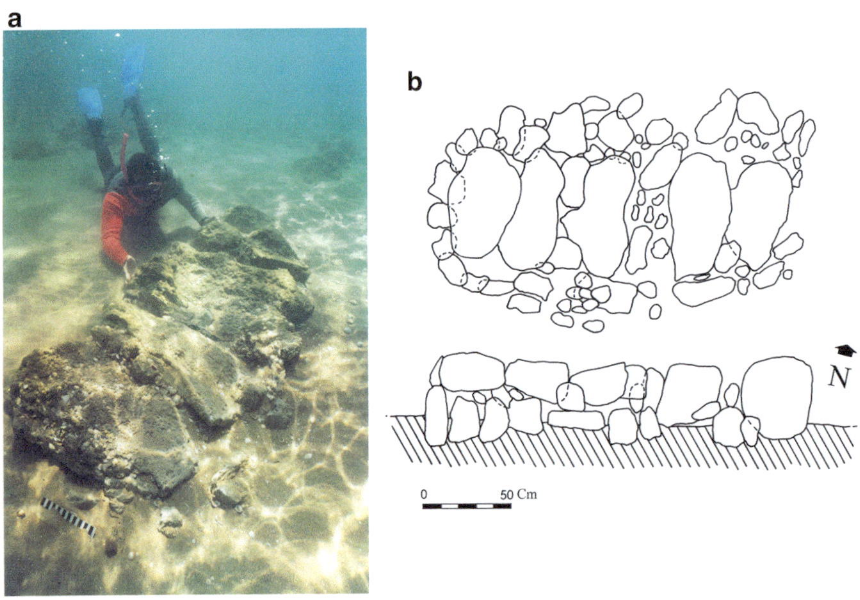


**Fig. 7.15** Neve-Yam: **a** Stone-built cist grave, **b** drawing and section of the cist grave (E. Galili)

**Botanical Remains** Three large concentrations of charred seeds, possibly associated with the burials, were recovered in the southern part of the site. Concentration #1 (Fig. 7.12a, no. 6) comprised about 300 cc of seeds mixed with clay, all identified as domesticated lentils (*Lens culinaris* var. *microsperma*) (Galili et al. 2009). The seeds varied in size and some were infested by pests, probably a

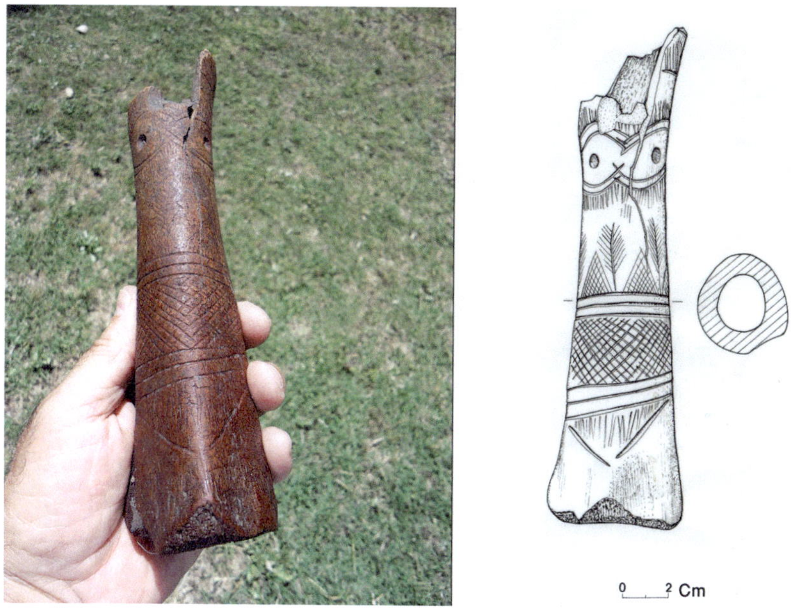


**Fig. 7.16** Neve-Yam: anthropomorphic bone Figurine (E. Galili and L. Zeiger)

**Fig. 7.17** Neve-Yam: anthropomorphic Figurine made of greenstone, 5.7 cm high (Israel Museum)

**Fig. 7.18** Neve-Yam: a concentration of charred legumes found near burial H3 (Concentration #3). (J. Galili)

beetle of the genus *Bruchus*. Concentration #2 (Fig. 7.12a, no. 9) comprised about 100 cc of grains mixed with clay, including seeds of various shapes and sizes that were concentrated in a hearth built of burnt mud bricks. About 90% of the grains were of domesticated barley (*Hordeum sativum* = *H. vulgare* L.). In addition, several seeds of edible plants and wild plants were recovered: domesticated emmer (*Triticum dicoccum*), one seed of *Vicia* sp., a few seeds of the Liliaceae family and several unidentified seeds. Concentration #3 comprised about 300 cc of seeds free of clay (Fig. 7.12a, no. 39, Fig. 7.18), and was found c. 1.5 m east of burial H3. This sample consisted of several species of pulses apparently domesticated, including (by frequency): pea (cf. *Pisum Vicia narbonensis*), vetch (*Lathyrus cercula*), horse bean (*Vicia faba* var. *minor*), lentil (*Lens culinaris* var. *microsperma*), domesticated flax (*Linum usitatissimum*), and remains of a few wild plants or weeds of Galium, of the genera Lolium and Liliaceae. The presence of field weeds and wild plants in the seed assemblages seems to indicate harvesting in cultivated fields. Barley and the pulses are harvested in the spring, while wheat is usually harvested in early summer. The under-representation of barley may be of significance, hinting at a spring event.

**Faunal Remains** A small faunal assemblage (NISP = 91 identified bones) was recovered during the 1960s salvage excavation and from collections along the sea-shore by Wreschner, and was described by Horwitz (1988). The subsequent 1989–1885 surveys yielded a more substantial faunal collection (NISP = 380 identified bones) but the remains were hand-collected during dives and so primarily consisted of large, relatively complete elements (Horwitz and Ducos 2005; Horwitz et al. 2006). Both collections indicate an animal economy based on four domestic herd animals: sheep (*Ovis aries*), goat (*Capra hircus*), cattle (*Bos taurus*) and pigs (*Sus scrofa*). The few remains of wild species that were recovered are mountain gazelle (*Gazella gazella*), badger (*Meles meles*) and perhaps wild boar (*Sus scrofa ferus*), attesting to the continuation of hunting, but also to its relative unimportance as a subsistence activity. Two-thirds of the ungulates were immature, a kill-off pattern indicative of herd management aimed at meat procurement. Analysis of cut marks indicates that animals were butchered and consumed on-site using a variety of stone tools (Greenfield et al. 2006). The scanty fish remains may indicate that the role of fishing was limited, or it may result from the fact that smaller-sized bones, such as fish bones, were missed because of recovery by hand-collection during dives.

#### 7.2.8.1 The Neve-Yam Burial Ground

The burials were concentrated in a relatively small area, 40 × 70 m. The cist graves (Figs. 7.11A, 7.12A, 7.15a, b, and 7.19) were oriented in an east–west direction and consisted of an oval burial chamber lined with undressed stones covered by large stone slabs. Altogether, 11 graves were located, of which

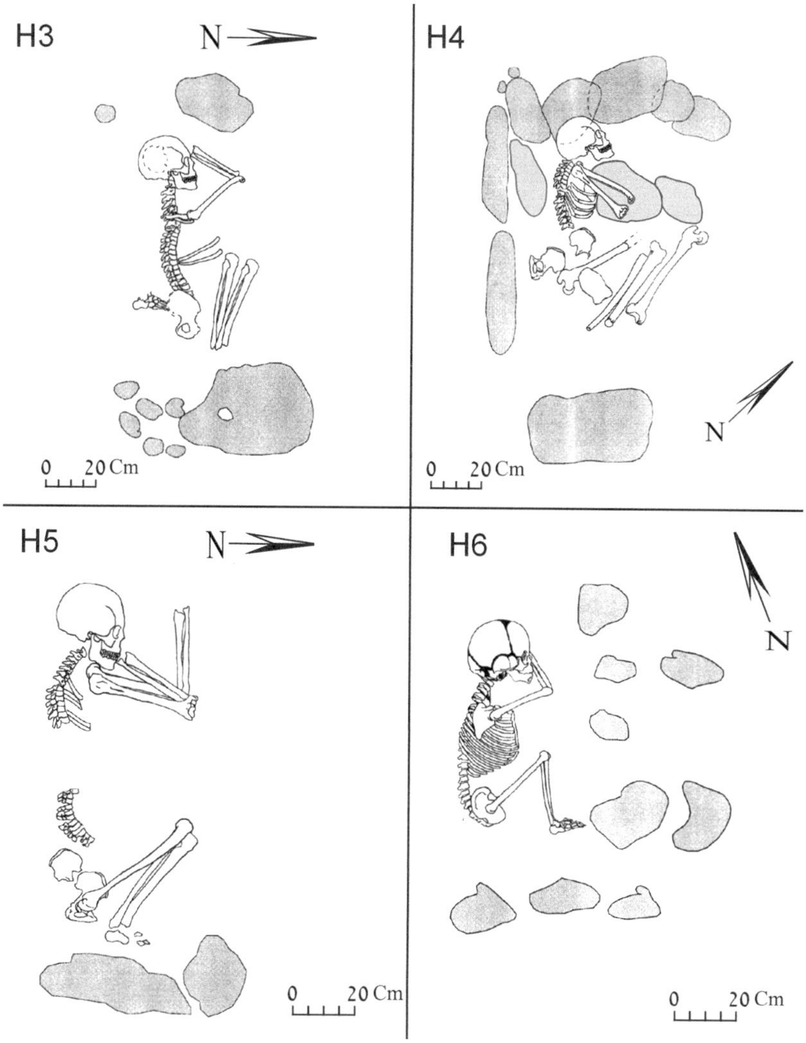


**Fig. 7.19** Burial positions of skeletons from Neve-Yam south site (After Galili et al. 2009)

two were covered cists, four were uncovered cists, two consisted of a few undressed stones in the clay, and three were pits dug into the clay. Some cist graves were eroded by sea action with only the skeletons and a few stones remaining. Four such eroded graves were excavated. No grave offerings were recovered within the excavated graves. However, the three charred seed assemblages found in the vicinity of the graves and the nearby hearths and paved surfaces, may be related to the burials and/or ritual activities, e.g., ceremonial meals that took place near the graves.

**Human Skeletal Remains** The sample consists of an estimated 15 individuals (Galili et al. 2009). Six skeletons, either primary or disturbed burials, were recovered from the graves. Nine additional individuals were represented by scattered bones. Four skeletons were buried in a fully flexed position, and four were partly flexed (Fig. 7.19). Eight children aged up to 10 years, one adolescent (10–18 years) and six adults were identified. Among the adults were three females, one male and two individuals of unknown gender. No charred bones or group burials were found. An unusual pattern of tooth attrition (oblique wear) was noted in one of the adult individuals suggesting use of the teeth as a working tool (Galili et al. 2009).

**Fig. 7.20** The Tmanun islet looking east and the location of the infant burial

#### 7.2.8.2 The Neve-Yam Baby Burial on the Tmanun Islet

During October 2014, burial remains were identified on the rocky Tmanun islet located opposite the site of Neve-Yam, presently c. 140 m off shore (located at 34° 55′ 32.7″ E, 32° 40′ 39.2″ N) (Figs. 7.11 and 7.20). The remains were embedded in a mass of conglomerate (60 × 40 × 30 cm) adhering to the Kurkar bedrock, some 60–100 cm above the present sea level. The human remains included a skull cap (calvarium), parts of a jaw with three teeth, long bones and fragments of ribs of a 1–2 year-old infant (Fig. 7.21). They were associated with a broken fragment of a bowl with an elongated nozzle. It is proposed that the burial took place in the PN period, when the sea level was lower by c. 7–8 m and the Tmanun islet was a prominent sandstone hill on the coast, partly covered by sand and facing the open sea.

## 7.3 Discussion

**The Development of the Traditional Mediterranean Subsistence Economy** At the end of the tenth millennium BP, a mixed mode of subsistence that included farming, animal husbandry, hunting and fishing evolved among the indigenous coastal inhabitants on the Levantine coast. This is shown by the archaeological remains recovered from the submerged PPNC site of Atlit-Yam (Galili et al. 2002, 2004), and possibly also from the terrestrial coastal sites of Ras Shamra in Syria and Ashqelon in Israel (Van Zeist and Bakker-Heeres 1984; Helmer 1989; Perrot and Gopher 1996; Garfinkel and Dag 2008). Galili et al. (2004) have noted that this represents the earliest appearance of the Mediterranean-Levantine fishing village. It is suggested that, subsequently, this coastal-adapted subsistence system continued to evolve locally in the PN, and also spread westward throughout the Mediterranean basin via sites on the Anatolian coast, e.g., Mersin-Yumuktepe on the Cilician coast, Coşkuntepe in western

**Fig. 7.21** Skull calvarium of a 1–2 year-old infant from Tmanun Islet near Neve-Yam

Anatolia and Karağaçtepe on the Gallipoli Peninsula (Lichter 2005; Takaoğlu 2005) and on to countries and areas located further to the west (Galili et al. 2002, 2004).

Overall, the subsistence mode of the submerged PN settlements on the palaeo-coast resembles that of their terrestrial counterparts (e.g., Gopher and Gophna 1993; Horwitz et al. 2002; Horwitz 2012) but with the addition of marine resource exploitation, primarily of fish.

Exceptional finds in the submerged Wadi Rabah sites are the installations and artefacts associated with olive oil extraction. In contrast to the botanical assemblage from PPNC Atlit-Yam, in most PN sites thousands of olive stones were recovered. This change may be seen as the beginning of intensive exploitation of olives for human consumption. Olive oil extraction, probably from wild olives, seems to be a PN innovation, first attested to in these submerged Wadi Rabah sites (Galili et al. 1997). Later still, during the fourth millennium cal BC, more plants were introduced into the subsistence economy such as the domesticated grape vine, which enabled the production of wine in the Levant (Zohary and Hopf 2000). By 3000 cal BC, the commonly termed 'traditional Mediterranean diet' had evolved (Galili et al. 2002, 2005), a complex defined by Butzer (1996) as based on cereal agriculture (primarily of wheat and barley), cultivation of fruit trees (olives, grapes, almonds and figs), vegetable gardening and husbandry of domestic caprines (sheep and goat) for meat as well as secondary products (milk and wool), while cattle were exploited as beasts of burden and for traction, as well as for other purposes.

**Material Culture in the Submerged Sites** In terms of material culture—lithic tools and technology, ground-stone artefacts and ceramics—there are no marked differences between the submerged Wadi Rabah sites and their terrestrial counterparts in the southern Levant (e.g., Gopher and Gophna 1993). However, one of the notable innovations on the coast is the construction of water wells built of wooden branches and stones. These are considered some of the earliest known wooden constructions. In contrast, the wells at the submerged PPNC site of Atlit-Yam, which pre-date the PN wells by ~1000 years, were lined only with undressed stones (Galili and Nir 1993; Galili and Rosen 2011a, Galili et al. Chap. 6).

**Fig. 7.22** Drawing illustrating the difference in the intensity of sub-soil excavations for pits, refuse, burials, wells etc. in a permanently occupied settlement (*lower*) and a temporary camp site (*upper*)

**Mortuary Practices and the Emergence of Separate Graveyards** Gopher and Eshed (2012) suggested that the burials found outside dwellings in many of the PN sites in the southern Levant were a product of socioeconomic and cultural changes. They referred to the two-dimensional distribution. The vertical dimension however, is associated with intensive excavation beneath the soil surface to create burial pits, which played a major role in settled Neolithic farming communities. Even during the Pre-Pottery Neolithic, sedentary agriculturists dug into the sub-surface intensively and systematically (Bar-Yosef 2001; Bar-Yosef and Belfer-Cohen 1991). The excavation work required would have resulted in considerable soil disturbance often within settlements or adjacent to them. Aside from the cultivation of fields, the appearance of extensive food storage facilities such as silos and pits after 9500 BP (Kuijt 2000) would also have disturbed the local sub-soil. In the PN, the digging of water wells, house foundations, and refuse and storage pits is included in this 'invasion' of the sub-surface. These combined factors must have greatly increased the number of intentional penetrations of the sub-surface in and immediately around settlements, resulting in the exposure and disturbance of older burials within sites settled for generations, centuries or even millennia (Fig. 7.22). This may have provoked frequent but unintended conflict between the use of ground for burials and their use by the living for cultivation, building, storage, refuse disposal and so on. Several forces may thus have influenced the separation of burial areas from residential ones as evident in Neve-Yam:

(a) Practical: hygienic and aesthetical considerations of smell, diseases etc.

(b) Lack of space: competing needs, especially the need to keep burials free from disturbance by other activities

(c) Social: at home the deceased ancestors and graves would have been the responsibility of the family, while in a common graveyard they are also a communal responsibility

(d) Ideological/symbolic: taboos and issues of purity versus impurity—the dead may not be pure and are thus not suitable to reside with the living but belong to the next world and should be kept apart from residential areas (Bartel 1982; Fahlander and Oestigaard 2008; Kuijt 2008).

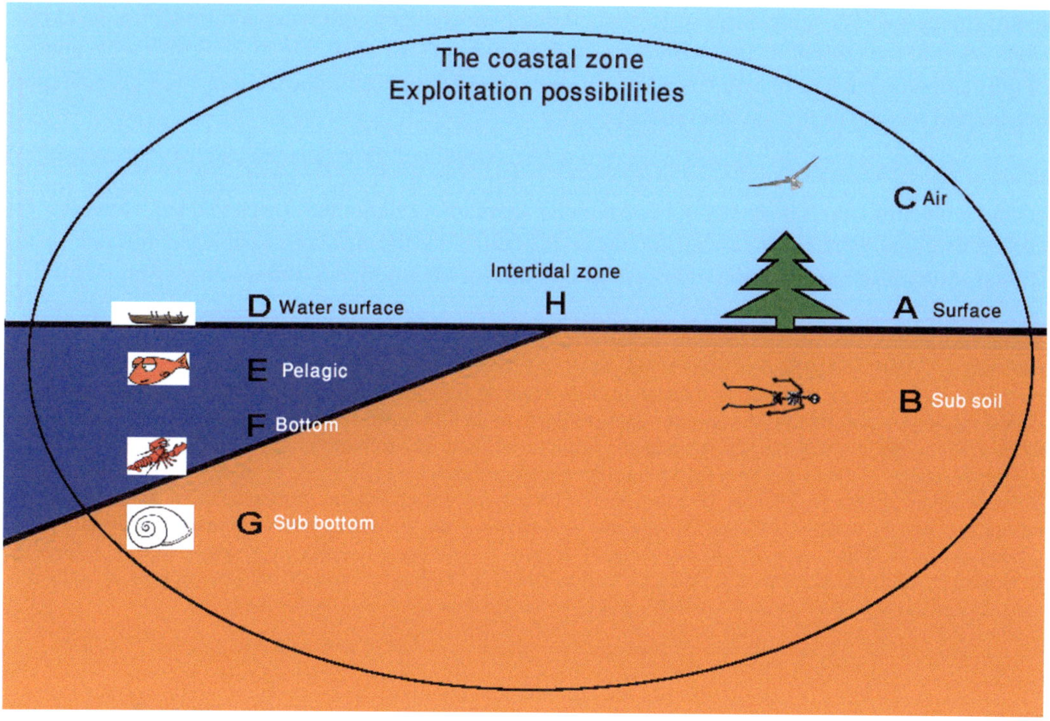


**Fig. 7.23** Schematic diagram showing the diversity of economic resources available in the coastal zone

Thus, the well-preserved submerged settlements discovered off the Carmel coast enable us to trace the significant changes that took place in burial practices from the late PPN to the late PN periods. A unique feature of these submerged PN fishing villages is the evidence for a separate burial ground. In the earlier PPNC site of Atlit-Yam, burials lack grave structures, and are scattered all over the site (Galili et al. 2005; Eshed and Galili 2011). In PN Neve-Yam, the graves are stone-built cists, a form known from the very beginning of the PN in terrestrial sites such as Byblos (Dunand 1973), Sha'ar Hagolan in Israel (Stekelis 1972) and Tabaqat Al-Buma in Jordan (Banning 1995; Banning et al. 1996,). The significance of the Neve-Yam burials is their organized pattern and concentration in a separate area of the site, set apart from the residential area. The presence of an organized burial area with standardized grave types outside the dwelling zone, accompanied by ritual and symbolic activities, should be considered as a burial ground or cemetery (Galili et al. 2009). The presence of agricultural products, perhaps representing offerings or leftovers of ritual or ceremonial feasts close to the graves, may reflect the later well-known practice of offering part of the harvest to the ancestors and gods as reflected in local archaeology, regional myths and biblical sources. Separate burial grounds with all these features are unknown in other PN sites in the region. The motivation to develop separate burial grounds may have been associated with the necessity to resolve 'territorial friction' between the living and the dead over the use of sub-surface space. During the late Chalcolithic period this mode of burial became the most common burial practice, and it is common in many human societies up to the present-day.

**The Abandonment of Sites** The earlier PPNC site of Atlit-Yam had been abandoned due to a global rise in Holocene sea-level from −16 to −8 m below sea level (See Galili et al. Chap. 6, Fig. 6.6). This sea-level rise resulted in salinization of PPNC water wells, and flooding of fields and homes. Relative to the PN sites, Atlit-Yam lies in deeper water off the Carmel coast, implying that post-PPNC settlements shifted inland and were established on the new coastline further to the east. However, the currently submerged PN settlements were also gradually abandoned following additional sea-level rise, from −8 m to the present level (Galili et al. 2008, Galili et al. Chap. 6, Fig. 6.7). Post-PN occupation of this region on the new coastline was less intensive, perhaps because the Carmel coast became quite narrow and marshy as the sea level rose.

The ever-changing coastal environment required ongoing adaptation, with changes of settlement location and subsistence strategies. Artificially raising the fresh water level of Atlit-Yam wells by adding a layer of stones is an example of such an adaptation (Galili and Rosen 2011a). However, the coastal dwellers were compensated for these difficulties by the diverse possibilities offered by the coastal zone for the exploitation of terrestrial and marine environments and resources (Fig 7.23).

**Acknowledgments** We wish to thank EU COST Action TD0902 SPLASHCOS, the National Geographic Research Foundation, CARE Archaeological Foundation, MAFCAF Foundation, and Sandy and Joseph Lepelstat, the Honor Frost foundation and the Dan David foundation their financial support. The Israel Antiquities Authority, the Israel Prehistoric Society, and Haifa University are acknowledged for their financial and institutional support of the underwater excavations. We also wish to express our appreciation to the divers and archaeologists who participated in the underwater excavations; to the many researchers who subsequently contributed to the study of the finds; S. Ben-Yehuda and B. Galili for the drawings; J. Galili, G. Arbel and I. Grinberg for the underwater photographs; M. Rasovsky for the treatment of organic material; I Carmi, D. Segal and E. Boaretto of the Weizmann Institute for radiocarbon dating; and I. Hershkovitz, M. Weinstein-Evron, N. Liphshitz, M. Kislev, A. Hartman, A. Gopher, D. Nadel, R. Barkai, A. Shifroni, D. Gersht, H. Michelin, K. Sharvit, I. Zohar, O. Lernau and H. K. Mienis for research on different aspects of the submerged sites and to O. Bar-Yosef, Y. Garfinkel and U. Avner for their comments on the discassion.

## References

Banning EB (1995) Herders or homesteaders? A Neolithic farm in Wadi Ziqlab, Jordan. Biblic Archaeol 58:2–13

Banning EB, Rahimi D, Siggers J, Ta'ani H (1996) The 1992 season of excavations in Wadi Ziqlab, Jordan. Annu Dep Antiquities Jordan 40:29–49

Bartel R (1982) A historical review of ethnological and archaeological analysis of mortuary practices. J Anthropol Archaeol 1:32–35

Bar-Yosef O (2001) From sedentary foragers to village hierarchies: the emergence of social institutions. In: Runciman G (ed) The origin of human social institutions. Proceedings of the British Academy, vol. 100, pp 1–38

Bar-Yosef O, Belfer-Cohen A (1991) From sedentary hunter-gatherers to territorial farmers in the Levant. In: Gregg SA (ed) Between bands and states. Cent Archaeol Investig, Carbondale, pp 181–202

Butzer KW (1996) Ecology in the long view: settlement histories, agrosystemic strategies, and ecological performance. J Field Archaeol 23:141–150

Dunand M (1973) Fouilles de Byblos, Tome V. Maisonneuve, Paris

Eshed V, Galili E (2011) Palaeodemography of southern Levantine Pre-Pottery Neolithic populations: regional and temporal perspectives. In: Pinhasi R, Stock J (eds) Human bioarchaeology of the transition to agriculture. Wiley, Chichester, pp 403–428

Fahlander F, Oestigaard T (2008) The materiality of death: bodies, burials, beliefs. In: Fahlander F, Oestigaard T (eds) The materiality of death, BAR International Series, vol 1768. Archaeopress, Oxford, pp 1–16

Galili E (2004) Submerged settlements of the ninth to seventh millennia BP off the Carmel coast. Unpublished PhD dissertation, Tel Aviv University (in Hebrew)

Galili E, Inbar M (1987) Underwater clay exposures along the Israeli coast, submerged archaeological remains and sea-level changes in the northern Carmel coast. Horiz Geogr 22:3–34. (in Hebrew)

Galili E, Nir Y (1993) The submerged pre-pottery Neolithic water well of Atlit-Yam, northern Israel, and its palaeoenvironmental implications. The Holocene 3:265–270

Galili E, Rosen B and Boaretto E (2007) Artifacts made of plant material, Kfar-Samir, submerged Neolithic site off the south coast of Haifa. Hadashot Archeologiot 119

Galili E, Rosen B (2011a) Submerged Neolithic settlements off the Mediterranean Carmel coast of Israel and water mining in the Southern Levant. Neo-Lithics 2:47–52

Galili E, Rosen B (2011b) Submerged Neolithic settlements off the Mediterranean coast of Israel. In: Benjamin J, Bonsall C, Pickard C, Fischer A (eds) Submerged prehistory. Oxbow, Oxford, pp 272–286

Galili E, Rosen B (2013) Atlit Yam – preliminary report. Hadashot Archaeologiot 125

Galili E, Schick T (1990) Basketry and a wooden bowl from the pottery Neolithic submerged site of Kefar Samir. Mitekufat Haeven, J Israel Prehist Soc 23:142–151
Galili E, Sharvit J (1994) Evidence of olive oil production from the submerged site at Kfar Samir, Israel. Mitekufat Haeven, J Israel Prehist Soc 26:122–133
Galili E, Weinstein-Evron M (1985) Prehistory and paleoenvironments of submerged sites along the Carmel coast of Israel. Paléorient 11:37–52
Galili E, Weinstein-Evron M, Ronen A (1988) Holocene sea-level changes based on submerged archaeological sites off the northern Carmel coast in Israel. Quat Res 29:36–42
Galili E, Weinstein-Evron M, Zohary D (1989) Appearance of olives in submerged Neolithic sites along the Carmel coast. Mitekufat Haeven, J Israel Prehist Soc 22:95–97
Galili E, Weinstein-Evron M, Hershkovitz I, Gopher A, Kislev M, Lernau O, Horwitz LK, Lernau H (1993) Atlit-Yam: a prehistoric site on the sea floor off the Israeli coast. J Field Archaeol 20:133–156
Galili E, Stanley DJ, Sharvit J, Weinstein-Evron M (1997) Evidence for earliest olive-oil production in submerged settlements off the Carmel coast, Israel. J Archaeol Sci 24:1141–1150
Galili E, Sharvit J, Nagar A (1998) Nevé-Yam – underwater survey. Excavations Surv Israel 18:35–36
Galili E, Rosen B, Gopher A, Horwitz LK (2002) The emergence and dispersion of the Eastern Mediterranean fishing village: evidence from submerged Neolithic settlements off the Carmel coast, Israel. J Mediterr Archaeol 15:167–198
Galili E, Lernau O, Zohar I (2004) Fishing and marine adaptations at Atlit-Yam, a submerged Neolithic village off the Carmel coast, Israel. Atiqot 48:1–34
Galili E, Eshed V, Gopher A, Hershkovitz I (2005) Burial practices at the submerged PPNC site of Atlit-Yam, northern Coast of Israel: what do they tell us about the final phase of the Pre-Pottery Neolithic Culture? BASOR 339:1–19
Galili E, Horwitz LK, Hershkovitz I, Eshed V, Salamon A, Zviely D, Weinstein-Evron M, Greenfield H (2008) Comment on "Holocene tsunamis from Mount Etna and the fate of Israeli Neolithic communities" by Maria Teresa Pareschi, Enzo Boschi, and Massimiliano Favalli. Geophys Res Lett 35:L08311. doi:10.1029/2008GL033445
Galili E, Eshed V, Rosen B, Kislev ME, Simchoni O, Hershkovitz I, Gopher O (2009) Evidence for a separate burial ground at the submerged Pottery Neolithic site of Neve-Yam, Israel. Paléorient 35:31–46
Galili E, Rosen B, Orrelle E, Yaroshevich A, Horwitz LK (2016) Symbolic artefacts from the submerged Neolithic village of Neve-Yam and contemporaneous Israeli sites as fossil directeur for the Wadi Rabah culture. Israel Explor J 67:129–150
Garfinkel Y, Dag D (eds) (2008) Neolithic Ashkelon, Qedem, vol 47. The Hebrew University, Jerusalem
Gopher A, Eshed V (2012) Burials and human skeletal remains from Nahal Zehora II in PN perspective. In: Gopher A (ed) Village communities of the Pottery Neolithic period in the Menashe Hills, Israel. Archaeological investigations at the sites of Nahal Zehora, Emery and Claire Yass Publications in Archaeology 29, vol III. Tel Aviv University, Tel Aviv, pp 1389–1412
Gopher A, Gophna R (1993) Cultures of the eighth and seventh millennia BP in the southern Levant: a review for the 1990s. J World Prehistory 7:297–353
Greenfield H, Galili E, Horwitz LK (2006) The butchered bones from Newe Yam. Mitekufat Haeven, J Israel Prehist Soc 36:173–200
Helmer D (1989) Le développement de la domestication au Proche-Orient de 9500 à 7500 BP: les nouvelles données d'El Kowm et de Ras Shamra. Paléorient 15:111–121
Hershkovitz I, Galili E (1990) 8000 year-old human remains on the sea floor near Atlit, Israel. Hum Evol 5:319–358
Horwitz LK (1988) Bone remains from Neve Yam, a pottery Neolithic site off the Carmel coast. Mitekufat Haeven, J Israel Prehist Soc 21:99–108
Horwitz LK (2012) The pottery Neolithic fauna from Stratum I. In: Valla F (ed) Les Fouilles de la Terrasse d'Hayonim (Israël). 1980–1981 et 1985–1989, Memoires et Travaux de Centre de Recherche Français de Jerusalem, vol 10. De Boccard, Paris, pp 131–148
Horwitz LK, Ducos P (2005) Counting cattle: trends in Neolithic Bos frequencies from the southern Levant. Rev Paléobiol 10:209–224
Horwitz LK, Galili E, Sharvit J, Lernau O (2002) Fauna from five submerged Pottery Neolithic sites off the Carmel coast. Mitekufat Haeven, J Israel Prehist Soc 32:147–174
Horwitz LK, Lernau O, Galili E (2006) Fauna from the pottery Neolithic site of Neve-Yam. Mitekufat Haeven, J Israel Prehist Soc 36:139–171
Kuijt I (2000) Keeping the peace: ritual, skull caching, and the community integration in the Levantine Neolithic. In: Kuijt I (ed) Life in Neolithic farming communities. Social organization, identity and differentiation. Kluwer Academic, New York, pp 137–162
Kuijt I (2008) Demography and storage systems during the Southern Levantine demographic transition. In: Bocquet Appel JP, Bar Yosef O (eds) The Neolithic demographic transition and its consequences. Springer, Netherlands, pp 287–313

Lichter C (2005) Western Anatolia in the late Neolithic and early Chalcolithic: the actual state of research. In: Lichter C (ed) How did farming reach Europe? Byzas, vol 2. Ege Yayinlan, Istanbul, pp 59–74
Olami Y (1984) Prehistoric Carmel. Israel Exploration Society, Jerusalem
Perrot I, Gopher A (1996) A Late Neolithic site near Ashkelon. Israel Explor J 46:145–166
Ronen A, Olami Y (1978) Atlit Map. Israel Dep Antiquities, Jerusalem
Stekelis M (1972) The Yarmukian culture of the Neolithic period. Magnes, Jerusalem
Takaoğlu T (2005) Coşkuntepe: an early Neolithic quern production site in NW Turkey. J Field Archaeol 30(4):419–433
Van Zeist W, Bakker-Heeres JAH (1984) Archaeobotanical studies in the Levant 2: neolithic and Halaf levels at Ras Shamra. Palaeohistoria 26:151–170
Wreschner EE (1977a) Neve-Yam, a submerged Late Neolithic settlement near Mount Carmel. Eretz Israel 13:260–271
Wreschner EE (1977b) Sea level changes and settlement location in the coastal plain of Israel during the Holocene. Eretz Israel 13:277–282
Zohary D, Hopf M (2000) Domestication of plants in the old world, 3rd edn. Oxford University Press, Oxford

# Chapter 8
# Hjarnø Sund: An Eroding Mesolithic Site and the Tale of two Paddles

**Claus Skriver, Per Borup, and Peter Moe Astrup**

**Abstract** Hjarnø Sund is one of 37 submerged archaeological sites that have been recorded in the 15-km-long Horsens Fjord. It is now located in a water depth of 0.5–2 m. Until recently, the gyttja deposits containing the archaeological material were covered with a protective layer of sand. Recent erosion by water currents has exposed the archaeological material and prompted detailed investigation before further damage occurs. Methods of investigation include underwater excavation and sediment coring, which have revealed a rich collection of material ranging in age from the late Kongemosen to the Middle Ertebølle. The deposits include two shell-midden layers dominated by oysters and cockles, which contain concentrations of flint artefacts and bones of fish and mammals, and layers of gyttja, from which artefacts of bone, antler and wood have been recovered. Of particular interest are decorated antler axes, one of which was found hafted on a wooden handle, and two wooden paddles decorated with distinctive painted designs. An extensive cultural deposit is still protected by overlying sediments, and would repay further excavation, especially given the threat of destruction by erosion.

## 8.1 Introduction

Since the mid-1970s, a number of submerged settlements dating from the Mesolithic period have been excavated in Denmark. Among the best known and best documented are those at Tybrind Vig (Uldum 2011; Andersen 2013), Ronæs Skov (Andersen 2009), Blak (Sørensen 1996) and Møllegabet (Skaarup and Grøn 2004), but Argusgrunden (Fischer 1987) and Næbbet (Skaarup 1983) also deserve mention. Well-preserved organic material was encountered at these sites, providing a much needed insight into artefact types of wood, bone and antler, for example dug-out boats, paddles, shafts for tools and implements, bows, arrows and leisters. These organic materials have survived due to the anoxic environment, which results in the absence of bacteria that would otherwise cause degradation and decay. The prevailing conditions have also favoured the preservation of plant macro-remains, such as nutshells, seeds and cones, and analysis of these can provide information on the use of plant resources, both for food and the production of tools and equipment such as twine and cordage of lime bast. Submerged settlements have therefore a great research potential in terms of the actual artefacts that are preserved as well as material for a range of environmental analyses. With time, these studies can

C. Skriver (✉) • P.M. Astrup
Moesgaard Museum, 8270 Højbjerg, Denmark
e-mail: cs@moesgaardmuseum.dk; pma@moesgaardmuseum.dk

P. Borup
Horsens Museum, 8700 Horsens, Denmark
e-mail: muspb@horsens.dk



G.N. Bailey et al. (eds.), *Under the Sea: Archaeology and Palaeolandscapes of the Continental Shelf*, Coastal Research Library 20, DOI 10.1007/978-3-319-53160-1_8

yield a more detailed and diverse picture of the exploitation by Mesolithic people of the resources available in the surrounding landscape and, as a consequence, of Mesolithic society itself.

Unfortunately, many submerged Mesolithic settlements are currently undergoing active erosion as a result of several different factors (Skriver et al. in press), and one that has been the focus of particular attention is so-called 'eelgrass death'. Eelgrass (*Zostera marina*) prefers a muddy seabed, where its roots form an entangled mass that lies like a protective blanket within the sediment and prevents this from becoming exposed (Rasmussen 1977; Fischer 2011). In the 1930s, a disease epidemic struck, which led to eelgrass disappearing from large parts of the seabed, leading to heavy erosion and significant sand movement (Rasmussen 1977). Subsequently, pollution has also become a contributory factor to the turbid water that reduces the amount of sunlight reaching the plants, thereby preventing them from photosynthesising efficiently and ultimately causing die-back (Krause-Jensen and Rasmussen 2009).

This chapter deals with the settlement at Hjarnø Sund, which is currently also undergoing erosion. The cause here is probably not eelgrass death, but more likely changes in local currents that have resulted in the disappearance of the protective sand layer. The find-bearing gyttja deposits have become exposed and eroded by wave action, and artefacts of organic material preserved within the gyttja have been washed out and degraded.

## 8.2 Topography

The settlement of Hjarnø Sund is located on the western shore of the small island of Hjarnø in Horsens Fjord (Figs. 8.1and 8.2). Horsens Fjord is a c. 15 km-long fjord arm that extends inland westwards from the east coast of Jutland to the town of Horsens. The large number of prehistoric settlements along the coast, some now located below water, shows that the fjord was a good place to live in the Mesolithic. Remains of 37 submerged settlements have been recorded to date (Fig. 8.3).

The Hjarnø Sund settlement has a classic location relative to the fishing place model developed by Anders Fischer (Fischer 1993, pp. 19–24). It lies alongside a tidal channel that runs between the island and the mainland. Static fish traps could have been set up here, and hunting in the extensive mainland forests was only a short 600 m trip by dug-out boat.

Due to the rise in sea level that took place in Atlantic times, the settlement now lies under water, at a depth of 0.5–2 m at high water, with parts of it becoming fully exposed at extreme low water.

## 8.3 History of Investigation

The existence of the settlement has been known for many years; the first archival notes date from 1957, and the locality has long been a favoured collecting site for amateur archaeologists due to its easy access and richness in artefacts. Flint artefacts, in particular, have been recovered, but there have also been finds of antler and bone, including a human femur and cranium, as well as several antler axes.

In 2008, it was observed that gyttja deposits had become exposed in an area previously covered by sand. Within these, artefacts of wood and antler could be observed, including several antler axes as well as the cranium of a dog (Fig. 8.4). This led to the local museum being contacted and, in a subsequent joint initiative, Moesgård Museum, Horsens Museum and the University of Aarhus applied to the Danish Agency for Culture for funding to mount a small investigation of the site in 2010. This was granted and the results of this investigation led to a further three investigations in subsequent years

**Fig. 8.1** The location of Horsens Fjord and the island of Hjarnø

(again funded by the Danish Agency for Culture). Further to this, the site has been regularly monitored in order to recover artefacts that have been washed out of the gyttja during storms (Fig. 8.5).

## 8.4 Method

Investigations at the site were carried out partly by divers, with areas being excavated on a grid network using an induction dredge. In order to gain an overview of the extent of the site, both vertically and horizontally, as well as the composition of the sediments, the relative rise in sea level and the

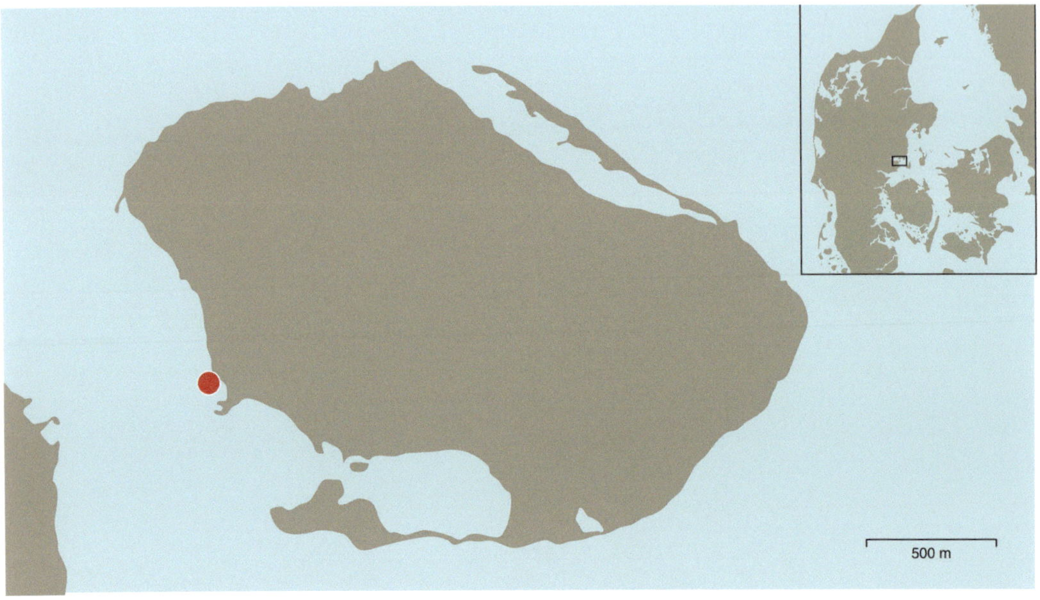


**Fig. 8.2** Location of the settlement of Hjarnø Sund on the island of Hjarnø

**Fig. 8.3** Map showing Mesolithic sites and finds in Horsens Fjord

likely settlement structures and features, a number of sediment cores were also taken manually in parallel with the excavation work. In the first year, sediment cores were taken at 20 m intervals in order to provide an immediate overview of the situation. Subsequently, a sampling strategy was implemented whereby the cores were taken at 2 m intervals in areas that appeared to be of interest based on the first series of wider-spaced cores.

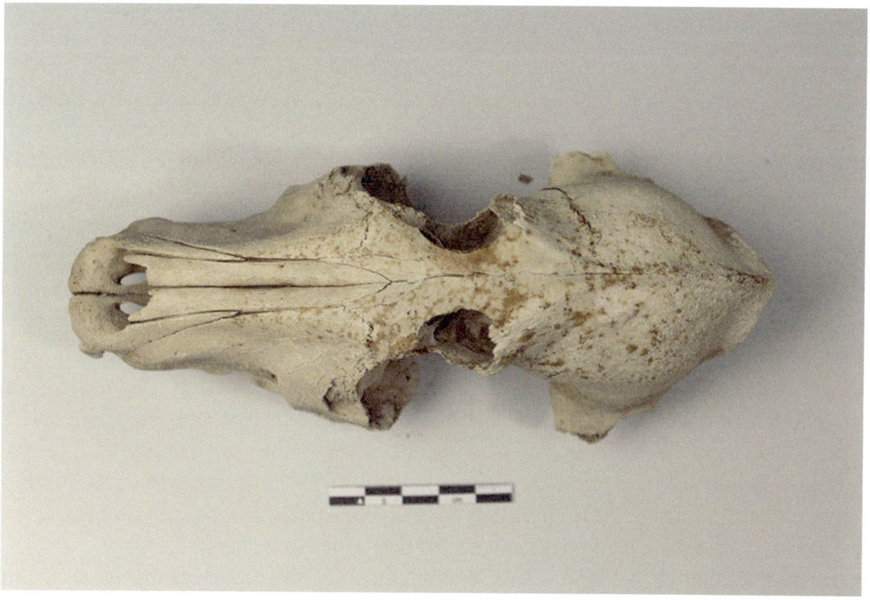

**Fig. 8.4** Cranium of a dog found in 2008 (Photo P Borup)

**Fig. 8.5** Wooden artefacts in the eroding gyttja (Photo P Borup)

Following excavation, the plant macro-remains and the bones recovered from the site were subjected to analysis, and radiocarbon dates were obtained for selected layers and a small number of artefacts of wood and bone.

## 8.5 Site Description

The investigations carried out to date show that, in addition to deep gyttja deposits, there are at least two shell deposits or middens at the site (Fig. 8.6). The first of these was found accidentally when sand dredging in order to reach the original clay surface. It was covered by up to 10 cm of sand and contained organic material, such as charcoal and bones of both fish and mammals, in addition to worked flint artefacts, including 11 transverse arrowheads. The shells were of oysters (*Ostrea edulis*), cockles (*Cerastoderma edule*) and periwinkles (*Littorina littorea*). The cockle and oyster shells proved not to be paired, as probably would have been the case had they died in a natural shell bank. Sediment data from cores indicate that the shell midden covers an area of at least 120 $m^2$, but only 2 $m^2$ has so far been excavated. The second shell deposit or midden excavated was not covered by sand but was visible on the seabed and undergoing heavy erosion. A 10 m-long trench through this midden, together with the core data, showed that it was situated in a steeply sloping area. At the top of this slope, the shells lay directly on clay, whereas lower down they lay over gyttja. This feature was stratified, being comprised of a layer of oyster shell up to 20 cm-thick, which at some point had become covered with a thin (c. 5 cm thick) layer of cockle shells. Like the first midden, it also contained flint artefacts, in addition to an abundance of fish bones and charcoal.

Given the composition of the shell layers, and the quantity and nature of the artefacts present, the shell deposits are considered to be the result of human activity rather than representing a natural shell bank. An area of 23 $m^2$ of these layers has been excavated, but it is still uncertain whether they are actual kitchen middens, with activity areas, or a refuse dump layer in a near-littoral zone. One possibility, however, is that they represent a kitchen midden that has been redeposited as a consequence of a subsequent rise in sea level. Three radiocarbon dates obtained from the first midden suggest that there were at least two periods when shell midden material accumulated, one during the Late Kongemose/Early Ertebølle culture and the other during the Middle Ertebølle culture. The relationship between the shell middens and the gyttja deposits has not yet been satisfactorily resolved, but the

**Fig. 8.6** Shell midden during excavation (Photo J Frederiksen)

sediment core data suggest that the earliest accumulation of shell material took place close to the area where the gyttja is located.

The shell deposits in the first midden later became inundated due to a rise in sea level, after which a primary gyttja deposit developed over them. They therefore lie protected beneath the gyttja and a sand/pebble layer and are not currently subject to erosion, unlike the second midden, which is clearly eroding.

## 8.6 The Finds

In addition to the flint artefacts mentioned above, there are also finds of wood and antler. Antler artefacts mainly comprise pressure flakers and axe heads. Nine axes have been recovered to date, all of the early type with the handle hole by the burr, which can be dated typologically to the Kongemose and Early-Middle Ertebølle cultures. Three of the axes were found with a wooden handle in situ and four are ornamented (Fig. 8.7). Decorated antler axes are known from several other Mesolithic localities in Denmark (Andersen 2013).

Two wooden paddles deserve special attention. These were found within a few minutes of each other and only 5 m apart. They lay exposed on the gyttja and it seemed obvious that they had been uncovered recently and would have been washed away in the course of a very short time (days or weeks depending on weather conditions) had they not been discovered (Fig. 8.8a, b).

One of the paddles has two thirds of the blade preserved: this was originally c. 20 cm long, c. 24 cm wide and c. 0.8 cm thick. Cut marks on the surface show that the paddle may have been used as a chopping board at some time. The other paddle is preserved in its full length and has half of its blade intact. It measures 103 cm in length, of which the blade takes up 21 cm. The blade is c. 22 cm wide and only 1 cm thick. This paddle has been radiocarbon dated to c. 4700–4540 cal BC. (Lab. no. AAR 16091), i.e. corresponding to the Middle Ertebølle culture.

**Fig. 8.7** Antler axe with wooden handle (Photo P M Astrup)

**Fig. 8.8** **a**, **b** Wooden paddles as found in the eroding gyttja (Photo - Peter Moe Astrup)

In spite of the presumed contemporaneity of the two paddles, they are not completely identical in form. One is of the type with a 'heart-shaped' paddle blade (Tybrind Vig type), while the other is of the 'spade-shaped' Satrup type (Andersen 2013, 169). These two types of paddle blade have been found at several western Danish Ertebølle settlements and there is one previous record from further west in Horsens Fjord, at Haldrup Strand, c. 7 km from Hjarnø (unpublished).

Despite degradation of their surfaces, both paddles show traces of paint on the side that was in contact with the gyttja deposits and had therefore not been exposed. The motifs were not completely identical, but the composition and painting technique are approximately the same. The lower part of both blades is painted solid black and above this are three horizontal parallel lines. On one paddle these lines are 0.6 cm wide and curve upwards from each edge towards the middle of the blade. On the other, they take the form of three separate c. 1.3 cm wide bands (Fig. 8.9a, b).

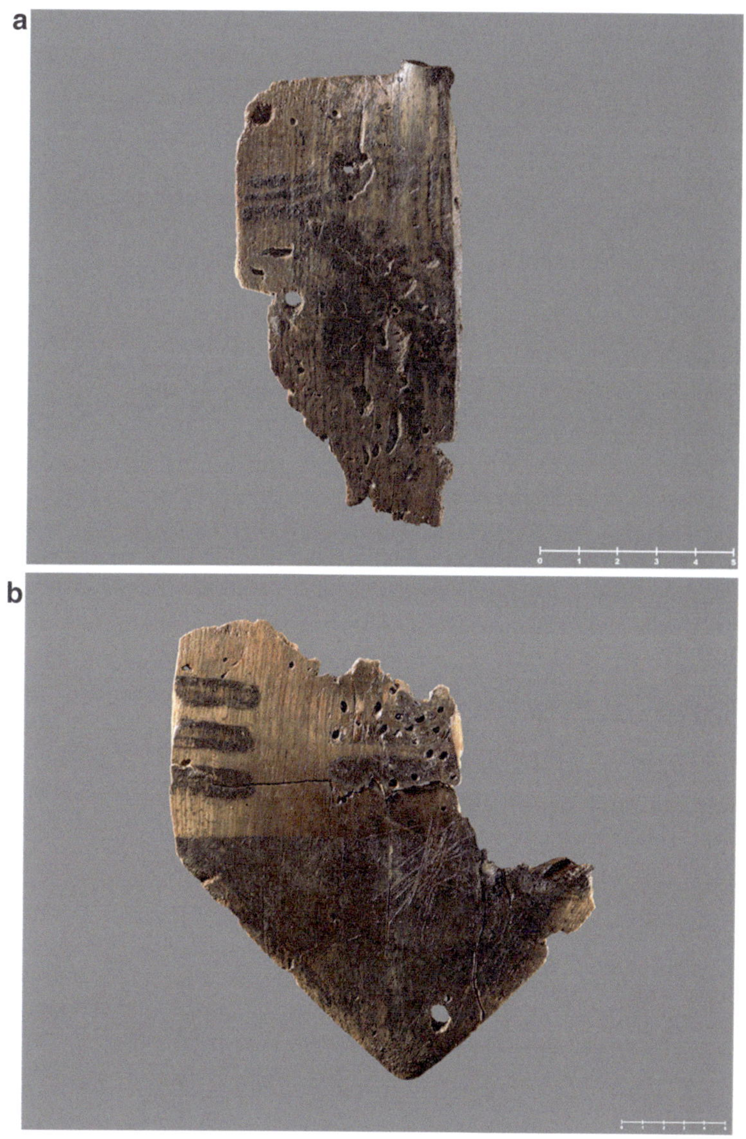


**Fig. 8.9 a, b** The two ornamented paddles from Hjarnø (Photo D Butler)

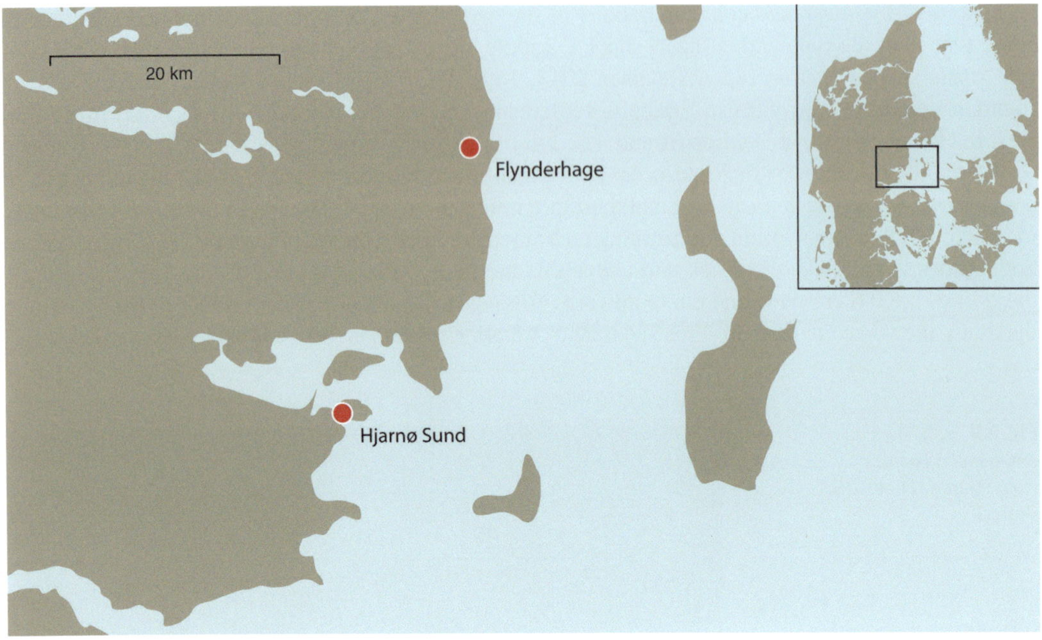


**Fig. 8.10** Location of the settlements of Flynderhage and Hjarnø Sund

Decorated paddles from the Ertebølle culture are not an unknown phenomenon in Denmark, but they are far from a common find. Some years ago, a paddle with very similar decoration to that on one of the Hjarnø paddles was found during an excavation of a kitchen midden at Flynderhage in Norsminde Fjord (c. 25 km from the Hjarnø site) (Fig. 8.10). This paddle has four narrow black lines painted above a similar solid black lower blade. Only the number of lines appears to differ in the motifs evident on these two paddles (Fig. 8.11).

The best-known decorated paddle blades are, however, from Tybrind Vig in western Funen. Four of the 13 finds of paddles from this site have varying ornamentation in the form of fine patterns made up of dots and lines executed in sunken relief in which traces have been found of an ochre-rich pigment. There are therefore clear differences between the paddles from the Horsens Fjord area and those from Tybrind Vig, both in terms of motif and technique (Fig. 8.12).

## 8.7 Art or Identity?

What was the significance of this ornamentation? The decoration evident on the paddles is typically classified as 'prehistoric art'. This is a broad term that also includes items such as small amber figures, ornamented pottery vessels and decorated antler artefacts.

As for its significance, there has been some discussion of its role as decoration on magical objects (Jensen 2001, 213–221) and as a symbol related to an owner or tribe (Andersen 2013, 184).

Based on his anthropological studies, M. Wobst, who has focussed on relations between style and information, considers that 'artefacts' which travel long distances are also those which carry symbols bearing a message relating to group affiliation and which are involved in the processes relating to the maintenance of boundaries (Wobst 1977, 330).

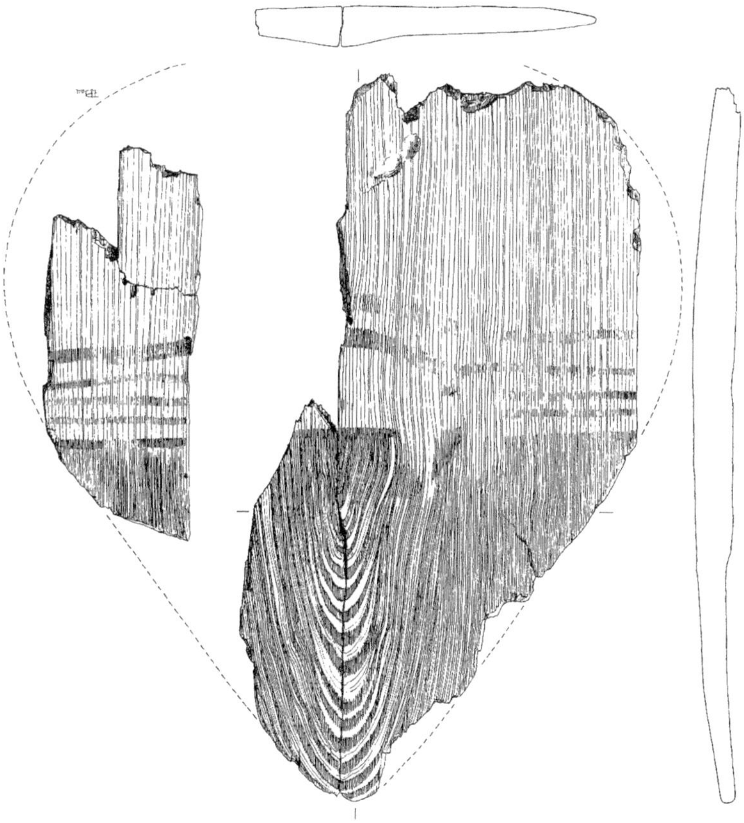

**Fig. 8.11** Ornamented paddle blade from Flynderhage. After Andersen (2013) (Drawing F Bau)

Paddles are clearly an artefact type that, potentially, travels far and wide across the landscape, being used during voyages in fjords and out on the open sea. In this respect, it is remarkable that two almost identically painted paddles were found c. 25 km apart, at Hjarnø and Flynderhage, while the paddle blades from Tybrind Vig (located 50 km south of Hjarnø Sund), although typologically identical to the Hjarnø and Flynderhage paddles, are painted in ways that are completely different both technically and pictorially. The question is whether this situation arises from territorial or kinship circumstances, or is simply a consequence of the fact that these two Mesolithic communities had different ways of expressing themselves in artistic terms. The ornamented antler objects should also be taken into consideration in this respect. Does the decoration they bear relate to the group, the family or perhaps the individual? At present, however, we are unable to cast further light on this intriguing aspect of Ertebølle society.

## 8.8 Conclusions and Future Perspectives

The Hjarnø Sund investigations have produced a wealth of fascinating and useful results. The primary aim was to gain an overview of the extent and morphology of the site, and of the erosion taking place, as well as to assess its scientific potential and, on this basis, to evaluate how best to approach work at the locality in the future. As demonstrated here, the results show that the site encompasses a large body of archaeological material, probably of unique content (the kitchen/shell midden), which is unfortunately being destroyed by erosion.

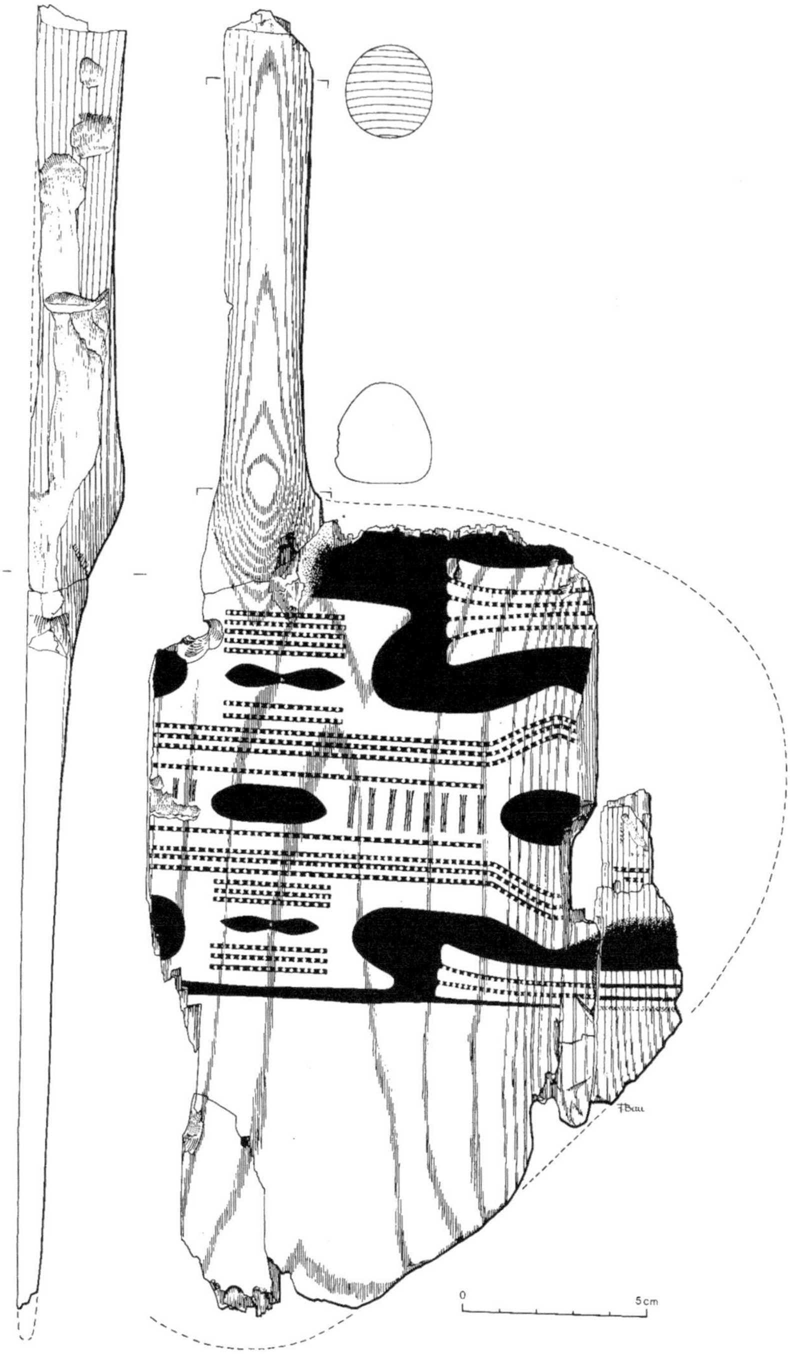


**Fig. 8.12** Ornamented paddle blade from Tybrind Vig. After Andersen (2013) (Drawing F Bau)

Given the presence of preserved organic material, it seems very likely that unique artefacts of wood, antler and bone will be uncovered at the site in the future. Similarly, the preserved botanical remains can be expected to provide information on the use of the landscape and the exploitation of plant resources for food and other purposes.

Previous investigations of submerged settlements with well-preserved organic material, where these kinds of artefacts survive, provide hope that preservation will also be possible in the future and underline the importance of either preserving these erosion-threatened sites or subjecting them to scientific investigation before the evidence disappears forever.

## References

Andersen SH (2009) Ronæs Skov. Marinarkæologiske undersøgelser af kystboplads fra Ertebølletid, Jutland archaeological society publications, vol 64. Jutland Archaeological Society, Højbjerg

Andersen SH (2013) Tybrind Vig. Submerged Mesolithic settlements in Denmark, Jutland archaeological society publications, vol 77. Jutland Archaeological Society, Højbjerg

Fischer A (1987) Argusgrunden – en undersøisk boplads fra jægerstenalderen. In: Kristiansen K, Fischer A (eds) Fortidsminder og kulturhistorie, Antikvariske studier, vol 8. Skov- og Naturstyrelsen, København

Fischer A (1993) Stenalderbopladser i Smålandsfarvandet. En teori afprøvet ved dykkerbesigtigelse. Miljøministeriet. Skov- og Naturstyrelsen, København

Fischer A (2011) Stone age on the continental shelf: an eroding resource. In: Benjamin J, Bonsall C, Pickard C, Fischer A (eds) Submerged prehistory. Oxbow Books, Oxford, pp 298–310

Jensen J (2001) Danmarks Oldtid. Stenalder 13.000–2.000 f. Kr. Gyldendal, København

Krause-Jensen D, Rasmussen MB (2009) Historisk udbredelse af ålegræs i danske kystområder. Danmarks Miljøundersøgelser, Aarhus Universitet – Faglig rapport fra DMU 755

Rasmussen E (1977) The wasting disease of eelgrass *Zostera marina* and its effect on environmental factors and fauna. In: Mcroy CP, Helfferich C (eds) Seagrass ecosystems. A scientific perspective. Marcel Dekker, New York, pp 1–51

Skaarup J (1983) Submarine stenalderbolpadser i Det sydfynske øhav. In: Kristiansen K, Tønnesen A (eds) Fortidsminder og Bygningsbevaring, Antikvariske studier, vol 6. Fredningsstyrelsen, København

Skaarup J, Grøn O (2004) Møllegabet II. A submerged Mesolithic settlement in southern Denmark, BAR international series, vol 1328. Archaeopress, Oxford

Skriver C, Galili E, Momber G (in press) Threats to the submerged cultural heritage. In: Fischer A, Pedersen L, Bailey G (eds) Oceans of archaeology

Sørensen SA (1996) Kongemosekulturen i Sydskandinavien, Egnsmuseet Færgegården

Uldum O (2011) The excavation of a Mesolithic double burial from Tybrind Vig, Denmark. In: Benjamin J, Bonsall C, Pickard C, Fischer A (eds) Submerged prehistory. Oxbow Books, Oxford, pp 15–20

Wobst HM (1977) Stylistic behaviour and information exchange. In: Cleland CE (ed) For the director, research essays in honor of James B. Griffin, Anthropological papers, vol 61. Museum of Anthropology, University of Michigan, Ann Arbor

# Chapter 9
# Fished up from the Baltic Sea: A New Ertebølle Site near Stohl Cliff, Kiel Bay, Germany

**Julia Goldhammer and Sönke Hartz**

**Abstract** In the coastal waters of Kiel Bay near the village of Strande (district of Rendsburg-Eckernförde, Schleswig-Holstein) divers unexpectedly came across trunks of fallen oak trees at 6 m depth, which led to the discovery of a new submerged late Mesolithic site. During a preliminary excavation in 2012, research divers uncovered a well-preserved coastal site, consisting of several organic sediment and silt layers with a large number of stone artefacts and organic finds. Wooden objects, plant remains, bones of several marine and freshwater fish, marine and terrestrial mammals, water birds and fragmented human bones were found. Tree ring dating, radiocarbon dates of leister prongs and human bones, and the artefact inventory pinpoint the site to the pre-pottery Ertebølle phase (5450–4750 cal BC). Sites of this time period are of particular interest as they are still rare in the south-western Baltic Sea area, where only very few sites have been examined in detail. To evaluate the extent of the organic sediment and silt layers and their potential for preserving more finds, a survey project was executed in summer 2014 over a wider area around the excavation trenches. This established a high potential for the recovery of additional finds and structures in the surrounding area, and further investigations at Strande are planned.

## 9.1 Introduction: The Discovery

The Strande site LA 163 was discovered by two divers in October 2011 when they spotted tree trunks in the middle of a sea grass meadow at 6 m water depth (Figs. 9.1 and 9.2). Under a layer of sand between the tree trunks they found flint artefacts and pieces of antler. The divers reported their discovery to Dr. Sönke Hartz (State Archaeological Museum of Schleswig-Holstein). Inspections of the recovered finds and underwater photographs indicated that the site dates to the Late Mesolithic Ertebølle culture. Ertebølle sites are known from the western and south-western Baltic. Named after the site 'Ertebølle' in the Danish Limfjord (Madsen et al. 1900) it describes a hunter-gatherer society living in close connection to the coast and exploiting marine and terrestrial resources. During the Late

J. Goldhammer (✉)
Lower Saxony Institute for Historical Coastal Research, Niedersächsisches Institut für historische Küstenforschung, Viktoriastr. 26/28, 26382 Wilhelmshaven, Germany

Bohusläns Museum, Museigatan 1, 45119 Uddevalla, Sweden
e-mail: julia.goldhammer@gmail.com

S. Hartz
State Archaeological Museum of Schleswig-Holstein, Stiftung Schleswig-Holsteinische Landesmuseen Schloss Gottorf, Schlossinsel 1, 24837 Schleswig, Germany
e-mail: soenke.hartz@schloss-gottorf.de



G.N. Bailey et al. (eds.), *Under the Sea: Archaeology and Palaeolandscapes of the Continental Shelf*, Coastal Research Library 20, DOI 10.1007/978-3-319-53160-1_9

**Fig. 9.1** Discovery of four collapsed tree trunks and organic sediment layers with late Mesolithic finds in autumn 2011. On this picture approximately 2 m of each tree is visible (Photograph: G. Lorenz)

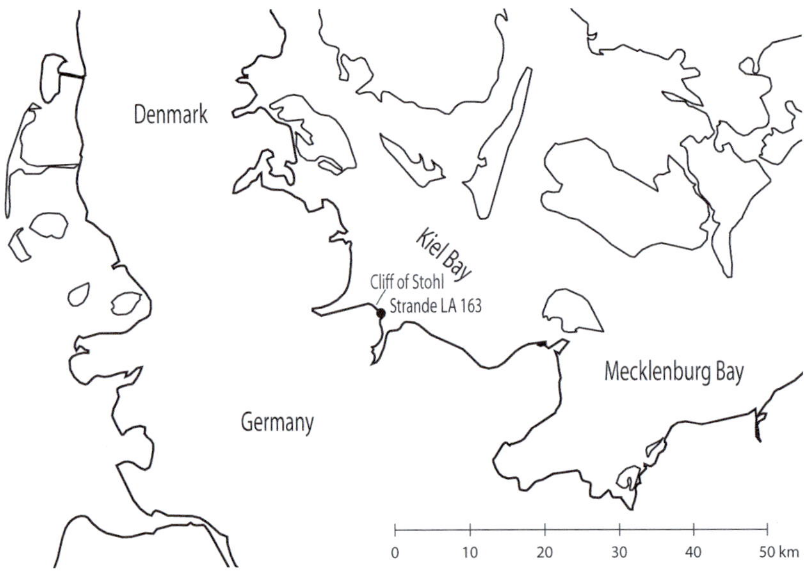


**Fig. 9.2** Geographical setting of the Strande site (Illustration: J. Goldhammer)

Mesolithic, Ertebølle groups were in contact with Neolithic groups from the North German Plain and were influenced by groups from the Fennoscandian area to introduce pottery (Piezonka 2008).

Around the Strande site some single finds were discovered earlier by people walking on the beach or diving in this area. Such finds were not expected, because the shallow coastal waters offshore of Stohl Cliff have been continuously exposed to waves, making the discovery of a well preserved site with in situ stratigraphy a surprise. In order to gain a more comprehensive overview of the site and its potential, in June 2012 test trenches were excavated by scientific divers and students from Kiel

University, coordinated by the authors and supported by the State Archaeological Department of Schleswig-Holstein.

## 9.2 Mesolithic Subsistence: The Excavated Area

The preliminary excavation in Summer 2012 uncovered a well-preserved coastal site, consisting of organic sediment and organic silt layers with a large number of stone artefacts and organic finds (Goldhammer and Hartz 2013; Glykou et al. 2014: 91–95). An area of 5 m² around the tree trunks was excavated, and test pits with a dredge and coring with a hand auger were carried out (Fig. 9.3). The stratigraphy is characterized by organic layers on Pleistocene till. Different layers of organic

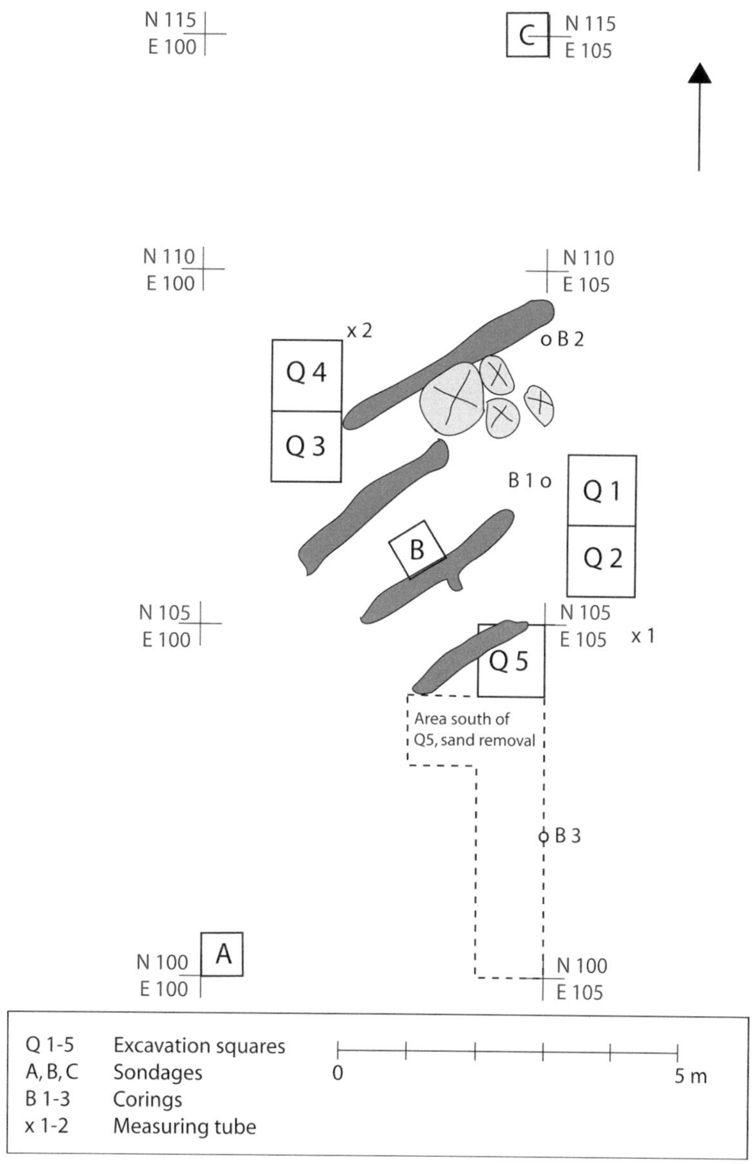


**Fig. 9.3** Sketch of 2012 excavation trenches around the tree trunks (Illustration: J. Goldhammer)

**Fig. 9.4** Trench 2 (Squares 3 and 4), part of Profile E 101, N107-109, 120 cm are displayed (Photograph: G. Lorenz)

sediments on top of the till were found: they differ in their consistency and the organic components. Most of the finds were found in the upper organic sediment layer. Above that, organic silt was deposited. In the lower deposits, molluscs and shell fragments were observed of common periwinkle (*Littorina littorea*), the common cockle (*Cerastoderma edule*) and the blue mussel (*Mytilus edulis*) (Fig. 9.4).

The organic finds included wooden objects, plant remains, bones of marine and freshwater fishes, marine and terestrial mammals and water birds. Archaeozoological analyses identified most of the fish bones as cod (*Gadus morhua*). The small size of the cod bones (indicating individuals of 20–30 cm in length) and the spectrum of the other fish species like Atlantic mackerel (*Scomber scombrus*), garfish (*Belone belone*), European eel (*Anguilla anguilla*), viviparous eelpout (*Zoarces viviparus*), short-spined sea scorpion (*Myoxocephalus scorpius*), Righteye flounders (Pleuronectidae) and carp (Cyprinidae) give evidence for shallow water near the site. Some fish bones were charred, which may indicate fishing and roasting of the catch at or near the site.

Moreover, cut marks on roe deer bones (*Capreolus capreolus*) also suggest food preparation at this site. Tools made of red deer (*Cervus elaphus*) antler were found. Two artefacts could be interpreted as a punch, a tool used for the production of lithic blades. A tool made from the tusk of a wild boar (*Sus scrofa*) shows use-wear and might have been used for fur or leather processing. A wild boar fibula was shaped to a bone point. Other terrestrial mammal bones originate from domestic dog (*Canis lupus cf. familiaris*), wildcat (*Felis silvestris*) and bank vole (*Clethrionomys glareolus*). Besides fish, the bone inventory also contains other marine and freshwater species like beaver (*Castor fiber*), seal (*Phocidae*), otter (*Lutra lutra*), and birds including duck, geese, swan (Anatidae), red-necked grebe (*Podiceps grisegena*), red-throated diver (*Gavia stellata*), and other unidentified species. The wide range of terrestrial and marine animals shows that Ertebølle people hunted land animals like wild boar, roe deer and red deer but also exploited marine food sources.

Of particular interest are the fragmented human bones that were found in the lowest layer above the Pleistocene till: two jawbone pieces with molar teeth and two single teeth were recovered (Fig. 9.5). Human-bone samples were taken and prepared for analysis of $^{13}C$ and $^{15}N$ isotopes and ancient DNA (aDNA), as well as for radiocarbon dating. An osteological examination by Peter Tarp, from the team of Prof. Dr. J.L. Boldsen at Odense University, Denmark, showed that one of the teeth belongs to a

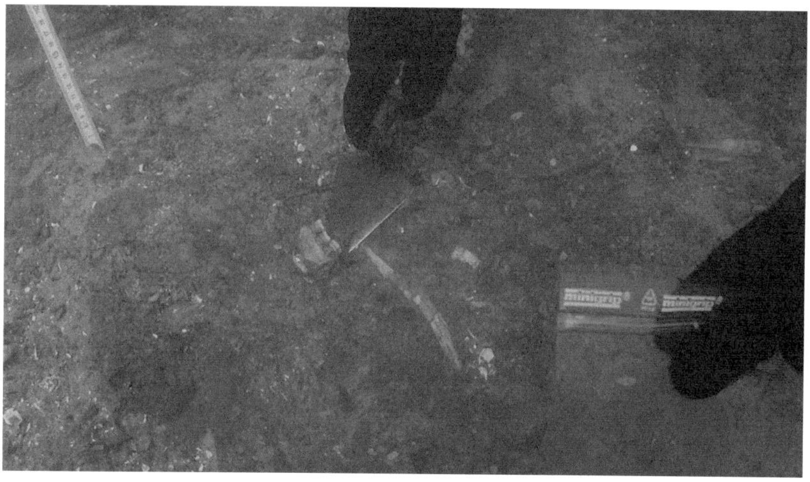

**Fig. 9.5** Recovery of a human jawbone fragment with teeth found at a depth of 6 m under water (Photograph: C. Howe)

different individual from the jaw bone pieces. The jaw bone is thought to belong to a female in her 20s. The second individual is likely older.

The analysis of the plant remains revealed a wide range of trees, such as hazel (*Corylus*), lime (*Tilia*), oak (*Quercus*), whitethorn (*Crataegus*), alder (*Alnus*), elm (*Ulmus*), and apple (*Malus*). Hazelnut shells and acorns were frequently found. Several of the hazelnut shells were charred, which can be interpreted as a sign of roasting, and thus indicate their use as food. Interestingly, in one of the excavated squares, a large bracket fungus was found.

The numerous lithics show the production and use of different kinds of flint tools. Core axes, borers, blade tools and transverse arrowheads were found. In particular, waste material and edge rejuvenation flakes gives evidence for the production and reshaping of core axes near the site. Some younger flake axes only appeared in blunted or patinated examples. These tools were collected near the excavation area from the upper sand layer. The condition and find circumstances of core and flake axes are important for the typological dating of the site. Besides lithics, wooden objects were also found. During survey dives and the excavation, composite tool parts like leister prongs were found in and under the covering sand layer. Small wooden fragments likely of a lime tree log boat were also detected on the surface (Fig. 9.6, determined by Dr. S. Klooß, Kiel, Germany).

Tree-ring dating, radiocarbon dating, and the lithic inventory place the site in the pre-pottery Ertebølle phase (~5450–4750 cal BC, Hartz and Lübke 2006). Sites from this period are rare in the south-western Baltic Sea area, where only a few sites, such as Jäckelberg-Nord and Rosenfelde, have been excavated and examined in detail. The trees found at Strande probably fell down around the year 5390 BC. Four samples of three tree trunks were taken by divers and analysed by Dr. K.-U. Heußner from the Dendrochronology Lab of the German Archaeological Institute (DAI) in Berlin, Germany (Table 9.1).

A leister prong (Fig. 9.7), found below the upper sand layer, dates to 5215–5020 cal BC (6160 ± 30 BP, KIA46461). In contrast, a radiocarbon date for one human-bone fragment at 5470–5340 cal BC (6380 ± 35 BP/6455 ± 30 BP, KIA 48437) appears too old. This discrepancy may be explained by the reservoir-effect as the individual's nutrition was mainly based on marine food as indicated by the $\delta^{13}C$ value of $-16.26 \pm 0.36/-13.47 \pm 0.41$ (Fischer et al. 2007). Hazelnut shells from the same layer as the human-bone fragment date to 5215–4981 cal BC (6140 ± 40 BP, Poz-64205). Interestingly, a second leister prong, which was also discovered under the upper sand layer, dates slightly younger to 4990–4795 cal BC (6000 ± 40 BP, KIA46711) and thus extends the chronological range of the site. The radiocarbon dates match the typological dating of the artefacts.

**Fig. 9.6** Probable fragment of a dugout canoe made of lime wood (*Tilia*), determined by Dr. S. Klooß, Kiel, Germany (Photograph: S. Hartz)

**Table 9.1** Dendrochronological dating of wooden samples from Strande LA 163. Analyses by Dendrochronology Lab of DAI Berlin, Germany. The outer edges of all samples represent probably the boundary between corewood and sapwood

| Lab.Nr. | Type of wood | Sample | Beginning | End | Date of death | Remark |
|---|---|---|---|---|---|---|
| C | | | | | | |
| 66133 | Quercus | Tree 1/1 | 5487 BC | 5413 BC | 5393 BC | ±10 |
| 66134 | Quercus | Tree 1/2 | 5490 BC | 5412 BC | 5392 BC | ±10 |
| 66135 | Quercus | Tree 2 | 5482 BC | 5014 BC | 5396 BC | Around/after |
| 66136 | Quercus | Tree 4 | 5468 BC | 5400 BC | 5380 BC | Around/after |

S. Hartz and H. Lübke suggested a four-phase chronology for the Late Mesolithic in the south-western Baltic (Hartz and Lübke 2005, 2006; Hartz et al. 2014). In summary, the older phases, Jäckelberg and Rosenfelde, are characterised by lack of ceramics and only a few flake axes in the later phase. The two younger phases, Jarbock and Timmendorf, are characterised by the production and use of pottery and the frequent appearance of flake axes. At Strande LA 163 no pottery and flake axes

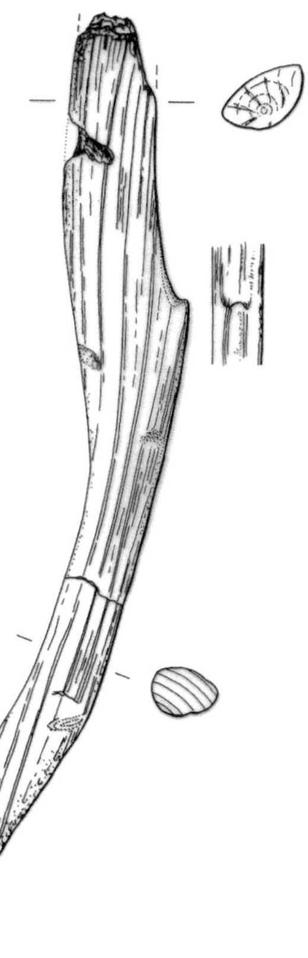

**Fig. 9.7** Leister prong made of Maloideae wood, determined by Dr. S. Klooß, Kiel, Germany (Drawing: J. Freigang)

were found in the cultural layers. Flake axes are only known from the covering sand layer, and they are patinated with worn edges indicating erosion and displacement from a younger settlement phase, but it is not possible to determine from the condition of the artefact its age or provenance. Taken together, the find layers of Strande can be dated with confidence to ~5200–4800 cal BC, and fit well with dates for the pre-pottery Ertebølle phases of 5450–4750 cal BC (Hartz and Lübke 2005, 2006; Hartz et al. 2014).

## 9.3 Potential and Perspectives: First Results of New Prospection

The excavation results suggest that the find layers pinch out to the east and north, though the sediment layers become thicker to the west and to the south. To investigate the possible existence of more extensive deposits and archaeological material, we carried out additional survey around the 2012 excavation area in August 2014 with four scientific divers and 160 survey dives.

Starting from the four fallen tree trunks, the surrounding area was inspected using transects. A 50 m-long yellow belt strap was set up with heavy sinkers attached to each end. This belt was first extended from the excavation area to the west, and then transferred to the south and so on until all cardinal directions were examined. Every 5 m along the transect belt, we dug a sondage. We used a diver propulsion vehicle to flush away the covering sand and expose underlying sediments, measured the thickness of the covering sand layer, and described, photographed and mapped the position of the sondage and the type of underlying sediment. Furthermore, we documented any artefacts present with particular attention to the degree of patination and edge sharpness of flint artefacts. From this information we were able to identify sediments with the potential to yield further prehistoric finds and structures.

The result of the prospection work gives a clear picture. The distribution of sediment types gives definitive indications of the degree of erosion or protection of material and the nature of the former landscape around the excavation trenches. Sediments remaining in the western, south-western and southern sectors comprise sandy layers with an organic component as well as organic silt layers (Fig. 9.8). Moreover, flint artefacts with sharp edges were recovered, indicating that more material in a good state of preservation may be found in these layers.

Additional locations with organic sediments were identified in the wider area, for example 150 m east of the fallen tree trunks and around 400 m to the south. It remains to be resolved whether the finds washed up on the present-day shore originate from the Strande LA 163 site or from other spots in the vicinity.

Seven years of research in the Wismar Bight following the accidental discovery of a single Mesolithic site (Hartz et al. 2014) have demonstrated the value of closer inspection using geoarchaeological surveys in revealing further settlement remains. So far, ten sites from various Mesolithic to early Neolithic time periods have been identified (Hartz et al. 2014, p 80, Fig. 1). The area of the Strande site in Kiel Bay seems to have a comparable high potential to reveal further sites from different periods.

## 9.4 Site and Surroundings: Landscape and Sea Level Rise

The 7th to 5th millennium BC is characterized by substantial changes of the landscape around the Baltic Sea basin. Due to sustained sea level rise, land was lost and people had to adapt to new coastlines. The Ancylus Lake stage was the last freshwater stage of the Baltic basin (Andrén et al. 2011). At that time Kiel Bay was still dry terrain. Until recently, geologists thought that the Ancylus Lake drained via a large river into the Kattegat. However, the picture seems to be more complex (Feldens and Schwarzer 2011). New findings suggest that a saltwater inflow through the Kattegat and the Great Belt occurred at around 7500 BP (ca. 6400 cal BC, Rößler et al. 2009) resulting in rising sea level and a gradual landward shift of the coastline.

From Mecklenburg Bay a transgression contact is known from 5500 cal BC. It was identified by 'a sudden start of records from pollen and macro fossil taxa of marine or perimarine origin' at Timmendorf-Nordmole II at a depth of 583 cm below present sea level (Hartz et al. 2014, pp. 115–119).

The Strande LA 163 site lies in a wide area of shallow water with a water depth of 4–6 m. The 2012 excavation described here also revealed that the sediment layers to the south and west are thicker than in the east and north. This might be explained by an uneven Pleistocene clay surface with depressions and elevations.

Taken together, these observations suggest that the site was originally located on the shore of a shallow water body, such as a bight or a lagoon, but additional geological and geophysical investigations will be needed to corroborate this reconstruction of the prehistoric landscape.

## 9.5 Conclusion

Further investigations are planned to determine when the site was used by coastal hunter-fisher societies and whether there is evidence of multiple independent visits. The inundation of land in the Baltic Sea area changed the human habitat within a few generations and forced people to move further

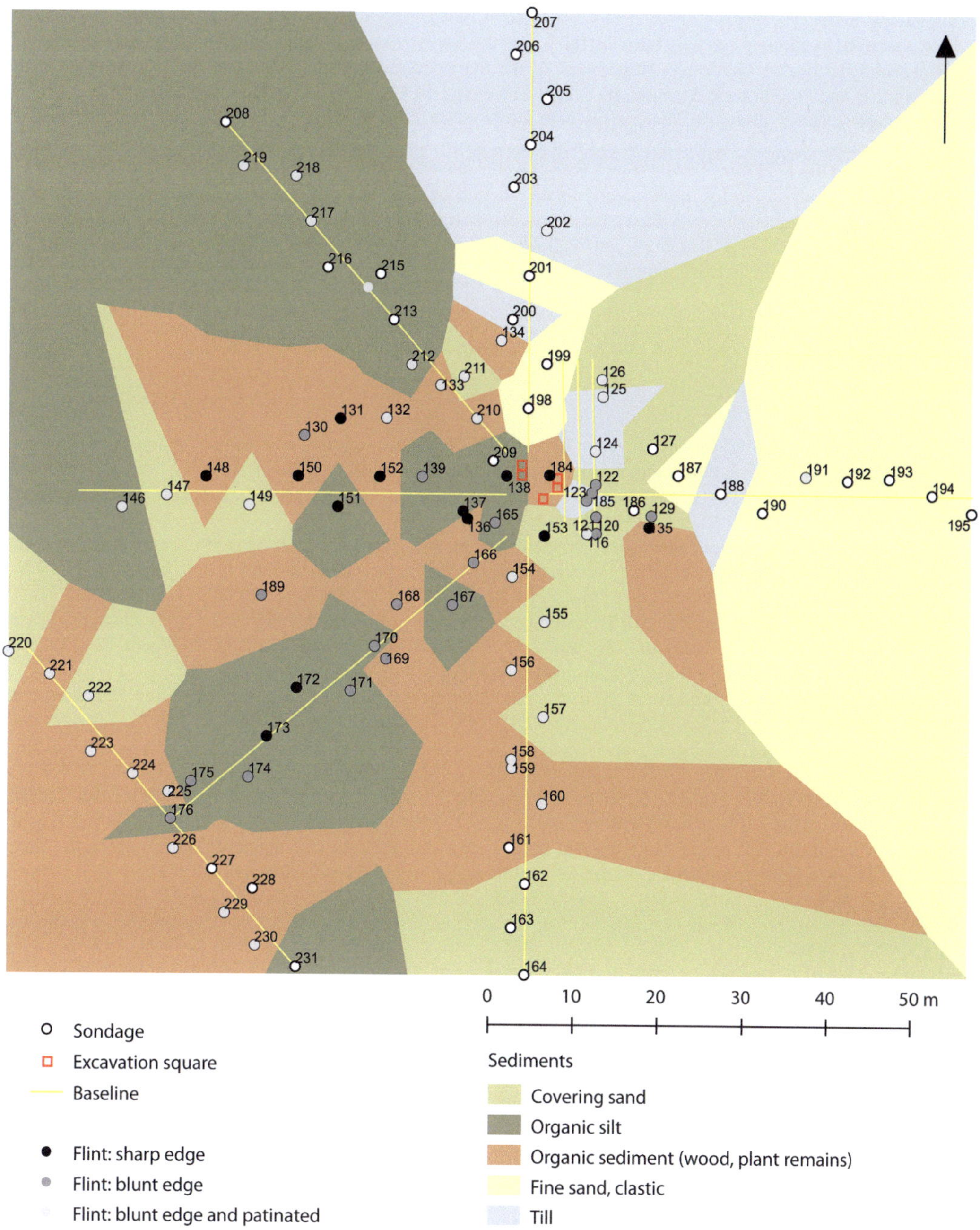


**Fig. 9.8** Assumed distribution of sediment types below covering sand modelled with Voronoi cells (Illustration: J. Goldhammer, R. Kiepe, NIhK)

inland. Once new geological and geomorphological analyses have been undertaken, the Strande site could become a case study of the way humans adapted to a changing coastal landscape. Furthermore, there is considerable potential to locate additional Mesolithic remains in the area around Strande LA 163. With new insights into the way of life in Kiel Bay it will be possible in due course to compare technology, economy and diet to the regions of Wismar Bay and the Danish Islands and thus to reconstruct pathways and networks of contact in the late hunter-gatherer period.

**Acknowledgements** The prospection of 2014 was funded by the German Science Foundation (DFG), the excavation of 2012 was enabled by funding and help from the following: the State Archaeological Department of Schleswig-Holstein; the Archaeological Society Schleswig-Holstein e. V.; the State Archaeological Museum of Schleswig-Holstein; the Institute of Pre- and Protohistoric Archaeology at Kiel University; the Centre for Scientific Diving, Kiel University; and the Sailing Centre, Kiel University. We thank all these institutions and the following individuals and organisations: Rolf and Gerald Lorenz, Matthias Wehkamp, Jonas Enzmann, Feiko Wilkes, Jana Ulrich, Karl Bruhns, Jörn Sievert, Florian Huber, Christian Howe, Philipp Schubert and Robert Lehmann of Submaris, Petra Amelow, Torsten Janßen, Henry Skorna, AMLA (Study group for maritime and limnic archaeology), all students and collegues involved in the 2012 excavation, Senckenberg am Meer Wilhelmshaven, the Workgroup of Prof. M. Wahl at Geomar Kiel, the Centre for Scientific Diving, University of Oldenburg, and the Stickenhörn and Schilksee Marinas. We also thank Geoff Bailey and Jonathan Benjamin for help in revising the text.

## References

Andrén T, Björck S, Andrén E, Conley D, Zillén L, Anjar J (2011) The development of the Baltic Sea basin ruring the last 130 ka. In: Harff J, Björck S, Hoth P (eds) The Baltic Sea basin. Springer, Heidelberg

Feldens P, Schwarzer K (2011) The Ancylus Lake stage of the Baltic Sea in Fehmarn Belt: indications of a new threshold? Cont Shelf Res 35:43–52. doi:10.1016/j.csr.2011.12.007

Fischer A, Olsen J, Richards M, Heinemeier J, Sveinbjörnsdottir AE, Bennike P (2007) Coast-inland mobility and diet in the Danish mesolithic and neolithic: evidence from stable isotope values of humans and dogs. J Archaeol Sci 34:2125–2150

Glykou A, Goldhammer J, Hartz S (2014) Versunkene Steinzeit – taucharchäologische Untersuchungen an der Schleswig-holsteinischen Ostseeküste. In: Huber F, Kleingärtner S (eds) Gestrandet, Versenkt, Versunken – Faszination Unterwasserarchäologie. Wachholtz, Neumünster

Goldhammer J, Hartz S (2013) Versunkene Steinzeit – endmesolithische Funde vor der Steilküste Stohl. Jb Eckernförde 71:151–162

Hartz S, Lübke H (2005) Zur chronostratigraphischen Gliederung der Ertebølle-Kultur und frühesten Trichterbecherkultur in der südlichen Mecklenburger Bucht. Bodendenkmalpflege Mecklenburg-Vorpommern, Jb 2004 52:119–143

Hartz S, Lübke H (2006) New evidence for a chronostratigraphic division of the Ertebølle culture and the earliest funnel beaker culture on the southern Mecklenburg Bay. In: Kind C-J (ed) After the ice age – settlements, subsistence and social development in the mesolithic of Central Europe. Proceedings of the International Conference 9th to 12th of September 2003, Rottenburg/Neckar, Baden-Württemberg, Germany. Theiss, Stuttgart

Hartz S, Jöns H, Lübke H, Schmölcke U, Carnap-Bornheim CV, Heinrich D, Klooß S, Lüth F, Wolters S (2014) Prehistoric settlements in the south-western Baltic Sea area and development of the regional stone age economy. Ber RGK 92(2011):77–210

Madsen AP, Müller S, Neergaard C, Petersen CGJ, Rostrup EV, Steenstrup KJ, Winge H (1900) Affaldsdynger fra Stenalderen i Danmark. Undersøgte for Nationalmuseet, Paris

Piezonka H (2008) Neue AMS-Daten zur frühneolithischen Keramikentwicklung in der Nordosteuropäischen Waldzone. Estonian J Archaeol 12:67–113

Rößler D, Lemke W, Moros M (2009) Reconstruction of the Littorina transgression in the Western Baltic Sea. In: Harff J, Lüth F (eds) SINCOS I – sinking coasts. Geosphere, ecosphere and anthroposphere of the Holocene Southern Baltic Sea. Bericht der RGK

# Chapter 10
# Investigations of Submerged Palaeoshorelines in the Kiel Fjord

**Frederick Feulner**

**Abstract** Underwater diver surveys have revealed remains of the root system of a submerged alder forest in front of a ferry pier in Kiel-Friedrichsort (Germany). This offers the rare chance to date material from a submerged forest recovered in situ from the inner Kiel fjord, which gives a $^{14}C$ result of 5200–5000 cal BC, and this in turn gives a date for a period after which the freshwater lakes of the fjord mixed with the transgressing seawater of the Baltic Sea.

## 10.1 Introduction

This chapter presents information on underwater investigations in the Kiel Fjord that have mapped the marine and submarine sediments and provided dateable material giving a date immediately preceding the local transition from freshwater to marine conditions.

At the southwestern end of the Baltic Sea, the wedge-shaped Kiel Fjord was formed by glacial processes during the last Ice Age and cuts 15 km deep into the moraine landscape. At the northern end at Friedrichsort, the fjord narrows from about 3 km width to only 1 km (Fig. 10.1). The greater part of the shoreline in the fjord has been altered, dredged and overbuilt by industrial and harbour structures. Therefore, discovering datable remains of prehistoric origin proves to be difficult. Beneath the surface a homogeneous picture emerges. The shallow water in front of the beach (about −2 m in depth) extends for up to 80 m near the ferry pier, whilst further to the west (shipyard) and to the east (factory quay) the submerged topography is steeper. The shallow water substrate consists of sand and gravel subsoil above organic layers, covering hard glacial till. Beyond the shallow water zone the bathymetry, between −7 and −9 m plunges steeply. Here, the seabed is mainly covered with thick mud sediments.

In this chapter I present information on the remains of a submerged forest, the conditions that have resulted in the exposure of the material, the methods of diver survey used to investigate the location, the dating of tree stumps, and the implications for the transition of the inner Fjord from a freshwater to a marine environment.

F. Feulner (✉)
Department of Archaeology, University of York, The King's Manor, York Y01 7EP, UK
e-mail: freder@palaeome.org


G.N. Bailey et al. (eds.), *Under the Sea: Archaeology and Palaeolandscapes of the Continental Shelf*, Coastal Research Library 20, DOI 10.1007/978-3-319-53160-1_10

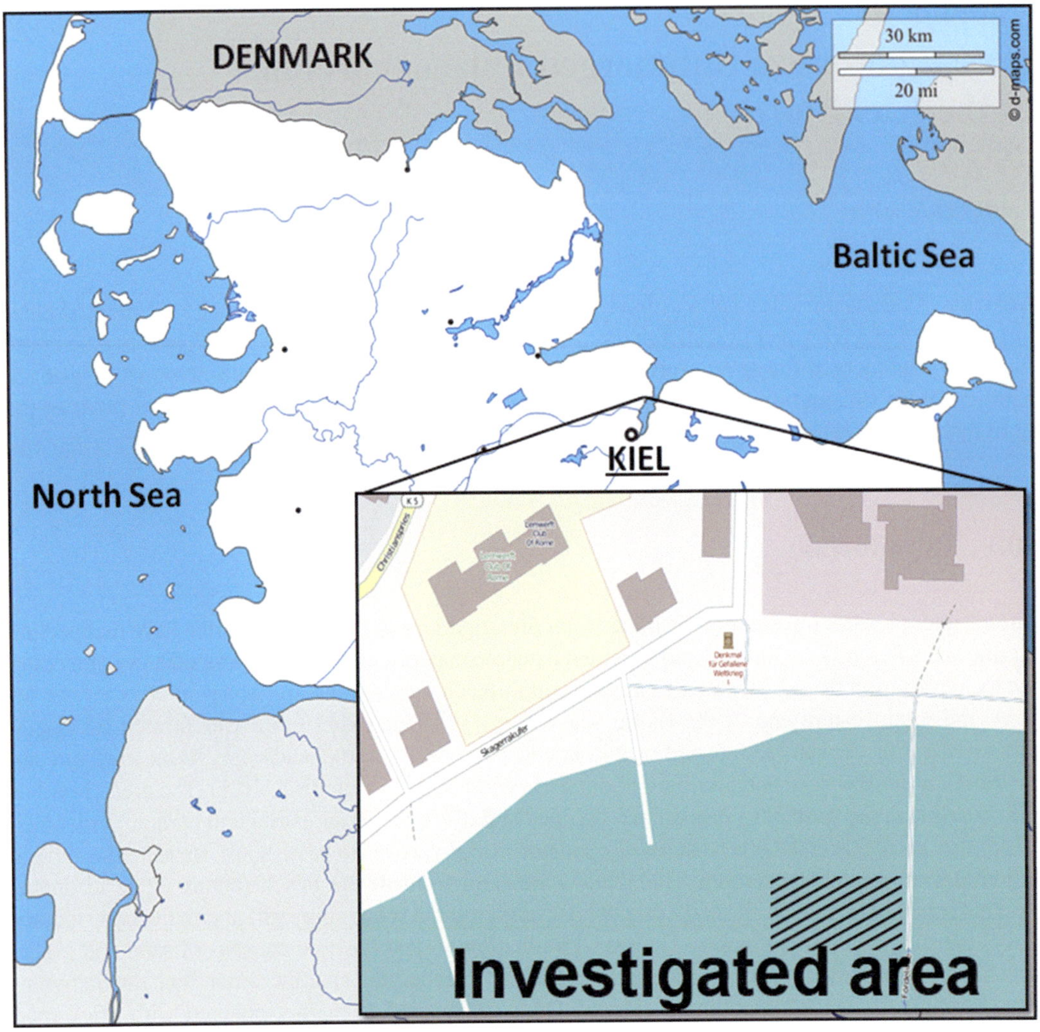


**Fig. 10.1** Map of Schleswig-Holstein showing detailed inset map of the Kiel-Friedrichsort site

## 10.2 The Site

In the area west of the passenger ferry pier (54°23′20.95″N, 10°10′41.22″E), I observed roots sticking out of the glacial till for the first time in 2004, and reported this to the authorities. In accordance with the Archäologisches Landesamt Schleswig-Holstein, the site was designated 'Kachel 1627 LA 2 Friedrichsort' (LA = Landesaufnahme). Since then I have regularly dived on the area, and monitored and documented visible features on the seabed. The submerged landscape near the ferry pier drops sharply from about −2 to −7 m, forming a jagged cliff. Near the cliff the uppermost covering of sand and gravel makes way for an organic soil about 30 to 70 cm thick, and beneath this there is glacial subsoil. At about −3.5 m depth, numerous plant roots start to appear, sticking out from the glacial till. These extend along the scarp over a distance of approximately 10 m in an east–west direction and about 3 m from the beginning of the upper edge of the scarp down to an exposed depth of −6 m with a further extension to be expected. Depending on changing circumstances associated with currents,

**Fig. 10.2** Divers setting out the measuring grid on the site

wind direction or ship traffic, the organic layer was variously covered with sand or completely and openly visible.

## 10.3 Survey Methods

Due to the changing nature of this site and the difficult circumstances associated with submerged obstructions and stretches of shallow water, it was not possible to obtain high resolution bathymetry by side-scanning sonar. Therefore a diver-supported survey was carried out. A measuring grid was set up in an east–west and south–north orientation, taking the end of the jetty pier as the zero-point (Fig. 10.2). Westward the grid stretched out for 40 m, and northward for 20 m. Depth measurements were taken by digital depth gauge at 5 m-intervals in shallow water (above −4 m) and calibrated against mean sea level (Figs. 10.3 and 10.4). In the deeper parts the number of measurement points taken was denser to allow a better mapping of the escarpment.

The wood residues were analyzed by S. Klooß, Kiel, and classified as roots of alder (*Alnus*). Gray and black alder (*Alnus incana* and *A. glutinosa*) grow primarily in wetland areas, along river banks and streams and also on nutrient-poor soils. Therefore, they are considered to be extremely competitive pioneer plants. Alders have a well-developed root nodule system and lack the usual vigorous lateral roots, which is immediately apparent at the Kachel 1627 LA 2 Friedrichsort. Although alders are moisture-loving, they may still be vulnerable to prolonged flooding of the stem base. This makes it possible to obtain a terminus post quem for the flooding of the inner Kiel fjord.

We owe it to the constant remodelling and erosion of the seabed by ferry traffic that these alder forest remains are now visible and can be examined. The local ferries with a capacity of 300 persons approach this pier multiple times per day (Fig. 10.5). The propellers of the ferries are situated at a depth of about −2.40 m beneath the hull and exert a force of up to 250 KW on the submarine scarp during berthing and departure manoeuvres. With every manoeuvre the erosion process continues, loosening parts of the seafloor. The immediate area has been abraded to form a semi-circular-shaped depression and a submarine cliff of more than 1 m has been formed. The eroded material is carried away by the stern propeller flow towards deeper areas, where it spreads out, and also by the bow

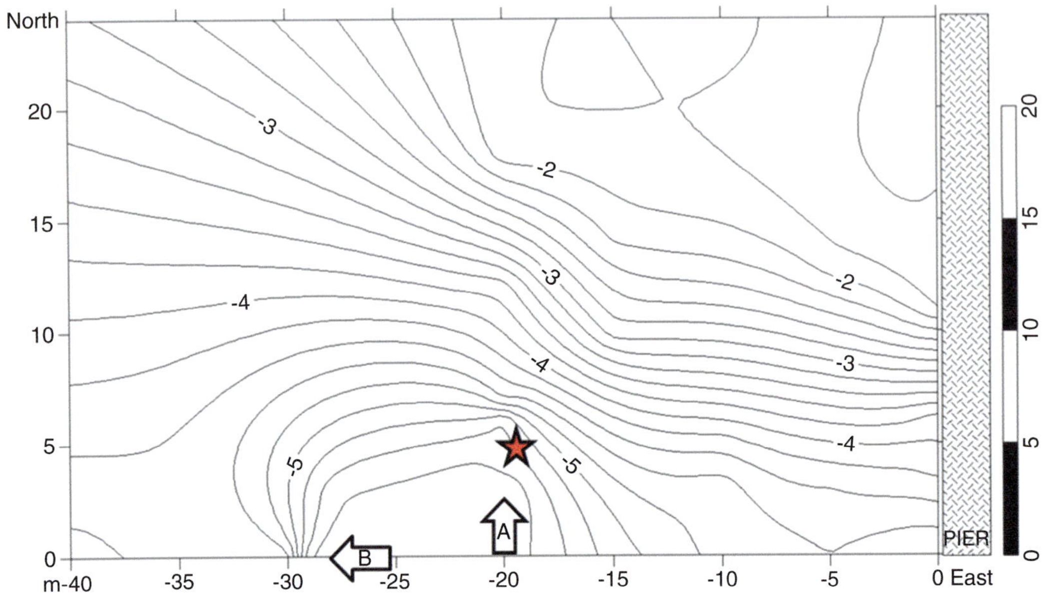


**Fig. 10.3** Contoured map of the investigated area. The star marks the sampling site, *arrows* mark profiles. Depths are in metres below present-day mean sea level

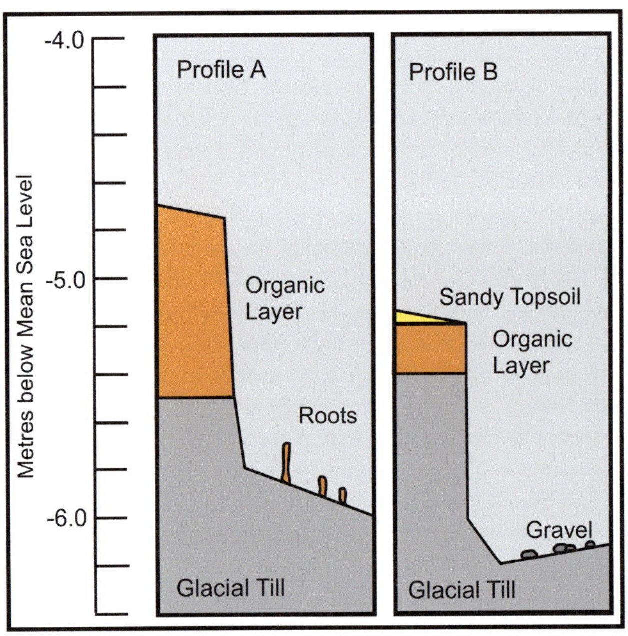


**Fig. 10.4** Profiles of submerged palaeo-landscape taken at two coordinates (**a** and **b** on Fig. 10.3)

**Fig. 10.5** Location of site showing the passenger ferry approaching the pier

thrusters into the shallow water zone to the north. On the eastern side of the ferry pier, a teardrop-shaped ejection zone has formed, in the direction parallel to the industrial compound, which raises the seafloor to a level of −2 m below sea level (Figs. 10.6 and 10.7).

## 10.4 Dating

A wooden root sample of about 3 cm thickness was taken at a central location (about 20 m west of the pier) from an eroded area in a calibrated depth of −5.9 m below sea level (since it is a root, it can be assumed that the former dry surface was located about 1–1.5 m higher) and passed on to the Leibniz Laboratory at the Christian-Albrechts-University of Kiel for AMS age determination. Sample KIA 50022 was controlled by microscope for any impurities and a suitable amount of material was selected for dating. The material was extracted by 1% HCl, 1% NaOH at 60 °C and again with 1% HCl. It was combusted at 900 °C in a quartz ampulla filled with CuO and silver wool. The generated $CO_2$ was reduced with $H_2$ at 600 °C to graphite. The iron-graphite was mounted on a specimen holder for AMS-measuring. The $^{14}C$ concentration in the sample results from a comparison of the simultaneously determined $^{14}C$, $^{13}C$ and $^{12}C$ values with the $CO_2$-standard (oxalic acid II) and a blank sample. The conventional $^{14}C$ age was subsequently calculated according to Stuiver and Polach (1977) and corrected for isotope fractionation ($^{13}C/^{12}C$ relation). This delta$^{13}C$ value incorporates the effects that occur during the graphitization and inside the accelerator mass spectrometer and therefore is not compatible with delta$^{13}C$ values measured with a $CO_2$ mass spectrometer. The uncertainty of the $^{14}C$ results includes the statistics, the stability of the AMS device and the uncertainty of the subtracted zero effect. For the first two factors the laboratory observed the counting statistics as well as the distribution of measured intervals and used the larger value. The sample yielded sufficient carbon

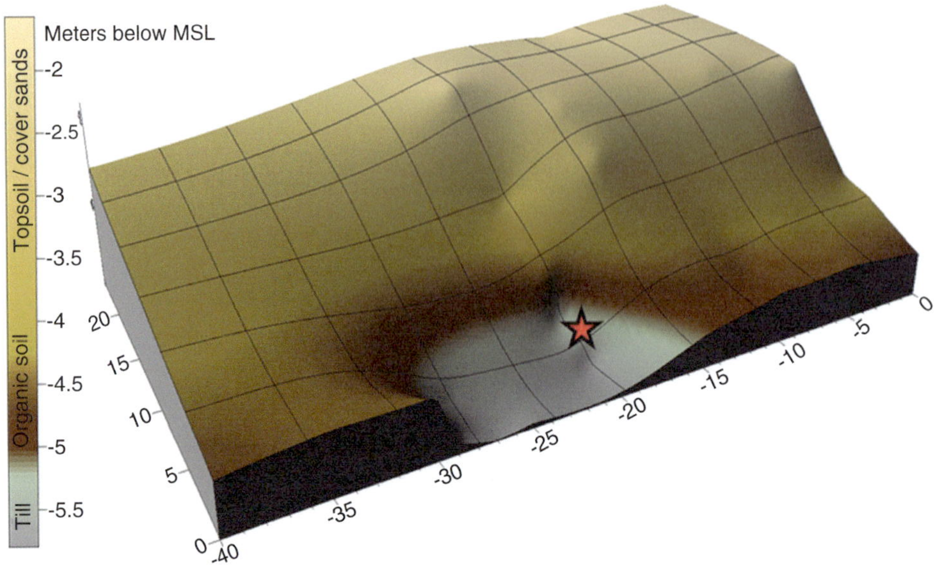


**Fig. 10.6** 3D-visualisation of Friedrichsort site, looking north. *Yellow*: sandy topsoil; *brown*: organic subsoil; *grey*: glacial till. The star marks the sampling site. The x and y axes show the horizontal distance in metres, the z axis depth in metres below mean sea level

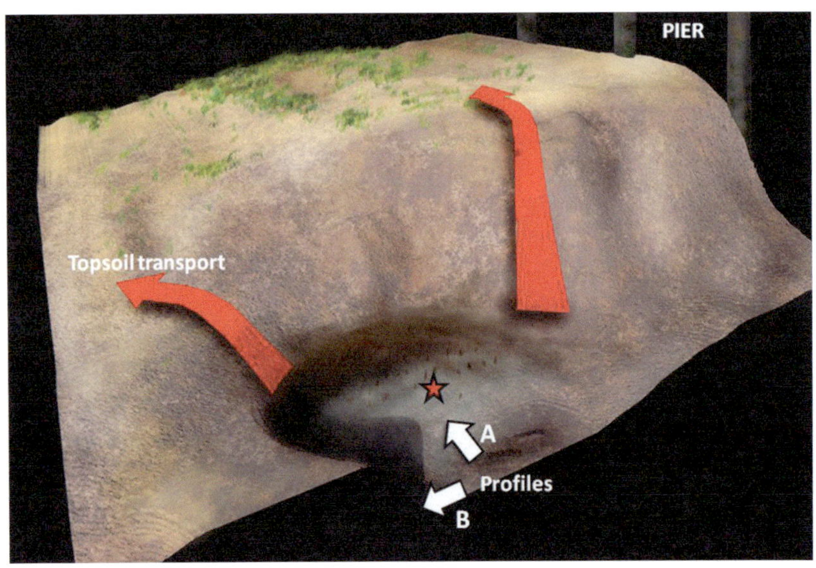


**Fig. 10.7** Visual representation of soil movement and how it has helped to expose the underlying terrestrial deposit. **a** and **b** mark the profiles shown in Figs. 10.3 and 10.4

(2.6 mg C), and the delta $^{13}$C value is within the normal range for organic samples and therefore valid (−30.01 ± 0.15 delta $^{13}$C ‰).The transformation into absolute age was done with the OxCal V4.2 program (Bronk Ramsey 2009) and the calibrating software IntCal 13 (Reimer et al. 2013) (Figs. 10.7 and 10.8).

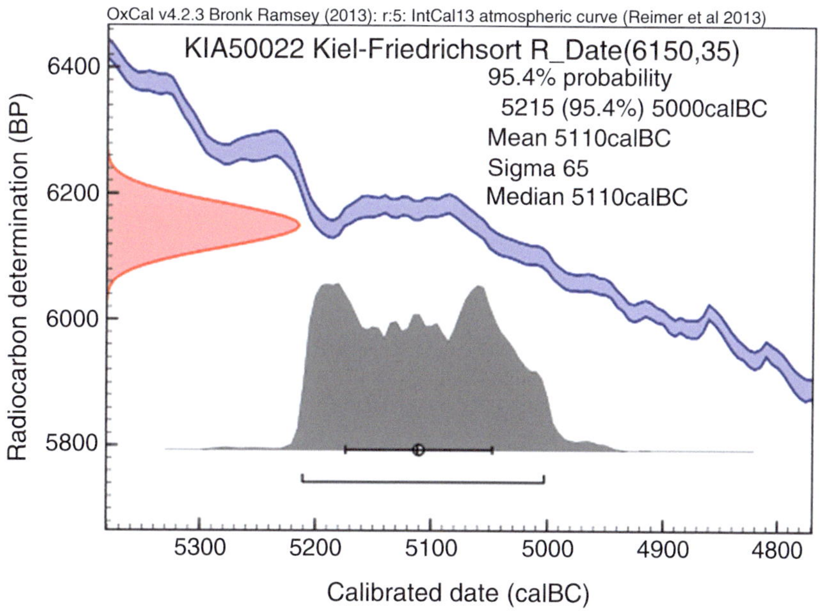


**Fig. 10.8** Calibrated date taken from the root of an alder tree (*Alnus* sp) sample KIA50022, as provided by Leibnitz Laboratory CAU Kiel. Calibration based on OxCal v4 2.3 and IntCal13 (Reimer et al. 2013)

## 10.5 Discussion

The results of the radiocarbon dating give an age of 5200–5000 cal BC (KIA 50022, 6151 ± 34 BP), which corresponds to the beginning of the pre-pottery Ertebølle culture. This falls into the main period of the Littorina transgression phase of the Baltic. The sea-level rose from about −10 to −3 m within a relatively short time-span (ca. 6500–5000 cal BC), flooding former dry land areas. Based on the current evidence, we can expect the prehistoric fertile dry-land horizon to be at about −5 m, with the roots reaching further down into the till layers, which is supported by a 0.2–0.7 cm-thick visible organic layer interspersed with mud and floral remains. As no tree stump has been found in situ so far, it is only possible to sample the root system of the tree, which tends to reach deep into the glacial till. I hypothesize that the tree(s) grew near the shore of a freshwater body, which was flooded by the transgressing Baltic Sea at −4 m at about 5000 cal BC (See also Goldhammer and Hartz, Chap. 9). Fischer (1997, p. 35) points out that different tree species have different requirements regarding location, moisture and salinity. Alder is much more resistant to water-level rise than, for example, certain oak or lime trees. Dörfler et al. (2009, p. 178) report that preserved radiant alder roots can be a good indicator for flooding events. Following the general mean graph of the south-western Baltic region as established by Labes (2002, Fig. 114) the sea-level for this time period should be at about −4 to −3.5 m below present-day MSL, which correlates well with the observations at Kiel-Friedrichsort. However, Jakobsen (2004, p. 75) estimated for parts of the Oldenburg valley a sea-level of −4.70 m MSL for 5100 cal BC, which would mean a sea-level rise of approximately 30 cm per 100 years until about 4750 cal BC.

Although there are many radiocarbon dates available from submerged/wetland prehistoric sites and sunken forests in the western Baltic (Labes 2002; Lübke 2005), none have so far been recovered from the inner Kiel Fjord itself. A number of $^{14}C$ dates have been acquired during analyses of dredged finds of the nearby eponymous site of Kiel-Ellerbek (Mestorf/Weber 1904; Mestorf 1905) by Schwabedissen, dating to the middle/late Ertebölle period, intermixed with early Neolithic finds

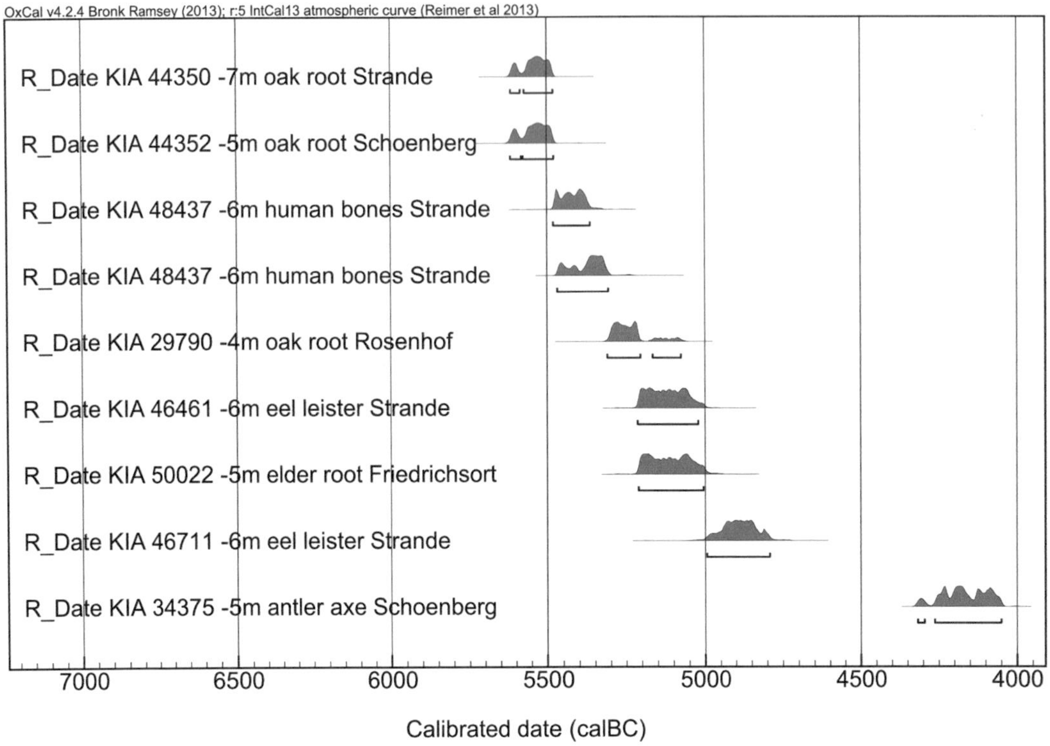


**Fig. 10.9** Calibrated dates of prehistoric submerged sites in the Kiel Bay

(ca. 4600–3400 cal BC, cf. Schwabedissen 1994). However, these old published dates should be used with care and need recalibration. In the vicinity of Friedrichsort, a growing number of submerged locations with evidence of prehistoric finds or palaeobotanic material were discovered and studied in recent years. About 12 km to the north-east, another site was investigated at LA 7 Schönberg-Kalifornien in 2007. At a depth of −4 to −5 m, tree stumps were found in situ but the sample did not yield enough information for a reliable dendrochronological date. Another oak tree at −5 m MSL was dated at 5573–5479 cal BC (KIA 44352, 6586 ± 41 BP, Glykou et al. 2014). However, a T-shaped antler axe found nearby was dated at 4317–4050 cal BC (KIA 34375, 5342 ± 34 BP, Hartz 2009; Glykou et al. 2014). In 2012 a further submerged prehistoric site was discovered 8 km to the north of the Friedrichsort site near Bülk (LA 163 Strande) in 6 m depth, which dates to the early Ertebölle period (Goldhammer and Hartz, Chap. 9). Here, a number of trees were dendrochronologically-dated to about 5390 cal BC, an oak tree at −7 m MSL was radiocarbon-dated to 5570–5480 cal BC (KIA 44350, 6588 ± 35) while two eel leister prongs were radiocarbon-dated to 5215–5020 cal BC (KIA 46461, 6160 ± 30 BP) and 4990–4795 cal BC (KIA 46711, 6000 ± 40 BP). Human remains from that site were dated at 5470–5340 cal BC (KIA 48437, 6380 ± 35/6455 ± 30 BP), but this did not take any possible reservoir effects into account (Goldhammer and Hartz 2012, 2013; Glykou et al. 2014). At LA 12 Siggeneben-Süd in the Oldenburg valley, an oak tree at about −3.5 to −3 m has been dated ca. 5210–4780 cal BC, while another one from LA 58 Grube-Rosenhof (also Oldenburg valley) at −4.1 m yielded a date of 5310–5200/5100 cal BC (KIA 29790, 6248 ± 31BP, MSL, Goldhammer 2008). The key dates from the Kiel Bay are summarized in Fig. 10.9.

## 10.6 Conclusions

To conclude, we can say that this area resembles a submerged palaeoshoreline of the inner Kiel fjord, with a top layer of sand protecting an organic layer that was once part of the prehistoric landscape. Although this site was and still is heavily influenced by anthropogenic and natural erosion, it has provided an in-situ radiocarbon date giving a terminus after which the flooding of this Baltic fjord occurred. Further steps to improve knowledge of this site and the development of sea-level rise in this area are planned, including a more detailed scientific analysis of the organic soil.

**Acknowledgements** I am indebted to Sönke Hartz and the Archaeological State Museum, Schleswig, who helped with funding the dating of the forest and to Stefanie Klooß, Kiel University, for identifying the plant remains. I would also like to thank the members of the Scientific Diving Association NGO, especially Steffen Häuser (Kiel) and Jörg Ewersen (Schleswig) for their help documenting this area.

## References

Bronk Ramsey C (2009) Bayesian analysis of radiocarbon dates. Radiocarbon 51(1):337–360

Dörfler W, Jacobsen O, Klooß S (2009) Indikatoren des nacheiszeitlichen Meeresspiegelanstiegs der Ostsee. Eine methodische Diskussion am Beispiel der Ostseeförde Schlei, Schleswig-Holstein. In: Müller U, Kleingärtner S, Huber F (eds) Zwischen Nord- und Ostsee 1997–2007. Zehn Jahre Arbeitsgruppe für maritime und Limnische Archäologie (AMLA) in Schleswig-Holstein. UPA 165. Habelt, Bonn, pp 171–186

Fischer A (1997) Drowned forests from the stone age. In: Pedersen L, Fischer A, Aaby B (eds) The Danish storebaelt since the ice age: man, sea and forest. Storebaelt, Copenhagen, pp 29–36

Glykou A, Goldhammer J, Hartz S (2014) Versunkene Steinzeit – taucharchäologische Untersuchungen an der Schleswig-holsteinischen Ostseeküste. In: Huber F, Kleingärtner S (eds) Gestrandet, Versenkt, Versunken – Faszination Unterwasserarchäologie. Wachholtz, Neumünster, pp 76–95

Goldhammer J (2008) Untersuchungen zur Stratigraphie, Fundverteilung und zum Fundspektrum der mittleren Ertebøllekultur in Ostholstein: Die Nachgrabungen in Grube-Rosenhof aus den Jahren 2001 und 2002, mit Beiträgen von P. Frenzel, J. Ewersen und S. Klooß. UPA 163, Habelt, Bonn 2008

Goldhammer J, Hartz S (2012) Steinzeitliche Siedlungsreste am Ostseegrund – Ein endmesolithischer Fundplatz in der Kieler Bucht. Archäologische Nachr Aus Schleswig-Holstein, pp 26–29

Goldhammer J, Hartz S (2013) Versunkene Steinzeit – endmesolithische Funde aus der Ostsee vor der Steilküste von Stohl. Jahrb Heimatgemeinschaft Eckernförde eV 2013:151–162

Hartz S (2009) Steinzeit am Ostseegrund. Submarine archäologische Forschungen vor Schönberg-Kalifornien. Jahrbuch für Heimatkunde in Kreis Plön 39: 33–45

Jakobsen O (2004) Die Grube-Wesseker Niederung (Oldenburger Graben, Ostholstein): Quartärgeologische und geoarchäologische Untersuchungen zur Landschaftsgeschichte vor dem Hintergrund des anhaltenden postglazialen Meeresspiegelanstiegs. Doctoral dissertation, Christian-Albrechts-Universität zu Kiel

Labes S (2002) Meeresspiegelschwankungen an der südwestlichen Ostseeküste während der letzten 8000 Jahre. Ein Überblick über archäologische und geologische Untersuchungsergebnisse. Unpublished Diploma thesis, Christian-Albrechts-Universität zu Kiel

Lübke H (2005) Spät-und endmesolithische Küstensiedlungsplätze in der Wismarbucht – Neue Grabungsergebnisse zur Chronologie und Siedlungsweise. Bodendenkmalpflege Mecklenburg, Jahrb 2004 (52):83–110

Mestorf J (1905) Einstmalige Wohnstätten in der Kieler Förde. Heimat 15:77–82

Mestorf J, Weber KA (1904) Wohnstätten der älteren neolithischen Periode in der Kieler Föhrde. Bericht des Kielers Museums vaterländischer Altertümer, Kiel.

Reimer PJ, Bard E, Bayliss A et al (2013) IntCal13 and Marine13 radiocarbon age calibration curves 0–50,000 years cal BP. Radiocarbon 55(4):1869–1887

Schwabedissen H (1994) Die Ellerbek-Kultur in Schleswig-Holstein und das Vordringen des Neolithikums über die Elbe nach Norden. In: Hoika J, Meurers-Balke J (eds) Beiträge zur frühneolithischen Trichterbecherkultur im westlichen Ostseegebiet. 1. Internationales Trichterbechersymposium in Schleswig vom 4 bis 7 März 1985. Wachholtz, Neumünster, pp 361–401

Stuiver M, Polach HA (1977) Discussion: reporting of 14C data. Radiocarbon 19(3):355–363

# Chapter 11
# Submerged Settlement in the Öresund, Western Scania, Southernmost Sweden

**Lars Larsson**

**Abstract** In order to obtain information on coastal settlement during the Early Mesolithic, marine archaeological investigations on the Swedish side of Öresund (the Sound) have been carried out since the 1970s. Early Mesolithic sites have been registered, the depths of which varied between −20 and −6 m. A number of settlement sites have been located close to the submerged outlet of a river in the central part of the present Öresund. One of these sites is partially covered in peat, situated at a depth of −7 to −8 m, and has been dated to about 7000 cal. BC. Finds of bones and worked wood make this a very important find location. A number of sites have been found during various activities along the former shoreline. The results of these investigations provide the basis for an argument that coastal settlement during the Boreal was equally as intensive as that which is well documented in the hinterland. The choice of settlement sites on the coast was dictated by the same factors as in the Late Mesolithic, a period when the hinterland became less attractive for settlement because of dense afforestation and the transformation of former lakes into bogs.

## 11.1 Introduction

Öresund is the sound that divides present-day Sweden from Denmark. It is about 100 km long, 30 km wide in the south, and narrowest in the north where it is 4 km wide (Fig. 11.1). The depth varies from more than 25 m to about 7 m below present sea level, with the shallowest part in the south. The region is of particular interest in relation to the study of submerged landscapes and prehistoric archaeology. It has been the focus of underwater investigations from an early period since the 1970s, and has produced a number of finds that illustrate the potentials and limitations of underwater research. There are also a number of well-known Mesolithic settlements on the present-day coastline and its adjacent hinterland, and this combination of offshore and onshore evidence provides a distinctive insight into the relationship between changing patterns of settlement and subsistence and the changing configuration of coastlines and environments associated with the transition from glacial to postglacial conditions in northern Europe.

Human settlement began at about 14,000 years ago, when the region first became accessible following de-glaciation. At the beginning of this period, most of the Sound, particularly the southern part, was still dry land, comprising a low-lying landscape with many freshwater lakes. Subsequently it underwent dramatic changes in palaeogeographic configuration and palaeoenvironment associated with postglacial climate change and sea-level rise. The lowlying basin was progressively inundated by seawater from the North Sea, becoming at first a marine bay opening to the North with numerous islands, most of which are now submerged, and finally becoming a throughway between the North

L. Larsson (✉)
Institute of Archaeology and Ancient History, LUX, Lund University, Box 192, SE-221 00 Lund, Sweden
e-mail: Lars.Larsson@ark.lu.se



G.N. Bailey et al. (eds.), *Under the Sea: Archaeology and Palaeolandscapes of the Continental Shelf*, Coastal Research Library 20, DOI 10.1007/978-3-319-53160-1_11

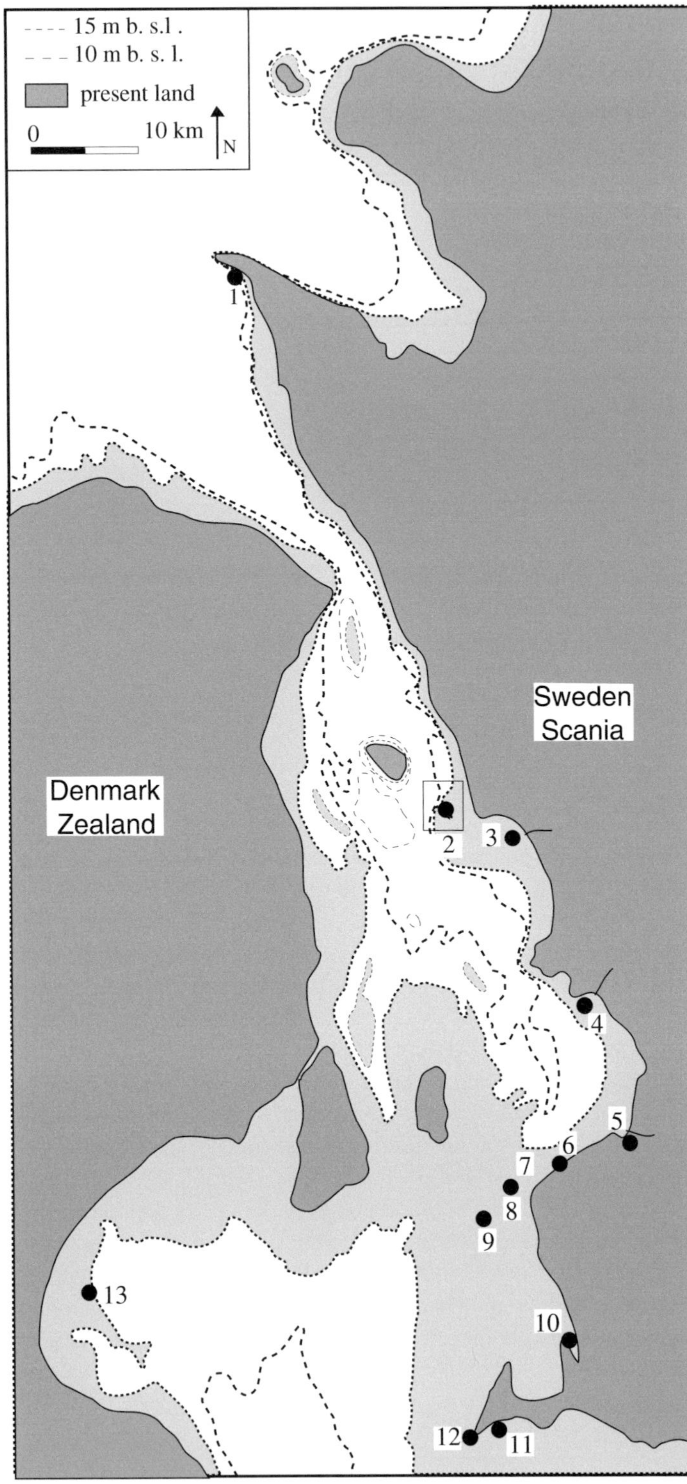


**Fig. 11.1** The Öresund Sound with bottom contours and sites mentioned in the text. *1* Kullen, *2* Pilhaken, *3* Saxån, *4* Lödde kar, *5* Segebro, *6* Malmö habour, *7* Limhamn, *8* Lernacken, *9* Knaggen, *10* Foteviken, *11* Falserbo, *12* Måkläppen, *13* Solrød strand

**Table 11.1** Prehistoric chronological subdivisions in southern Scandinavia

| Uncal BP | Chronozones | Cultures | Phases | Cal BC |
|---|---|---|---|---|
| 5000 | Subboreal | TRB | Neolithic | 4000 |
| 6000 | Atlantic | Ertebølle culture | Late Mesolithic | 5000 |
| 7000 | Atlantic | Kongemose culture | Middle Mesolithic | 6000 |
| 8000 | Preboreal and Boreal | Maglemose culture | Early Mesolithic | 7000 |
| 9000 | Younger Dryas | Ahrensburg culture | Late Palaeolithic | 8000 |
| 10,000 | | | | 9000 |

Sea and the Baltic as sea level reached its present level about 5,000 years ago. On land, progressive afforestation and transformation of freshwater lakes to bogs altered the opportunities for settlement and subsistence.

The aim of this chapter is to summarise the evidence for underwater finds in the Öresund, to highlight some of the problems of underwater investigation and interpretation, and to use the evidence to assess the changing nature of coastal settlement and its relationship to the hinterland. Despite the fact that shorelines for the earlier part of the prehistoric sequence are now submerged, underwater finds, as well as indications from the on-land archaeological record of contacts with the contemporaneous coastline and exploitation of marine resources, provide many insights into the nature of early coastal settlement. Several different and partially overlapping cultural and chronological labels are used to describe the sequence, and the relationship between them is shown in Table 11.1.

## 11.2 Late Palaeolithic and Early Mesolithic

Due to considerable rise in sea levels during the Late Boreal and the Atlantic periods, there is very little surviving evidence of coastal settlement from the Late Glacial and the Early Postglacial, apart from individual finds which have been observed in connection with sand extraction. These include actual artefacts, such as a worked flint and bones from the sea floor off Solrød strand on the Danish side (Fig. 11.1) at a depth of between 6 and 10 m below sea level (Vang Petersen and Johansen 1995), with a worked reindeer bone dated at 12,140 ± 100 BP (AAR-1036) indicating settlement remains from the Hamburgian culture.

At a depth of about 10 m just off Falsterbo on the Swedish side of the southernmost side of the sound (Fig. 11.1), reindeer antlers and bones of giant deer have been found during dredging for a canal. Radiocarbon dating of the bone of a giant deer provided a date of 11,220 ± 70 BP (LuS 6140).

This evidence of Late Palaeolithic activity in the southern Öresund can be linked to the supposed migration routes of reindeer moving from present-day north-western Germany to Zealand and further to the north-east (Vang Petersen and Johansen 1995). A land bridge existed across the southern Öresund between present-day Denmark and Sweden during certain periods of the Late Palaeolithic and the entire Early Mesolithic. During other periods of the Late Palaeolithic the sound might have been narrow and, like the large Siberian rivers (Klein 1982), could have been crossed by swimming animals. This area might, therefore, have been a strategic location for mass slaughtering in conjunction with migrations undertaken by herd animals such as reindeer and probably other migrating animals like horse (Larsson 2008). This land bridge exhibits a dynamic history, however, and existed only during certain periods, including from 11,300 to 10,900 BP and from 10,300 until around 8,000 BP (Larsson 1994; Björck 1995).

Evidence relating to coastal settlement during the Preboreal and large parts of the Boreal (11200–7000 BP) is missing from Öresund because the relevant coastlines are now submerged. Coastal sites are better exemplified on the west coast of Sweden, where older shorelines have been elevated by isostatic uplift following deglaciation (Nordqvist 1995; Kindgren 1995). These sites mostly lack good conditions of faunal preservation and are known only from the presence of stone artefacts, so that we do not have reliable evidence about the degree of reliance on marine resources. Nevertheless, the majority are located on old shorelines and the large number of sites indicates that coastal settlement was of major importance compared to the hinterland.

In the southernmost part of Sweden, settlement in the Early Mesolithic is well documented by several inland sites close to and within the larger bogs, which were at that time lake areas attractive for fishing, hunting and gathering (Larsson 1978). On these sites, at a distance of 20–80 km from the coast, most or all of the lithic material is flint, the source of which is in the coastal area to the west. This means that the distribution networks between coast and inland areas were very active.

That permanent coastal settlement existed during a late part of the Boreal period is indicated by a find of a human skull off the Danish coast of the Öresund. It is dated to 7270 BC (K-5099), and a $^{13}C$ value of $-14.7$ ‰ indicates a strongly marine diet (Fischer 1997).

Even though the sea level was lower than today, no other land connections were formed across the Sound. There are only three islands of any size today in Öresund, although the northern and central part of the sound was a deep arm of the North Sea with a true archipelago during the Early Mesolithic. A rapid rise in the water level during the Late Boreal caused the bay to enlarge at the same time as most of the islands were submerged. A number of submerged areas with layers of peat and gyttja on the bottom of Öresund show that this area, before it became a marine bay, contained a number of freshwater basins, which were gradually transformed into lagoons before being finally inundated. The area contained several biotopes which made it attractive for settlement. But so far no survey has been conducted around these former lakes.

## 11.3 Settlement Sites Associated with Submerged Rivers

Zoologists in the 1930s found some worked flints offshore of the south Swedish town of Landskrona when sampling the sea bottom (Rausing and Larsson 1977). Most of the Late Mesolithic settlement sites that are still situated in the present-day coast zone are found beside the outflows of rivers. Therefore, the area between present-day Landskrona and the island of Ven should be a suitable area for finding submerged settlements from earlier periods of the Mesolithic (Fig. 11.1). The bottom topography shows a river channel from the present-day outflow of the Saxån River some kilometres south of the town, and the submerged river can be followed though the harbour of Landskrona and at least half-way out to the island of Ven (Fig. 11.2).

During a scraping test in 1975, some worked flints indicated the existence of settlement remains. In order to obtain further information on coastal settlement formed during an earlier part of the Mesolithic, a marine archaeological investigation on the Swedish side of Öresund was carried out in 1979 (Larsson 1983). This was concentrated on the submarine channel corresponding to the prehistoric course of the river Saxån (Fig. 11.2). Along this river, both surveys and excavation have revealed a number of rich sites from the Middle and Late Mesolithic close to a deep bay (Karsten and Knarrström 2003).

From the study of sea charts, it was possible to trace the former course of the river, as well as submarine elevations and depressions. Areas of particular interest in a submarine context were noted. Areas to the west as well as to the east of the former river outlet seemed to be of special interest. These became the object of sampling from a boat with a special scraper used by marine biologists. The scraped bottom material was then sieved on board the boat. With the aid of divers, certain areas were also surveyed using an underwater sled towed by the boat. At least three Early Mesolithic sites were recorded during this preliminary investigation, the depths of which varied between 20 and 6 m below

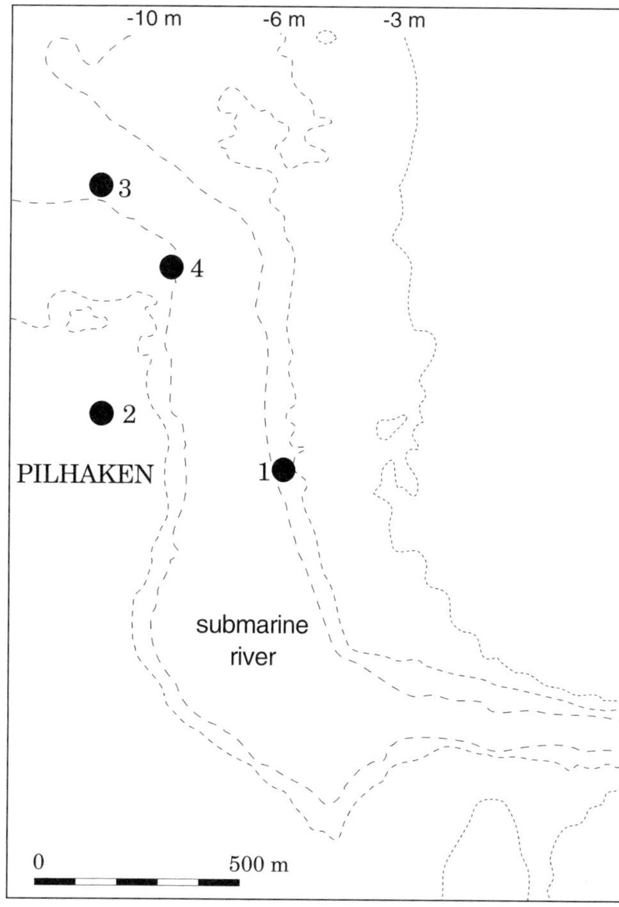


**Fig. 11.2** The location of sites close to the submerged river channel of the Saxån outside the town of Landskrona

sea level (Fig. 11.2). The somewhat raised bank named Pilhaken to the west of the former river outlet revealed more concentrations of artefacts than the eastern bank.

In the early 1990s the planning of a bridge across the Öresund prompted further underwater investigations. In 1992, the Pilhaken bank off Landskrona was used for field training by a Danish team. A further site, located on Pilhaken on the outermost part of the former river mouth, was discovered. This site, which is partially covered in peat, was situated at a depth of 7–8 m and yielded artefacts and samples of oak and hazelnut dated to 8120 ± 90 and 7945 ± 75 uncal. BP (AAR-1225 and T-10667) (Fischer 1993). The site was located directly adjacent to the clearly marked course of the river, which is about 300 m wide and a maximum of 14 m deep at this point. Extensive dredging further inshore and busy shipping movements presumably played a part in the severe erosion of the outer reaches of the navigable channel. The muddy layer could be observed as a horizon in the steeply sloping submarine channel of the river (Fig. 11.3). The fact that the deposit nearest the course of the river channel is exposed to continuous erosion meant that a trial investigation was urgently needed in order to obtain broader information about the extent of the find-bearing layer and the degree of destruction due to erosion—not only the erosion that is taking place now, but also that which may be expected to occur in the near future.

An investigation of the site was accordingly carried out in 1995 (Larsson et al. 1996; Larsson 1999). The location of the site, exposed to winds and currents, made the investigation work extremely difficult and time-consuming. The manner in which the excavation was conducted involved excavating a trench from the edge of the submarine course of the river inland as far as the deposits that had

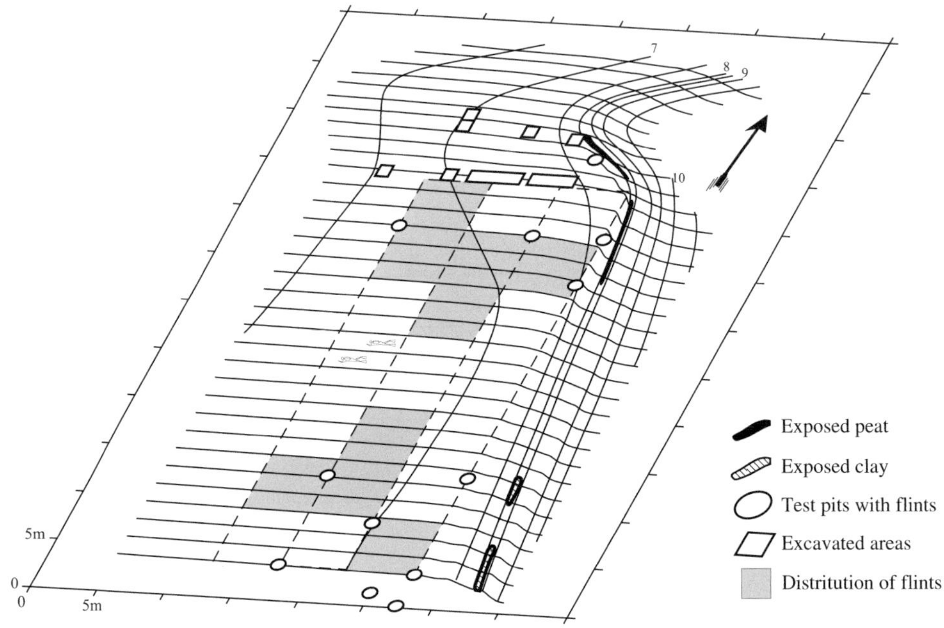


**Fig. 11.3** The location of the Pilhaken 4 site

been more or less dry land during the period of settlement (Fig. 11.3). The water was pumped out with large-capacity nozzles, and the resulting back suction was used to collect material recovered from the sea bed in fine-mesh nets at the end of the nozzle. The deposits comprise alternate layers of mud and sand with a combined maximum thickness of 0.5 m. A thicker layer of peat was present closest to the lowest sand layer. An abundance of tree branches was also present in the organic layers. Significant quantities of common mussels that had dug down into the bottom made the investigation more difficult and were a major cause of disturbances.

The majority of the flint artefacts were found in the lower part of the peat layer in direct contact with the underlying sand, and the layer of peat could be followed for a distance of at least 20 m (Fig. 11.3). The bottom was severely eroded further away from the former river channel, which meant that the find-bearing layer had been completely washed away in that area. The finds consist of flakes, blades and micro-blades and a smaller number of implements including scrapers, a burin and a burin spall (Fig. 11.4). The previous underwater investigation of the site in 1992 had revealed a number of blades and micro-blades, and a small number of bones from roe deer, red deer and aurochs were also found. The finds include a worked branch and sticks with traces of fire. The investigation was sufficiently extensive to give an idea of the extent of the find-bearing layer away from the course of the river, indicating a settlement that could have extended for some tens of metres along the course of the river. The layers of peat were deposited in a comparatively well-protected basin, possibly in a part of the delta that may have included the former mouth of the Saxån River.

A survey of the shallower waters further up the former river channel, from about −5 m to the present shoreline, was conducted by diving and swimming. Worked flints were found, but the distribution of these finds and the traces of rolling on the flints indicate that serious erosion has destroyed the sites in this area.

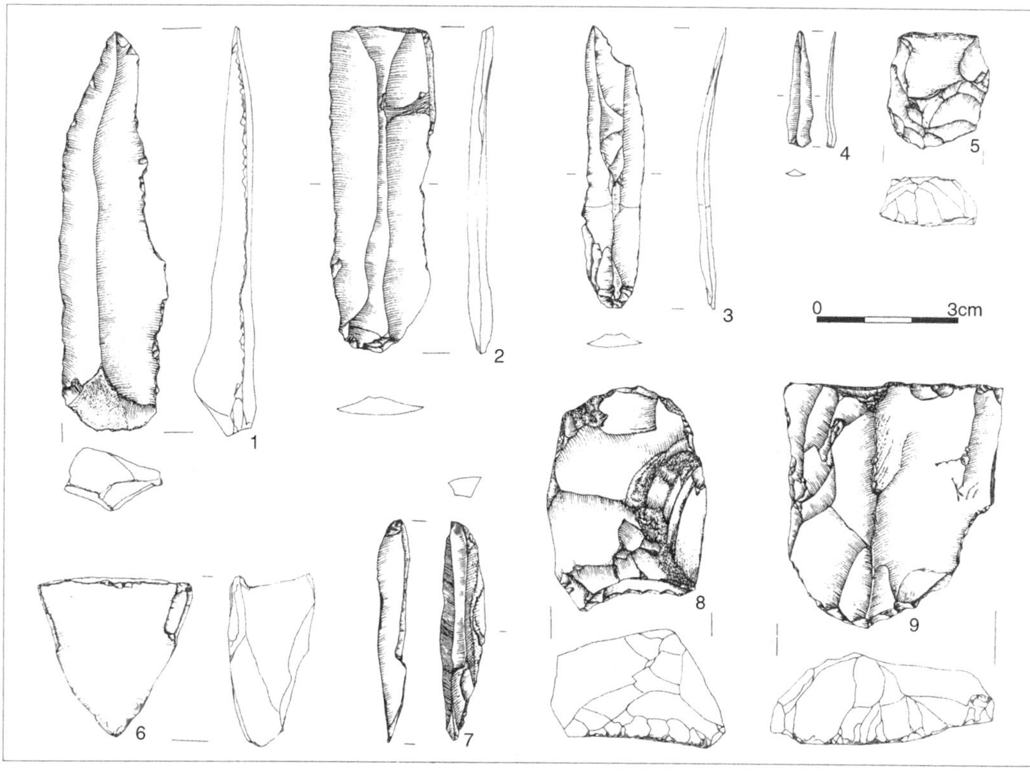


**Fig. 11.4** Flint artefacts from Pilhaken 4: blades, *1–3*; microblades, *4*; scrapers, *5*, *8–9*; burin, *6*; burin spall, *7* (Drawings by Björn Nilsson)

## 11.4 Other Submerged Sites

Surveys of the seabed in the southern part of the Sound in the vicinity of the location of the former land bridge between Denmark and Sweden were carried out in 1992–1994 in conjunction with prospection associated with the building of a new bridge connecting Denmark and Sweden over the Öresund (Dencker et al. 1994) (Fig. 11.1). In the course of its transformation from a terrestrial landscape to a waterway in the previous millennia, the land connection in the south part of the Sound totally disappeared, and the shallowest parts of the seabed were subjected to severe erosion by strong currents. The limestone bedrock is fully exposed in certain parts, providing some idea of the heavy action of the currents. There are also less exposed areas with sand and clay. It was hoped that settlement remains would be found there. A thorough trial investigation was accordingly carried out off Lernacken from 0.5 to 2 km offshore. Trial trenches of material were excavated using the suction method described above. Smaller areas of the seabed were also examined by divers by hand. The loose layers in the test trenches varied in thickness from a few centimetres to 1 m. Evidence of modern material, such as the remains of modern vegetation down to a considerable depth, indicates that the area has been subject to considerable transformation. Flint artefacts were found in a significant number of test pits, indicating evidence of a former settlement. However, none of these flints could be confirmed with certainty as in situ finds. Most also bore traces of rolling and salt water patination, but some were in fresh conditions indicating either that undisturbed find-bearing layers had been eroded quite recently, or that they may still exist in particularly well-protected areas. However, it was not possible to identify such a find location during the trial investigation. The find material includes both blades and micro-blades dating from a late part of the Boreal and the earlier part of the Atlantic, i.e., a late period of the Maglemose culture and the earliest part of the Kongemose culture (Fischer 1993). The finds emerged at a depth of between 9 and 5 m. The roots of trees descending into the clay at the base of the trial pits to a depth of approximately 7 m have been dated to 7800 BP, i.e., at some time in the Atlantic (Fischer 1996). The inundation threshold in Öresund lies at this level, which indicates that significant quantities of salt water only found their way into the Baltic Basin after that time.

Besides these underwater archaeological investigations, a number of other finds suggest settlement below the present sea level. In the northernmost part on the Swedish side, Öresund ends in Cape Kullen, with a rocky shoreline that shelves very steeply under water. On the southern side flints have been found at a depth of about 10 m. Core axes indicate settlement from the Preboreal. However, given the steep topography, these finds might have been eroded out from the deposits on the present-day shoreline and re-deposited on the sea bed.

Most site indications are found south of the above mentioned Pilhaken. Flints were found during a survey of the shallow water just off the present mouth of the river Saxån (Fig. 11.1). These were at a depth of less than 1 m below sea level and thus may well date from the early Atlantic. An indication of the date is given by wood associated with muddy sediments recovered at a depth of −5 m during dredging of the harbour at Landskrona just north of the present outflow of the river Saxån, with a radiocarbon date of 8100 ± 100 BP (Mörner 1969).

At Lödde Kar (Fig. 11.1), where there is an early Medieval loading platform at the mouth of the Lödde River, worked flints have been found associated with a submarine bog. This lies at a depth of −3 m and should also be dated to the transition from the Boreal to the Atlantic (Larsson 1983).

At Segebro (Fig. 11.1), immediately to the North of Malmö, extensive deposition of sand during the Late Atlantic and Sub-Boreal resulted in preservation of the remains of a settlement of Atlantic age (Fig. 11.1). The occupation layer is situated at a shallow depth of between 0 and −1 m, in spite of the fact that the settlement was situated hundreds of metres beyond the shoreline in the outer part of a river delta (Larsson 1982b). Dates range between 7390 ± 80 and 6970 ± 90 BP. Comparing other south Scandinavian settlements from an early part of the Kongemose culture, the more recent date is the most probable. Measurement of $^{13}C$ in grey seal and human remains gave values of −16.7 ‰ and −15.7 ‰, respectively, which shows that the inhabitants had a considerable intake of marine food. Grey seal is important among the bone finds, as is cod (Lepiksaar 1982).

As a result of expansion within the present harbour of Malmö, on the south-eastern side of Öresund (Fig. 11.1) a large area of the sea bed totalling 32,000 m$^2$ was exposed before construction (Hammarstrand Dehman 2009), and uncovered parts of the original land surface, with fallen tree trunks and tree stumps still in place at a level of −2 to −1 m. According to dates of an oak trunk and dendrochronological analysis, a mixed, dense forest was still standing at 6200–6100 cal. BC (Hardevik et al. 2008). After 120–140 years, the area became wetter and oak was replaced by alder. This change occurred within 11–75 years depending on sampling points. Artefacts were found at the original surface level but no real settlement site. Fish traps were found as evidence of fishing activities on the sea bed after the area had been inundated. The traps, one almost intact and two represented by fragments of wicker cages, are cylindrical in form and made of small branches bound close together with roots, and are dated to the interval 5750–5550 cal. BC. Based upon these dates, the sea rose to the present level somewhat later than 6000 cal. BC.

In the course of dredging work to extend the harbour at Limhamn (Fig. 11.1) at the end of the nineteenth century, extensive settlement remains in the form of flint implements and bone, including a slotted bone point and human bones, were found about 300 m from the shore. These objects were found at a depth of approximately −2 m and belong to the Kongemose culture. Wood recovered from peat layers at a depth of −8 m in the harbour at Limhamn has been dated to between 7990 ± 160 and 7895 ± 115 uncal. BP (Persson 1962).

A site known as Knaggen (Fig. 11.1) was found in conjunction with general surveying for sites in the vicinity of the line of the proposed Öresund Bridge (Fischer 1997). The site is situated 2 km off the coast. Small deposits of peat have been preserved by the protective presence of large rocks. The find location is at a depth of −8 m. Worked flints were found in the peat but only a single microlithic implement. These finds indicate that the site belongs to the middle part of the Maglemose culture, and it has been dated to 8340±90 and 8190±100 uncal. BP (AAR-1527 and AAR-1526).

Flint artefacts, including blades, were discovered at Foteviken, a deep bay on the southern part of the west coast of Scania (Fig. 11.1) in the course of investigating an early Medieval barrier. These flints were found at a depth of about −3 m (Ingelman-Sundberg and Söderhielm 1982).

Amber is abundant on the beaches of the southernmost part of Öresund close to the small towns of Skanör and Falsterbo. While collecting amber pieces, an amber craftsman who is also an amateur archaeologist has collected bones and worked flints, mostly found on the western side of the small island of Måkläppen (Larsson and Brost 2011) (Fig. 11.1). Several human bones were identified. Radiocarbon dating shows that most of them are from the Iron Age and the Middle Ages, probably representing peoples drowned when shipwrecked. However, two finds date to the Mesolithic: a femur bone dated to 7100±50 uncal BP (LuS 6148) and a large part of a skull dated to 6095±50 uncal BP (LuS 6533). The worked flints that were washed onto the beach are blades that can be related to the late Maglemose and Kongemose cultures. The femur bone probably belongs to the late Maglemose, while the skull dates to an early stage of the Ertebølle culture. The bone might belong to a settlement site or grave that could have been destroyed by currents and washed up on the beach. The skull dates to the period when transgressions reached higher than present sea level and settlements as well as graves might have been eroded, thus ending up on the sea bed and later being washed ashore. Both bone finds have a $^{13}$C value of −14.5 ‰, indicating that marine food was of primary importance.

That heavy erosion of organic layers on the sea bottom is in progress is evident from the pieces of peat that are washed up in large quantities on the beach during winter storms.

## 11.5 Coast-Inland Connections and Submerged Sites

A small number of finds indicate connections between inland sites and the now submerged coastal settlements. The traditional interpretation is either that the hinterland was seasonally exploited by the settlers from the coastal region or that coastal and inland settlements coexisted with contacts between them (Larsson 1980).

During the Early Mesolithic, coastal as well as inland areas were permanently populated. The forest grew denser during the Atlantic period, causing the population of large animals to decrease. At the same time several of the lakes, attractive for fishing and hunting, developed into bogs. Therefore, most of the permanent inland settlements disappeared, and the inland area became a resource area used only seasonally from settlements based on the coast.

Peat extraction has been active for several decades at the bog of Rönneholms Mosse in central Scania, about 40 km from the present coast in western Scania. During the past decade, peat extraction has been combined with intensive surveying. Two find categories are related to coastal settlement. A number of small, unworked flint stones found in rows in a couple of bog camps have been interpreted as net sinkers (Larsson and Sjöström 2011, 2013). This kind of raw material is not found in the area but common in the shore zone. Another find is nine shells with perforations, probably a necklace lost or deposited in the shallow lake. The shells are of *Nassarius reticulatus*, a marine gastropod mollusc (Larsson and Sjöström 2011, p. 460).

A perforated tooth from a grey seal was found at a late Maglemosian site, Ageröd I:HC, a shoreline settlement at the bog of Ageröds Mosse belonging to the same bog complex as Rönneholms Mosse mentioned above from central Scania (Larsson 1978).

When dealing with submarine settlement in southern Scandinavia we should include a special category of sites, namely those that have been submerged because of transgressions during the late

Atlantic when the sea reached 3–4 m above present sea level. During the subsequent Sub-Boreal when the sea retreated, these sites were situated once again on firm land after being submerged for hundreds or thousands of years. Such a site is Skateholm II with a settlement and cemetery situated on a low island in a lagoon dated at about 5200 cal. BC, which was submerged and re-exposed above water at least a couple of millennia later (Larsson 1993). Such sites might provide information about the effects of waves and currents in different types of terrain. As for Skateholm II, the refuse layer was totally eroded away while the occupation layer was less affected and the graves well preserved.

## 11.6 Conclusions

The results of these investigations support an argument that coastal settlement during the Boreal was equally as intensive as that well-documented in the hinterland. The location of archaeological material on the outer estuaries of river courses or in bays points to locations very similar to the coastal settlements from the Late Mesolithic (Larsson 1984, 1988). This indicates that the choice of settlement site was dictated by the same factors in both periods. Whether these Early Mesolithic settlements were as extensive and of as permanent a character as their later variants is a question to which we are still unable to provide a satisfactory answer, because the material recovered so far has suffered erosion, so that the original status of sites cannot be determined with any certainty.

Also, organic material is almost entirely absent from the find locations investigated at Pilhaken, so that it is impossible to gain an insight into the importance of marine food in the subsistence strategy. Nevertheless, the resulting impression is of an extensive site, and this appears to be very different from lake-edge finds dating from the same period, which suggest small settlements comprising a single hut (Larsson and Sjöström 2011).

## References


Björck S (1995) Late Weichselian/early preboreal development of the Öresund Strait; a key area for northerly mammal immigration. In: Larsson L (ed) The earliest settlement of Scandinavia and its relationship with neighbouring areas. Acta Archaeologica Lundensia, Series in 8o, Nr. 24: 123–134

Dencker J, Hansen J-S, Petersen S-B (1994) Marinarkeologiske rekognosceringer efter fredede vrag og fortidsminder i Prøvegravnings- og Prøveomraderne. Miljøministeriet, Skov- og Naturstyrelsen, København

Fischer A (1993) Marinarkreologiske førundersøgelser forud for etablering af en fast Øresundsforbindelse. Stenaldersbopladser pa bunden af Øresund. Miljøministeriet, Skov- og Naturstyrelsen, Hørsholm

Fischer A (1996) Naturvidenskaplige undersøgelser. In: Bartholin T, Christensen K, Dencker J, Fischer A, Gulliksen S, Schou Hansen J, Bydal R, Bødtker Petersen S, Söderhielm P (eds) Marinarkaeologiske rekognosceringer efter fredede vrag og fortidsminder i nøjbrotraceet. Skov- og Naturstyrelsen, København, pp 41–46

Fischer A (1997) Stenalderbopladser på bunden af Øresund. Afprøvning af en model. Del 2, det central og sydlige Øresund. Miljø- og Energiministieriet, Skov- og Naturstyrelsen, København

Hadevik C, Hammarstrand Dehman K, Serlander D (2008) Malmö C Nedre – i samband med byggandet av tunnel och ny stationsbyggnad. Malmö stad RAÄ 20 Skåne län. Malmö Kulturmiljö Enheten för Arkeologi Rapport 2008: 030, Malmö

Hammarstrand Dehman K (2009) Våtmarksarkeologiska undersökningar i Malmö på senare år. Fornvännen 2009 Nr. 1: 1–16

Ingelman-Sundberg C, Söderhielm P (1982) Vikingaskepp Foteviken. Marinarkeologisk undersökning i Foteviken, Skåne

Karsten P, Knarrström B (2003) The Tågerup excavations. Skånska spår – arkeologi längs Västkustbanan. Riksantikvarieämbetet. Avdelningen för arkeologiska undersökningar, UV-syd, Lund

Kindgren H (1995) Reindeer or seals? Some late palaeolithic sites in Middle Bohuslän. In: Larsson L (ed) The earliest settlement of Scandinavia and its relationship with neighbouring areas. Acta Archaeologica Lundensia, Series in 8o, Nr. 24: 191–205

Klein DR (1982) Distribution and status of wild reindeer in the Sovjet union. J Wildl Manag 46(3):728–733
Larsson L (1978) Ageröd I:B – I:D. A study of early Atlantic settlement in Scania. Acta Archaeologica Lundensia 4 (12), Lund.
Larsson L (1980) Some aspects of the Kongemose culture of southern Sweden. Pap Archaeological Inst Univ Lund 1979–1980:5–22
Larsson L (1982) Segebro. En tidigatlantisk boplats vid Sege ås mynning. Malmöfynd 4, Malmö
Larsson L (1983) Mesolithic settlement on the sea floor in the strait of Öresund. In: Masters PM, Flemming NC (eds) Quaternary coastlines and marine archaeology. Towards the prehistory of land bridges and continental shelves. Academic, New York, pp 283–301
Larsson L (1984) Skateholm project. A late Mesolithic settlement and cemetery complex at a south Swedish bay. Pap Archaeological Inst Univ Lund 1983–1984:5–38
Larsson L (1988) The use of the landscape during the Mesolithic and Neolithic in Southern Sweden. Archeology en Landschap. Bijdragen aan het gelijknamige symposium gehouden op 19 en 20 oktober 1987, ter gelegenheid van het afscheid van H.T. Waterbolk, Groningen, pp 31–48
Larsson L (1993) The Skateholm project: late Mesolithic coastal settlement in southern Sweden. In: Bogucki P (ed) Case studies in European prehistory. CRC Press, Boca Raton, pp 31–62
Larsson L (1994) Earliest settlement in southern Sweden. Late Paleolithic remains at Finjasjön, in the north of Scania. Curr Swed Archaeology 2:115–177
Larsson L (1999) Submarine settlement remains on the bottom of the Öresund strait, southern Scandinavia. In: Bintz P, Thevénin A (eds) L'Éurope des derniers chasseurs. Èpipaléolithique et Mésolithique. Actes du 5e colloque international UISPP, commission XII, Grenoble, 18–23 septembre 1995. Editions du Comité des Travaux Historiques et Scientifiques, Paris, pp 327–334
Larsson L (2008) Horse hunters during the deglaciation of southern Scandinavia. In: Sulgostowska Z, Tomaszewski AJ (eds) Man – millennia – environment. Studies in honour of Romuald Schild. The Institute of Archaeology and Ethnology, The Polish Academy of Sciences, Warszawa, pp 99–106
Larsson L, Brost L (2011) Uppspolad forntid. Strandfynd av människor och deras redskap på Måkläppen. Hilstorisk tidskrift för Skåne, Halland och Blekinge 2011, nr. 4, Ale, pp 25–31
Larsson L, Sjöström A (2011) Bog sites and wetland settlement during the Mesolithic: research from a bog in central Scania, southern Sweden. Archäologisches Korrespondenzblatt 4:457–4722011, Heft
Larsson L, Sjöström A (2013) Mesolithic research in the central part of Scania, southern Sweden. In: Johanson K, Tõrv M (eds) Man, his time, artefacts, and places. Collection of articles dedicated to Richard Indreko, Muinasaja Teadus 19. University of Tartu, Institute of History and Archarology, Tartu, pp 487–513
Larsson L, Ljungkvist J, Theander C (1996) Submarine settlement on the bottom of the Öresund strait, southern Scandinavia. NewsWARP 19:8–10
Lepiksaar J (1982) Djurrester från den tidigatlantiska boplatsen vid Segebro nära Malmö I Skåne (Sydsverige) in Larsson, L, Segebro. En tidigatlantisk boplats vid Sege ås mynning, vol 4. Malmöfynd, Malmö, pp 105–134
Mörner NA (1969) The late quaternary history of the Kattegatt Sea and the Swedish West Coast. Sveriges Geologiska Undersökning, ser. C, n° 640
Nordqvist B (1995) The Mesolithic settlement of the West Coast of Sweden with special emphasis on chronology and topography of coastal settlements. In Fischer A (ed) Man and sea in the Mesolithic. Coastal settlement above and below present sea level. Oxbow Monograph 53. Oxford, 185–196
Persson G (1962) En transgressionslagerföljd från Limhamn. Geol Fören Stockh Förh 8:47–55
Rausing G, Larsson L (1977) Pilhaken, en stenåldersboplats under Öresund. Historisk tidskrift för Skåneland 1977, Nr. 2, Ale, pp 1–3
Vang Petersen P, Johansen L (1995) Tracking late glacial reindeer hunters in eastern Denmark. In: Larsson L (ed) The earl iest settlement of Scandinavia and its relationship with neighbouring areas. Acta Archaeologica Lundensia, Series in 8o, Nr. 24: 75–88

# Part III
# Underwater Landscapes and Archaeology

# Chapter 12
# Prospecting for Holocene Palaeolandscapes in the Sound of Harris, Outer Hebrides

**Andrew Bicket, Genevieve Shaw, and Jonathan Benjamin**

**Abstract** The Outer Hebrides Coastal Community Marine Archaeology Pilot Project (OHCCMAPP) considered a range of themes across the Outer Hebrides including Maritime History and Transport, Marine Resource Exploitation and Submerged Prehistory potential. This paper introduces the past landscape component, introducing palaeogeographic reconstructions for the Sound of Harris for the mid-Holocene as proxy scenarios for Mesolithic seascapes. These scenarios are based on publicly-available bathymetric datasets, community-informed field investigations and published sources. The implications for the interpretation of the terrestrial archaeological record are discussed with a focus on maritime connections and distribution of intertidal land and by association coastal resources such as shellfish. Areas of potential are identified for future investigation. The influence of sea-level rise on the coastal configuration is considered for the Mesolithic, in particular the positive impact of increased intertidal zone area and increased penetration into the interior of the landscape by boat. A major seaway is indicated linking the Atlantic to the Minch close to the Harris coast in the early Holocene, which provides a direct context for interpreting the Mesolithic (and later periods) at Northton, Harris within a maritime framework.

## 12.1 Introduction

The Sound of Harris, located in the Outer Hebrides of north-west Scotland separates the Isle of Harris (Fig. 12.1) and the Island of North Uist to the south by a broad seaway interspersed with a series of islands, islets, intertidal sandbars and skerries. The Sound of Harris links the Atlantic Ocean and The Minch (the strait between mainland Scotland and the northern Outer Hebrides) and the western seaways of the British Isles (Fig. 12.2). The western seaways—a network of fjords and small seas linking a number of island groups—have been postulated as key maritime routes since at least the Neolithic period and probably earlier (Garrow and Sturt 2011). Archaeological evidence from the Mesolithic is

A. Bicket (✉)
WA Coastal & Marine, 7–9 North Saint David Street, Edinburgh EH2 1AW, UK

Department of Geography, Loughborough University, Loughborough LE11 3TU, UK
e-mail: a.bicket@wessexarch.co.uk

G. Shaw
AOC Archaeology Group, Edgefield Road Industrial Estate, Loanhead, Midlothian EH20 9SY, UK
e-mail: genevieve.shaw@aocarchaeology.com

J. Benjamin
Department of Archaeology, Flinders University, PO Box 2100, Adelaide 5001, Australia
e-mail: jonathan.benjamin@flinders.edu.au


G.N. Bailey et al. (eds.), *Under the Sea: Archaeology and Palaeolandscapes of the Continental Shelf*, Coastal Research Library 20, DOI 10.1007/978-3-319-53160-1_12

**Fig. 12.1** Aerial photograph of the south coast of Harris looking east from Northton (*bottom left*) to Leverburgh. Skye can be seen on the horizon (*top right*) (©Wessex Archaeology 2011, Photo: J. Benjamin)

preserved widely within these island groups from Ireland, the Isle of Man, Inner Hebrides and Outer Hebrides. The recently re-characterised and re-investigated organic and lithic remains at Northton on the south coast of Harris (Gregory et al. 2005; Simpson et al. 2006; Bishop et al. 2010) represent some of the most distant (geographically) early prehistoric archaeological remains in the British Isles dating to around 6000 BC. There is potential for an earlier phase around 7000 BC (Bishop et al. 2010) with indications of anthropogenic activity long-suggested across the archipelago at this time from palaeoenvironmental records (e.g. Tipping 1996, 2004; Edwards 1996, 2004). Submerged peats have previously been recovered from contexts around Pabbay in the Sound of Harris and dated by Ritchie (1985) from around 3 m below 'mean sea level' broadly contemporaneous with the activity at Northton providing some local-to-regional scale data-points (sensu Bradley et al. 2011) for establishing that mid-Holocene sea levels were lower than present. Within this context coastal land which was previously available for human exploitation and now under water (i.e. submerged palaeolandscapes) may have been preserved, and consideration of this geography and its potential for in situ archaeological material is important for understanding the early prehistory in the area (Sturt et al. 2013; Benjamin et al. 2014).

Prospecting for sites and palaeolandscapes in submerged or intertidal locations can be very challenging. Palaeogeographical reconstructions based upon Glacial Isostatic Adjustment (GIA) models (e.g., Lambeck 1991; Peltier et al. 2002; Shennan and Horton 2002; Bradley et al. 2011) and other techniques aimed at identifying palaeo-shorelines (e.g. Smith et al. 2006) are often regional in scale ($10^3$ km$^2$) or greater, with limited use for resolving human-scale features ($10^{0-1}$ km$^2$) or local coastline configurations ($\geq 10^2$ km$^2$) that can be examined using archaeological techniques. Relative sea-level (RSL) reconstructions (used to test GIA models) are available on the basis of data points that include regional information, for example from the Harris coast of the Hebrides (Ritchie 1985; Jordan et al.

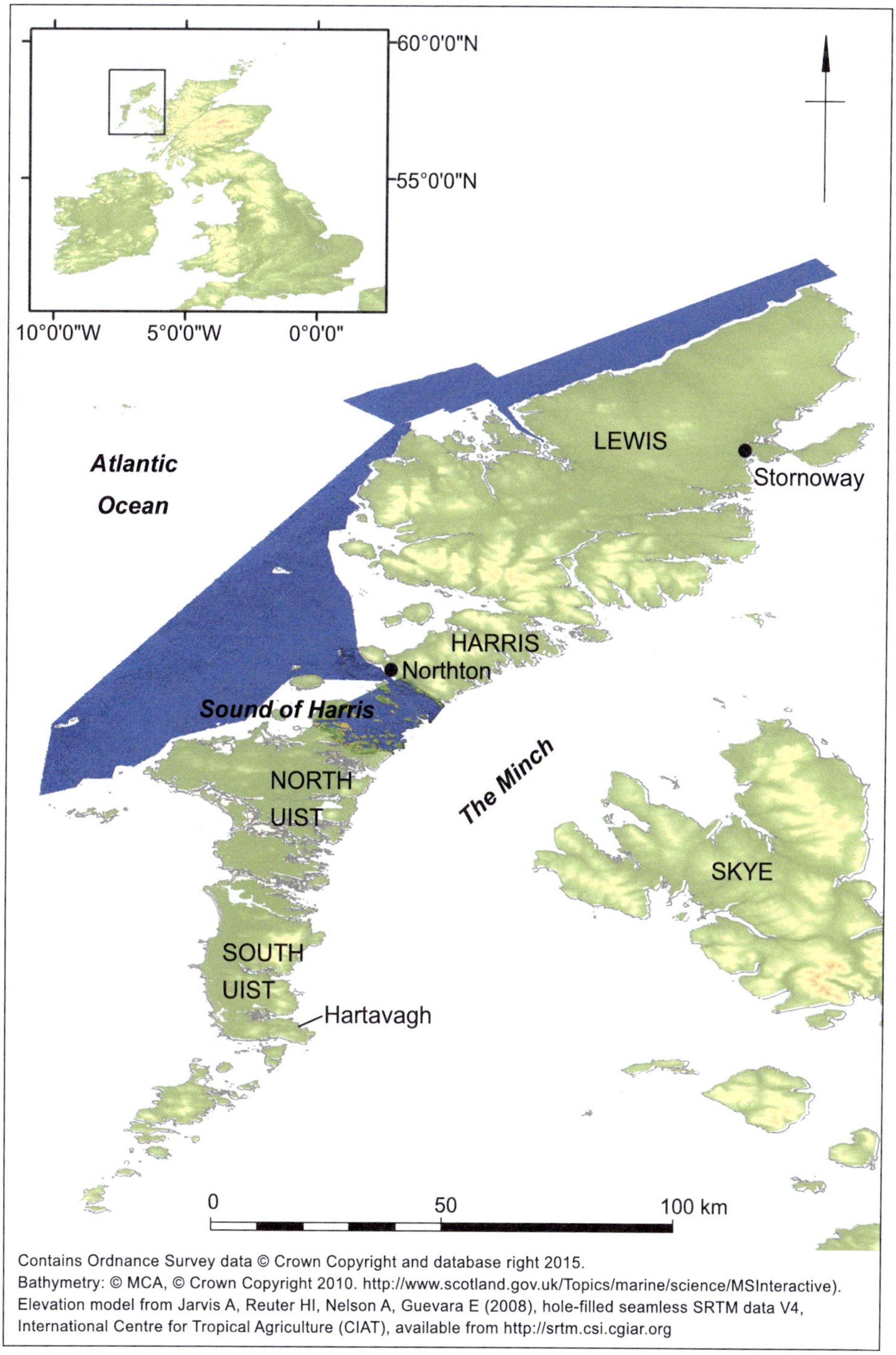


**Fig. 12.2** Location of study area and bathymetric survey data coverage used within this assessment (Contains OS Open Data Crown Copyright 2015; Bathymetry: © MCA, © Crown Copyright 2010. http://www.scotland.gov.uk/Topics/marine/science/MSInteractive). Elevation model from Jarvis et al. (2008), hole-filled seamless SRTM data V4, International Centre for Tropical Agriculture (CIAT), available from http://srtm.csi.cgiar.org

2010; Bradley et al. 2011), and can provide important local context for prospecting for submerged prehistory. However the spatial scale is limited, as highlighted in the substantial variation of modelled relative sea-level from around the British Isles (e.g. Shennan and Horton 2002; Bradley et al. 2011). In our experience when prospecting for new sites and palaeoenvironmental resources, published RSL models are difficult to deploy on the ground at specific sites around Scotland unless sea level index points and RSL models constrained by them are locally available. Clearly, producing a bespoke palaeoenvironmental sea-level reconstruction in every study area is not feasible either in terms of logistics or of data likely to be locally available. The artefactual and palaeoenvironmental samples, the targets for early prehistory prospection, are also small, fully or partly buried and potentially ephemeral, i.e., challenging even for the trained archaeological eye to observe during field survey. To date, chance finds dominate the record of prehistoric artefacts recovered from the continental shelf around the British Isles, consisting primarily of artefacts and faunal material recovered from fishermen's nets (Wessex Archaeology 2004) or occasionally from geotechnical samples (Long et al. 1986; Wessex Archaeology 2009; http://www.splashcos-viewer.eu). Archaeologically investigated sites are rare (i.e., Bouldnor Cliff, Momber 2011; Momber et al. 2011; Area 240, Tizzard et al. 2014; Bicket et al. 2014).

A further limiting factor is the availability of geophysical and geotechnical datasets from offshore and nearshore locations to provide a basis for archaeological assessments of palaeolandscape potential. However, with the development of open-access archiving and data portals, availability is more driven by the geographic distribution of existing data rather than problems of access to the data itself (MAREMAP/INSPIRE etc.). Substantial baseline datasets relevant to submerged prehistory exist for the last 1 million years of human activity in the southern North Sea, the British Isles and the English Channel through initiatives such as the Regional Environmental Characterisations (RECs[1]) undertaken through the Marine Aggregates Levy Sustainability Fund (MALSF) or the Marine Aggregates Regional Environmental Assessments (MAREAs[2]). With the end of funding streams such as the ALSF in 2011 (Dellino-Musgrave et al. 2009; Flatman and Doeser 2010; Bicket 2011, also see UK chapters in Evans et al. 2014), development-led assessments of the seabed have become the dominant source of baseline data (Wessex Archaeology 2013; Bicket et al. 2014).

Freely-available geophysical datasets can be obtained through government bodies, such as Marine Scotland,[3] the Maritime and Coastguard Agency (MCA) Civil Hydrography Programme and UK Hydrographic Office (UKHO) (see Wessex Archaeology 2011[4]). However, the software required to process and analyse them is typically expensive and requires specialist training and experience of undertaking assessments for archaeological purposes. Within this overall context, where opportunities exist for the assessment of these available datasets, they do in some cases provide sufficiently detailed spatial resolution to be of use for archaeological prospection (Wessex Archaeology 2011). The recently completed Outer Hebrides Coastal Communities Marine Archaeology Pilot Project (OHCCMAPP) (Benjamin and Hale 2012; Benjamin et al. 2014) provided a thematic basis for examining marine resource exploitation, maritime history and transport, and submerged prehistory within the Outer Hebrides as a whole; the Sound of Harris was one of the study areas under examination (Fig. 12.2).

In this paper we develop a set of prospection models for submerged prehistory in the Outer Hebrides based on the thematic baseline developed by OHCCMAPP, focusing particularly on marine resource exploitation and the intertidal zone in relation to maritime transport and available evidence for Postglacial and Holocene sea-level change. This is framed by the current span of the archaeological record in Scotland—c. 14,000 years—from the Later Upper Palaeolithic to the present (Ballin et al.

[1] http://www.marinealsf.org.uk/

[2] http://www.marine-aggregate-rea.info/

[3] http://www.gov.scot/Topics/marine/science/MSInteractive

[4] http://www.wessexarch.co.uk/projects/marine/scotland/historic-scotland-marine-data-audit

2010). We present palaeogeographic scenarios for the earliest archaeological period in the Outer Hebrides, i.e., the Mesolithic (c. 7000 BC) with a particular focus on the Sound of Harris, and discuss the implications for the interpretation of maritime activity, subsistence and land use, and for future prospection of archaeological material.

## 12.2 Methods

In order to provide a basis for hypothesis building, prospection and testing, palaeogeographic models incorporating a variety of parameters are valuable tools in lieu of known submerged sites or reported materials. The key variables are reconstructions or models of past relative sea level, an archaeological framework which defines the periods of interest, seabed bathymetry and ideally sub-bottom geophysical survey data in areas of sediment accumulation, and estimates of palaeo-shorelines and palaeogeographic configurations. Models of this type can and have been developed for Pleistocene palaeolandscapes (e.g., Cohen et al. 2011; Hijma et al. 2012; Tizzard et al. 2014) and Holocene ones (e.g. Gaffney et al. 2009; Westley et al. 2011, 2014; Bates et al. 2013).

### *12.2.1 Geophysical Datasets*

During Year 1 of the OHCCMAPP, marine bathymetry datasets were identified that, when processed, would be publicly available for research purposes. These are: north-west Lewis multibeam bathymetry (produced by Marine Scotland and British Geological survey, BGS), Sound of Harris bathymetric LiDAR (MCA 2004) and multibeam bathymetry (UKHO), and Stornoway Approaches & Loch Erisort (Marine Scotland) multibeam bathymetry. The Stornoway Approaches & Loch Erisort dataset was judged to be too far offshore for easily assessing palaeolandscape potential, and the north-west Lewis dataset has been shown also to be too far offshore to directly inform current estimates of early Holocene palaeolandscapes within the parameters described below. The north-west Lewis coastal study area benefits from a combined dataset of offshore and inshore multibeam bathymetry surveys gathered by Marine Scotland (MS) and the British Geological Survey (BGS) extending into East Loch Roag. This inshore area is of particular archaeological interest for examining prehistoric palaeolandscape potential in the Islands due to its proximity to major archaeological areas such as Calanais and associated palaeoenvironmental records which hint at human activity around 8,000 years ago (Edwards 1996, 2004).

The focus of this paper is OHCCMAPP Study Area 7 (Benjamin et al. 2014) (Fig. 12.2). The bathymetry is mainly LiDAR-based and as such abuts seamlessly with the coastline. The north-western portion extending towards Toe Head and Northton, and to the northern coast of Pabbay, is based on multibeam data (the southern coasts of Pabbay are the source areas for the submerged peats investigated in Ritchie (1985)). The Sound of Harris geophysical datasets and local palaeoenvironmental control are therefore of great value for assessing the potential for submerged prehistoric archaeology. The local archaeological record is represented by the multi-period archaeological remains at Northton, with Mesolithic, Neolithic, Bronze Age and Iron Age material, and also material from historical periods (Simpson et al. 2006).

**Table 12.1** Data management parameters used for palaeogeographic models for periods outlined in Table 12.2

| Parameter | Parameter | Source | Notes |
|---|---|---|---|
| Tidal range | ±2.5 m | National Oceanographic Centre | Approximation based on HAT/LAT Stornoway:[a] |
| Bathymetry Survey Chart Datum – OD Stornoway | −2.93 m | MCA (2004) | OD Stornoway is an approximation of mean sea level developed from historical measurements[b] |
| | | | Chart Datum is derived from local measurements at Bays Loch and Leverburgh[c] |

[a]http://www.pol.ac.uk/ntslf/tgi/portinfo.php?port=stor.html
[b]http://www.ordnancesurvey.co.uk/gps/legacy-control-information/levelling-datum
[c]http://webarchive.nationalarchives.gov.uk/20121107103953/http://www.dft.gov.uk/mca/sound_of_harris.pdf

**Table 12.2** Relative seal-level values underpinning palaeogeographic scenarios, c. 7000–6000 BC

| Model | Bathymetry mean RSL, m OD | Source |
|---|---|---|
| Mesolithic A | −10 | Gregory et al. (2005), Shennan et al. (2006), and Bishop et al. (2010) |
| Mesolithic B | −8 | Peltier et al. (2002), Gregory et al. (2005), Bishop et al. (2010), and Bradley et al. (2011) |
| Mesolithic C | −5 | 'high-stand scenario', Gregory et al. (2005), Smith et al. (2006), and Bishop et al. (2010) |
| Mesolithic D | −15 | 'low-stand scenario'. Lambeck (1993), Peltier et al. (2002), Gregory et al. (2005), and Bishop et al. (2010) |

### 12.2.2 Data Management

The datasets have been referenced to Ordnance Datum (original geophysical surveys are corrected to Chart Datum) to permit calibration to existing relative sea-level models (e.g., Shennan et al. 2006; Bradley et al. 2011) for the Mesolithic period (summarised in Table 12.1).

The assessment of the bathymetry focussed on identifying major submerged landforms including valleys, partially infilled palaeo-channels, inundated islets, palaeo-shoreline configurations and areas of inundated coastal land. The resolution of the multibeam bathymetry is c.1 m cell-size and the LiDAR data approximately 1.5–3 m cell-sizes. This is sufficient for providing geomorphological assessments in support of ground-testing and national research objectives. Digital Elevation Terrain was analysed in software package Fledermaus. Geotiffs with a cell-size of c. 3 m were exported for further analysis in ArcGIS 10, combined with archaeological datasets and field observations.

### 12.2.3 Modelling Parameters

Several parameters have been defined to produce modelled scenarios for the Mesolithic period (Tables 12.1 and 12.2). The addition of a general tidal range parameter has been included to align the analysis with the existing objectives of OHCCMAPP, particularly theme of Marine Resource Exploitation. Instead of focussing upon a single value for mean sea level and therefore producing a defined palaeo-shoreline, the resulting models are aimed at identifying environmental zones. With a tidal range currently of around 5 m in this part of the British Isles, the 'intertidal range' is more representative of the coastal environment. It also provides a useful buffer that can 'absorb' some of the uncertainty associated with local datum points and RSL models. In the scenarios presented here an estimate is made of the terrestrial, intertidal and fully submerged areas, based on the parameters in Table 12.1. This may be of particular use for developing prospection models that incorporate resource gathering behaviours in prehistory (as well as during later periods). The use of generalised zones may also inhibit the over-interpretation of the palaeogeographic reconstructions.

Submerged peats in the vicinity of Quinish and Forvath, Pabbay, and Manish beach, Ensay (both outside the geophysical survey coverage) provide some indication of local resources for future study (Ritchie 1980, 1985). Calibration of the Pabbay dates using the IntCal 09 dataset (Reimer et al. 2009) indicates submerged peats at depths of around 2.8 m below 'mean tide level' and some 80 m seaward of the low tide mark from Quinish (Ritchie 1985). These date to 7530–6450 cal. BC (2σ). Intertidal peat from Quinish beach dates to 3100–2900 cal. BC at around 0.5 m RSL (Fig. 12.2). The older dates are comparable to recent dates from Northton (Gregory et al. 2005; Bishop et al. 2010) and provide limiting points for regional relative sea level models (Shennan et al. 2006) (Fig. 12.3).

Based upon the earliest dates for the earliest Mesolithic stratigraphy at Northton, of c. 7060–6650 cal. BC (Gregory et al. 2005; Bishop et al. 2010), we use the c. −10 m RSL relative to Ordnance Datum (OD) at c.7000 BC (Table 12.2, Mesolithic A). Further iterations of the model are based on RSLs of −8 m OD (Mesolithic B), a 'high-stand sea-level' scenario at −5 m OD (Mesolithic C), and a 'low-stand sea-level' scenario at −15 m OD (Mesolithic D) which generalise the potential effects of lower sea levels (Table 12.2). The −15 m reconstruction is comparable to the 'Storegga shoreline' of Smith et al. (2006, Fig. 12.2a). See also an overview of the regional variation in Holocene RSL models in Jordan et al. (2010).

We appreciate the effects of isostasy and local crustal submergence or uplift, but we have not currently incorporated these into the model due to a lack of local and regional values for Holocene timescales (cf. Jordan et al. 2010). Models for the UK indicate that the study area is on the fringes of isostatic readjustment (Gehrels 2010; Smith et al. 2011).

## 12.3 Results

As Garrow and Sturt (2011) have recently discussed, the enduring features of the palaeogeography are the maritime configuration of islands and relatively sheltered and interconnected seas (Fig. 12.1). At a national and regional-scale they have discussed the feasibility of Late Mesolithic and Early Neolithic maritime connectivity. With the suite of palaeogeography models developed for the Sound of Harris we can apply these concepts directly to our study area.

Figures 12.3, 12.4, 12.5, and 12.6 present four palaeogeography scenarios for the Sound of Harris using different bathymetric contours (Table 12.1). Models A (Fig. 12.3) and B (Fig. 12.4) are based upon the most recent RSL models although they are constrained by few (but local) data points for the Outer Hebrides (see Bradley et al. 2011, Hebrides (7) model). The most striking feature of the bathymetry datasets is the shallow rocky seabed that characterises most of the central and southern areas of the Sound, although it is overlain by considerable expanses of sand bars and sand waves between Berneray and North Uist. Ideally, we would include reconstructions of the land surfaces buried beneath marine sands, but we do not have the necessary geophysical data to achieve this. Instead, we rely on results from similar coastal environments elsewhere (Westley et al. 2014), which demonstrate that bathymetric data provide a good approximation of the submerged topography for the purposes of general palaeogeographic reconstruction.

The largest and most enduring bathymetric feature highlighted by the models is the open channel running SE–NW along the Harris coast passing by Leverburgh and Northton, directly linking the Minch with the Atlantic coast; except for Model D which requires a high tide to flood the intertidal land west of Leverburgh (Fig. 12.6). A certain level of shelter may have been afforded by Chaipaval,

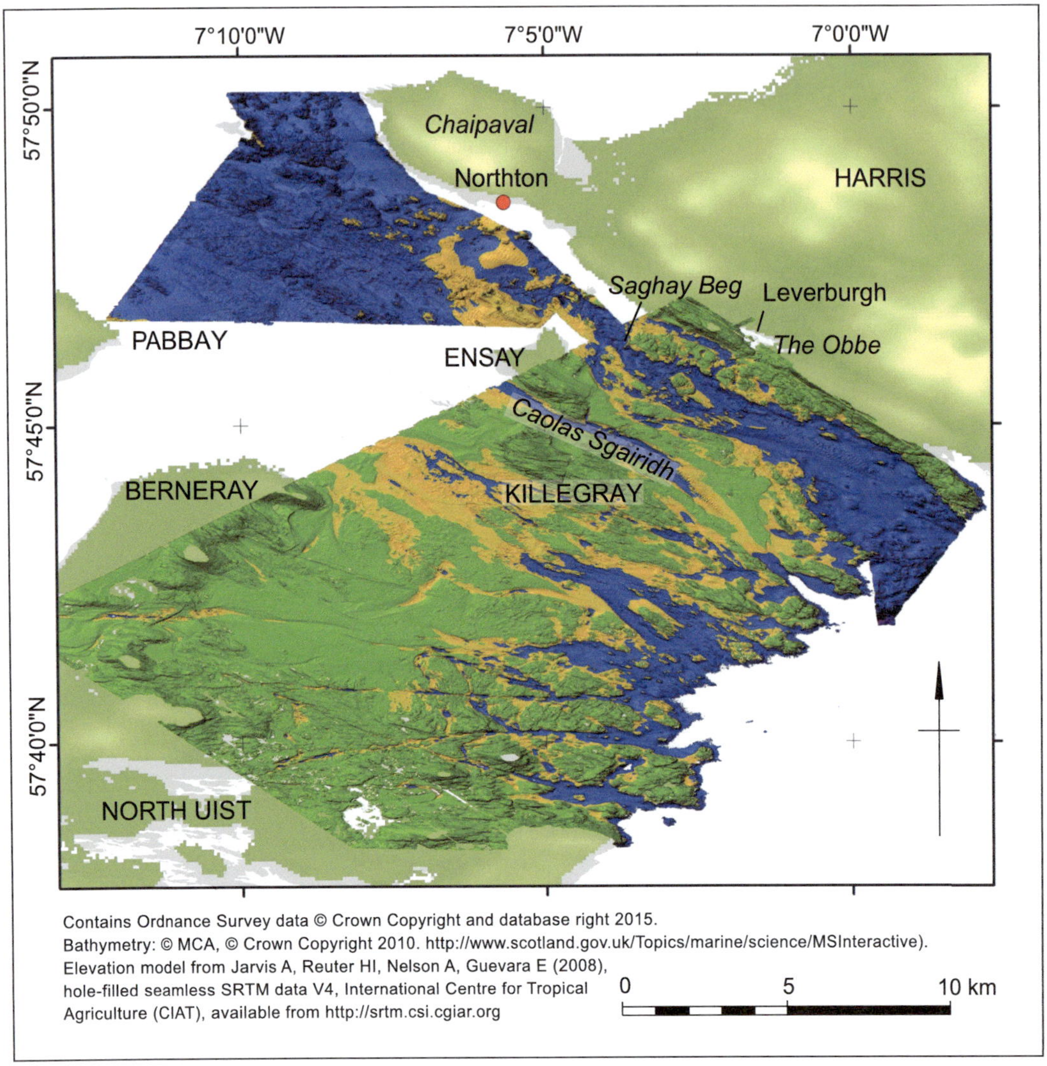


**Fig. 12.3** Scenario A, Mesolithic period palaeogeography for the Sound of Harris c. 7000–6000 BC (Contains OS Open Data Crown Copyright 2015; Bathymetry: © MCA, © Crown Copyright 2010. http://www.scotland.gov.uk/Topics/marine/science/MSInteractive). Elevation model from Jarvis et al. (2008), hole-filled seamless SRTM data V4, International Centre for Tropical Agriculture (CIAT), available from http://srtm.csi.cgiar.org

Ensay and Killegray and it seems feasible from these models that a direct seaway existed in the early-to-mid Holocene through the Sound of Harris when sea levels were lower than present.

Due to the significant volumes of seabed sediment in the South-west of the study area, uncertainties in the configuration of the coastline exist, but these are reduced in areas of exposed bedrock. Sea-level rise resulting from key events in the northern hemisphere in the early Holocene, in particular the breaching of Lake Agassiz and the transfer of huge volumes of water into the North Atlantic (see Weninger et al. 2008, Table 4; Wanner et al. 2011), is roughly contemporaneous with the Mesolithic artefact layers at Northton (c. 6500–6000 BC) and would have significantly contributed to landscape inundation of critical importance to contemporaneous Mesolithic populations (Smith et al. 2011). At this time the Northton site would have been several metres above the high tide.

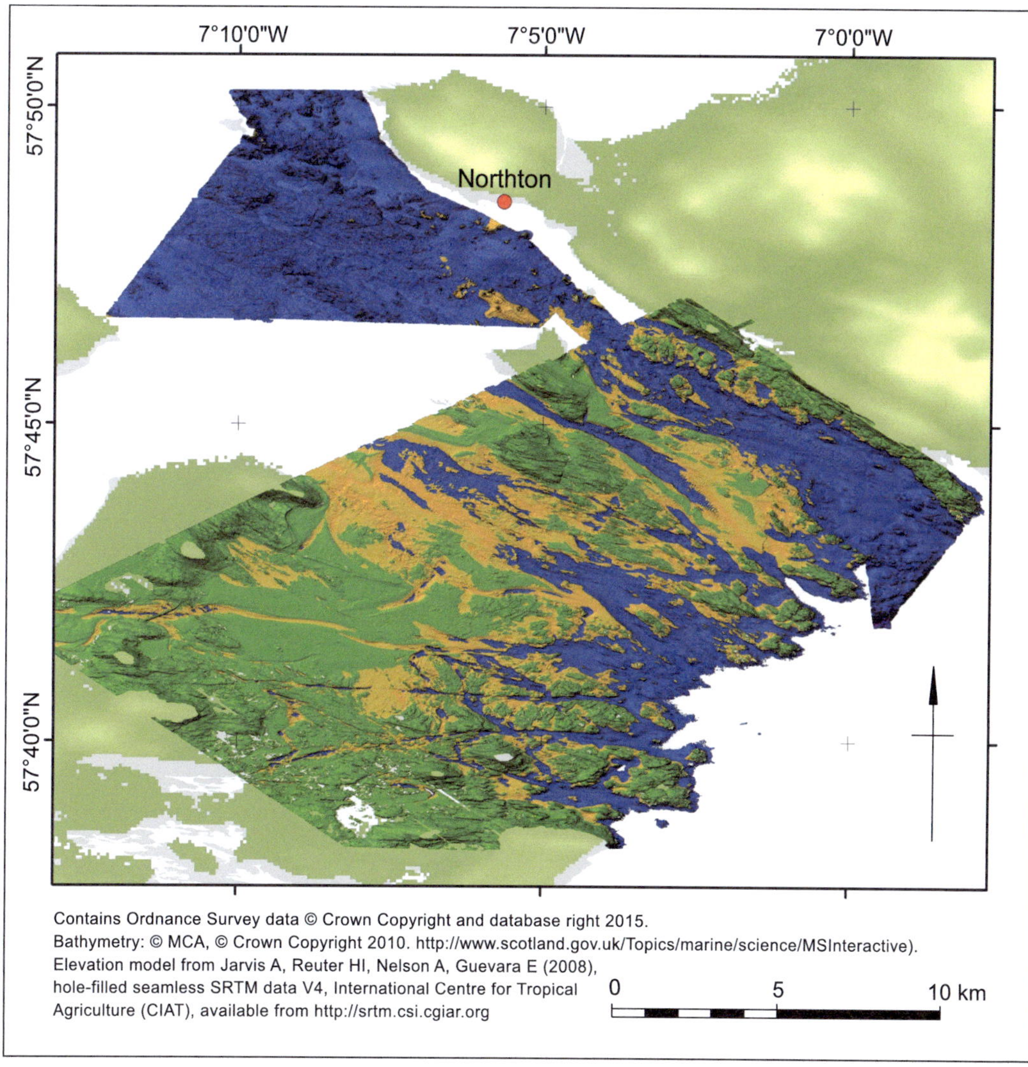


**Fig. 12.4** Scenario B, Mesolithic period palaeogeography for the Sound of Harris c. 7000–6000 BC (Contains OS Open Data Crown Copyright 2015; Bathymetry: © MCA, © Crown Copyright 2010. http://www.scotland.gov.uk/Topics/marine/science/MSInteractive). Elevation model from Jarvis et al. (2008), hole-filled seamless SRTM data V4, International Centre for Tropical Agriculture (CIAT), available from http://srtm.csi.cgiar.org

With a relatively low sea level, as in Model D, the configuration of the coast around the time of Mesolithic activity preserved at Northton would have created an intertidal zone distributed very unevenly, with a paucity on the Minch coasts and a much more extensive intertidal area in the north-west around Northton and approaching Pabbay. Perhaps this palaeogeographical configuration reflects a preference of the Northton people for exploiting coastal resources linked to the largest available areas of intertidal land. As sea level 'rises' throughout Models B, A and C, the area of intertidal increases significantly, becoming more evenly distributed around the Sound of Harris, and at high tides permitting a number of maritime transport routes between the Minch and Atlantic coasts. The Models also suggest that a subsistence strategy incorporating maritime travel and marine resources such as shellfish and seals was increasingly favoured by the more extensive intertidal areas and other palaeogeographic changes associated with early Holocene sea-level rise in the Sound of Harris.

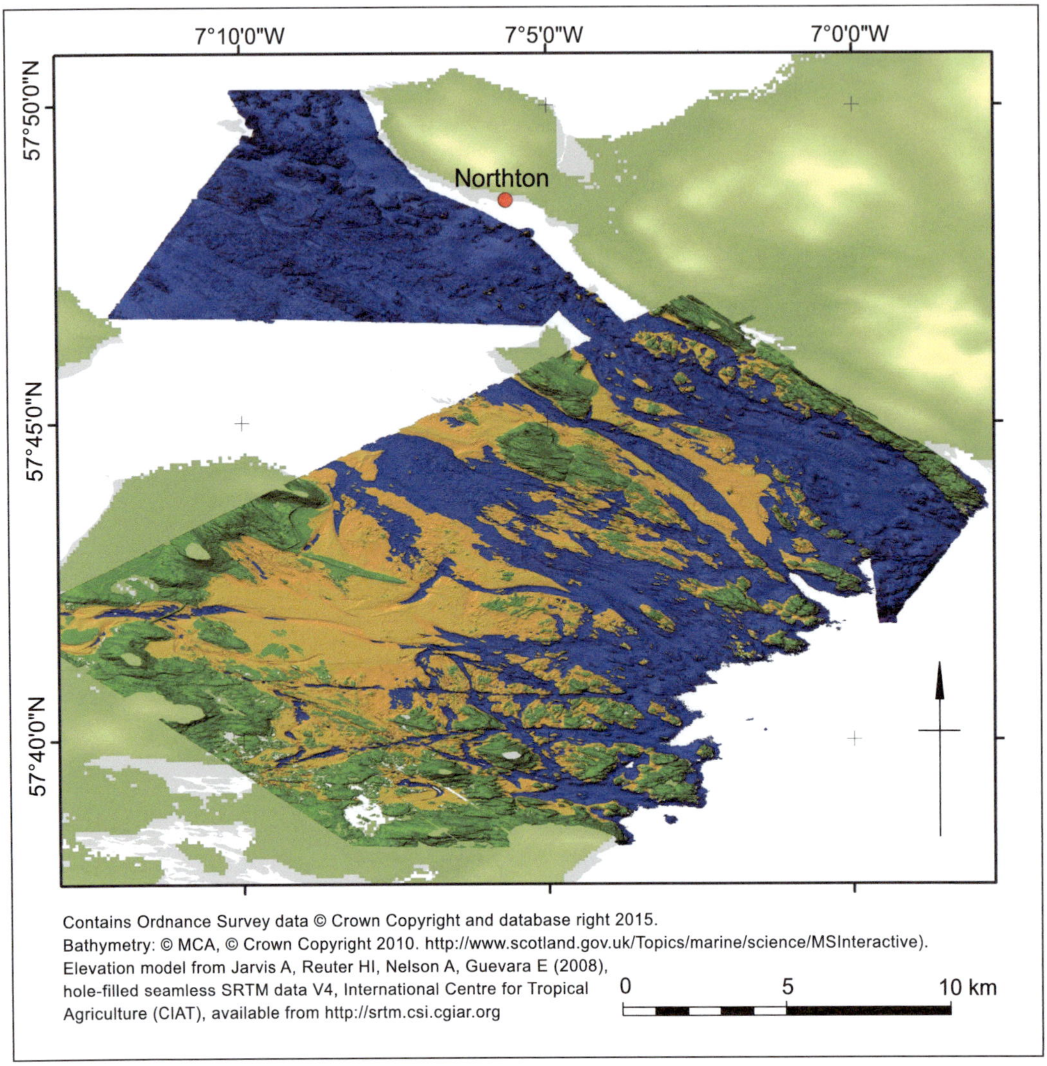


**Fig. 12.5** Scenario C, Mesolithic period palaeogeography for the Sound of Harris c. 7000–6000 BC (Contains OS Open Data Crown Copyright 2015; Bathymetry: © MCA, © Crown Copyright 2010). http://www.scotland.gov.uk/Topics/marine/science/MSInteractive). Elevation model from Jarvis et al. (2008), hole-filled seamless SRTM data V4, International Centre for Tropical Agriculture (CIAT), available from http://srtm.csi.cgiar.org

The Storegga tsunami that affected the east and north coasts of Scotland at around 6100 BC and had a wider impact on the North Sea basin including the area often referred to as 'Doggerland' (Coles 1998) is generally argued to have had a catastrophic impact on coastal populations (Weninger et al. 2008; Smith et al. 2011). However, the process of inundation associated with the breaching of Lake Agassiz was slower, occurring over a period of months, and the significant loss of coastal dry land may have been compensated in coastal regions such as the Sound of Harris by a substantially increased extent of intertidal resources together with greater penetration of seaways into the interior, arguably a favourable and varied set of conditions for Mesolithic subsistence strategies which were spreading around the coasts of the northern British Isles at this time (Waddington 2007). At a higher relative sea level scenario of around −5 m, this 'benefit' becomes tempered by a loss of land (coastal or otherwise)

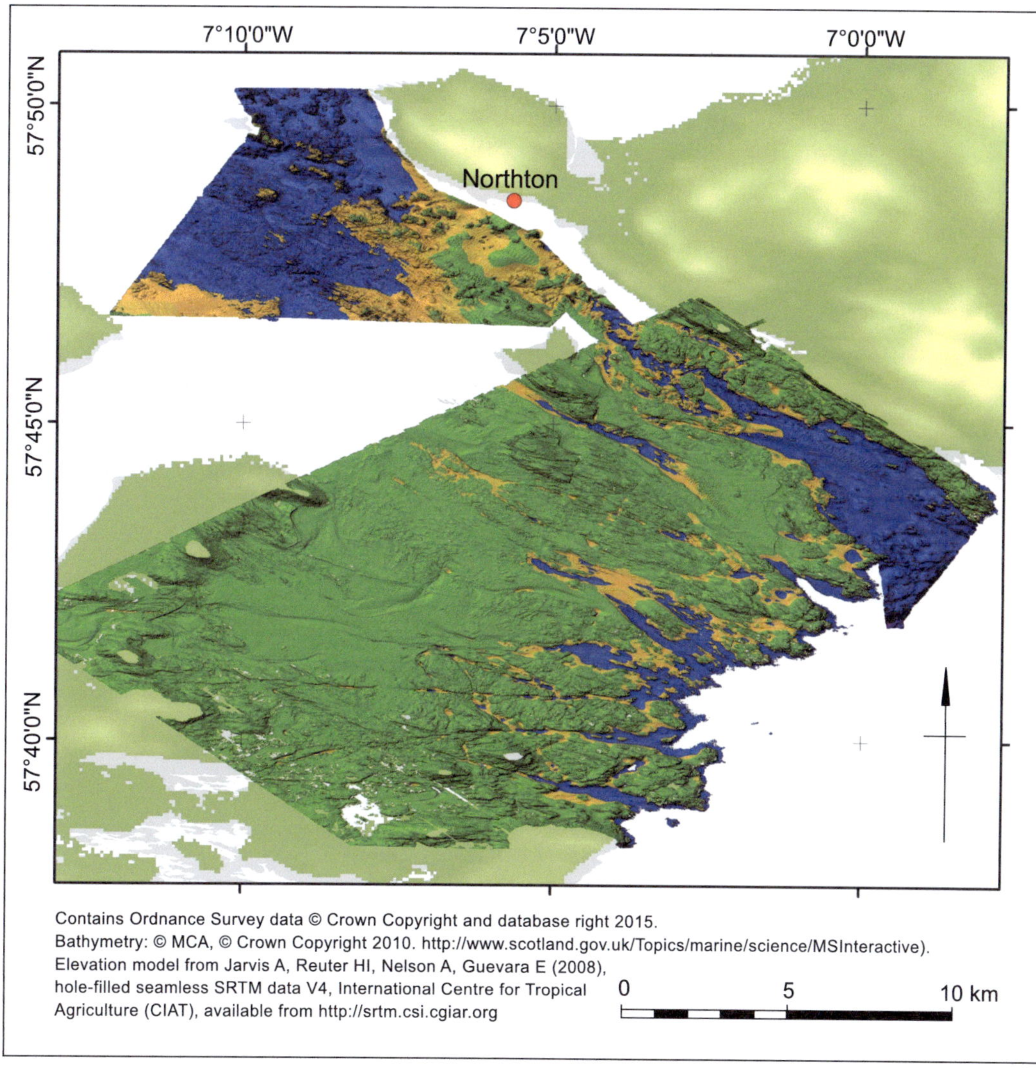


**Fig. 12.6** Scenario D, Mesolithic period palaeogeography for the Sound of Harris c. 7000–6000 BC (Contains OS Open Data Crown Copyright 2015; Bathymetry: © MCA, © Crown Copyright 2010. http://www.scotland.gov.uk/Topics/marine/science/MSInteractive). Elevation model from Jarvis et al. (2008), hole-filled seamless SRTM data V4, International Centre for Tropical Agriculture (CIAT), available from http://srtm.csi.cgiar.org

but with significantly increased connectivity of permanent seaways between the Minch and Atlantic coasts (Table 12.2).

From a taphonomic perspective, the relative increase in the rate of marine transgression following rapid meltwater pulses such as that associated with the breach of Lake Agassiz, may lead to favourable conditions for the preservation of archaeological material; inundation is relatively quick and coastal archaeological and peat deposits are impacted by tides and waves for a shorter period of time than would be the case with eustatic sea-level changes generated by general ice-cap melting alone. These factors are key considerations for future prospection and field testing in submerged contexts in the Outer Hebrides. Conditions suitable for in situ site preservation do clearly exist; examples of submerged peat from the Sound of Harris have been preserved since at least 7500 BC (Ritchie 1985)

(Fig. 12.3), with some new examples of early Holocene peat deposits recently reported from the east coast intertidal zone on South Uist[5] (Benjamin et al. 2014).

A distinctive feature of the models are the bathymetric deeps formed on fault lines, particularly the strait between Ensay and Killegray, Caolas Sgairidh, but also on smaller faults off the North Uist coast (Fig. 12.7) mapped by the BGS (Fettes et al. 1992). The faults add considerable variation to the modelled biotopes. Substantial stretches of the intertidal zone provide marine resources such as molluscs and embayments with narrow mouths provide locations for fish traps (cf. Fischer 2011). Intertidal shelves and beaches may also provide seasonal opportunities for hunting seals. A significant element

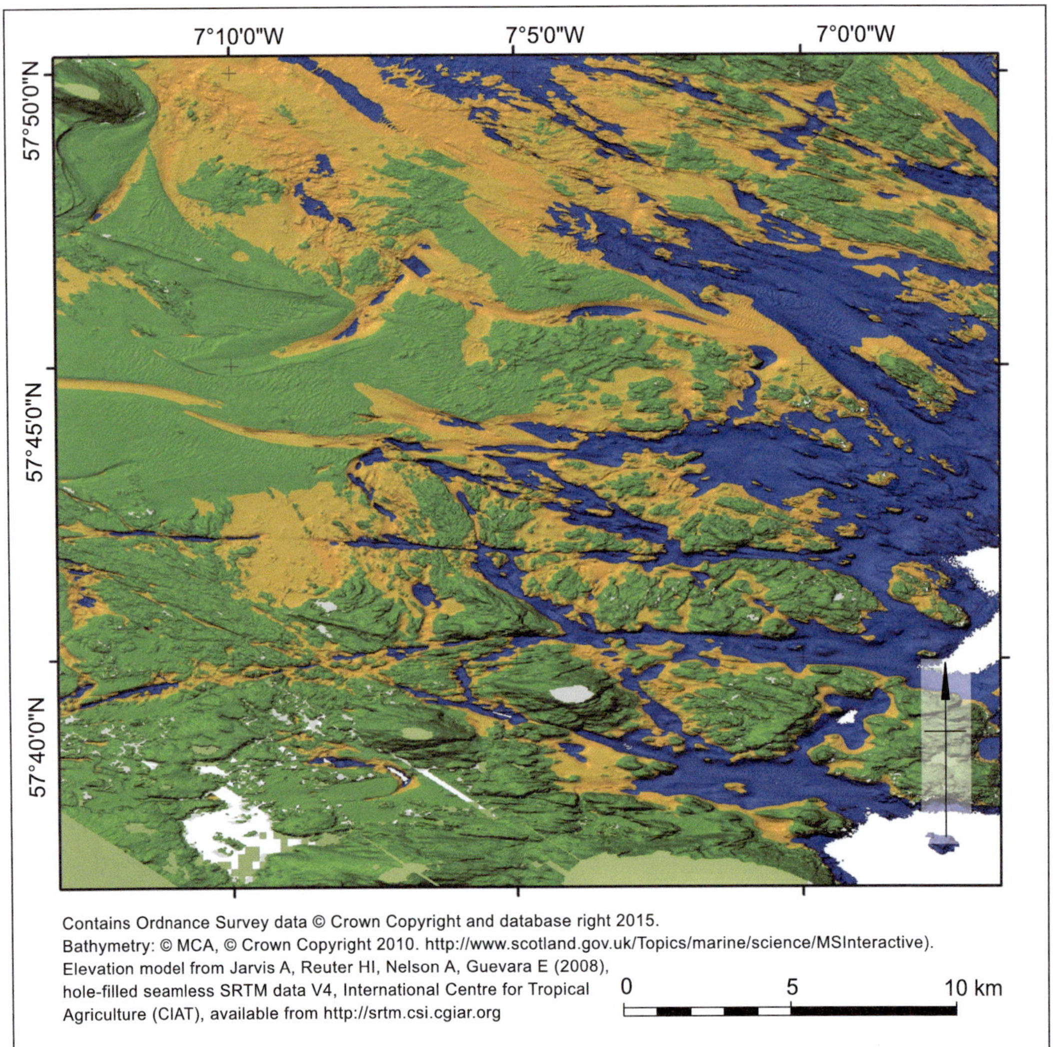


**Fig. 12.7** Influence of fault lines on maritime accessibility within the Sound of Harris, North Uist coast (Contains OS Open Data Crown Copyright 2015; Bathymetry: © MCA, © Crown Copyright 2010. http://www.scotland.gov.uk/Topics/marine/science/MSInteractive). Elevation model from Jarvis et al. (2008), hole-filled seamless SRTM data V4, International Centre for Tropical Agriculture (CIAT), available from http://srtm.csi.cgiar.org

[5] Dated peats from the intertidal zone at Hartavagh, South Uist: Tr1-4/50 (SUERC-42567/GU28481): 9643 ± 27 bp (9229–9119 cal BC (52%); 9007–8915 cal BC (36%)). Tr1-4/55 (SUERC-42568/GU28482):10135 ± 28 bp(9896–9745 cal BC (55.5%); 10027–9908 cal BC (32.6%)), calibrated using OxCal 4.1.7 using IntCal 09 curve.

for building hypotheses about Mesolithic activity is the maritime connectivity between different areas indicated by the palaeogeographic reconstructions. The faults form linking seaways that afforded ease of access and efficient, direct transportation, by boat, and this is especially significant in view of the evidence for the use of boats and resources used in their construction in the Mesolithic of western Scotland Bonsall et al. (2013).

## 12.4 Discussion

### *12.4.1 Maritime Technology*

Hypotheses about Mesolithic access to the Outer Hebrides c. 7000 BC depend on a number of factors, notably transport across open, deep water by boat, whether by eastern or southern routes (Waddington 2007). Boats would have been required to cross the Minch from mainland Scotland throughout the Holocene (Sturt et al. 2013). If seasonal return voyages were necessary to avoid winter conditions (which may have been colder than today, Garrow and Sturt 2011), the technology would have to have been well enough developed to ensure that boats could be repaired or replaced with the resources available locally (Bonsall et al. 2013). Trees suitable for logboat building may have been limited in a birch-dominated mid-Holocene Outer Hebrides, with pollen records of <3% *Pinus* and *Quercus* (Tipping 1996; Edwards 2004). Land mammals with large hides suitable for making coracles or currachs would also arguably have been relatively restricted. Seals would have offered an alternative, particularly Grey Seals, which provide larger hides than Harbour Seals., Grey Seals are easily accessible during the moulting and breeding seasons, and they are also reportedly widely distributed in the Islands, at least in modern conditions (Scottish Natural Heritage 2011).

A range of seafish may have been an element of the Mesolithic subsistence in the Outer Hebrides. Midden evidence for marine resources from early prehistory could imply fishing in deep water or offshore locations (Pickard and Bonsall 2004), but, equally, fish species that have been sought in more recent times using boats and long lines and nets could in the past also have been retrieved occasionally from shallow water requiring less advanced methods (Bonsall et al. 2013). Therefore the presence of offshore species, such as cod, in the Mesolithic midden deposits at Northton is not surprising.

Although we have emphasised the marine focus here, we should also note the abundant distribution of freshwater lochs, lochans, rivers, streams, springs and brackish estuaries adjacent to the marine environment across the Islands as critically important features of the landscape of relevance to the archaeological record (Benjamin and Hale 2012; Benjamin et al. 2014).

### *12.4.2 Why Northton?*

Access by boat during the Mesolithic was clearly possible, but may have been seasonal, infrequent and of limited extent. Despite its attractive location for the use of boats, it could be argued that the preservation of Mesolithic material at Northton may simply reflect preservation bias in aeolian dune contexts. However, based on our palaeogeography models, perhaps the more pertinent question is, 'why not Northton?' The enduring feature of all the models of the Mesolithic palaeogeography along the northern fringe of the Sound of Harris is a wide marine seaway, the Obbe, passing Rodel, Leverburgh and Ensay and linking the Minch to the Atlantic. With wide, low-lying coastal plains to the west fringed by the higher ground of Uist and Berneray and the Ensay and Killegray Hills, the narrow "Strait of Saghay Beg" provided direct access through the Outer Hebrides during the mid-Holocene. In addition, Northton, located in the lee of Chaipaval (365 m) would have afforded clear views west and north across the Atlantic and south and east across the complex network of intertidal islands and narrow seaways.

### *12.4.3 Prospecting for Sites and Palaeolandscapes*

It is clear from our palaeogeographical reconstructions and the available datasets and RSL models, that there are distinct areas within the Sound of Harris that may preserve in situ sediments, in locations which may have been intertidal or terrestrial environments since the Mesolithic (Fig. 12.2). Similar palaeogeographic approaches in Ireland and Northern Ireland have successfully been used to locate in situ artefactual material (Westley et al. 2014). The regional potential for submerged early prehistoric sites is indicated by in situ early Holocene lithics from around the Inner Sound, particularly at Clachan Harbour, Raasay off the east coast of Skye (Ballin et al. 2010) Also, recent examinations of the intertidal zone at Upper Loch Torridon on the western coast of the Scottish mainland have located thousands of lithics at Lub Dubh-Aird (Hardy et al. 2016).

Due to the pattern of currents that pass through the Sound of Harris, there is considerable spatial variation in the accumulation of seabed sediments. This is most notable when comparing the large sand banks and sand waves in the southwest between Berneray and North Uist, and the exposed bedrock to the east of the bathymetric survey area. The north-west to south-east axis of currents between the Atlantic and the Minch leads to a pattern of seabed sediment accumulation in the lee of headlands and islands, typically on the south-east side. The distribution of these leeward sediment accumulations provides a target for prospecting for preserved Holocene intertidal or terrestrial sediments.

The palaeogeography models provide an initial basis for prospecting for early prehistoric material by locating inundated peat deposits, and these in their turn may offer palaeoenvironmental data and dates and possibly the establishment of sea level index or limiting points. On the north coast of the Sound of Harris the landscape around Leverburgh provides a focus. It is located on an ecotone associated with a marine environment and large bodies of freshwater lochs within a system of valleys that link the east, west and south coast of Harris. A submarine shelf which appears to have accumulated sediment in the lee of a rocky headland also provides a target for investigating submerged prehistoric material (Fig. 12.2). The seaward edge of the shelf is sharply defined in the bathymetry, in a similar way to the Bouldnor Cliff formation in the Solent, southern England (Momber 2011). This may indicate erosion by currents passing through the Sound of Harris that have excavated a channel exposing earlier terrestrial deposits with the potential for revealing archaeological material. Sheltered coves with large areas that have preserved submarine sediments also exist on Ensay, Killegray, and Berneray with preserved submerged peat recovered from Pabbay (Ritchie 1985). Diving on these locations is difficult, but they could also be investigated by anchored platforms for recovering geotechnical cores and possibly by small, shallow-draft boats at high-tide for sub-bottom geophysical surveys.

**Acknowledgements** The authors would like to thank Historic Scotland for funding the OHCCMAPP. The support of various members of the communities of Grimsay, Locheynort, Stornoway, Manish and Lochboisdale greatly aided the project team in the field. Marine Harvest at Lochboisdale is thanked for logistical support and transport to and from Hairteabhagh. Philip Robertson and Rod McCullough at Historic Scotland are thanked for supporting the radiocarbon dating assay and geophysical analyses and the project in general. Various staff members made significant contributions to the field and reporting efforts of the OHCCMAPP from within the project partner organizations, in particular Abby Mynett, John McCarthy, Simon Davidson (WA Coastal & Marine), George Geddes, Dave Cowley and Kevin Grant (RCAHMS). Additionally, we thank all those who reported finds sites and stories to the OHCCMAPP team for their time and interest.

## References

Ballin TB, Saville A, Tipping R, Ward TAM (2010) An upper Palaeolithic flint and chert assemblage from Howburn farm, south Lanarkshire, Scotland: first results. Oxf J Archaeol 29(4):323–360
Ballin TB, White R, Richardson P, Neighbour T (2011) An early Mesolithic stone tool assemblage from Clachan harbour, Raasay, Scottish Hebrides. Lithics 31:94–104
Bates MR, Nayling N, Bates R, Dawson S, Huws D, Wickham-Jones C (2013) A multi-disciplinary approach to the archaeological investigation of a bedrock-dominated shallow-marine landscape: an example from the bay of firth, Orkney, UK. Int J Naut Archaeol 42(1):24–43. doi:10.1111/j.1095-9270.2012.00360.x
Benjamin J, Hale A (2012) Marine, maritime, or submerged prehistory? Contextualizing the prehistoric underwater archaeologies of inland, coastal, and offshore environments. Eur J Archaeol 15:237–256
Benjamin J, Bicket A, Anderson D, Hale A (2014) A multi-disciplinary approach to researching the intertidal and marine archaeology in the outer Hebrides, Scotland. J Island Coast Archaeol 9:400–424
Bicket A (2011) Submerged prehistory: research in context. Marine aggregates levy sustainability fund (MALSF), Science monograph series no. 5. Wessex Archaeology, Edinburgh
Bicket A, Firth A, Tizzard L, Benjamin J (2014) Heritage management and submerged prehistory in the United Kingdom. In: Evans AM, Flatman JC, Flemming NC (eds) Prehistoric archaeology on the continental shelf: a global review. Springer, Heidelberg, pp 213–232
Bishop RR, Church MJ, Rowley-Conwy PA (2010) Excavations at Northton, Western Isles of Scotland, 2010; data structure report. Department of Archaeology, Durham University
Bonsall C, Pickard C, Groom P (2013) Boats and pioneer settlement: the Scottish dimension. Nor Archaeol Rev 46(1):87–90
Bradley SL, Milne GA, Shennan I, Edwards R (2011) An improved glacial isostatic adjustment model for the British isles. J Quat Sci 26:541–552
Cohen KM, MacDonald K, Joordens JCA, Roebroeks W, Gibbard PL (2011) The earliest occupation of north-West Europe: a coastal perspective. Quat Int 271:79–83. doi:10.1016/j.quaint.2011.11.003
Coles B (1998) Doggerland, a speculative survey. Proc Prehist Soc 64:45–81
Dellino-Musgrave V, Gupta S, Russell M (2009) Marine aggregates and archaeology: a golden harvest? Conserv Manag Archaeol Sites 11(1):29–42
Edwards KJ (1996) A Mesolithic of the western and northern isles of Scotland? Evidence from pollen and charcoal. In: Pollard T, Morrison A (eds) The early prehistory of Scotland. Edinburgh University Press, Edinburgh, pp 23–38
Edwards KJ (2004) Palaeoenvironments of the late upper Palaeolithic and Mesolithic periods in Scotland and the North Sea area: new work, new thoughts. In: Saville A (ed) Mesolithic Scotland and its neighbours. Society of Antiquaries of Scotland, Edinburgh, pp 55–72
Evans AM, Flatman JC, Flemming NC (eds) (2014) Prehistoric archaeology on the continental shelf: a global review. Springer, New York
Fettes DJ, Mendum JR, Smith DI, Watson JV (1992) Geology of the outer hebrides. BGS, HMSO, London
Fischer A (2011) Stone age on the continental shelf: an eroding resource. In: Benjamin J, Bonsall C, Pickard C, Fischer A (eds) Submerged prehistory. Oxbow, Oxford, pp 298–310
Flatman J, Doeser J (2010) The international management of marine aggregates and its relation to maritime archaeology. Hist Environ 1(2):160–184
Gaffney V, Fitch S, Smith D (2009) Europe's lost world: the rediscovery of Doggerland, CBA research report 160. Council for British Archaeology, York
Garrow D, Sturt F (2011) Grey waters bright with Neolithic argonauts? Maritime connections and the Mesolithic-Neolithic transition. Antiquity 85(327):59–72
Gehrels WR (2010) Late Holocene land- and sea-level changes in the British isles: implications for future sea-level predictions. Quat Sci Rev 29(13–14):1648–1660
Gregory RA, Murphy EM, Church MJ, Edwards KJ, Guttmann EB, Simpson DDA (2005) Archaeological evidence for the first Mesolithic occupation of the western isles of Scotland. The Holocene 15(7):944–950
Hardy K, Benjamin J, Bicket A, McCarthy J, Ballin T (2016) Scotland's Intertidal Prehistory: Lub Dubh Aird, a raw material and knapping site in Upper Loch Torridon. Proceedings of the Society of Antiquaries of Scotland 145
Hijma MP, Cohen KM, Roebroeks W, Westerhoff WE, Busschers FS (2012) Pleistocene Rhine-Thames landscapes: geological background for hominin occupation in the southern North Sea region. J Quat Sci 27:17–39
Jordan JT, Smith DE, Dawson S, Dawson AG (2010) Holocene relative sea-level changes in Harris, outer Hebrides, Scotland, UK. J Quat Sci 25:115–134
Lambeck K (1991) Glacial rebound and sea-level change in the British isles. Terra Nova 3(4):379–389
Lambeck K (1993) Glacial rebound of the British isles. II. A high- resolution, high-precision model. Geophys J Int 115:960–990

Long D, Wickham-Jones CR, Ruckley NA (1986) A flint artefact from the northern North Sea. In: Roe DA (ed) Studies in the upper Palaeolithic of Britain and North Western Europe, British archaeological reports international series, vol 296. BAR, Oxford, pp 55–62

MCA (2004) Hydrographic lidar survey – sound of Harris HI 1086: specification v1.2. Hydrography, meteorology and ports branch. http://www.dft.gov.uk/mca/sound_of_harris.pdf

Momber G (2011) Submerged landscape excavations in the Solent, southern Britain: climate change and cultural development. In: Benjamin J, Bonsall C, Pickard C, Fischer A (eds) Submerged prehistory. Oxbow, Oxford, pp 85–98

Momber G, Tomalin D, Scaife R, Satchell J, Gillespie J (eds) (2011) Mesolithic occupation at Bouldnor cliff and the submerged prehistoric landscapes of the Solent, CBA research report, vol 164. Council for British Archaeology, York

Peltier WR, Shennan I, Drummond R, Horton BP (2002) On the postglacial isostatic adjustment of the British isles and the shallow viscoelastic structure of the earth. Geophys J Int 148:443–475

Pickard C, Bonsall C (2004) Deep-sea fishing in the European Mesolithic: fact or fantasy? Eur J Archaeol 7:273–290

Reimer PJ, Baillie MGL, Bard E, Bayliss A, Beck JW, Blackwell PG, Bronk Ramsey C, Buck CE, Burr GS, Edwards RL, Friedrich M, Grootes PM, Guilderson TP, Hajdas I, Heaton TJ, Hogg AG, Hughen KA, Kaiser KF, Kromer B, McCormac FG, Manning SW, Reimer RW, Richards DA, Southon JR, Talamo S, Turney CSM, van der Plicht J, Weyhenmeyer CE (2009) IntCal09 and Marine09 radiocarbon age calibration curves, 0–50,000 years cal BP. Radiocarbon 51(4):1111–1150

Ritchie W (1980) The beach, dunes and machair landforms of Pabbay, sound of Harris. In: Ranwell DS (ed) Sand dune Machair 3. Institute of Terrestrial Ecology, Cambridge, pp 13–19. http://nora.nerc.ac.uk/7906/1/Machair3.pdf

Ritchie W (1985) Intertidal and sub-tidal organic deposits and sea-level changes in the Uists, outer Hebrides. Scott J Geol 21:161–176

Scottish Natural Heritage (2011) http://www.snh.gov.uk/about-scotlands-nature/species/mammals/marine-mammals/seals/

Shennan I, Horton B (2002) Holocene land- and sea-level changes in great Britain. J Quat Sci 17(5–6):511–526

Shennan I, Bradley S, Milne G, Brooks A, Bassett S, Hamilton S (2006) Relative Sea-level changes, glacial isostatic modelling and ice-sheet reconstructions from the British isles since the last glacial maximum. J Quat Sci 21:585–599

Simpson D, Murphy E, Gregory R (2006) Excavations at Northton, Isle of Harris, British archaeological reports British series 408. Archaeopress, Oxford

Smith DE, Fretwell PT, Cullingford RA, Firth CR (2006) Towards improved empirical isobase models of Holocene land uplift for mainland Scotland, UK. Philos Trans R Soc A Math Phys Eng Sci 364(1841):949–972

Smith DE, Harrison S, Firth CR, Jordan JT (2011) The early Holocene Sea level rise. Quat Sci Rev 30(15–16):1846–1860

Sturt F, Garrow D, Bradley S (2013) New models of north west European Holocene palaeogeography and inundation. J Archaeol Sci 40:3963–3976. doi:10.1016/j.jas.2013.05.023

Tipping R (1996) Microscopic charcoal records, inferred human activity and climatic change in Mesolithic of northernmost Scotland. In: Pollard T, Morrison A (eds) The early prehistory of Scotland. Edinburgh University Press, Edinburgh, pp 39–61

Tipping R (2004) Interpretative issues concerning the driving forces of vegetation change in the early Holocene of the British isles. In: Saville A (ed) Mesolithic Scotland and its neighbours. Society of Antiquaries of Scotland, Edinburgh, pp 45–53

Tizzard L, Bicket AR, Benjamin J, De Loecker D (2014) A middle Palaeolithic site in the southern North Sea: investigating the archaeology and palaeogeography of area 240. J Quat Sci 29(7):698–710

Waddington C (ed) (2007) Mesolithic settlement in the North Sea basin: a case study from Howick, north-East England. Oxbow, Oxford

Wanner H, Solomina O, Grosjean M, Ritz SP, Jetel M (2011) Structure and origin of Holocene cold events. Quat Sci Rev 30(21–22):3109–3123

Weninger B, Schulting R, Bradtmöller M, Clare L, Collard M, Edinborough K, Hilpert J, Jöris O, Niekus M, Rohling EJ et al (2008) The catastrophic final flooding of Doggerland by the Storegga slide tsunami. Doc Praehist 35(Neolithic Studi):1–24

Wessex Archaeology (2004) Artefacts from the Sea. Unpublished report ref: 51541.05

Wessex Archaeology (2009) Seabed prehistory. doi:10.5284/1000050

Wessex Archaeology (2011) Scottish marine historic environment data audit: sources for the enhancement of the coastal and marine historic environment record. Unpublished report for Historic Scotland, ref: 76680.01

Wessex Archaeology (2013) Audit of current state of knowledge of submerged palaeolandscapes and sites. Unpublished Report ref: 84570.02: http://www.english-heritage.org.uk/publications/audit-current-state-knowledge-submerged-palaeolandscapes-sites/

Westley K, Bell K, Quinn R, Plets R (2011) Investigating submerged archaeological landscapes: a research strategy illustrated with case studies from Ireland and Newfoundland, Canada. In: Benjamin J, Bonsall C, Pickard C, Fischer A (eds) Submerged prehistory. Oxbow, Oxford, pp 65–73

Westley K, Plets R, Quinn R (2014) Holocene paleo-geographic reconstructions of the Ramore head area, Northern Ireland, using geophysical and geotechnical data: paleo-landscape mapping and archaeological implications. Geoarchaeology 29(6):411–430

# Chapter 13
# Early Holocene Landscape Development and Baltic Sea History Based on High-Resolution Bathymetry and Lagoonal Sediments in the Hanö Bay, Southern Sweden

**Anton Hansson, Svante Björck, Hans Linderson, Mats Rundgren, Björn Nilsson, Arne Sjöström, and Dan Hammarlund**

**Abstract** The Baltic basin has experienced extensive water-level fluctuations since the last deglaciation. During two occasions of lower than present water levels, c. 11,700–10,200 and 9800–8000 cal BP, areas along the present-day coast of the Hanö Bay in south-eastern Sweden were exposed and pine-dominated forests were established. Around the mouth of the Verkeån River at the Haväng site, remains of this landscape occur in the form of organic-rich lacustrine deposits and well-preserved stumps and trunks of pine trees, reaching depths of 21 m and distances from the present coastline of around 3 km. This study aims at refined reconstructions of the dynamic Early Holocene environment and shore-level displacement to increase the understanding of how Mesolithic people exploited the landscape. Stratigraphic analyses were performed on a sediment sequence obtained from an organic-rich deposit situated at 8.3 m water depth, aided by detailed bathymetric surveys. Radiocarbon dates obtained from the 3.6 m long sequence, supported by pollen stratigraphic correlation, indicate deposition during the period 9000–8600 cal BP with an unusually high sediment accumulation rate. As indicated by a consistently high organic matter content, stable C/N ratios and a general lack of coarse mineral matter, the sediments were deposited in a low-energy environment. Our preliminary interpretation is that the organic-rich deposit was formed in a highly productive oxbow lake, connected via a shallow threshold to the Verkeån River, only allowing fine-grained, fluvially transported particles to reach the depositional environment. As indicated by numerous Mesolithic artefacts in the area, the dynamic landscape at the Haväng site with its rich fishing waters and access to fresh water must have been attractive to Mesolithic hunter-gatherers. People had to adapt to the changing water levels and climatic conditions during the Early Holocene, and traces of their presence are preserved as a consequence of the accumulation of organic-rich fluvial deposits around the river mouth.

A. Hansson (✉) • S. Björck • H. Linderson • M. Rundgren • D. Hammarlund
Department of Geology, Lund University, Sölvegatan 12, 22362 Lund, Sweden
e-mail: anton.hansson@geol.lu.se; svante.bjorck@geol.lu.se; hans.linderson@geol.lu.se; mats.rundgren@geol.lu.se; dan.hammarlund@geol.lu.se

B. Nilsson • A. Sjöström
Department of Archaeology and Ancient History, LUX, Lund University, Box 192, 221 00 Lund, Sweden
e-mail: bjorn.nilsson@ark.lu.se; arne.sjostrom@ark.lu.se



G.N. Bailey et al. (eds.), *Under the Sea: Archaeology and Palaeolandscapes of the Continental Shelf*, Coastal Research Library 20, DOI 10.1007/978-3-319-53160-1_13

## 13.1 Introduction

The Baltic basin has undergone a dynamic development since the deglaciation of the Scandinavian Ice Sheet, which is reflected in the present landscapes around the Baltic Sea. At the Haväng site in the Hanö Bay, south-eastern Sweden, an ancient submerged landscape with remains of pine trees, organic-rich sediments and Mesolithic artefacts is preserved.

At 11,700 years ago, at the end of the Younger Dryas, the melting of the ice margin that had previously dammed the Baltic Ice Lake led to a catastrophic 25 m lowering of the water level in just a few years (Björck 2008). Low-lying areas in south-central Sweden became the outlet for the newly formed Yoldia Sea. New land appeared around the present-day coasts of the southern Baltic Sea as a consequence of the regression. A land bridge between Denmark and Sweden was formed and led to large-scale colonization of plants and animals in southern Sweden. The open landscape remaining from the Younger Dryas was first colonized by birch and shrubs followed by pine a few hundred years later (Gaillard 1984). Later again, hazel became established and the pine-hazel forest supported large mammals such as wild boar and red deer (Liljegren and Lagerås 1993). In the shallow Hanö Bay a pine-dominated forest was established when vast areas of the present-day sea floor became land with well-drained sandy soils exposed during the Yoldia regression (Björck and Dennegård 1988).

As the isostatic uplift was more rapid in the recently deglaciated more northerly parts of Scandinavia, the Yoldia Sea outlets in south-central Sweden gradually became shallower, leading to the Ancylus transgression that started around 10,700 cal BP (Björck 2008). The rising water reclaimed the coastal landscapes of the southern Baltic formed during the Yoldia Sea low stand, and the pine forest in the Hanö Bay was rapidly flooded. Alder, elm, oak and lime immigrated to southern Sweden in rapid succession around 10,000–9000 cal BP (Gaillard 1984), altering the landscape from a shrub-dominated ecosystem to a dominance of more closed forests (Liljegren and Lagerås 1993).

Around 10,200 cal BP, the dammed Ancylus Lake found a new southerly outlet, the Dana River, through Darss Sill, Mecklenburg Bay, Fehmarn Belt and the Great Belt Strait (Björck 2008). The establishment and threshold erosion of the Dana River led to a second regression and the start of the Initial Littorina Sea Stage. At c. 9800 cal BP restricted amounts of brackish water started to enter the Initial Littorina Sea through the Dana River (Andrén et al. 2011). Organic-rich lacustrine sediments situated at about 5–10 m water depth at the Haväng site have been dated to 9500–8800 cal BP, during the Initial Littorina Sea, showing that parts of the Hanö Bay became land again (Gaillard and Lemdahl 1994). The Öresund strait opened at c. 8500 cal BP as global sea level rise exceeded the rate of isostatic rebound in southern Sweden and large amounts of saline water flowed into the Baltic Basin, defining the Littorina Sea Stage proper (Andrén et al. 2011, see also Larsson, Chap. 11). The rapid Littorina transgression led to drowning of the forested landscape at the Haväng site, and since c. 8000 cal BP this region has been permanently covered by the Baltic Sea.

Wood remains and organic deposits have been observed at the Haväng site during times of extremely low water level, and local fishermen have reported stumps and roots caught in their nets (Gaillard and Lemdahl 1994), some of which have yielded calibrated radiocarbon ages in the range of 11,200–10,400 cal BP (Håkansson 1968, 1972, 1974, 1976, 1982). In the early 1960s a pollen study was performed on an organic sediment sampled at the Haväng site and described as a compacted gyttja, indicating an age of about 9000 cal BP (Nilsson 1961). More recently Björck et al. (1990) studied the Late Quaternary stratigraphy in the Hanö Bay, whereas Gaillard and Lemdahl (1994), based on sampling by sport divers, performed macrofossil analyses of the organic deposits at depths of 5–15 m at the Haväng site. These organic deposits were interpreted as being formed between beach ridges or sand dunes during the Littorina transgression. Remains of submerged landscapes of corresponding ages from the Ancylus Lake and the initial Littorina Sea stages have been found in Poland (Uścinowicz et al. 2011), Denmark (Fischer 1997a) and Germany (Lampe 2005).

Due to the water level changes since the last deglaciation, many of the coastal Mesolithic sites in Scandinavia (Fischer 1995) and the North Sea (Gaffney et al. 2009) are now submerged. Often the state of preservation at these sites is exceptional (Fischer 1995, see also Uldum et al. Chap. 5, Skriver et al. Chap. 8). Previous surveys of the Haväng site have yielded artefacts such as sea-polished flint flakes, indicating the presence of Mesolithic activity (Hansen 1995). Furthermore, a pecked axe was found on the shore, washed up from a possible settlement below the present water level (Hansen 1995). Recent investigations by marine archaeologists have revealed additional traces of human presence in the area, with findings dated to around 10,500 cal BP and 9000 cal BP (Holmlund et al. Chap. 4).

The submerged forest landscape at the Haväng site is the focus of a joint geological and archaeological project based at Lund University, aimed at refined reconstructions of the dynamic Early Holocene environment and shore-level displacement and at improved understanding of how people exploited the changing landscape. As part of this effort, sedimentological, stratigraphic and dendrochronological studies have been performed, in order to reconstruct the local environmental development and the transgression rates of the Ancylus Lake and the Initial Littorina Sea. The geological results will be used as a basis for refined archaeological interpretations.

In this study we present chronological and pollen stratigraphic data obtained from an organic-rich sediment sequence, which form the basis for preliminary interpretations of the landscape development at the Haväng site.

## 13.2 Study Location

The Haväng site is situated in south-eastern Sweden on the coast of the Hanö Bay (Fig. 13.1). The bedrock in the area consists mainly of Mesozoic and Tertiary limestone, sandstone and clay (Kumpas 1978). The Quaternary stratigraphy in the Hanö Bay consists of a clayey diamicton overlying the bedrock, generally followed by a thin covering layer of sand and clay (Björck et al. 1990). From the coastline down to about 20 m water depth, a layer of sand with occasional organic deposits covers the diamicton.

The Verkeån River, which is a small stream with a mean annual flow of c. 1 $m^3\ s^{-1}$ (SMHI 2015), has its outlet on the coast at the Haväng site. Outside the river mouth in the Hanö Bay, organic-rich deposits are found, forming lake basins, river beds and beach-parallel ridges to a distance of about 750 m from the present coastline (Figs. 13.1 and 13.2). The organic-rich sediments appear as positive landforms, the thickest about 3 m above the seafloor with erosional channels perpendicular to the coastline (Fig. 13.3). About 1.3 km from the coast, the organic-rich sediments re-appear as a river bed formation, continuing to c. 3 km from the coastline at 21 m depth. Apart from these organic-rich sediments, the seafloor is characterized by submerged tree remains (Fig. 13.4). In situ pine trunks and stumps are scattered across the landscape down to a depth of 21 m. The longer trunks are mostly fallen and lie on the sea floor whereas some shorter stumps are still standing upright. In some cases bark and branches remain on the trees. Some trunks protrude at an angle from the edges of the organic-rich deposits.

## 13.3 Methods

The sediment sequence used in this study was sampled with a vibrocorer from the vessel *R/V Ocean Surveyor* at 8.3 m water depth in May 2009 from one of the organic-rich sediment deposits approximately 650 m east of the coast at the Haväng site (Figs. 13.1 and 13.3). The position of the sample site is N 55°43″44.90″; E 14°12″19.43″. The retrieved sequence is 3.6 m long and was split into 1 m sections.

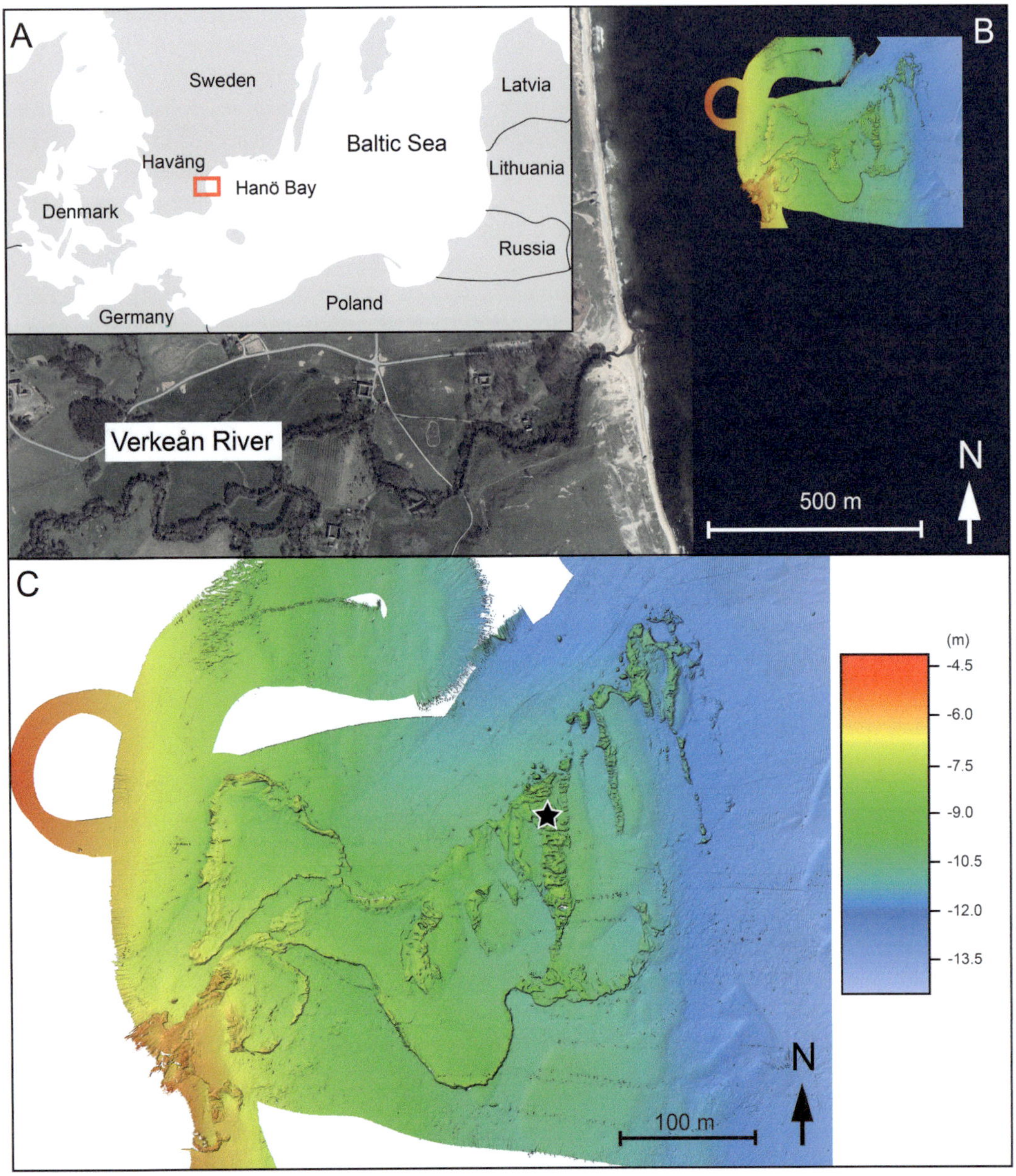


**Fig. 13.1** (**a**) Map showing the location of the Haväng site (*red box*), situated in the Hanö Bay, south-eastern Sweden. (**b**) Blow-up of the area contained within the *red box* showing an aerial photograph of the Haväng site. The Verkeån River is seen in the southern part of the landscape. North-east of the river mouth outside the coast a bathymetrical map shows the location of the submerged landscape. © Lantmäteriet [License No. I2014/00579]. (**c**) Close-up of the submerged landscape showing the meander-shaped and beach-parallel organic-rich ridges. The *star* indicates the sampling point of the sediment sequence

The bathymetrical data were obtained with a MultiBeam Echo Sounder system with 1–3 cm horizontal resolution and 5 cm vertical resolution. The measurements were performed in the summer of 2013.

Eight AMS radiocarbon dates were obtained on terrestrial macroscopic plant remains at the Radiocarbon Dating Laboratory, Lund University. The samples were calibrated using the INTCAL13

Fig. 13.2 Photograph of an organic-rich deposit rising about 50 cm above the sea floor. The sediment is tilted indicating the erosion of a supporting river valley. The photograph is facing north (Photo Arne Sjöström)

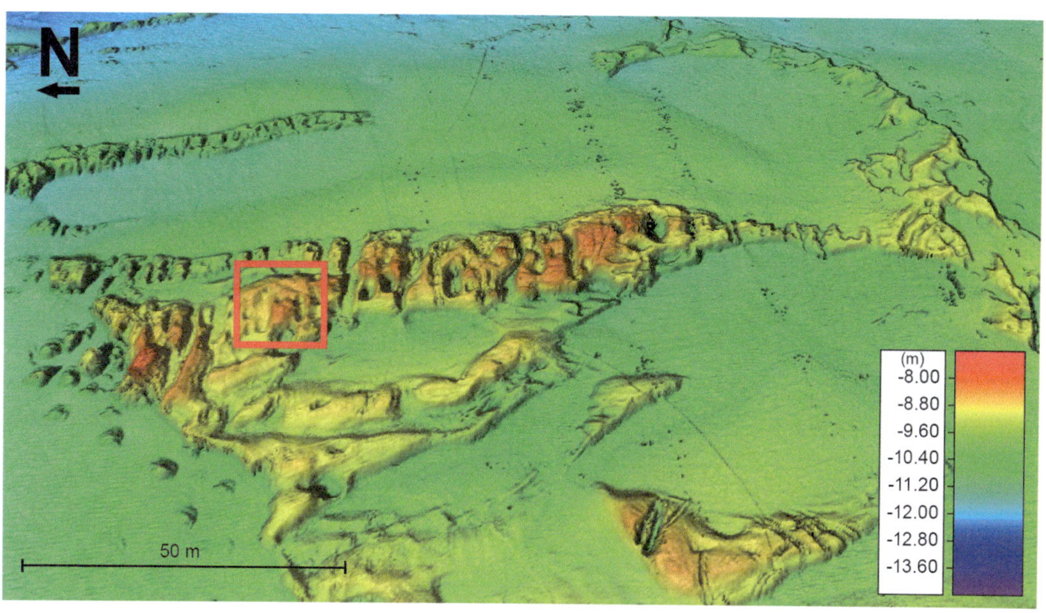


Fig. 13.3 Close-up of the sampled organic-rich sediment ridge at 8.3 m depth. The sediment sequence was obtained from the area enclosed by the *red box*. Erosional channels can be seen in east-west direction, perpendicular to the coastline

calibration data set (Reimer et al. 2013) and the software OxCal 4.2 (Bronk Ramsey 2009). A Poisson-process deposition model, assuming younger ages upwards in the sequence, based on seven of the radiocarbon dates was constructed with a K-value of 0.5 $cm^{-1}$ using the software OxCal 4.2 (Bronk Ramsey 2009) in order to assess the sedimentation rate of the sequence (Bronk Ramsey 2008).

**Fig. 13.4** Photograph of a rooted pine stump at the Haväng site (Photo Arne Sjöström)

Twenty samples with a volume of 1–2 $cm^3$ were retrieved for pollen analysis and prepared according to the standard methodology of Berglund and Ralska-Jasiewiczowa (1986). *Lycopodium* spore tablets were added to the samples in order to estimate pollen concentrations (Stockmarr 1971). At least 500 pollen grains per sample were counted. The main purpose of the pollen analysis was to assess the validity of the radiocarbon dates and therefore the focus was set on pollen from tree taxa. The identification keys in Fægri and Iversen (1989) and Moore et al. (1991) were used for pollen determination.

Samples for loss-on-ignition analysis were collected at every 5 cm throughout the sequence to estimate the organic matter content. The samples were placed in crucibles and weighed after heating to 105 °C and 550 °C, respectively (Bengtsson and Enell 1986). Carbon and nitrogen elemental contents were determined at 19 selected levels using a Costech ECS 4010 elemental analyser for assessment of the character and origin of the organic matter. Elemental C/N ratios were converted to atomic ratios by multiplying by 1.167.

Magnetic susceptibility was measured at every 5 mm throughout the sequence with a Bartington Instruments MS2E1 surface scanning sensor connected to a Tamiscan-TS1 automatic logging conveyor in order to assess the general grain size of the sediment sequence.

## 13.4 Results

Seven of the radiocarbon dates gave calibrated ages in the range of c. 9000–8600 cal BP throughout the sequence (Table 13.1). In contrast, the uppermost sample gave an unrealistically young age, possibly due to contamination by recent carbon. An age model based on seven radiocarbon dates shows deposition of 338 cm of sediments over a time span of c. 300–400 years (Table 13.1), yielding an average sediment accumulation rate of 0.85–1.1 cm per year.

The entire 360 cm sediment-sequence can be described as a dark, homogeneous clayey and silty gyttja. The sediments, which are virtually devoid of macroscopic plant remains and visible mineral grains, are markedly compact and erosion resistant. The organic matter content of the sediment

**Table 13.1** Radiocarbon dates, and calibrated and modelled ages from the retrieved sediment sequence. Depth refers to depth of the sediment

| Depth (cm) | Lab. no | Material | $^{14}$C years BP | Cal. age BP (1σ interval) | Modelled ages (1σ interval) |
|---|---|---|---|---|---|
| 8.5–10.5 | LuS-10677 | Bark and needle fragments, insect fragment, moss stems | 710 ± 50 | 566–694 | – |
| 15–19 | LuS-10880 | Bark and needle fragments, insect fragment, moss stems | 7900 ± 35 | 8628–8767 | 8631–8780 |
| 15–19 | LuS-10962 | Bark and needle fragments, insect fragment, moss stems | 8000 ± 45 | 8779–8996 | 8760–8985 |
| 39–41 | LuS-10676 | Bark and needle fragments, insect fragment, moss stems | 8025 ± 55 | 8779–9010 | 8652–8795 |
| 199–201.3 | LuS-10675 | Bark and needle fragments, insect fragment, moss stems | 8040 ± 115 | 8664–9088 | 8816–8911 |
| 328.5–329.8 | LuS-10674 | Bark and needle fragments, insect fragment, moss stems | 7950 ± 55 | 8660–8977 | 8958–9007 |
| 335.5–336.5 | LuS-10673 | Bark and needle fragments, insect fragment, moss stems | 8020 ± 120 | 8650–9028 | 8963–9013 |
| 349–353 | LuS-10881 | Bark and needle fragments, insect fragment, moss stems | 8085 ± 45 | 8983–9091 | 8975–9027 |

sequence varies between 45% and 55% (Fig. 13.5). Between 0 and 156 cm, the organic matter content remains below 50% whereas the lower half of the sequence, 156–360 cm, shows organic matter content values above 50%. Minimum values occur at 21 cm and 31 cm, with values of 43% and 41%, respectively. The magnetic susceptibility shows values that are consistently between $2 \times 10^{-5}$ and $4 \times 10^{-5}$ m$^3$ kg$^{-1}$ (Fig. 13.5). Peaks occur at 34 cm and 71.5 cm with values of $10.3 \times 10^{-5}$ m$^3$ kg$^{-1}$ and $11 \times 10^{-5}$ m$^3$ kg$^{-1}$, respectively. The total organic carbon content (TOC%) varies between 20% and 30% throughout the sequence whereas the nitrogen content (N%) varies between 1% and 2%. The atomic C/N ratio is between 19.5 and 21 throughout the sequence (Fig. 13.5).

The pollen record was divided into two stratigraphic zones (Fig. 13.6). Zone 1 consists of the lower and middle parts of the stratigraphy, 350–88 cm. This zone is dominated by the tree taxa *Alnus*, *Betula*, *Corylus* and *Pinus*. *Ulmus* and *Quercus* are also common in this zone. The total pollen concentration varies in the range of 400,000–600,000 grains/cm$^3$. Zone 2 consists of the uppermost part of the stratigraphy, 88–10 cm, and is defined by the appearance of *Tilia* pollen grains. Zone 2 is dominated by *Alnus*, *Betula*, *Corylus* and *Pinus* and a slight increase in *Filipendula* can be seen. *Ulmus* and *Quercus* are still common throughout zone 2. The total pollen concentration decreases to less than half of the values recorded in zone 1, showing values in the range of 100,000–200,000 grains/cm$^3$.

## 13.5 Discussion

### *13.5.1 Sediment Properties and Accumulation Rate*

The magnetic susceptibility shows values typical of lake sediments with a mineral content predominantly in the clay and silt fractions. Coarser material such as sand would give susceptibility values at least two orders of magnitude lower than obtained from our sediment sequence (Björck et al. 1982). The peaks seen in the susceptibility data probably do not reflect short term changes in the depositional environment and could instead be the result of individual, highly magnetic, mineral grains close to the

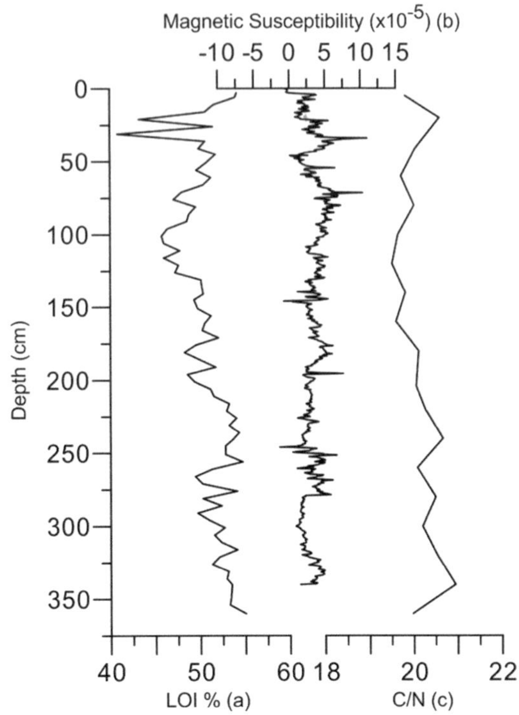


**Fig. 13.5** Graph showing (**a**) the organic matter content, (**b**) the magnetic susceptibility and (**c**) C/N atomic ratios throughout the sediment sequence. Organic content is expressed in percent and the magnetic susceptibility in a dimensionless unit

susceptibility sensor (Hambach et al. 2008). Organic matter content values are consistently between 45% and 55% throughout the sediment sequence, indicative of a nutrient-rich and productive environment. This is 10–20% higher organic matter content than reported from a previous study of similar organic-rich sediments at the Haväng site (Gaillard and Lemdahl 1994). The relatively stable values for organic matter content indicate that the depositional environment remained rather constant during the time of sedimentation, and the largely uniform C/N ratios around 20 suggest a stable balance between aquatic productivity and fluvial supply of fine-grained terrestrial organic matter (Meyers and Lallier-Verges 1999).

The radiocarbon dates obtained from the sediment sequence yielded indistinguishable ages, apart from the uppermost date that gave an age of 694–566 cal BP. This anomalous age strongly indicates contamination by younger organic material. Most likely the sample was contaminated during the sampling procedure as the compact sediments would not have been susceptible to intrusion of younger organic material before sampling. The remaining dates all gave ages of 9000–8600 cal BP, dating the sediment succession to the Initial Littorina Sea stage, when the water level was at least 10 m below the present (Gaillard and Lemdahl 1994). A theoretical explanation for these indistinguishable ages would be that the sediment succession tipped over at some point in time, but this hypothesis can be rejected based on the pollen stratigraphy, which indicates undisturbed sedimentation. The radiocarbon calibration curve exhibits an atmospheric $^{14}C$ plateau at 9000–8600 cal BP (Reimer et al. 2013), which explains some of the encountered dating difficulties. The modelled ages obtained from the age model are still in the span of 9000–8600 cal BP but are distinguishable from each other (Table 13.1). Two radiocarbon dates were obtained from the level of 15–19 cm (Table 13.1), and the calculated sediment accumulation rate based on both dates indicates a sediment accumulation rate of either 0.85 cm or 1.1 cm per year, respectively. As both rates must be considered correct we can only conclude that the sedimentation rate was in the span of 0.85–1.1 cm per year, which is one order of magnitude higher than what is usually seen in Swedish lakes, where values around 0.5–1 mm per year are common (Björk 2010).

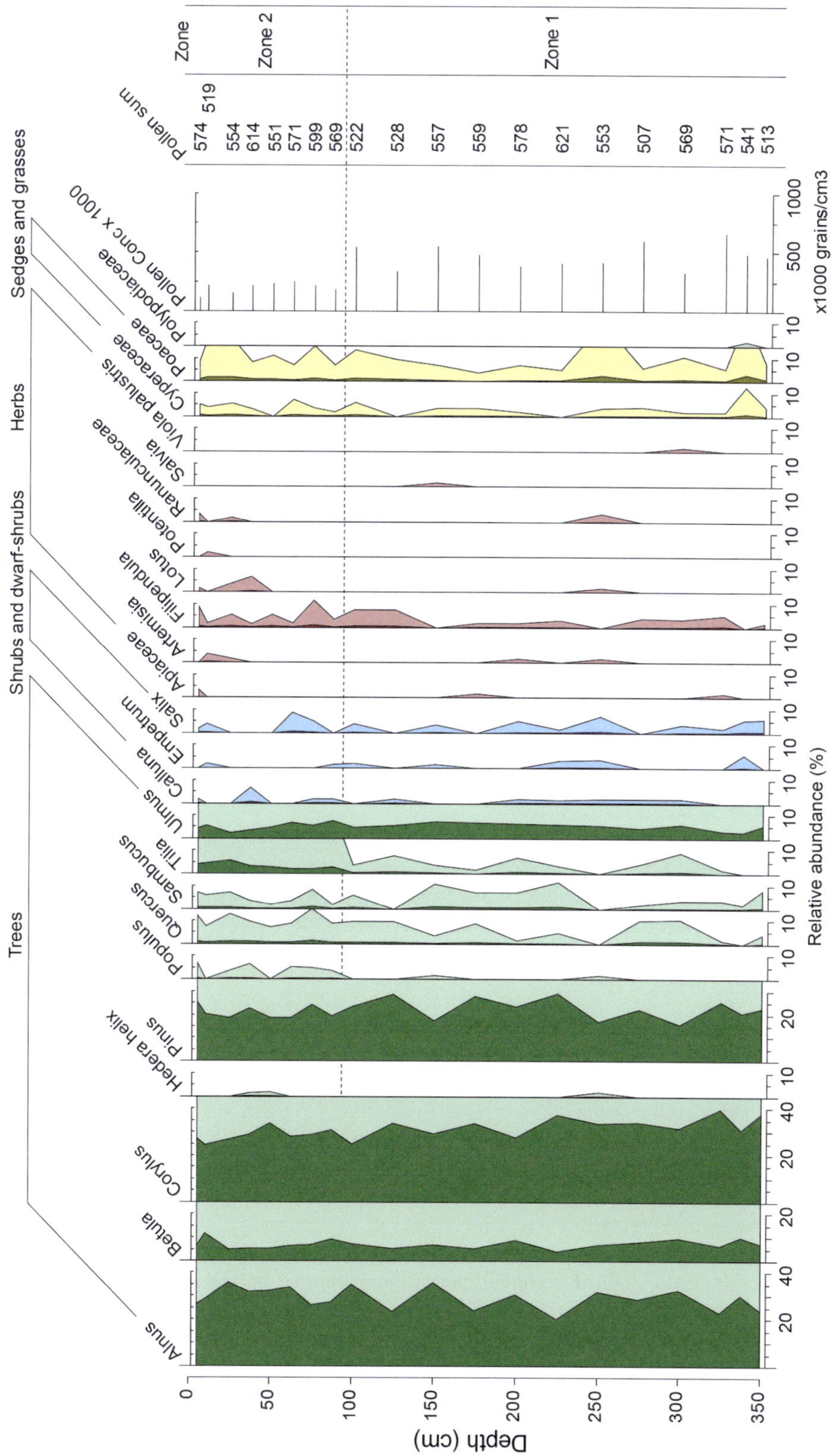


**Fig. 13.6** Pollen diagram of the analysed sediment sequence. The sequence is dominated by Pinus, Corylus, Betula and Alnus. The stratigraphy is divided into two zones based on the absence or presence of Tilia. Pollen counts and diagram by Anton Hansson

*Tilia* pollen first appears at 88 cm (Fig. 13.6). A pollen diagram from the Krageholmssjön Lake, approximately 40 km southwest of the Haväng site, in southernmost Sweden shows that *Tilia* first appeared at 9200 cal BP and was established in the landscape shortly after 9000 cal BP (Gaillard 1984). The pollen analysis therefore confirms the radiocarbon dates of the Haväng sediment sequence. Because the sampled basin was most likely connected to the Verkeån River, a substantial part of the pollen may have arrived by fluvial transport and therefore reflects vegetation in areas upstream.

The pollen concentration depends on the influx of pollen grains and the sediment accumulation rate (Fægri and Iversen 1989). The substantial decrease in pollen concentration slightly above 1 m occurs simultaneously with the appearance of *Tilia* pollen grains. Whether these events are connected cannot be determined at this point, but it is possible that the decreased concentration stems from an increased inflow of particles that led to dilution of the pollen content in the sequence. A decreased inflow of pollen grains seems unlikely as the forest vegetation experienced generally favourable climatic conditions during this time period (Wanner et al. 2008).

As coarse mineral grains are absent in the sediment sequence, input of material from the local sandy substrate from episodic events such as storms can be eliminated. The basin was apparently protected from the sea and influenced exclusively by fine-grained fluvial material and organic production within the basin. Probably, a shallow and productive environment maintaining high rates of organic deposition in the basin in combination with an ample supply of fine mineral matter from the river contributed to the extremely high sediment accumulation rate. The mineral matter consists mostly of clay and fine silt, which would indicate that the basin was partly cut off from the river by a shallow threshold that only allowed suspended material to enter and settle. The compact appearance of the sediments is most likely a result of compaction by sand that was deposited on top of the organic-rich sediment sequence during the Littorina Sea and then subsequently eroded away.

### *13.5.2 Depositional Environment*

A compelling scenario is that the Haväng site has been subject to several depositional and erosional episodes, as the shoreline has passed across the study site four times during the Early Holocene. Therefore the depositional landforms might not have been uniformly preserved, and what is seen today on the seafloor is partly an erosional landscape. Additionally, older sediments from the Yoldia Sea stage that are found farther from the coastline may also be hidden under the Initial Littorina Sea sediments investigated in this study.

Based on the bathymetry (Fig. 13.1), the investigated area can be divided into two parts. The inner part consists of meander-shaped basins confined by edges of organic-rich material. Outside these deposits the bathymetrical map is dominated by elongated beach-parallel organic-rich ridges. It is still unclear whether these ridges originate from the same depositional event or if they were deposited in succession. Further radiocarbon dating of the other ridges is needed in order to clarify this development and to interpret more confidently the formation and subsequent denudation of the landscape. However, given the overlap of the radiocarbon ages discussed above, it may prove impossible to distinguish between the two hypotheses.

The beach-parallel, organic-rich ridges described in this study have previously been interpreted as protective beach ridges in a shallow lagoonal environment formed behind them during the Initial Littorina Sea stage (Gaillard and Lemdahl 1994). According to this model, new beach ridges were formed as the Littorina transgression progressed, producing new basins closer to the coast. However, the study by Gaillard and Lemdahl (1994) was carried out without access to bathymetric data from the area, and the maps produced by divers missed the meander-shaped ridges seen in the present bathymetrical map (Fig. 13.1). Our results provide strong sedimentological evidence against the beach-ridge formation hypothesis. The presence of beach ridges would most likely have caused aeolian transport into the adjacent basins, but no sand is observed in the sampled sediment sequence.

Our interpretation is that the investigated sediment sequence was deposited in an oxbow lake partly cut off from the meandering river, with a shallow threshold allowing only suspended material to settle together with the organic matter produced within the basin. A steadily rising water level in the oxbow lake due to the Littorina transgression probably contributed to the rapid sediment accumulation and kept the water depth in the basin fairly constant. The deposit was then covered by sand and compacted during the Littorina transgression. As the water level regressed down to the present level the superimposed sand was eroded away and the present morphology of the organic-rich deposit was shaped by currents and wave action.

Further radiocarbon dates of both tree remains and organic deposits from different depths in the study area will provide increased insight into the age relationship between the meander-shaped and ridge-shaped deposits, as well as the age relationships within the ridges, and thereby contribute to a better understanding of how the submerged landscape at the Haväng site was formed.

The southernmost part of the Baltic Basin has experienced a constant transgression during the Holocene (Uścinowicz 2006). In the Gulf of Gdansk, remains of trees and organic-rich sediments have been found at different depths. In the western part of the Gulf, at depths of 16–17 m below present, remains of deciduous trees have been dated to somewhat younger ages than the sediment sequence described here (Uścinowicz et al. 2011). This shows that the water level was about 6–7 m lower than at the Haväng site in the southernmost part of the Baltic Basin during the Initial Littorina Sea stage (Björck 2008). On the eastern side of the Gulf, in the Vistula Lagoon, there are trees at 2 m below the present sea level, which have been dated to 5600–3500 cal BP (Łęczyński et al. 2007). During this time the sea level at the Haväng site was approximately 3–5 m above the present (Berglund et al. 2005). In Danish waters, evidence of human presence in the form of Mesolithic coastal settlements is common (Fischer 1997b); in contrast, in the Baltic Sea findings of coastal settlements are less common, but there is considerable potential for discovery of submerged settlements in the southern Baltic Sea.

The coastal environment along the Hanö Bay was probably very attractive to groups of hunter-gatherers in the early Holocene. The combination of a reliable fresh water supply and rich fishing waters, both in the Verkeån River and in the Initial Littorina Sea, and the habitable river plain in the valley likely made the area around the Haväng site especially attractive, as shown by findings of fish traps and other remains of human activity, such as worked animal bones, hazel nuts and flints (Holmlund et al. Chap. 4). All archaeological findings so far are characterized as organic refuse deposited in the river, or in lagoons near the coast. The actual settlement areas are likely to have been eroded away during transgressions and by river-bank erosion. However, certain artefact concentrations indirectly indicate the presence of more intensively used settlement areas. So far, no remains of marine resources have been found, but since no excavations have been conducted this could reflect a methodological bias. People living at the Haväng site not only had to cope with a changing climate. They also had to adapt to the constant changes of the riverine landscape as both the meandering river and the water level fluctuations of the Ancylus Lake and Littorina Sea altered the path of the river and the configuration of the coastline.

## 13.6 Conclusions

The calibrated radiocarbon dates of 9000–8600 cal BP from the sediment sequence are confirmed by the pollen stratigraphy and the immigration of *Tilia*. The age span of the sequence indicates a very high sedimentation rate, which may be the result of deposition in a basin where the water level successively rose during the initial Littorina transgression. However, the lack of coarse mineral matter in the sediments suggests that the basin was only supplied with material from the Verkeån River and organic matter produced within the basin and that aeolian processes and wave action did not affect the basin.

The submerged landscape at the Haväng site can be divided into two parts, an inner with meander-shaped deposits and an outer consisting of beach-parallel ridges. Radiocarbon dates from the inner area are presently lacking and it is unclear whether the inner and outer areas were deposited simultaneously. The basin is interpreted as an oxbow lake that was partly cut off from the main river by a threshold that only allowed material in suspension to enter. It is evident that people occupied the Haväng site during the Early and Middle Mesolithic. Radiocarbon dates obtained on worked bones from mammals (c. 10,500 cal BP) and wooden fish traps used in the river (c. 9000 cal BP) give us a hint of the patterns of resource utilization at the river mouth. These human populations must have adapted to a dynamic landscape characterised by changes in the fluvial and coastal environment as well as changes in climate and water level.

The recent investigations at the Haväng site demonstrate a more complex formation history than previously described (Gaillard and Lemdahl 1994; Hansen 1995). The development of the landscape is not yet fully understood, but further archaeological and geological studies of the area will take place during the coming years. Dendrochronological studies of the wood remains and analyses of both stratigraphic sequences and discrete surface samples of organic deposits are ongoing. These investigations, in conjunction with proper archaeological excavations of the refuse layers, will contribute to a deeper understanding of the dynamics of this landscape, how it was affected by changes in water level, and how Early Mesolithic populations adapted to the changing landscape. The preservation of archaeological material is made possible by the accumulation of organic-rich sediments near the river mouth, and therefore similar near-coastal sites may have a high potential for archaeological studies.

**Acknowledgements** This project is funded by The Crafoord Foundation, the Royal Physiographic Society of Lund and the Knowledge Foundation. The crew of *R/V Ocean Surveyor* are thanked for help with the coring. We also thank MMT AB for bathymetry mapping, MARIS and Södertörn University for their part in this project, and Beesham Soogrim and Krister Jeppson for invaluable help during the diving expeditions. Finally, we thank the South Skåne Regiment of the Swedish armed forces for their cooperation during the field operations.

## References

Andrén T, Björck S, Andrén E, Conley D, Zillén L, Anjar J (2011) The development of the Baltic Sea basin during the last 130 ka. In: Harff J, Björck S, Hoth P (eds) The Baltic Sea basin. Springer, Heidelberg, pp 75–98. doi:10.1007/978-3-642-17220-5

Bengtsson L, Enell M (1986) Chemical analysis. In: Berglund BE (ed) Handbook of holocene palaeoecology and palaeohydrology. Wiley, Chichester, pp 423–451

Berglund BE, Ralska-Jasiewiczowa M (1986) Pollen analysis and pollen diagrams. In: Berglund BE (ed) Handbook of holocene palaeoecology and palaeohydrology. Wiley, Chichester, pp 455–484

Berglund BE, Sandgren P, Barnekow L, Hannon G, Jiang H, Skog G, Yu S-H (2005) Early Holocene history of the Baltic Sea, as reflected in coastal sediments in Blekinge, southeastern Sweden. Quat Int 130:111–139. doi:10.1016/j.quaint.2004.04.036

Björck S, Dennegård B (1988) Preliminary stratigraphic studies on the late Weichselian and Holocene development of the Hanö Bay, southeastern Sweden. Geogr Pol 55:51–62

Björck S (2008) The late quaternary development of the Baltic Sea. In: Team TBA (ed) Assessment of climate change for the Baltic Sea basin. Springer, Heidelberg, pp 398–407. doi:10.1007/978-3-540-72786-6

Björck S, Dearing JA, Jonsson A (1982) Magnetic susceptibility of late Weichselian deposits in southeastern Sweden. Boreas 11(1):99–111

Björck S, Dennegård B, Sandgren P (1990) The marine stratigraphy of the Hanö Bay, SE Sweden, based on different sediment stratigraphic methods. Geol Fören Stockh Förh 112(3):265–280

Björk S (2010) The evolution of lakes and wetlands. In: Eiseltová M (ed) Restoration of lakes, streams, floodplains, and bogs in Europe: principles and case studies. Springer, Heidelberg, pp 25–35. doi:10.1007/978-90-481-9265-6

Bronk Ramsey C (2008) Deposition models for chronological records. Quat Sci Rev 27(1–2):42–60. doi:10.1016/j.quascirev.2007.01.019
Bronk Ramsey C (2009) Bayesian analysis of radiocarbon dates. Radiocarbon 51(1):337–360. doi:10.2458/azu_js_rc.51.3494
Fægri K, Iversen J (1989) Textbook of pollen analysis, 4th edn. Wiley, Chichester
Fischer A (1995) An entrance to the Mesolithic world below the ocean. Status of ten years' work on the Danish sea floor. In: Fischer A (ed) Man and sea in the Mesolithic. Oxbow, Oxford, pp 371–384
Fischer A (1997a) Drowned forests from the stone age. In: Pedersen L, Fischer A, Aaby B (eds) The Danish Storebaelt since the ice age – man, sea and forest. A/S Storebaelt Fixed Link, Copenhagen, pp 29–36
Fischer A (1997b) People and the sea – settlement and fishing along the Mesolithic coasts. In: Pedersen L, Fischer A, Aaby B (eds) The Danish Storebaelt since the ice age – man, sea and forest. A/S Storebaelt Fixed Link, Copenhagen, pp 63–77
Gaffney V, Fitch S, Smith D (2009) Europe's lost world. The rediscovery of Doggerland. CBA research report 160. Council for British Archaeology, York
Gaillard M-J (1984) A palaeohydrological study of Krageholmssjön (Scania, South Sweden). Regional vegetation history and water-level changes. LUNDQUA Report 25: 40
Gaillard M-J, Lemdahl G (1994) Early-Holocene coastal environments and climate in Southeast Sweden: a reconstruction based on macrofossils from submarine deposits. The Holocene 4(1):53–68
Håkansson S (1968) University of Lund radiocarbon dates I. Radiocarbon 10:36–54
Håkansson S (1972) University of Lund radiocarbon dates V. Radiocarbon 14(2):380–400
Håkansson S (1974) University of Lund radiocarbon dates VII. Radiocarbon 16(3):307–330
Håkansson S (1976) University of Lund radiocarbon dates IX. Radiocarbon 18(3):290–320
Håkansson S (1982) University of Lund radiocarbon dates XV. Radiocarbon 24(2):194–213
Hambach U, Rolf C, Schnepp E (2008) Magnetic dating of quaternary sediments, volcanites and archaeological materials: an overview. In: Freund H (ed) E&G quaternary science journal special issue: recent progress in quaternary dating methods. E. Schweizerbart'sche Verlagsbuchhandlung, Stuttgart, pp 25–51
Hansen L (1995) Submerged Mesolithic landscapes. Preliminary results from the Hanö Bay, southern Baltic. In: Fischer A (ed) Man & sea in the mesolithic. Oxbow, Oxford, pp 409–414
Kumpas M (1978) Distribution of sedimentary rocks in the Hanö Bay and S of Öland, S Baltic. Stockh Contrib Geol 31:95–103
Lampe R (2005) Lateglacial and holocene water-level variations along the NE German Baltic Sea coast: review and new results. Quat Int 133–134:121–136. doi:10.1016/j.quaint.2004.10.005
Łęczyński L, Miotk-Szpiganowicz G, Zachowicz J, Uścinowicz S, Krapiec M (2007) Tree stumps from the bottom of the Vistula lagoon as indicators of water level changes in the southern Baltic during the late Holocene. Oceanologia 49(2):245–257
Liljegren R, Lagerås P (1993) Från mammutstäpp till kohage. Djurens historia i Sverige. Wallin & Dalholm, Lund
Meyers PA, Lallier-Verges E (1999) Lacustrine sedimentary organic matter records of late quaternary paleoclimates. J Paleolimnol 21:345–372
Moore PD, Webb JA, Collinson ME (1991) Pollen analysis. Blackwell Scientific Publications, Oxford
Nilsson T (1961) Ravlundafältets geologi. Skånes Natur 1961:73–106
Reimer PJ, Bard E, Bayliss A, Beck JW, Blackwell PG, Ramsey CB, Buck CE, Cheng H, Edwards RL, Friedrich M, Grootes PM, Guilderson TP, Haflidason H, Hajdas I, Hatté C, Heaton TJ, Hoffman D, Hogg AG, Hughen KA, Kaiser KF, Kromer B, Manning SW, Niu M, Riemer RW, Richards DA, Scott EM, Southon JR, Staff RA, Turney CSM, van der Plicht J (2013) IntCal13 and Marine13 radiocarbon age calibration curves 0–50,000 years cal BP. Radiocarbon 55(4):1869–1887. doi:10.2458/azu_js_rc.55.16947
SMHI (2015) Sveriges Meteorologiska och Hydrologiska Instituts vattenwebb, Norrköping http://vattenweb.smhi.se/station/. Accessed 15 Dec 2015
Stockmarr J (1971) Tablets with spores used in absolute pollen analysis. Pollen Spores 13:615–621
Uścinowicz S (2006) A relative sea-level curve for the polish southern Baltic Sea. Quat Int 145–146:86–105. doi:10.1016/j.quaint.2005.07.007
Uścinowicz S, Miotk-Szpiganowicz G, Krapiec M, Witak M, Harff J, Lübke H, Tauber F (2011) Drowned forests in the Gulf of Gdańsk (southern Baltic) as an indicator of the Holocene shoreline changes. In: Harff J, Björck S, Hoth P (eds) The Baltic Sea basin. Springer, Heidelberg, pp 219–231. doi:10.1007/978-3-540-72786-6
Wanner H, Beer J, Bütikofer J, Crowley TJ, Cubasch U, Flückiger J, Goosse H, Grosjean M, Joos F, Kaplan JO, Küttel M, Müller SA, Prentice IC, Solomina O, Stocker TF, Tarasov P, Wagner M, Widmann M (2008) Mid- to late holocene climate change: an overview. Quat Sci Rev 27:1791–1828. doi:10.1016/j.quascirev.2008.06.013

# Chapter 14
# Tributaries of the Elbe Palaeovalley: Features of a Hidden Palaeolandscape in the German Bight, North Sea

**Daniel A. Hepp, Ursula Warnke, Dierk Hebbeln, and Tobias Mörz**

**Abstract** Prior to postglacial global sea-level rise in the present North Sea area, Mesolithic hunters and gatherers were able to settle in the coastal lowland landscape between England, Germany and Denmark, commonly known as Doggerland. Regarding the reconstruction of this now drowned palaeolandscape, the German exclusive economic zone (EEZ) sector is still 'terra incognita'. Recent discoveries of two ancient fluvial systems, both of which were tributaries of the Elbe Palaeovalley, give new insights into the formation of the Mesolithic Doggerland landscape in the German EEZ. One of these fluvial systems developed during the last glaciation and connected the Dogger Hills with the Elbe Palaeovalley. The second river structure discovered in the south seems to be slightly younger and can be identified as the drowned extension of the modern Ems River.

## 14.1 Introduction

Nearly 10,700 years ago, Mesolithic people were able to cross an extensive coastal lowland which is now covered by the south-eastern North Sea. Archaeological and palaeoenvironmental data document that hunters and gatherers were present in the area between England, Germany and Denmark at this time, living in a landscape shaped by a system of rivers, lakes, birch groves, pine woods, fens and grasslands (Coles 2000; Fitch et al. 2005; Gaffney et al. 2007; Spinney 2008; Cohen et al. 2014). The present shallow water area, the so-called Dogger Bank, was a hill ridge and the exposed red sandstone rocks of Helgoland stood out against the surrounding plains, rather like Uluru (Ayers Rock) in Australia today. This submerged Mesolithic palaeolandscape is known as Doggerland (Coles 1998).

The final postglacial transgression drowned the prehistoric surface of Doggerland by shifting the coastline southwards towards its current position. Presently, the former Doggerland is covered by modern North Sea sands (Zeiler et al. 2000). There is, however, still considerable cultural heritage buried in the ancient sediments (Coles 2000; Spinney 2008; Gaffney et al. (n.d), Chap. 20). Despite

D.A. Hepp (✉) • D. Hebbeln • T. Mörz
MARUM – Center for Marine Environmental Sciences, University of Bremen,
Leobener Str. 8, 28359 Bremen, Germany
e-mail: dhepp@marum.de; dhebbeln@marum.de; tmoerz@marum.de

U. Warnke
German Maritime Museum – Institute of the Leibniz-Association (Deutsches Schiffahrtsmuseum – Institut der Leibniz-Gemeinschaft), Bremerhaven, Germany
e-mail: warnke@dsm.museum


G.N. Bailey et al. (eds.), *Under the Sea: Archaeology and Palaeolandscapes of the Continental Shelf*, Coastal Research Library 20, DOI 10.1007/978-3-319-53160-1_14

this, no German state heritage authority has responsibility in the German Exclusive Economic Zone (EEZ) outside its territorial waters (12 nautical miles). Moreover, this heritage is currently threatened by diverse and potentially destructive economic activities, for example, laying of submarine cables and pipelines, the construction of offshore wind farms, oil and gas production, offshore mining, engineering installations, fisheries and sea farming activity. Almost every inch of seafloor within the German EEZ is affected by such activities.

On the other hand, this intensive and extensive commercial exploitation of the seabed has been accompanied by numerous surveys of surface and sub-surface features, and these offer a vitally important source of scientific knowledge on the origin and development of this unique prehistoric landscape (Ward et al. 2014).

During recent decades especially, scientists from the UK and the Netherlands have re-discovered the Doggerland landscape in the British and Dutch sectors of the North Sea (Fitch et al. 2005, 2011; Gaffney et al. 2007, 2009; Spinney 2008; Cohen et al. 2014; van Heteren et al. 2014). Whilst the German EEZ comprises prominent parts of the late Pleistocene-Holocene drainage system of Doggerland, particularly the Elbe Palaeovalley or 'Elbe-Urstromtal', its presence and extent is still largely 'terra incognita'.

During surveys within the German EEZ running high resolution reflection seismics, two shallow palaeo-river structures draining into the Elbe Palaeovalley were discovered (Fig. 14.1). One is located in the northern part of the German EEZ, at the entry of the so-called Duck's Beak ('Entenschnabel'; study area 1, Fig. 14.2), the second runs through the southern part of the German EEZ, about 40 km north of the island of Juist (study area 2, Fig. 14.3). The two study areas are located about 170 km apart.

Although both palaeo-river systems drained into the Elbe Palaeovalley, they were affected in different ways by the Holocene marine transgression. This raises two questions. Firstly, how did each river system contribute to the formation of the late glacial-postglacial palaeolandscape? And secondly, how did they interact with the late Weichselian/early Holocene hydrogeological regime on the North Sea shelf in the course of sea-level rise?

## 14.2 Database

In both study areas, all seismic profiles are based on channel boomer seismic data. Boomer data were obtained using a boomer plate operating at an energy level of 300 J. The data were recorded using a short single-channel streamer with 20 hydrophones. This configuration provided a maximum shot rate of 3.5 shots per second, which offered a good horizontal resolution.

Study area 1 was sampled during one expedition in 2012 with a seismic grid of south–north and east–west seismic tracks using a spacing of approximately 300 m. For methodical reasons (namely the density of the seismic grid), study area 2 is sub-divided into a northern sector (surveyed during three expeditions from 2009 to 2011) and a southern sector (surveyed in 2013). For the northern sector, a dense grid of south–north and east–west seismic tracks was run with a spacing of approximately 200 m. The southern sector was added to trace the southern connection of the river structure towards the outer Wadden Sea. Due to the test character of this latter survey, and the fact that a Traffic Separation Scheme crosses the survey area, the resulting grid has a line spacing of only about 500 m. This difference in line spacing has to be taken into account in any further interpretations.

The seismic data were used to produce initial seismic profiles of the proposed Holocene surface and to define the location, form and dimensions of the palaeorivers. For the depth estimation of the base of the ancient valleys, the two-way-traveltime (TWT) was converted into metres using a propagation velocity of 1,600–1,650 m/s. The difference between the depth of the seafloor and the estimated base of the valley gives the depth of the valley in metres below sea floor (mbsf). The base of the valley was used to compute gridded maps based on a method of natural neighbour interpolation.

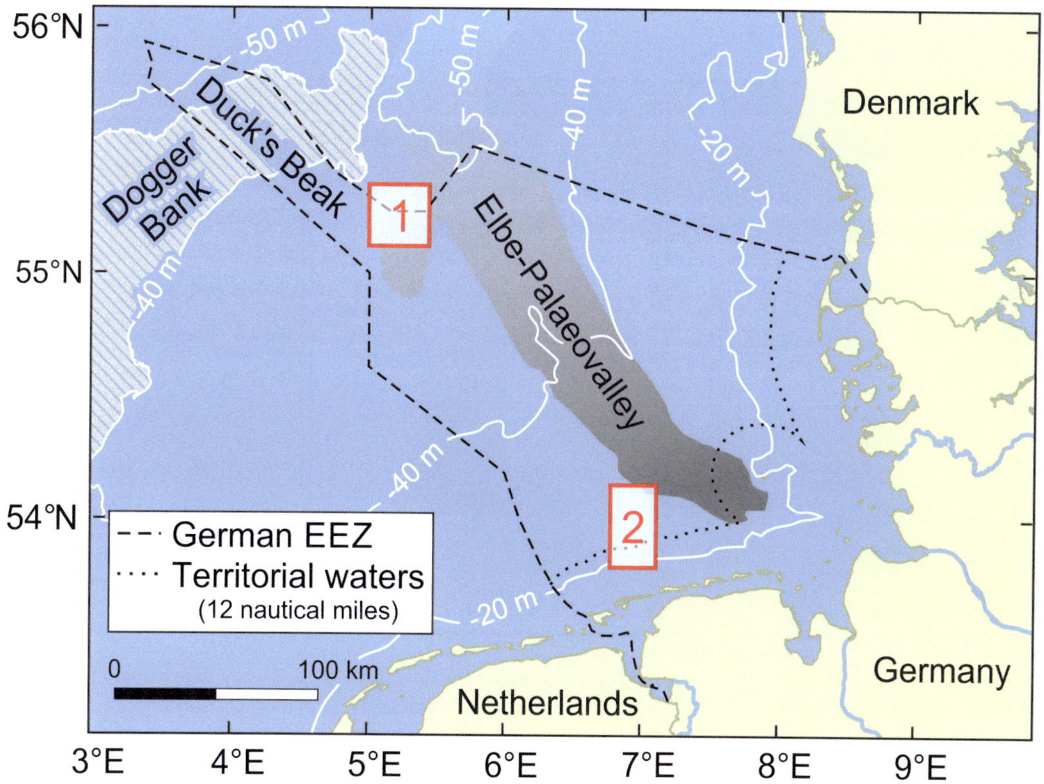


**Fig. 14.1** The southern North Sea, showing the locations of study area 1 (see Fig. 14.2), at the entry of the Duck's Beak ('Entenschnabel'), and study area 2 (see Fig. 14.3), which is about 40 km north of the island of Juist. *Shading* indicates the Elbe Palaeovalley and the Dogger Bank. The *dotted line* marks the border of German territorial waters (12 nautical miles) and the *dashed line* marks the German exclusive economic zone (EEZ). Contour lines for −50 m, −40 m, and −20 m (*white solid lines*) are with reference to the present sea level and represent hypothetical shorelines at about 10.6, 9.8, and 8.2 ka

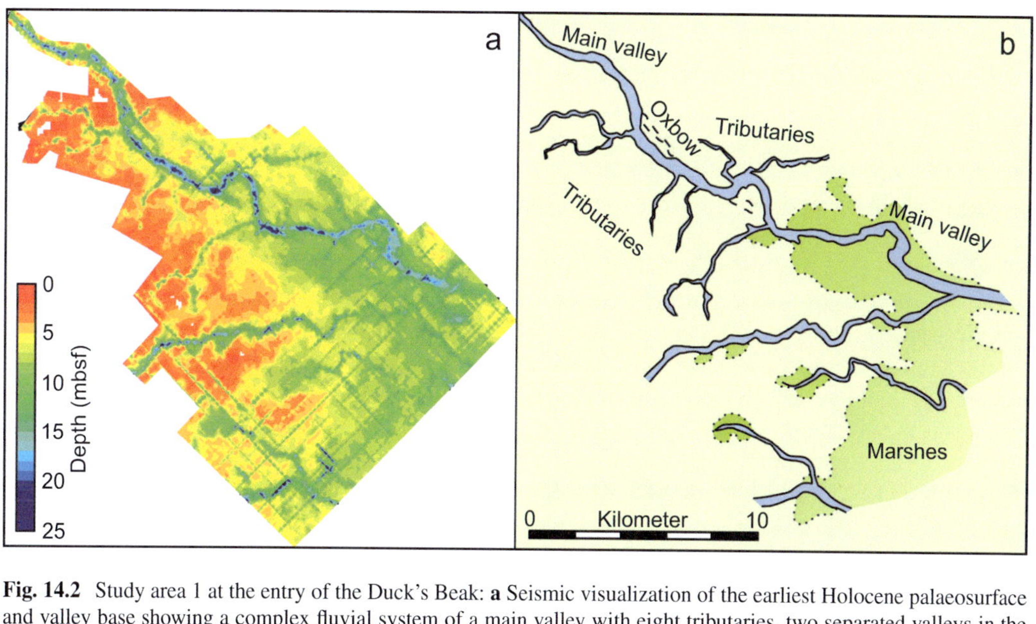


**Fig. 14.2** Study area 1 at the entry of the Duck's Beak: **a** Seismic visualization of the earliest Holocene palaeosurface and valley base showing a complex fluvial system of a main valley with eight tributaries, two separated valleys in the south-eastern part as well as marshes. **b** Interpretation of the seismic grid showing rivers (*blue*), Oxbows (*dashed line*) and the extension of marshes (*green and dotted line*)

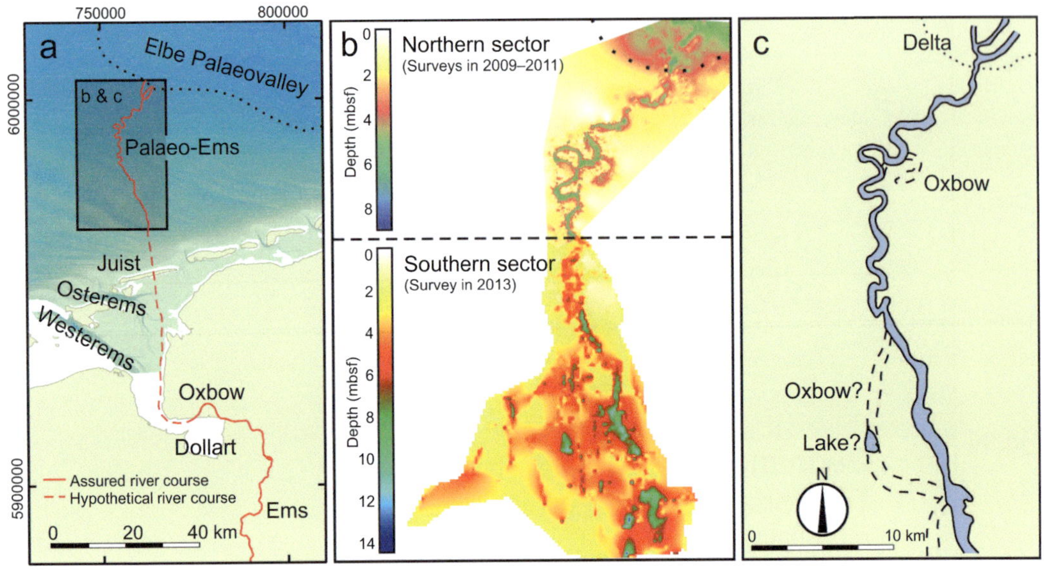


**Fig. 14.3** Study area 2, Paleo-Ems: **a** Assumed river course of the drowned Palaeo-Ems and the modern Ems River (*solid red lines*) in relation to the Elbe Palaeovalley (*dotted black line*). The hypothetical course (*dashed red line*) crosses today's barrier island Juist, which was not formed when the Palaeo-Ems existed. **b** Seismic visualization of the palaeosurface and valley base, and **c** interpretation of the seismic grids showing a meandering river (*blue*) with oxbows (*dashed lines*) and oxbow lakes (*blue*) and the delta of the Elbe Palaeovalley (*dotted line*). The seismic visualization is separated into a northern and a southern sector due to different qualities of the seismic grid (see text for explanation)

The degree to which the river valley structures meander—their sinuosity—was measured by the ratio of the real length of a river section to the linear distance between its endpoints (Leopold et al. 1964). Sinuosity ratios ≥1.5 indicate meandering structures.

## 14.3 Stratigraphic Interpretation

Due to the lack of sediment cores and geological ground-truth data in study area 1, the stratigraphy is obtained by interpretation of seismic profiles. We used characteristic reflectors to identify seismic packages which cover or intersect the substratum (Figs. 14.4 and 14.5). The horizontally stratified reflectors 'A' and 'B' represent the top and base of the modern sea floor. The package they comprise is about 1 m thick and is observed in both profiles. Reflector 'C' may be congruent with the base of the mobile sand deposits spatially distributed in the North Sea (Zeiler et al. 2000). The c. 3.5 m-thick seismic package between reflectors 'B' and 'C' is characterised by a more transparent horizontal stratification which increases in strength downward and may reflect a fining downward in grain size from middle and fine sands to silts. This package is only seen in the profile in Fig. 14.4. Strong horizontal seismic reflectors 'D' are indicative for clay and peat deposits at the base of the mobile sand deposits of 'C'. Reflector 'E' marks the irregular-shaped base of various depressions in both profiles (Figs. 14.4 and 14.5), which incise 5–10 m deep into the substratum. They show different shapes and internal seismic patterns but seem to belong to the same stratigraphic horizon. The infill 'C–E' in Fig. 14.4 shows a complex depositional body with irregular internal reflectors. These internal reflectors partly overly the seismic package 'C–D' and are partly cropped by the same seismic package. The base of the depression in Fig. 14.5 is more U-shaped and marked by a strong reflector indicating clay

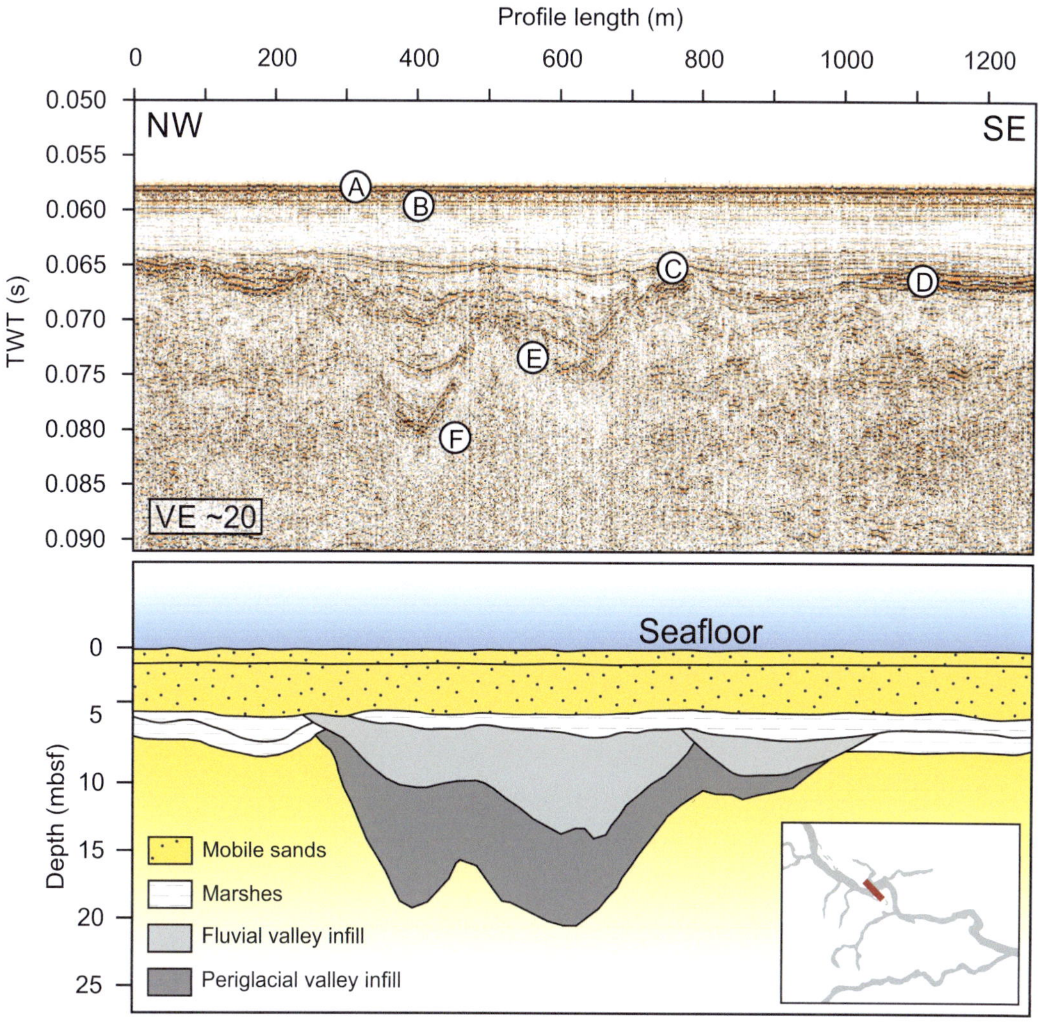


**Fig. 14.4** Study area 1. Seismic cross section and interpretation of the main valley showing two drainage phases: (1) A deep periglacial valley which is incised up to 20 m into the subsurface can be related to the late Weichselian glacial, and (2) a shallower, down to 8 m-deep fluvial valley with repeated dislocations of its bed. Prominent seismic reflectors used for the stratigraphic interpretation are marked with capital letters *A–F*

and peat deposits. The infill of this depression is horizontally stratified and has similar characteristics to package 'B–C' in Fig. 14.4. A specific characteristic observed in Fig. 14.4 is that the already described depression cuts another stratigraphically older depression, which incises more than 15 m deep into the substratum. The base of this older depression is barely identifiable as a continuous transparent zone 'F'. The infill of the depression is marked by some stronger internal reflectors.

Interpretation of seismic profiles in study area 2 (Figs. 14.6 and 14.7) is supported by visual descriptions of four cores (Fig. 14.8). The cores penetrate the entire valley infill from the seafloor to the base. The observed lithological succession is typical of this North Sea region. The river valley incises into glacio-fluvial sand deposits of the substratum. The base of the valley infill is determined by a 0.2–1 m-thick basal peat layer which is overlain by intertidal or marine deposits of clays or silty clays. The river valley is covered by modern or mobile sands, which is also evident in the horizontally

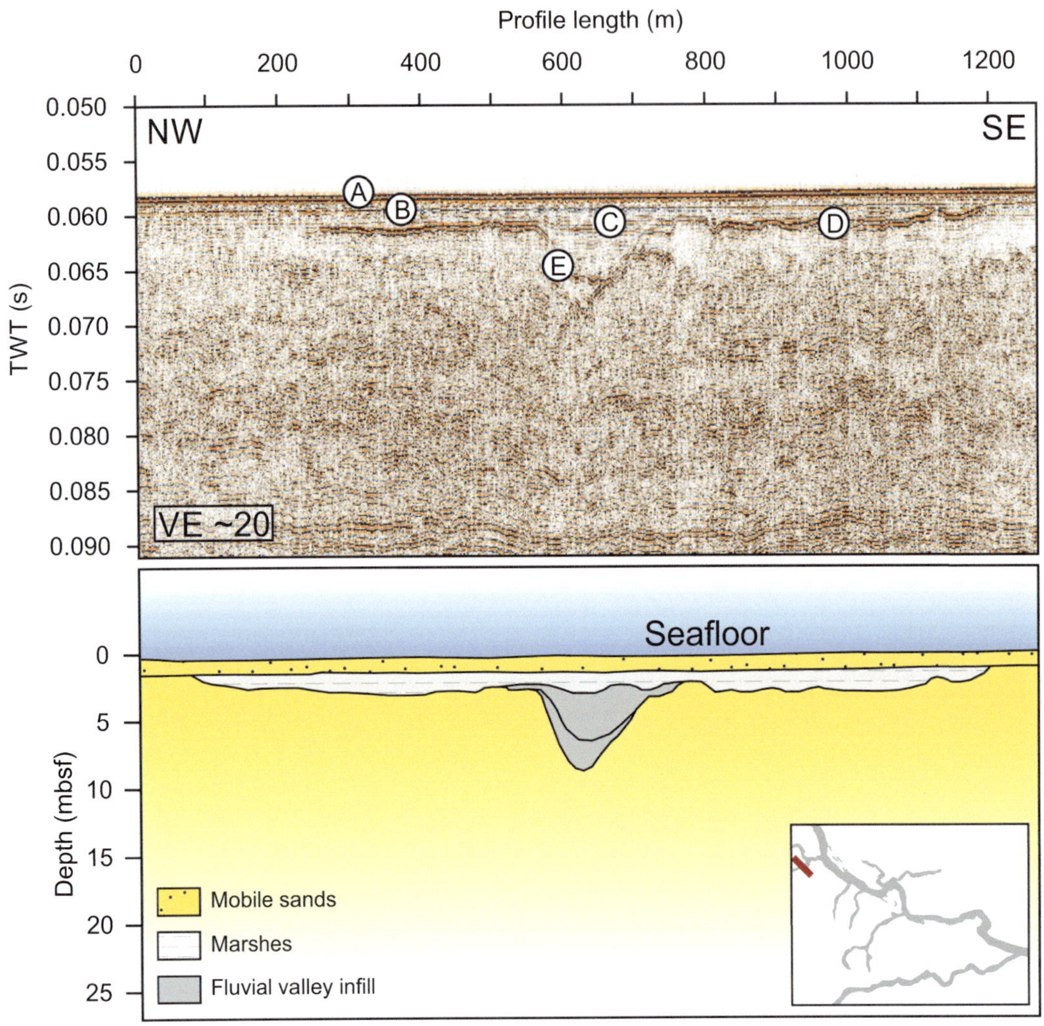


**Fig. 14.5** Study area 1. Seismic cross section and interpretation of a tributary to the main valley shows an 8 m deep, V-shaped valley body with a homogeneous fill, which is embedded in a very shallow 2 m-deep body interpreted as a marsh. Prominent seismic reflectors used for the stratigraphic interpretation are marked with capital letters *A–E*

stratified seismic reflectors 'A–B' (Figs. 14.6 and 14.7). The strong seismic reflectors at the valley base in Figs. 14.6 and 14.7 can be attributed to the peat layers observed in the cores.

## 14.4 Fluvial Systems Seen in the Geophysical Record

### *14.4.1 Study Area 1*

At the entry of the Duck's Beak ('Entenschnabel'), a system consisting of a deeper main valley with eight shallower tributaries (five on its southern side and two on its northern side) was observed in the seismic grid, running in a north-west to south-east direction from the Dogger Bank area toward the Elbe Palaeovalley (Fig. 14.2). Two separated valleys in the south-eastern part of the study area run

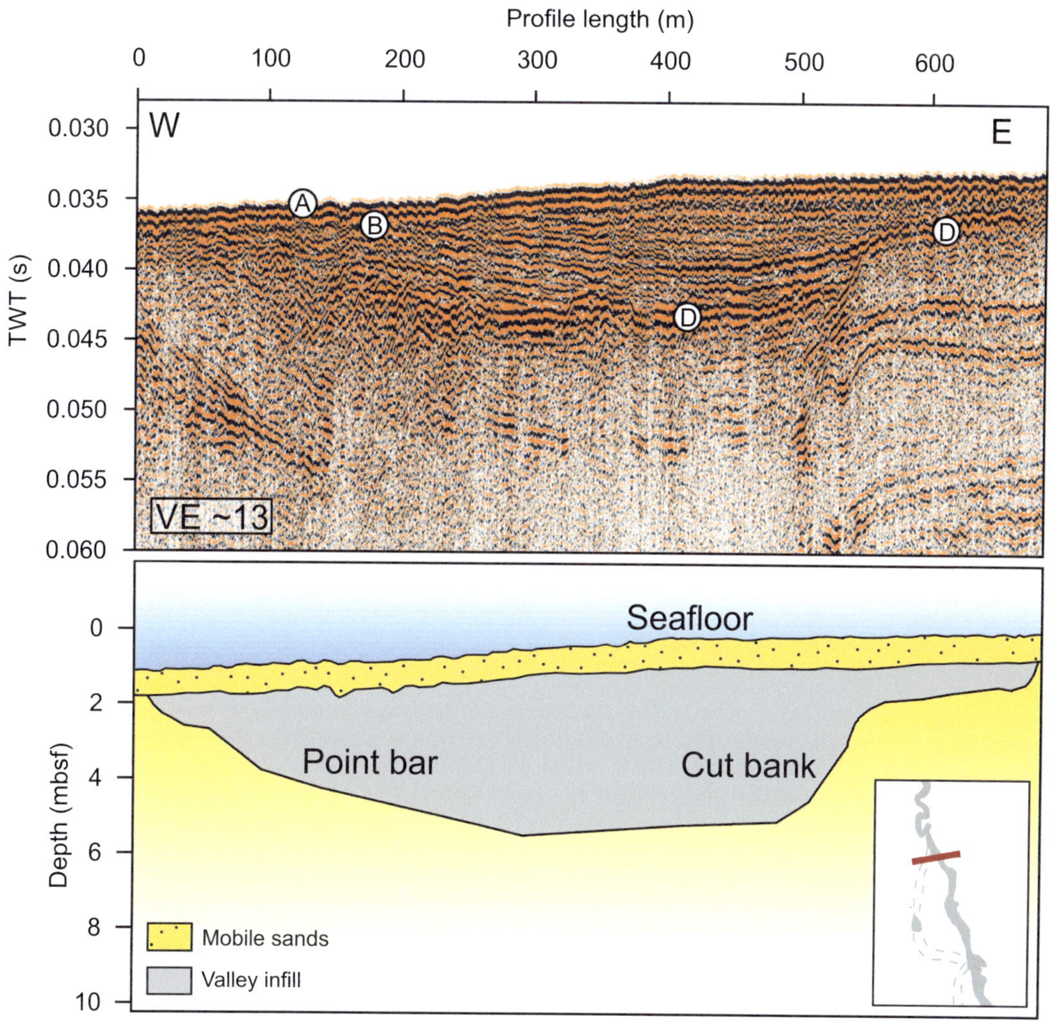


**Fig. 14.6** Study area 2. Seismic cross section and interpretation of the Palaeo-Ems (southern sector) showing pointbar and cut bank of a meandering river structure. Prominent seismic reflectors used for the stratigraphic interpretation are marked with capital letters *A–D*

more or less parallel to the main valley. The seismic visualization of the proposed Holocene base (Fig. 14.2) shows a slightly elevated area in the south-western and western part of the study area. This is interpreted as the origin of some of the tributaries. The north-eastern and eastern part of the study area is dominated by shallow depressions connecting the valleys. Strong seismic reflectors at the base of the depressions indicate clay or peat deposits. Hence, these depressions are interpreted as palaeo-marshes. Similar relationships have been documented in the Dogger Bank area for this time period by Fitch et al. (2005). The source of the main valley and its exit into the Elbe Palaeovalley are not covered by the existing surveys.

Inferred from seismic interpretation, the valleys are covered by a 3.5–5.0 m-thick unit of mobile sands. The greater thickness of the sand coverage compared to study area 2 is caused by the greater water depth and the associated reduced vulnerability to wave and current exposure on the shallow shelf. The water depths increase from 39 m in the South–east to 46 m in the North-west.

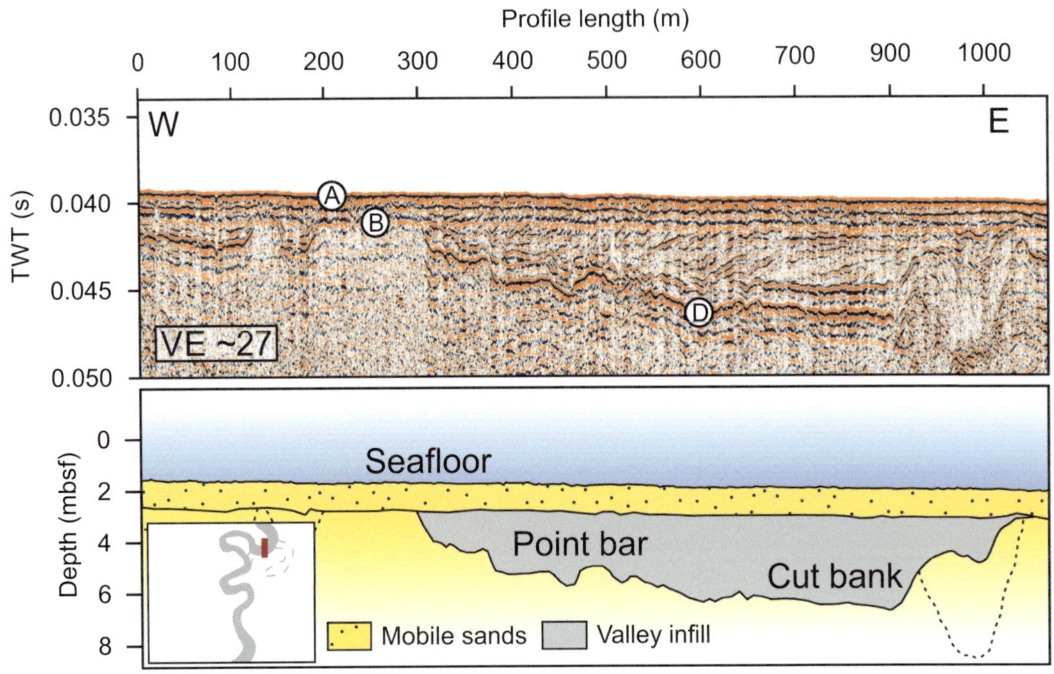


**Fig. 14.7** Study area 2. Seismic cross section and interpretation of the Palaeo-Ems (northern sector) cut bank, showing an undulating base to the pointbar and cut bank of a meandering river structure. On its eastern flank the valley appears to truncate another valley structure (*dashed line*), although this has not yet been confirmed. Prominent seismic reflectors used for the stratigraphic interpretation are marked with capital letters *A–D*

The main valley was mapped over a length of 30 km. The sinuosity ratio is 1.19 and is therefore indicative of a rather linear river. The valley is approximately 700 m wide and incises, with relatively steep flanks, down to 20 mbsf into the sandy sediments of a larger Pleistocene glacial and fluvio-periglacial accumulation plain. This proposed periglacial deeper valley was re-used by a shallower valley system whose morphology is indicative for a fluvial valley with repeated dislocations of its bed. However, without ground truth data from sediment cores, it is not clear if the actual infill deposits are of fluvial or tidal origin. During the process of drowning, the river valleys were occupied by tidal channels and we have no indications as to how severe the reworking of the fluvial sands has been. Reflections of tidal sand deposits and fluvial sand deposits are often quite similar. This can lead to misinterpretations, as is shown by van Heteren et al. (2014) for the Dutch sector of the North Sea.

The bed of this shallower river valley is located at a depth similar to that of the other river structures and tributaries in the study area (Fig. 14.2). These tributaries are up to 6 km long, 250 m wide, and incise 8–12 m below the seafloor. Their sinuosity is rather linear, with ratios varying between 1.09 and 1.27. An example is given in the seismic cross section of a tributary to the main river in Fig. 14.5. The valley is U-shaped, 200 m wide and as much as 8 m deep. Its infill is more homogeneous and shows horizontally stratified seismic reflectors.

## *14.4.2 Study Area 2*

About 40 km north of the island of Juist, a shallow south–north oriented buried valley with meanders, oxbows and possible oxbow lakes is observed on seismic grid profiles over a distance of nearly 50 km (Fig. 14.3). Water depths in this study area increase from South to North from 19 to 34 m. The valley

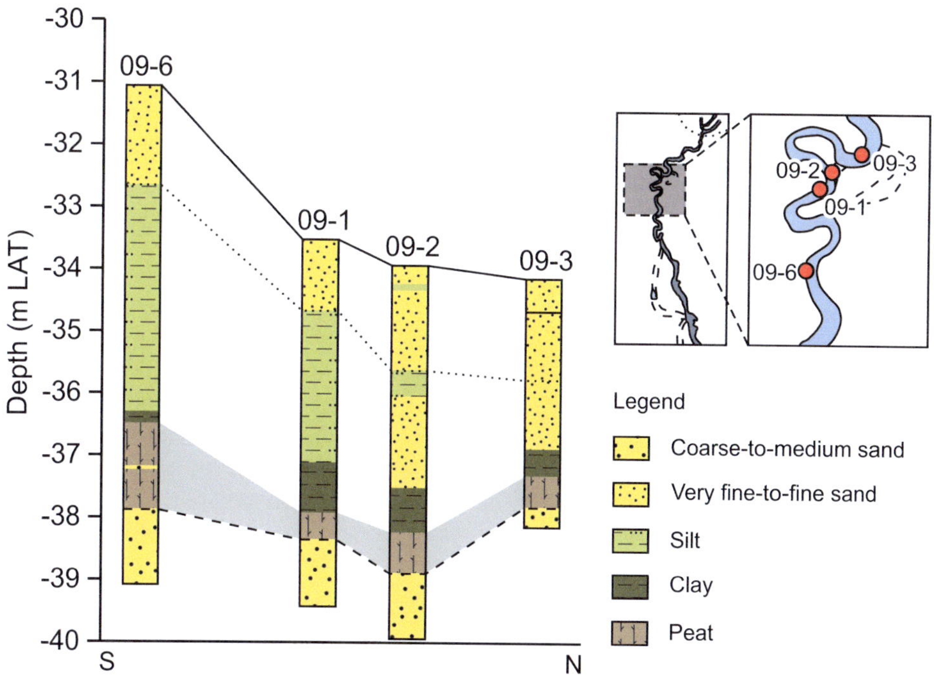


**Fig. 14.8** Study area 2. Sediment columns derived from visual core description of four cores along a virtual south to north transect in the northern sector. The top of the cores are corrected to m LAT depths. The base of the river valley infill (*dashed line*) is determined by a basal peat layer (*grey shading*) which is overlain by intertidal or marine deposits of clays or silty clays. The valley is covered by modern or mobile sands (*dotted line* = approximate depth of the valley top). The different *thicknesses* of the valley infill reflect core locations at the centre of the valley or at the valley flanks

is covered by a 0.8–2.0 m-thick unit of mobile sands. As with the valley system in study area 1, the morphology of the valley in study area 2 is indicative of a fluvial valley. Peat layers at the base of the valley overlain by clay or silty clay deposits lead to the inference that the system gradually drowned due to transgression and that the valley infill is largely of intertidal or marine origin.

The valley incises to a depth of 7–14 m into the Pleistocene glacial and fluvio-periglacial sands, which form the sub-seafloor, with depths decreasing from South to North. The width of the valley varies between 500 and 1,500 m with a generally increasing width from South to North. The north–south gradient over the total length is broadly consistent with the very shallow slope of the seafloor of 0.027%.

The shallow gradient may have caused the meandering course of the valley. At first view, the sinuosity ratio of the valley structure shows apparent differences between the northern and southern sector. The river course in the northern sector has sinuosity ratios between 1.61 and 2.15, which clearly indicates a meandering river course of about 24 km length. In contrast, sinuosity ratios in the southern sector range between 1.02 and 1.33 indicating a more linear river course of about 26 km length. However, the more linear river course in the southern sector is obviously an artefact owing to the less dense seismic coverage, which may be too coarse to resolve details of the river's sinuosity. The southern origin of the river valley remains undiscovered so far. However, derived from its location and course, we suppose that the ancient river is the drowned predecessor of the modern Ems River (Fig. 14.3a). The northern end of the river valley fans out into at least three valleys forming a delta at the western flank of the Elbe Palaeovalley.

Morphological features of meandering rivers like point bars and cut banks are well-developed in seismic profiles (Figs. 14.6 and 14.7). The valley base in the cross section is irregular due to several small adjoining valleys. Seismic reflectors within the infill are continuous and parallel to sub-parallel, showing a homogeneous, horizontally stratified infill (Fig. 14.6). In some seismic profiles, inclined reflectors indicate a different infill process across the western valley flank (Fig. 14.7).

## 14.5 Styles of Tributaries in the Elbe Palaeovalley

Studies in the British and Dutch EEZ subsurface suggest a diversity of styles of river systems related to the late Weichselian/early Holocene Doggerland palaeolandscape, depending on specific topographic settings and their change over time (Törnqvist 1993; Berendsen and Stouthamer 2000; Fitch et al. 2005; Gaffney et al. 2007; Hijma 2009). Data from the western Dogger Bank show that, during the Holocene, a complex meandering river system had developed consisting of a 600 m-wide palaeovalley with tributaries or distributary channels with moderately developed sinuosity, and associated lakes or marshes (Fitch et al. 2005). We infer that these were finally filled with tidal sediments due to transgression. The general fluvial morphology of study area 1 appears to have analogies with the earlier fluvial phase of this western Dogger Bank area in terms of a major river channel with low to moderate sinuosity, tributaries and associated marshes.

From our investigation, study area 1 seems to have experienced two phases of development with distinctive features forming in each case. The first phase consisted of a deep palaeovalley, which we suppose originated during or at the end of the last glaciation. Both the bottom of the periglacial valley and the plains of the early phases of the Elbe Palaeovalley are chronologically related, occurring at the same stratigraphic level (see Figge 1980). In the second phase, during a postglacial stage, the deeper valley was replaced by a valley system with tributaries of moderately developed sinuosity or distributary valleys and lakes. The valley fill of this younger system is associated with a higher stratigraphic level than the periglacial valley and may reflect a time when the Elbe Palaeovalley was already influenced by transgression. The estimated age for the second fluvial phase is based on the assumption of a coexistence of the fluvial systems in both study areas, whereas the Dogger Bank fluvial system overlies buried late Weichselian glacial tunnel valleys. The fluvial system ceased during the course of the North Sea transgression when this region was flooded not later than 9,800 years ago. This is based on the −40 m contour line (Fig. 14.1) and the regional sea level curve of Vink et al. (2007).

The river structure located in study area two to the South differs significantly from the valley system of study area 1. The river structure here is characterized by a clear meandering channel with lateral valley displacements running parallel to its flow direction and shifting sinuosities, but without any change in its dimension or meandering characteristics over time. Analogies in morphology, incision depth and dimensions are known from the late glacial/early Holocene Scheldt River (Kiden 1989, 1991) and from the early phase of the Rhine-Meuse system (Törnqvist 1993; Berendsen and Stouthamer 2000). The deposits of the late Weichselian river incision in the Scheldt are 7–8 m thick and the alluvial plain was about 1.8 km wide. The incision took place at about 13,000 BP, based on palynological studies and radiocarbon dating (Kiden 1991). According to Kiden (1989), the character of the valley experienced important changes after the first occurrence of peat layers at about 8,700 BP. After that time, a new river valley was incised into the already existing peat deposits of the alluvial plain. Several subsequent lateral migrations of the river course caused the formation of point bars whilst peat and underlying sediment deposits were eroded to form cut banks. Using a time-space model, Törnqvist (1993) illustrated a braided river system for the late glacial followed by an early Holocene meandering river system. From about 5,000 years ago the system changed to an alternating meandering anastomosing system associated with a rapid sea level rise.

The comparison of these findings from the Scheldt River and the Rhine-Meuse valley belt with the small incisions in the valley base of the river of study area 2, together with the formation of point bars and cut banks, suggests that the river of study area 2 may have changed its valley character from a meandering, late glacial river valley to a wide, Holocene alluvial plain with migrating meanders, point bars and peat growth as a consequence of progressive North Sea transgression.

The river systems described for the two study areas seem to have their origin in the latest glacial, when permanently frozen ground was still present. We infer that the conspicuous difference in morphology of the river systems described for the two study areas originates in differences in their source areas and regional differences in relation to sea-level change after the Last Glacial Maximum. The perimarine river of the Ems palaeovalley persisted as a single migrating river draining across exposed land about 8,200 years ago based on the –20 m contour line (Fig. 14.1) and the regional sea level curve, at a time when the northern river system in the Duck's Beak area was already drowned.

## 14.6 Conclusion

The postglacial landscape in today's North Sea area known as Doggerland was covered by a network of rivers, lakes, wetlands and huge drainage channels. Seismic records of two newly discovered structures lying within the German EEZ that once formed part of Doggerland provide evidence for a morphology of clearly riverine origin and a complex hydrogeological regime. However, the riverine meanders are filled for the most part with intertidal deposits, which points to the fact that the sedimentary infill was only deposited when the river had already been transformed to an estuary during the course of marine regression.

Both river structures are tributaries of the Elbe Palaeovalley, and both are assumed to have developed since the latest Weichselian. However, differences in topography and source, as well as timing and intensity of fluvial activity, caused significant differences in river morphology.

The northern river structure connected the Dogger Bank in the North-west with the western flank of the Elbe-Palaeovalley. The system was influenced very early on by the postglacial sea-level rise and developed from a single, deep valley, which may initially have originated during the last glacial, to a network of shallower rivers and tributaries. Their morphology is in various aspects similar to river structures known from the western part of the Dogger Bank.

The southern river structure is the drowned extension of the modern Ems River and can be distinguished as part of the tributary system feeding the southern head of the Elbe Palaeovalley, together with the Elbe, Weser and Eider Rivers. The river structure was, from the beginning, formed as a meandering river valley with repeated dislocations of its river bed. However, the Ems palaeovalley still existed whilst the northern river structure was already being drowned by the advancing North Sea.

The development of both river systems reflects the landscape evolution of Doggerland along the west flank of the Elbe Palaeovalley. This, in turn, has implications for the landscape configuration available to Mesolithic hunter and gatherers. In the initial phase, the northern system was already draining the Dogger Bank area while the adjacent last-glacial ice-sheet was still collapsing. However, the southern Ems palaeovalley seems to have developed during the postglacial phase. It is evident that we need further investigations and stronger ground truth data from sediment cores to improve our understanding about the role of such drainage systems in forming the Doggerland landscape within the German EEZ and their influence on changing patterns of Mesolithic settlement.

**Acknowledgments** We gratefully acknowledge Sebastian Feldmann, Lukasz Socko, Mike Belasus and Vanessa Wahlers for their excellent technical and scientific operation on board *LEV Taifun* and for assistance with the data processing onshore. We thank Mark Coughlan for his suggested improvements to the text, and Henk Weerts and Dimitris Sakellariou for constructive reviews that improved our manuscript. This study was funded through DFG-Research Center/Cluster of Excellence 'MARUM – The Ocean in the Earth System' in cooperation with the project 'Bedrohtes Bodenarchiv Nordsee' funded by the Federal Ministry of Education and Research (Germany). The study was further supported with data from several anonymous industry partners.

## References


Berendsen HJA, Stouthamer E (2000) Late Weichselian and Holocene palaeogeography of the Rhine-Meuse delta, the Netherlands. Palaeogeogr Palaeoclimatol Palaeoecol 161:311–335

Cohen KM, Gibbard PL, Weerts HJT (2014) North Sea palaeogeographical reconstructions for the last 1 Ma. Neth J Geosci 93:7–29

Coles BJ (1998) Doggerland: a speculative survey. Proc Prehist Soc 64:45–81

Coles BJ (2000) Doggerland: the cultural dynamics of a shifting coastline. In: Pye K, Allen SRL (eds) Coastal and estuarine environments: sedimentology, geomorphology and geoarchaeology, vol 175. Geological Society of London, Special Publication, London, pp 393–401

Figge K (1980) Das Elbe-Urstromtal im Bereich der Deutschen Bucht (Nordsee). Eiszeit Gegenw 30:203–211

Fitch S, Gaffney VL, Ramsey E, Kitchen E (2011) West coast palaeolandscapes survey, final report. Visual and Spatial Technology Centre and University of Birmingham, Birmingham

Fitch S, Thomson K, Gaffney V (2005) Late Pleistocene and Holocene depositional systems and the paleogeography of the Dogger Bank, North Sea. Quat Res 64:185–196

Gaffney VL, Fitch S, Smith D (2009) Europe's lost world: the rediscovery of Doggerland. CBA research report 160. Council for British Archaeology, York

Gaffney VL, Thomson K, Fitch S (eds) (2007) Mapping Doggerland: the Mesolithic landscapes of the southern North Sea. Archaeopress, Oxford

Hijma MP (2009) From river valley to estuary: the early-mid Holocene transgression of the Rhine-Meuse valley, The Netherlands. PhD thesis, University of Utrecht

Kiden P (1989) The late Holocene evolution of the perimarine part of the river Scheldt. In: Henriet JP, De Moor G, De Batist M (eds) The quaternary and tertiary geology of the southern bight North Sea. Belgian Geological Survey, Brussels, pp 173–184

Kiden P (1991) The late glacial and Holocene evolution of the middle and lower river Scheldt, Belgium. In: Starkel L, Gregory KJ, Thornes JB (eds) Temperate palaeohydrology: fluvial processes in the temperate zone during the last 15000 years. Wiley, Chichester, pp 283–299

Leopold LB, Wolman MG, Miller JP (1964, reprinted 1995) Fluvial processes in geomorphology. Dover Publication, New York

Spinney L (2008) The lost world. Nature 454:151–153

Törnqvist TE (1993) Holocene alternation of meandering and anastomosing fluvial systems in the Rhine-Meuse Delta (Central Netherlands) controlled by sea-level rise and subsoil erodibility. J Sediment Res 63:683–693

Van Heteren S, Meekes JAC, Bakker MAJ, Gaffney V, Fitch S, Gearey BR, Paap BF (2014) Reconstructing North Sea palaeolandscapes from 3D and high-density 2D seismic data: an overview. Neth J Geosci 93:31–42

Vink A, Steffen H, Reinhardt L, Kaufmann G (2007) Holocene relative sea-level change, isostatic subsidence and the radial viscosity structure of the mantle of Northwest Europe (Belgium, the Netherlands, Germany, southern North Sea). Quat Sci Rev 26:3249–3275

Ward I, Larcombe P, Firth A, Manders M (2014) Practical approaches to management of the marine prehistoric environment. Neth J Geosci:1–12

Zeiler M, Schulz-Ohlberg J, Figge K (2000) Mobile sand deposits and shoreface sediment dynamics in the inner German bight (North Sea). Mar Geol 170:363–380

# Chapter 15
# The Wadden Sea of North-West Germany: An Intertidal Environment of High Archaeological Research Potential

**Martina Karle and Julia Goldhammer**

**Abstract** This paper presents the first results of an inventory of archaeological remains in the Wadden Sea of northwest Germany. We have generated palaeogeographic maps on the basis of a geological reconstruction of Holocene landscape evolution from archive core data, and use these to identify areas where human traces have potentially been preserved during different time intervals. Currently, the cultural heritage is being increasingly threatened by coastal protection measures, dredging of waterways as well as offshore industrial activity. Recording of known sites, prospecting for new sites and the subsequent investigation and monitoring of both demonstrate the research potential of this region. These results will be instrumental in helping to recover information about the cultural archive before it is destroyed by erosion and to identify target areas for future research.

## 15.1 Introduction

The Wadden Sea refers to the southern sector of the North Sea. Today, it is a World Heritage Site famous for its natural fauna and flora and its marine ecosystems and comprises one of the largest unbroken areas of coastal wetlands and intertidal mudflats in the world, extending for about 500 km from the Netherlands in the West to the west coast of Denmark in the East. It represents a depositional coastline of coastal wetlands, intertidal mudflats and offshore sand barriers and islands that has been subject to continuous conditions of rising relative sea-level since the last glacial (Streif 1990b, 2004; Behre 2007).

Initially, the melting of the Scandinavian ice sheet caused a rapid sea level rise and consequently a continuous displacement of the coastline throughout the late glacial period and the early postglacial, drowning large tracts of the now-submerged landscape of Doggerland that were important zones of human settlement and subsistence activity. However, a more modest rate of relative sea-level rise has continued up to the present day, with highly dynamic processes of sediment accumulation and erosion that have steadily encroached on and partially buried land surfaces and associated evidence of human activities in the later prehistoric and historical periods.

M. Karle
Niedersächsisches Institut für historische Küstenforschung, Viktoriastr.26/28, 26382 Wilhelmshaven, Germany
e-mail: karle@nihk.de

J. Goldhammer (✉)
Lower Saxony Institute for Historical Coastal Research, Niedersächsisches Institut für historische Küstenforschung, Viktoriastr. 26/28, 26382 Wilhelmshaven, Germany

Bohusläns Museum, Museigatan 1, 45119 Uddevalla, Sweden
e-mail: julia.goldhammer@gmail.com



G.N. Bailey et al. (eds.), *Under the Sea: Archaeology and Palaeolandscapes of the Continental Shelf*, Coastal Research Library 20, DOI 10.1007/978-3-319-53160-1_15

In coastal lowlands, human occupation is closely tied to changes in regional water levels and resulting locations of waterways. Accordingly, landscape evolution and the occupation history of coastal zones are closely related and may therefore provide information on regional fluctuations in relative sea level. The tidal range and the extent of tidal inundation are the two most relevant factors in reconstructing transformations in the coastal landscape, understanding where material traces of past cultural activity are most likely to be preserved, and interpreting archaeological material in its landscape setting.

From 3000 BP there is growing evidence from archaeological data that people have occupied particular coastal lowlands and left their mark on the landscape (Kluiving et al. 2013). In the central Wadden Sea region only a few finds are known from this time period, because archaeological remains are likely to have been dispersed and mixed into the sediments of former land surfaces. If these remains can be detected and recovered, they should provide important evidence for transgressions and regressions as well as catastrophic storm events. Such former land surfaces are therefore unique archives for the exploration of cultural heritage (Flemming et al. 2014). Obtaining information about storm surges and other erosional processes and the resulting redistribution of sediments makes it possible to identify areas where former land surfaces can be detected. However, erosion also involves the risk of total destruction of these natural archives and any associated cultural remains and human settlements. Exposed finds on tidal flats are particularly vulnerable and therefore in need of immediate registration (Jöns et al. 2013).

The current state of research in the Wadden Sea (Jöns et al. 2013, Niederhöfer, pers. comm.) suggests that significantly more cultural remains should exist than are recorded in the latest site data base. Since summer 2012, the geoarchaeological project 'Settlement and Cultural History of the Lower Saxony Wadden Sea Area' has set out to record known sites and to search for as yet unknown settlement remains in the Lower Saxony Wadden Sea (Jöns et al. 2013). The area of investigation stretches from the Ems estuary in the west to the Elbe estuary in the east, including the estuaries, embayments and back-barrier tidal flats of the East Frisian coast (Fig. 15.1).

## 15.2 Research Questions

Since 2009, when UNESCO declared the Dutch and German parts of the Wadden Sea as a World Heritage Site, the general appreciation and respect for the Wadden Sea area has increased significantly. However, the main focus of public awareness and attention has been the unique natural phenomena of the region, with much less emphasis placed on the cultural heritage. One reason for this is that archaeological surveys in the Wadden Sea have been restricted to a few distinct locations, the known sites being mostly attributable to incidental finds. At the same time commercial exploitation (e.g., wind farms, commercial fishing, oil and gas production, aggregate extraction) in the offshore regions of the North Sea is expected to increase rapidly over the next decades. Also, nearshore sand placements, beach nourishment and other coastal protection measures will greatly expand in the future, so that the archaeological heritage is increasingly at risk.

To achieve sufficient appreciation of the drowned cultural heritage of the Wadden Sea and its consideration and integration in official conservation planning, the availability and dissemination of reliable data on the location and conservation status of archaeological finds and sites is of central importance. As one part of the project, data are collected for archaeological heritage conservation. These constitute an important basis for the ability to assess the impacts of any commercial activity affecting the cultural heritage. For the correct interpretation of the cultural material archived in the archaeological sources, the availability of data on the palaeogeography of each site and find area is of major importance (Jöns et al. 2013). Consequently, the investigation and reconstruction of the evolution of the Wadden Sea landscape is another key objective in the project presented here. By means of

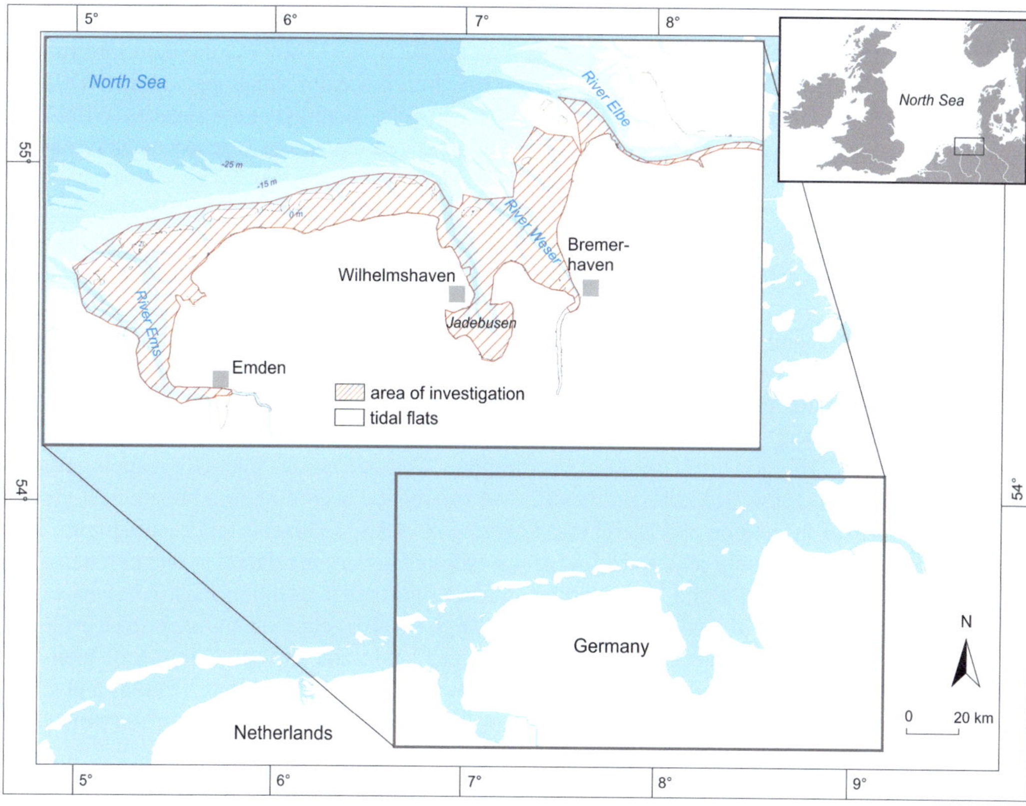


**Fig. 15.1** Map showing the area of investigation

palaeogeographic maps and supplementary data on morphodynamics resulting from sedimentary processes, it is possible to identify potential sites of human occupation where corresponding cultural remains may be found today.

Another objective of the project is to explore existing sites in greater detail in order to facilitate their cultural and historical classification, and this, in its turn, will contribute to a better understanding of the settlement history of north-western Germany.

New information on the long-term use of North Sea coastal areas and the varying strategies of the prehistoric and early historic populations in dealing with rising sea level will greatly expand our perception of human life along the southern North Sea coast.

## 15.3 Methods

The Wadden Sea, including its coastal marshes and islands, presents a large variety of landscapes, each with its own natural and cultural features in different periods. These landscape types can be understood and described as habitats having different characteristics and potential for human settlement and land use during the individual periods.

Available data about the palaeogeography of each site and find area is of central importance for the interpretation of archaeological sources with respect to the cultural history (Jöns et al. 2013; Goldhammer and Karle 2015). Data from sediment cores and seismic profiles, as well as geological, nautical and historical maps are currently being evaluated to obtain comprehensive information about the subsurface morphology of the seabed. The *Landesamt für Bergbau, Energie und Geologie (LBEG)* provides data from a variety of geoscientific fields with the aid of the Lower Saxony Soil Information System via the NIBIS® MAP SERVER (Niedersächsischen Bodeninformationssystems NIBIS®). This information forms a very extensive but, at the same time, heterogeneous dataset. Within the marsh belt of the Lower Saxony coastal area, over 30,000 sediment cores are documented, about 3,000 of which are from the intertidal area of the Wadden Sea and the islands. These data have been inherited from both regular geological research and from applied coring activities. Only a few datasets are supported by radiocarbon dates giving an absolute chronology of the depositional history. Therefore, quality and content of the core data differ widely and data have to be evaluated regarding their significance for mapping potential archaeological sites.

The investigation mainly focused on sedimentological aspects of the deposits to determine the depositional environment and related processes. The borehole database of the LBEG (Fig. 15.2) enables an integrated reconstruction of coastal evolution throughout the Holocene. The deposits were interpreted in terms of sedimentary successions, these being greatly enhanced by the detailed analyses of Barckhausen et al. (1977) and Streif (1990b), who proposed a classification system for mapping coastal plain deposits, in which the Holocene sediments were subdivided according to a hierarchical system based on the vertical succession and lateral interfingering of clastic sediments and peat deposits. This classification system accurately reflects Holocene depositional history in relation to rate of relative sea level change, sediment budget and accommodation space. Particular attention must be given to the depth and changing courses of tidal channels and waterways, so as to determine the areas to be excluded due to erosion. It is then possible to visualize the Holocene landscape units at various time intervals. Along with the evaluation of the preservation potential of archaeological artefacts, features and structures, an important framework is created for the sustainable management of the Wadden Sea which also incorporates the cultural heritage (Jöns et al. 2013; Goldhammer and Karle 2015).

Archaeological data found in databases like ADAB*web* ([A]llgemeine Denkmal[DA]ten[B]ank [web]-basiert, a web-based general heritage database) are systematically documented to help reconstruct the palaeolandscape and potential locations of human settlement. Selected finds were examined in the

**Fig. 15.2** Map of the East Frisian coastal area with locations of sediment cores situated in the Wadden Sea and the adjacent marsh land (data base: NIBIS® Kartenserver 2014 – Landesamt für Bergbau, Energie und Geologie, Hannover)

field to assess their preservation state. Besides archives of heritage offices and museums, aerial photographs provided by the National Park Administration Wadden Sea Lower Saxony were screened for traces of human activity and subsistence (Jöns et al. 2013; Goldhammer and Karle 2015).

## 15.4 Landscape Reconstruction, Archaeological Potential and Landuse Strategies

Knowledge of the morphology of the pre-Holocene surface that reflects the original topography at the beginning of the postglacial sea level rise is essential to understand the distribution of the various sediment successions on the coastal plain. The landscape prior to marine flooding shows complex drainage patterns of numerous rivers and their tributaries. These palaeovalleys were formed during the last glacial sea level lowstand. The analysis of the core database shows that there is a regular distribution of smaller channels in the western part, whereas deeply incised valleys of the Jade and Weser palaeorivers characterize the eastern part (Fig. 15.3, see also Hepp et al. Chap. 14).

The rising sea invaded the coastal lowland through palaeovalleys, which developed into estuaries with tidal channels and flats along their margins. As the groundwater level rose with the rising sea, the flooded area was fringed by freshwater marshes in which peat bogs could develop. This peat forms the basal part of the Holocene sedimentary succession and occurs everywhere unless reworked by late Holocene tidal channel migration. This basal peat is covered by a complex sequence of sediments associated with transgression that shifts landward with the rising sea level.

Environmental interpretation of core data is essential in documenting coastal changes because each tidal sedimentary environment has a specific relation to sea level. Moreover, these sedimentary environments (e.g., marine, estuarine, freshwater, terrestrial) form a dynamic system that can shift landward or seaward, with superimposition of one type of sediment over another in response to a number of controlling factors such as the rate of sea level rise, sediment supply and accommodation space. This dynamic character implies that, at any time in the Holocene, all the different types of sediments could have existed next to one another, sometimes even over short distances (Walther 1894; Barckhausen et al. 1977; Streif 1990a, 2004). However, the evolution of one environment into another cannot be generalized for the entire coastal plain. It is related to the position of the tidal channels, which act as local sediment suppliers and drainage networks (Baeteman 2013).

To interpret the geological data, cores and types of sediment successions in terms of an indicative map for potential archaeological areas, it is necessary to merge these with modern measurements of the morphodynamics of tidal channels and creeks provided by the NLWKN Norderney (Lower Saxony Water Management, Coastal Defence and Nature Conservation Agency). Erosion depth and lateral channel migration are the limiting factors for the preservation of archaeological finds and their relationship to the former land surface.

The distribution of archaeological finds and areas with the potential for discovery of new cultural material have been compiled into a composite map (Fig. 15.4). Nearly all known settlement finds are located within the areas with the potential for preservation of cultural material. Working on a systematic record of archaeological sites in the modern tidal flats highlights the difficulties in verifying aerial anomalies and locating sites described in old find reports. The main reason for this is the sediment dynamics, which bury settlement remains and thereby prevent their discovery. In this project, previously unknown sites were discovered and examined. This work has shown that the lack of surficial archaeological remains on the tidal flats cannot be interpreted as evidence for a lack of settlements. With the enormous sedimentation rate along the southern North Sea coast since the last ice age, archaeological remains are expected to be preserved at greater depth below the present seabed. The older the find, the deeper it will be buried, which explains the fact that most archaeological finds from the Wadden Sea of north-western Germany are not older than 2,000 years.

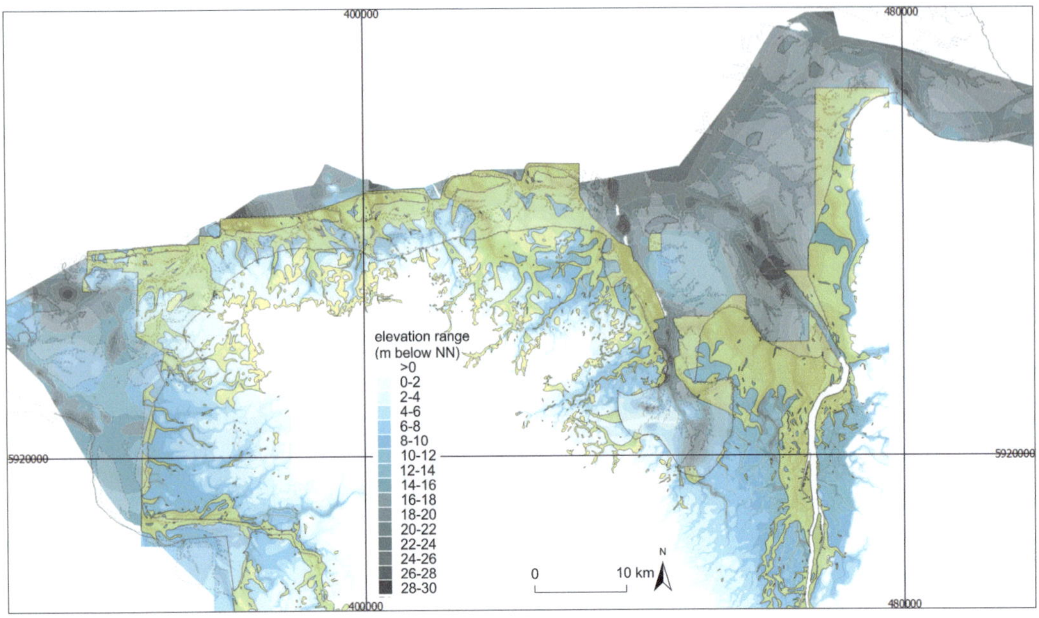


**Fig. 15.3** Digital elevation map of the pre-Holocene surface at 2 m intervals relative to NHN (German chart datum). *Yellow* areas depict zones of solely clastic deposition (Data base: Holozänbasiskarte, © Landesamt für Bergbau, Energie und Geologie, Hannover, 2012)

Nevertheless, people began to settle in the North Sea basin already in early Holocene times, although it is still unclear exactly when hunter-gatherers began to traverse the tundra-like landscape while hunting for game and gathering food. There are some older finds from the Pleistocene: Neanderthal bones from the Middeldiep off the Dutch coast (Peeters 2011) and Palaeolithic flint artefacts from the Great Yarmouth site (Area A240) off the English coast (Tizzard et al. 2011), which are evidence for earlier human activity. Various Mesolithic finds from the area around Brown Bank and Dogger Bank (van de Noort 2011) and from the Dutch coast (Peeters 2011) prove that Mesolithic communities used the North Sea basin as it changed from tundra to a more forested landscape (see also Gaffney et al. Chap. 20, Momber and Peeters, Chap. 21). As the sea level rose, people were forced to adapt to the changing coastline. In Neolithic times, when people became sedentary, the largest part of the basin, and also the slightly elevated Dogger Bank, must have become inundated (Shennan et al. 2000; Behre 2003), although Michelsberg Culture artefacts found in these areas indicate a human presence at that time (van de Noort 2011, see also Gaffney et al. Chap. 20). Remains of Stone Age settlements should be expected in the Wadden Sea region, although these are likely to have been covered by a thick sediment layer.

From the metal ages onwards until early modern times, the sea was an important economic factor, and sedentary societies with their settlements on or near the coast would have had to adapt to catastrophic storm surges as well as progressive inundation. Furthermore, the fertile marsh land was highly attractive for agriculture. Therefore, the coastal region was an important economic zone from an early period. Around 600 BC people started to build their houses on artificial dwelling mounds (Nieuwhof et al. 2013) to protect their houses and cattle during periods of unusually high tides and storm floods. From the twelfth century AD onwards, coastal societies built dikes to protect their marsh land and settlements. However, catastrophic floods like the Second Marcellus Flood in 1362, the Burchardi-Flood in 1634, and the Christmas Flood of 1717 demonstrate that the dikes of that time were unable to provide reliable protection against severe storm surges.

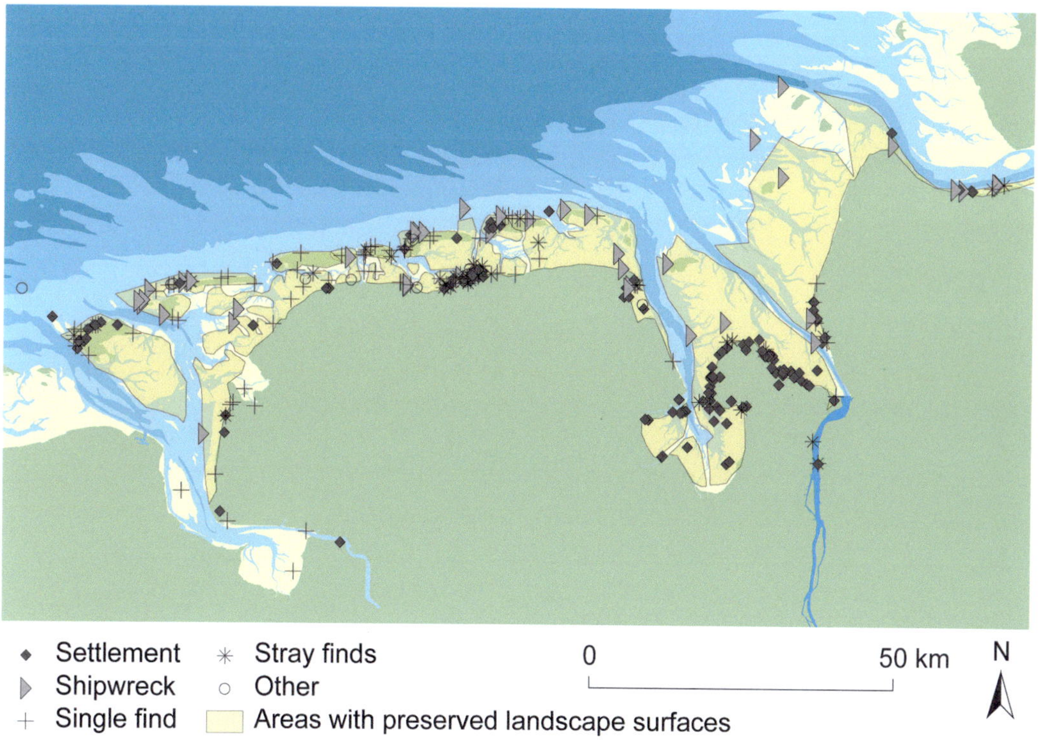


**Fig. 15.4** Potential areas and locations of archaeological finds in the Wadden Sea. Find spots located outside the areas of preserved landscape surfaces are in most cases shipwrecks or displaced single finds

In the course of this project, we found the remains of a farmstead which may have been destroyed by a storm flood. North of the Jade Bay, off the village of Horumersiel (District of Wangerland), broken bricks, animal bones, pottery sherds, clay pipes and lead fragments from window panes were found. Additional pottery sherds and fragments of a grindstone were discovered nearby (Fig. 15.5). The inventory, especially the pottery, dates this previously unknown dwelling to the sixteenth or seventeenth century AD (Goldhammer and Karle 2015).

## 15.5 Future Prospects

In the future, systematic surveys as well as a refinement of landscape reconstruction by filling in gaps in the subbottom data are necessary to pinpoint potential archaeological sites that may be endangered by the activities of coastal engineering and offshore renewable energy projects. Although this project succeeded in creating palaeogeographical maps for the whole area of interest, it turned out that a number of gaps in the data currently prevent a higher precision in time and space, especially for the last 2,000 years. Unfortunately, the sedimentary record is very complex due to multiple seaward and landward shifts of the coastline and it is thus impossible to reconstruct the palaeoenvironment and former positions of the coastline in detail without a high resolution grid of absolute dates and other indicative factors like diatoms or plant remains.

A monitoring strategy for the tidal flats in close cooperation with the Wadden Sea National Park Administration of Lower Saxony and the Lower Saxony State Office for Heritage is recommended to

**Fig. 15.5** Fragments of a grinding stone and pottery probably belonging to an inundated settlement in the tidal flats off Horumersiel, Germany, dated to Early Modern Times (Photo: Niedersächsisches Institut für historische Küstenforschung)

allow quick responses to find reports. Only instant reaction will allow a proper assessment of such sites before they are destroyed by erosion or commercial activities.

In conclusion, geological research has provided new data by which researchers have gradually gained a better understanding of the close relationships between sedimentary processes, sea level change, and landscape evolution in the wider region. At the same time, historical and archaeological research has also developed. Most significantly, ideas regarding the chronology of human occupation have changed dramatically. Therefore, further investigation will profit from a multidisciplinary approach to fill gaps in the record and reach a higher level of data quality through improved spatial and temporal resolution. Merging different local datasets will lead to a broader insight into the interplay of natural changes and human adaptation strategies.

**Acknowledgements** The project was funded by the Lower Saxony Ministry for Science and Culture. A large number of different institutions provided data, assistance and knowledge, and we are particularly grateful to: Nationalparkverwaltung Niedersächsisches Wattenmeer, Landesamt für Denkmalpflege Hannover, Fachdienst Informationssysteme und IT, Gebietsreferat Oldenburg, Ostfriesische Landschaft, Landesmuseum für Natur und Mensch Oldenburg, Archäologische Denkmalpflege Bad Bederkesa, Archäologische Denkmalpflege Landkreis Stade, Stadtarchäologie und Bodendenkmalpflege Cuxhaven, Museum Leben am Meer Esens, Landesamt für Bergbau, Energie und Geologie (LBEG), Hannover, Niedersächsischer Landesbetrieb für Wasserwirtschaft, Küsten- und Naturschutz (NLWKN) – Betriebsstelle Norden-Norderney, and Senckenberg am Meer, Wilhelmshaven. Finds were reported by a large number of interested non-professionals and we were supported in a massive way by our colleagues – we wish to thank them all heartily. Reviews by A. Wehrmann and J. Harff greatly helped to improve this work, and we also wish to thank B. Flemming for help in editing the English language.

## References

Baeteman C (2013) History of research and state of the art of the Holocene depositional history of the Belgian coastal plain. In: Thoen E, Borger GJ, de Kraker AMJ, Soens T, Tys D, Vervaet L, Weerts HJT (eds) Landscape or seascape? The history of the coastal environment in the North Sea area reconsidered. Brepols Publishers, Turnhout, pp 11–29
Barckhausen J, Preuss H, Streif H (1977) Ein lithologisches Ordnungsprinzip für das Küstenholozän und seine Darstellung in form von Profiltypen. Geologisches Jahrbuch, A 44:45–77
Behre KE (2003) Eine neue Meeresspiegelkurve für die südliche Nordsee. Probleme der Küstenforschung im südlichen Nordseegebiet 28:9–63
Behre KE (2007) A new Holocene Sea level curve for the southern North-Sea. Boreas 36:82–102
Flemming NC, Cağatay MN, Chiocci FL, Galanidou N, Jöns H, Lericolais G, Missiaen T, Moore F, Rosentau A, Sakellariou D, Skar B, Stevenson A, Weerts H (2014) Land beneath the waves: submerged landscapes and sea level change. a joint geoscience-humanities strategy for European continental shelf prehistoric research. In: Chu NC, McDonough N (eds) Position Paper 21 of the European Marine Board, Ostend, Belgium.
Goldhammer J, Karle M (2015) Geoarchaeological research in the Wadden Sea area of lower Saxony. Siedlungs- und Küstenforschung an der südlichen Nordseeküste 38:59–70
Jöns H, Karle M, Kleingärtner S (2013) Das Nordseebecken und der Wattenmeerraum als Forschungsgebiet. Methodische Überlegungen, Strategien und aktuelle Forschungsprojekte. Offa 69(70):71–80
Kluiving SJ, Lascaris MA, de Kraker AMJ et al (2013) Potential and use of archaeological and historical data for a reconstruction of the sea level curve of the last 3000 years in the coastal zone of the southern North Sea. Results of a case study. In: Thoen E, Borger GJ, de Kraker AMJ, Soens T, Tys D, Vervaet L, Weerts HJT (eds) Landscapes or seascapes? The history of the coastal environment in the North Sea area reconsidered. Brepols Publishers, Turnhout, pp 61–84
Nieuwhof A, Jöns H, De Langen G et al (2013) Leben mit dem Meer: terpen, wierden und wurten – Leven met de zee: terpen, wierden en wurten. In: Kegler JF (ed) Land der Entdeckungen. Die Archäologie des friesischen Küstenraums – lend van ondekkingen. De archeologie van het Friese kustgebied. Ostfriesische Landschaftliche Verlags- und Vertriebsgesellschaft mbH, Aurich, pp 146–155
Peeters H (2011) How wet can it get? Approaches to submerged prehistoric sites and landscapes on the Dutch continental shelf. In: Benjamin J, Bonsall C, Pickard C, Fischer A (eds) Submerged prehistory. Oxbow, Oxford, pp 55–64
Shennan I, Lambeck K, Falther R et al. (2000) Modelling western North Sea palaeogeographies and tidal changes during the Holocene. In: Shennan I and Andrews JE (eds) Holocene land–ocean interaction and environmental change around the North Sea. Geological Society of London, Special Publication 166: 299–319
Streif H (1990) Quaternary Sea-level changes in the North Sea, an analysis of amplitudes and velocities. In: Brosche P, Sundermann PJ (eds) Earth's rotation from eons to days. Springer, Berlin, pp 201–214
Streif H (1990) Das Ostfriesische Küstengebiet. Nordsee, Inseln, Watten und Marschen. Sammlung Geologischer Führer 57. Gebr. Borntraeger, Berlin
Streif H (2004) Sedimentary record of Pleistocene and Holocene marine inundations along the North Sea coast of lower Saxony, Germany. Quat Int 112:3–28
Tizzard L, Baggaley PA, Firth AJ (2011) Seabed prehistory: investigating palaeolandsurfaces with Palaeolithic remains from the southern North Sea. In: Benjamin J, Bonsall C, Pickard C, Fischer A (eds) Submerged prehistory. Oxbow, Oxford, pp 65–74
Van der Noort R (2011) North Sea archaeologies: a maritime biography, 10,000 BC to AD 1500. Oxford University Press, Oxford
Walther J (1894) Einleitung in die Geologie als historische Wissenschaft. Dritter Theil: Lithogenesis der Gegenwart. G. Fischer, Jena

# Chapter 16
# Sacred Landscapes and Changing Sea Levels: New Interdisciplinary Data from the Early Neolithic to the Present in South-Eastern Sicily

**Giovanni Scicchitano, Elena Flavia Castagnino Berlinghieri, Fabrizio Antonioli, Cecilia Rita Spampinato, and Carmelo Monaco**

**Abstract** Through the analysis of geomorphological processes coupled to archaeological time markers in one selected site – Ognina in south-eastern Sicily – this paper investigates ritual practices and sacred places associated with sea-level change and shoreline locations. The interdisciplinary approach adopted in this research also provides new data on relative sea-level change during the late Holocene, while at the same time bringing together diverse approaches and methods for the analysis of submerged landscapes. It also aims to act as a blueprint for future directions in this specific field.

## 16.1 Introduction

The concept of 'sacred landscape' encompasses multiple forms, meanings and belief systems, and includes relationships or actions that may have different values in different settings. New studies have brought many different perspectives to bear on the role and human significance of coastal environments, variously emphasising their significant heterogeneity and instability on a number of

G. Scicchitano (✉)
Dipartimento di Scienze Biologiche, Geologiche e Ambientali, Università di Catania, Catania, Italy

Studio Geologi Associati T.S.T., Via Galliano 157, Misterbianco, CT, Italy
e-mail: gianfrancoscicc@hotmail.com

E.F.C. Berlinghieri (✉)
Dipartimento di Scienze Biologiche, Geologiche e Ambientali, Università di Catania, Catania, Italy

Soprintendenza per i Beni Culturali e Ambientali di Siracusa, Catania, Italy
e-mail: elfcb@tiscali.it

F. Antonioli
Italian National Agency for New Technologies, Energy and Sustainable Economic Development (ENEA), Via Anguillarese 301, 00060 S. Maria di Galeria, Rome, Italy
e-mail: fabrizio.antonioli@enea.it

C.R. Spampinato
Dipartimento di Scienze Biologiche, Geologiche e Ambientali, Università di Catania, Catania, Italy

Portable Lab Geology and Engineering s.r.l – Academic Spinoff University of Catania, Catania, CT, Italy
e-mail: spampinatocecilia@hotmail.it

C. Monaco
Dipartimento di Scienze Biologiche, Geologiche e Ambientali, Università di Catania, Catania, Italy
e-mail: cmonaco@unict.it



G.N. Bailey et al. (eds.), *Under the Sea: Archaeology and Palaeolandscapes of the Continental Shelf*, Coastal Research Library 20, DOI 10.1007/978-3-319-53160-1_16

spatio-temporal scales (e.g. Westley and Dix 2006), their role as a focus for the development of ideas about 'maritime cultural landscapes' (Westerdahl 1992; Parker 2001), the impact of 'sea-level changes', originally developed in an archaeological context by Flemming (1968), the significance of 'submerged landscapes' (Benjamin et al. 2011), and the relationship between coastal landscapes and maritime activity (Farr 2006; Castagnino Berlinghieri 2011, and references therein). Some of these studies are concerned not only with the study of relationships between coastal landscapes and maritime activity in a practical sense, but also with the creation of ritual beliefs and social identity, and that is the emphasis in this chapter.

Archaeological sites in areas of small tidal range can also provide significant information on relative sea-level change during recent millennia, especially man-made coastal structures which required a precisely defined relationship to sea level at the time of construction for their successful functioning, for example fish traps or harbour walls. Along Mediterranean shores, in particular, there are many such remains extending over a long period that can be used to establish constraints on relative sea level (Schmiedt 1966, 1972; Flemming 1969; Caputo and Pieri 1976; Flemming and Webb 1986; Lambeck et al. 2004a; Anzidei et al. 2006, 2011a,b, 2013; Antonioli et al. 2007; Scicchitano et al. 2008, 2011; Auriemma and Solinas 2009; Lopresti et al. 2014). Relative sea level change during the Holocene results from the contribution of many factors: ice melting, glacio-hydro-isostasy and tectonics (Lambeck et al. 2004b, 2011). For those areas where it is possible to evaluate all of these factors, it is also possible to perform accurate reconstructions of the coastal palaeolandscape that may be of great significance to archaeological interpretation.

South-eastern Sicily between the town of Augusta and Siracusa (Fig. 16.1) is particularly rich in archaeological remains located along the present coastline and below it, ranging in date from the Neolithic to the C17th AD (ca. 8–0.4 ka cal BP), and spanning a period of relatively rapid Holocene sea-level change. This offers an unusual opportunity to analyse the relationship between sea-level change and human settlement activity, including its impact on ritual and belief systems.

Of particular interest are the archaeological remains on the small island of Ognina some 10 km south of Siracusa. The area is characterized by several phases dating from prehistory to the Byzantine age. Some of this evidence has already been discussed in detail as an indicator of relative sea-level change (Kapitaen 1970, 1991; Basile et al. 1988; Castagnino Berlinghieri 1993–1995; Scicchitano et al. 2008). There is also a substantial body of evidence to suggest that Ognina played a key role in the wider spectrum of Neolithic and Bronze Age interactions, which probably reflects a wide-ranging network of contacts including both social and ritual practices. A better understanding of these interactions demands more detailed examination of the location of the island in relation to the mainland and its modification over time, coupled with archaeological investigation of the interactions between local traditions and external influences.

Our aim in this chapter is to use geomorphological reconstructions of the coastal landscape of Ognina during the Early Neolithic, the Middle Bronze Age and the Byzantine period to throw light on the impact of sea-level change on the sacred significance of this coastal location and its ritual associations. For the geomorphological reconstructions, we use: (1) an accurate morphobathymetric and topographical dataset; (2) a model of late Holocene sea level rise that takes into account glacio-hydro-isostatic and eustatic components; and (3) biological and radiometric data.

## 16.2 Geological Setting

South-eastern Sicily is characterized by thick Mesozoic to Quaternary carbonate sequences and volcanics forming the emerged foreland of the Siculo–Maghrebian thrust belt (Grasso and Lentini 1982). This area, mostly constituted by the Hyblean Plateau, is located on the footwall of a large normal fault system which since the Middle Pleistocene has reactivated the Malta Escarpment (Bianca et al. 1999),

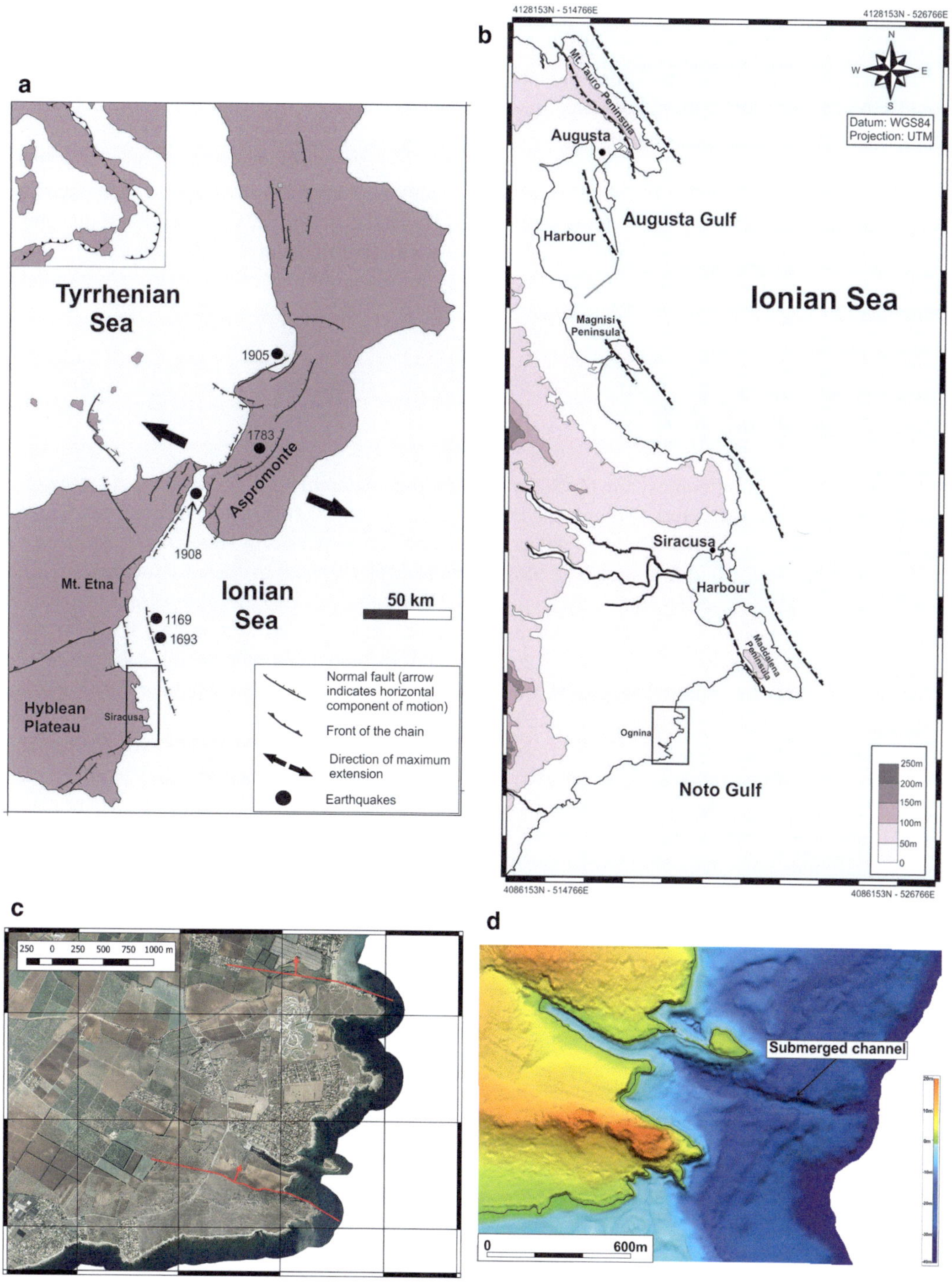


**Fig. 16.1** (**a**) Geodynamical setting of south-eastern Sicily and southern Calabria; (**b**) Structural setting between Augusta and Siracusa; (**c**) Map of Ognina area, red line corresponds to normal faults and arrows indicate the hanging wall; (**d**) Reconstructed DTM of the emerged and submerged area of Ognina

a Mesozoic boundary separating the continental domain from the oceanic crust of the Ionian basin (Scandone et al. 1981; Sartori et al. 1991; Hirn et al. 1997).

The physical setting of the Ognina archaeological site is characterized by the presence of three small promontories alternating with small pocket beaches (Fig. 16.1b). From a geological point of view, the outcropping rocks comprise massive Miocene calcarenites, gently dipping seaward, unconformably overlain by middle–late Quaternary terraced marine deposits composed of poorly cemented biocalcarenites (Di Grande and Raimondo 1982). The Ognina Island is a small island surrounded by very shallow water offshore of the northern promontory. To the south, a 600 m long natural channel occurs, marked today by a dock and quarry. This channel continues underwater extending from a depth of −3 m to −30 m (Fig. 16.1d) and probably represents the remnant of a Holocene river channel terminated by rapid sea level rise (Scicchitano and Monaco 2006; Scicchitano et al. 2010).

## 16.3 Archaeological Data

While providing a good context for the study of processes of prehistoric interactions, Ognina also exhibits evidence for ritual and ideology and evidence that it was a centre of social power and a meeting place of different cultures. The main archaeological evidence consists of various classes of vessels and goods along with Neolithic post-hole alignments and pits, as well as sacred structures (Bronze Age rock cut tomb, Byzantine church), which indicate that ritual activities were integral to the use of the site. Moreover, given that the fragments of prehistoric vessels found were mainly bowls or open shapes used for eating and drinking, this suggests functions associated with social aggregation and ritual. By viewing these data through the lens of the ritual society model (Baumann 1993; Barley 1995), it is possible to better understand the nature and role of ritual and elite communities. Recent excavations (Tanasi 2014: 109–112) undertaken nearby in the trenches excavated in 1964 by Bernabò Brea, confirm that the stratigraphy is characterized by layers dating from Neolithic to Byzantine periods. In this panorama, the value of some old finds will be discussed in the light of their probable importance in ritual performance, and some new connections with larger sacred landscapes will be established. In essence, recurring artefacts that are rich in symbolism and their related contexts suggest modes of ritual behaviour that imply an overarching, region-wide framework of shared ritual and beliefs.

The earliest evidence of human presence in Ognina goes back to about 6000 BC and is associated with artefacts of the Neolithic period, a period typically associated with farming and fixed settlements. Currently exposed remains on the island comprise a complex system of post holes indicating large compartmentalised structures, arranged in parallel alignments (Fig. 16.2), and resembling the ground plans of the Neolithic longhouses in Central Europe (cf. Pásztor and Barna 2014).The longest rows extend for over 40 m in the northern area while individual post holes are large, between 0.8 and 1.3 m in diameter.

The post-hole alignments on the island appear to delineate the foundations of timber-framed dwellings, and are also associated with shorter rows of smaller holes likely used as pits. These suggest the presence of a stable settlement established during the Neolithic period. While the island is characterized by scattered sherds of pottery and lithic tools, early strata of a deposit excavated in the 1960s (Bernabò Brea 1966), partially overlying a linear pattern of holes cut into the rock, revealed a sequence of Stentinello impressed wares, 5700–5200 BC, including fragments of large jars (Fig. 16.3a). A ditched settlement with scattered post-holes has been detected on the mainland facing the island (Lazzarini et al. 1965), but this has now been mostly destroyed by marine erosion. Historical black and white pictures taken in the year of discovery (Fig. 16.3b) clearly show a ditch-like feature cut into the bedrock (now heavily damaged by sea erosion). Earlier levels of the ditch-fill investigated in the 1960s revealed abundant sherds of incised and impressed ware of Stentinello type (Fig. 16.3c), thereby

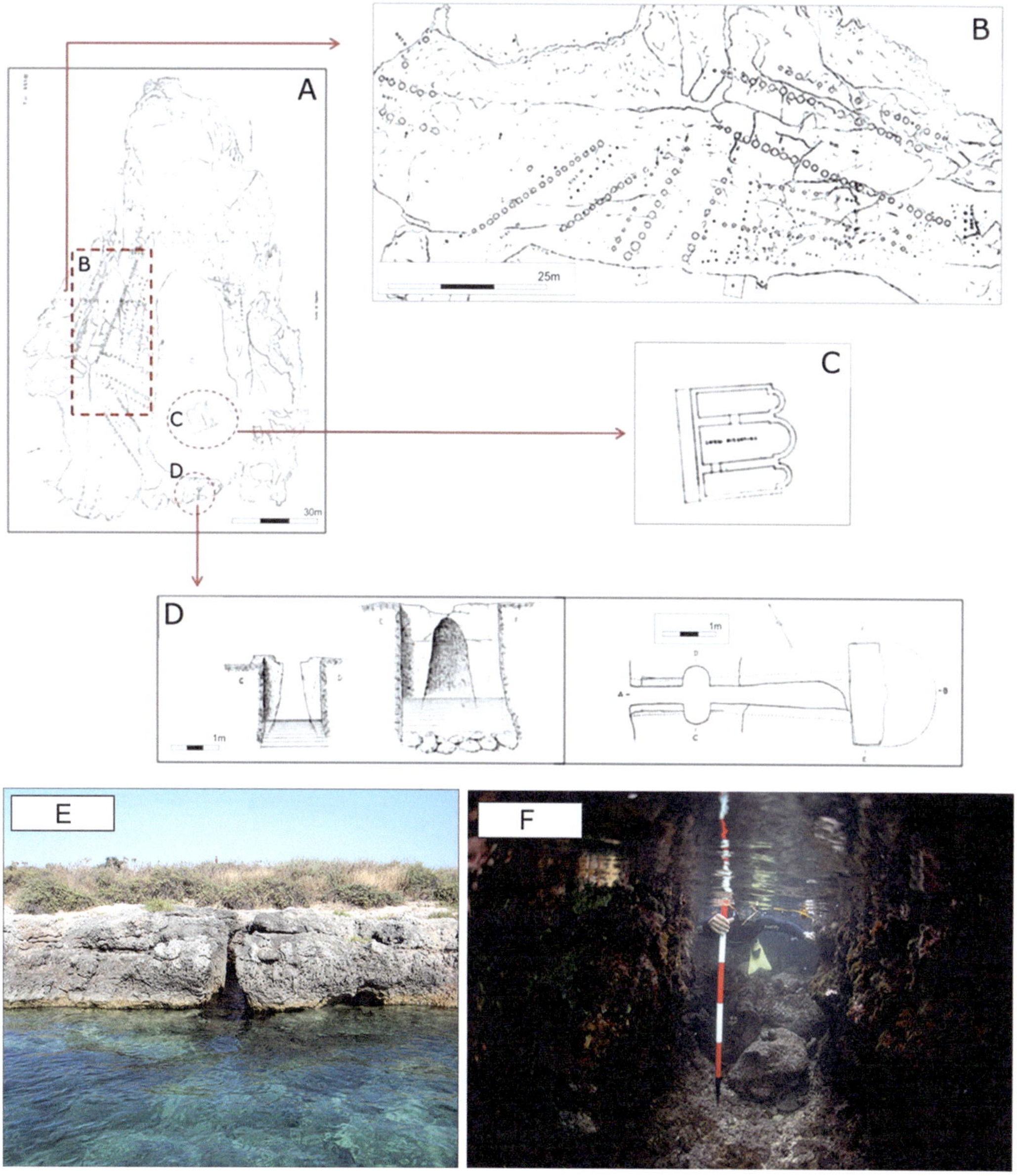


**Fig. 16.2** (**a**) Archaeological evidence on the island of Ognina; (**b**) Post-holes arranged in parallel in the northern sector of the island; (**c**) Remains of the triple-apsed Byzantine basilica; (**d**) Rock-cut chamber tomb and long rock cut dromos with two symmetrically opposed niches near the opening (Adapted from Bernabò Brea 1966); (**e**) Now-submerged entrance to the tomb; (**f**) Direct survey of the tomb

confirming that this part of the deposit dates to the early Neolithic and was contemporaneous with that on the island.

The position of the Ognina settlement on the highest part of the promontory, dominating its surroundings, may reflect some sort of power organisation, feasibly related to emerging power structures (cf. Castagnino Berlinghieri 2011).

Bronze Age life at Ognina has usually been connected with the Maltese world, more specifically with the Tarxien Cemetery phase. Ognina was first interpreted as a Maltese colony by Bernabò Brea

**Fig. 16.3** (**a**) Fragments of large jars of Stentinello impressed ware; (**b**) Ditch cut into the bedrock shown during 1960s excavation; (**c**) Sample of incised and impressed Stentinello ware from earlier levels of the ditch fill excavated in the 1960s. Archivio Storico Fotografico della Soprintendenza ai Beni Culturali e Ambientali di Siracusa

in his fundamental work first published in 1966. The same author later argued convincingly (Bernabò Brea 1976–1977) that there were dissimilarities between the finds from the Tarxien Cemetery and those found at Ognina. As he himself wrote (Bernabò Brea 1976–1977: 78, note 37):

> vi è senza dubbio una sensibile differenza nel repertorio decorativo delle tazze a orlo ingrossato costituente la classe denominata dallo Evans e dal Trump "Thermigreyware" e le rimanenti forme del complesso ceramico del Tarxien Cemetery, differenza data soprattutto dalla prevalenza nelle prime dei motivi a triangoli e a bande punteggiate (o a cerchielli impressi) che nelle seconde, pur non essendo del tutto assenti, sono alquanto più rare.
>
> There is definitely a noticeable difference between the decorative repertoire of cups with thickened rims belonging to the form called by Evans and Trump 'Thermigreyware' and the remaining shapes of the ceramic complex of the Tarxien Cemetery. Such a difference is especially marked by the prevalence of triangles and dotted bands (or small impressed circles) in the former, while in the latter case such decoration is very rare, although not entirely absent.

Nevertheless, his first hypothesis – of a Maltese colony established at Ognina – has continued to dominate the archaeological literature. This interpretation has been recently challenged by a re-examination of some Bronze Age pottery known as 'Thermi Ware' or 'grey wares' with dot-filled incised decoration (Palio 2008, see also Cazzella and Recchia 2013). According to Palio (2008) these vessels show very close parallels in fabric, decoration and shape with contemporaneous material in southern Italy (Laterza and Zurgi) as well as in the Aegean region (Olympia and Lerna AEIII). This re-evaluation suggests that Ognina was the final terminus of a route that led from the Aegean and the Balkans, across the Adriatic and over the Italian peninsula (Laterza and Zungri), as far as the island itself and then radiated inland (to Chiusazza), southwards to Vendicari and Malta and northwards to Thapsos, the Etna area and the Aeolian Islands. It is also useful to emphasise that even the well-documented class of Bronze Age funerary objects known as 'bone bossed plaques' seem to track the same itinerary of diffusion attested by Ognina-type pottery, with notable examples from Lerna, Altamura, Malta and Petraro (see bibliography in Procelli 1991).

This interpretation indicates a developed system of exchange transactions between Bronze Age communities in Ognina, the Aegean and southern Italy. The results of more recent campaigns undertaken at Ognina, published in a preliminary form (Tanasi 2014) on the basis of new data from laboratory analysis, suggest that 'Thermi ware' vessels were imported from centres of production in Malta as well as being locally produced at Ognina, whether by itinerants or by the native community exploiting their specialised knowledge.

This opens up a new scenario, which implies various long-distance exchange networks as a means of facilitating population movement and expansion, and reflects cultural interactions on a large scale with an important role played by the Bronze Age communities of Ognina. These are new hypotheses (Palio 2008; Tanasi 2014) for the study of the development of the economy of the Bronze Age involving movement of goods, knowledge and people in search of new territory, and call for further evaluation within the wider exchange network of the Mediterranean. Old and new investigations have revealed the presence of 'Thermi Ware' along with plain and decorated ceramics in small percentages in both the Castelluccio and Capo Graziano styles.

Despite the intriguing and exotic nature of the array of pottery present at Ognina, there is little evidence for Bronze Age settlement, apart from two areas with traces of hearths possibly related to huts, Also, we cannot exclude the possibility that at least some of the Early Neolithic structures could still have been in use during the Bronze Age, even if for a different purpose: perhaps for storage or for rubbish disposal.

In addition to the sparse evidence of structures, funerary architecture is also limited. There is no evidence for cemeteries or smaller clusters of tombs, as in well-documented Bronze Age burials in western Sicily (Leighton 1999 and references therein), but rather a single, atypical, chamber tomb, which combines the local tradition of rock cutting with some elaboration (Fig. 16.2c). It consists of a partially submerged rock-cut tomb carved into calcarenite (Fig. 16.2d,e), and represents one of the more interesting Bronze Age burial sites, possibly implying a social or political hierarchy.

**Table 16.1** Heights and ages of sea-level stands (relative to present sea-level) associated with Marine Isotope Stages 3 and 5 and the uplift rates implied by radiocarbon-dated balanid samples

| MIS stage | Age kyr | Height relative to modern sea level metres | Uplift rate |
|---|---|---|---|
| 5.5 | 125 | 6.3 | −0.02 |
| 5.3 | 100 | −20.86 | 0.24 |
| 5.1 | 80 | −18.67 | 0.27 |
| 3 | 60 | −48.02 | 0.85 |

The semi-circular chamber is approached by a sloping rock-cut dromos or pathway, 4.2 m long, with two symmetrically opposed niches close to the opening. Its original shape has been partially altered by complex geological factors so that its floor now lies at −1.20 m (corrected height −1.21 m) below the present sea level (Fig. 16.2f). As already demonstrated (Scicchitano et al. 2008), taking into account the functional height, the palaeo-sea level indicated for the Bronze Age must have been at ≤−1.81 m depth (Table 16.1).

From a structural viewpoint, Bernabò Brea (1966) pointed out certain affinities with the 'Allées couvertes' gallery graves widespread in southern France (Bernabò Brea 1966: 57; Procelli 1995: 23), while recently Tusa (1992: 368) and Cultraro (2000: 713–714) have suggested a derivation from types of Chambered Tombs of Middle Helladic Age in the Aegean. However, leaving aside its design and possible derivation, it is tempting to account for the tomb in terms of a social organization with a clear power structure producing a formalised leadership with ritual authority that dominated the wider religious and social landscape.

The solemnized space of the Ognina tomb, which one might suppose functioned as a sort of 'sanctuary', lacks the items that one might associate with the attempts of an elite to legitimise religious authority. This is due to the fact that the grave goods have been removed or destroyed by marine erosion. By the same token, the existence of some formalized ritual activity on the island is supported by the occurrence of special items brought to light in earlier excavations (Bernabò Brea 1966) which may be rooted in ritual factors and may lend weight to the hypothesis that the site had religious significance. In this respect, some valuable clues come from the re-evaluation of the deposit found in trench 'D' excavated by Bernabò Brea in 1964, which reveals an interesting assemblage including a circular hearth 70 cm in diameter (cut D5: Fig. 16.4a) and a range of pottery sherds, as well as ceremonial items such as a fragment of a violin-shaped figurine, a clay horn, a figurine of phallic shape, a small horn with incised dotted decoration, and other 'prestige' goods (Fig. 16.4b). One of the most distinctive artefact categories is a 'gynaecomorphic vessel' found as a stray item close to trench 'D,' which implies a some form of prestige or cult item (Fig. 16.4c).

The combination of all of these artefacts recovered in a restricted area suggests that ritual practices were a central function of the site and perhaps part of a widespread ceremonial process. At the same time, these and other components have parallels in other Bronze Age sites, such as the Manfria deposits (Orlandini 1962, Table 30) which are ritual in nature, and are sometimes associated with burial places, providing some degree of evidence for ritual symbolism. In this period, the rocky isthmus connecting the mainland and the island provided the settlement with sheltered leeward anchorages and beaching places (Kapitaen 1970; Kapitaen 1991; Castagnino Berlinghieri 1993–1995). As previously discussed (Scicchitano et al. 2008), these features are now still in place but submerged between −0.20 and −3.30 m relative to the present sea level.

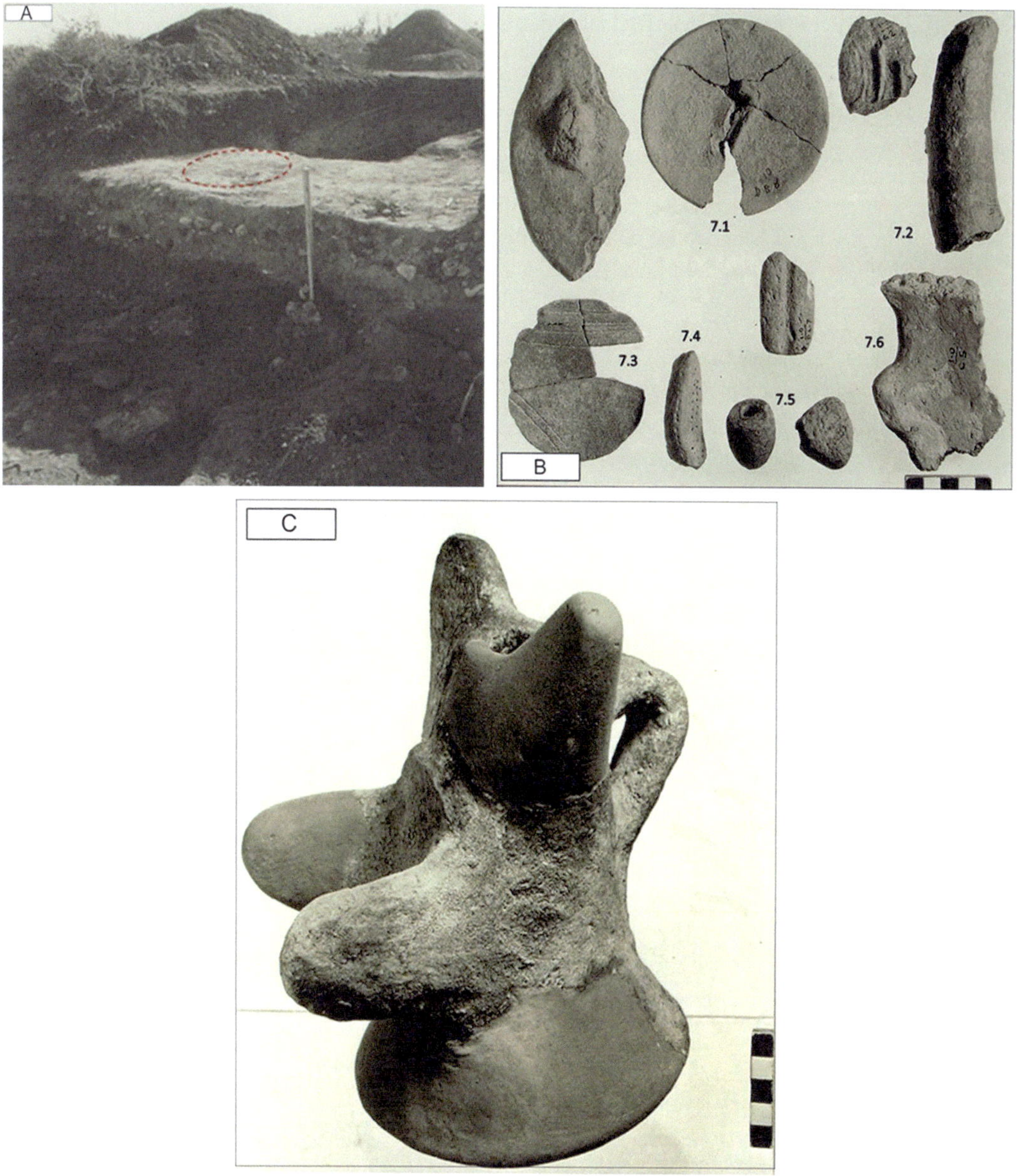


**Fig. 16.4** (**a**) Circular hearth 70 cm in diameter (cut D5) and stratigraphic sequence revealed in 1960s excavation; (**b**) Sample of ceremonial items: 7.1 dish-shaped artefact; 7.2 phallic figurine; 7.3 round object with parallel incised lines; 7.4 small horn with incised dotted decoration; 7.5 small loom weights; 7.6 fragment of violin-shaped figurine and other 'prestige' goods; (**c**) Stray find of 'gynaecomorphic vessel' near trench 'D,' a presumed prestige or cult item. Archivio Storico Fotografico della Soprintendenza ai Beni Culturali e Ambientali di Siracusa

## 16.4 Coastal Morphology and Sea-Level Change

### *16.4.1 Material and Methods*

In order to reconstruct the coastal palaeogeography of the Ognina site, all the components of relative sea level rise have been taken into account. The effects of ice melting and glacio-hydro-isostasy are taken from the model of Lambeck et al. (2011) (Fig. 16.5a).

This predicts sea level rise during the Holocene for 40 sites along the Italian shores (Fig. 16.5b). As regards the vertical deformation for this sector of Sicily, several authors estimated the amount of uplift by analyzing different markers. Considering several archaeological sites presently submerged, Scicchitano et al. (2008) proposed for the area of Ognina an uplift rate of about 0.4 mm/yr. during the Holocene. Dutton et al. (2009) and Spampinato et al. (2011) found similar results for the last 80 kyr, using U/Th dating of speleothems located inside submerged caves offshore of the Siracusa area. An accurate geomorphological survey, recently performed along the coasts of the Ognina area, revealed the presence of several biological markers such as balanids and serpulids (Fig. 16.6a), closely related to fluctuations of sea level.

Biological remains, mostly represented by balanids, have been discovered at heights ranging between 3.2 and 1.3 m above sea level (corrected for tide and pressure at time of survey). They are locally accompanied by serpulids and briozoa. Close to the southern Ognina Cape, these organic encrustations are uniformly developed and often associated with other morphological evidence of sea level stands such as tidal notches (Sample 2, Fig. 16.6b) or abrasion platforms. The lower sample (Sample 1) is located under the present body of a small beach (Fig. 16.6c), along a paleo-erosional platform, at an elevation of 1.3 m a.s.l. Table 16.2 shows the results of radiocarbon age determinations performed on balanid samples used to reconstruct the timing of sea level fluctuations.

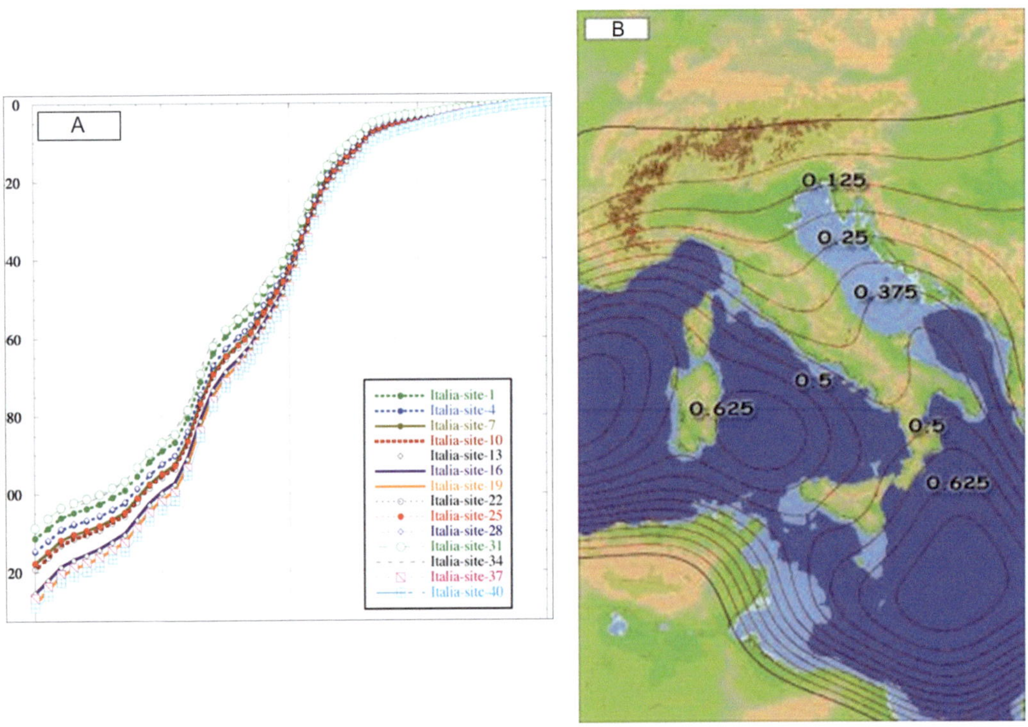


**Fig. 16.5** (**a**) Curves of sea level rise for 40 sites located along Italian coasts (X-axis: age in kyr; Y-Axis: sea level height in meters above present); (**b**) Values of isostatic deformation. After Lambeck et al. (2011)

**Fig. 16.6** (**a**) Location of palaeoshorelines showing close-up of samples of balanid shells recovered from them; (**b**) Beach deposit above ancient erosional platform showing position of balanid sample (*sample 1*); (**c**) Tidal notch found along the higher palaeoshoreline showing position of balanid shells (*sample 2*)

**Table 16.2** Radiocarbon ages of balanid samples from the Ognina coast

| Lab name | Sample name | Radiocarbon agr (BP) | δ13C‰ |
|---|---|---|---|
| LTL13385A | Sample 1 lower shoreline | 39,647 ± 760 | −17,8 ± 0,5 |
| LTL13386A | Sample 2 upper shoreline | 41,227 ± 1205 | −2,1 ± 0,3 |

In order to depict the impact of past sea-level change, it is necessary to have accurate topographical and bathymetric data against which to apply all the components of sea level rise. Topographical data have been extracted from a Digital Terrain Model (DTMl) with a cell size of 2 m (Fig. 16.7a) obtained from airborne LiDAR survey performed in 2008 for the government of Regione Sicilia. A multibeam echo sounder survey was performed in 2011 and 2014 (Fig.16.7b) to integrate the LiDAR

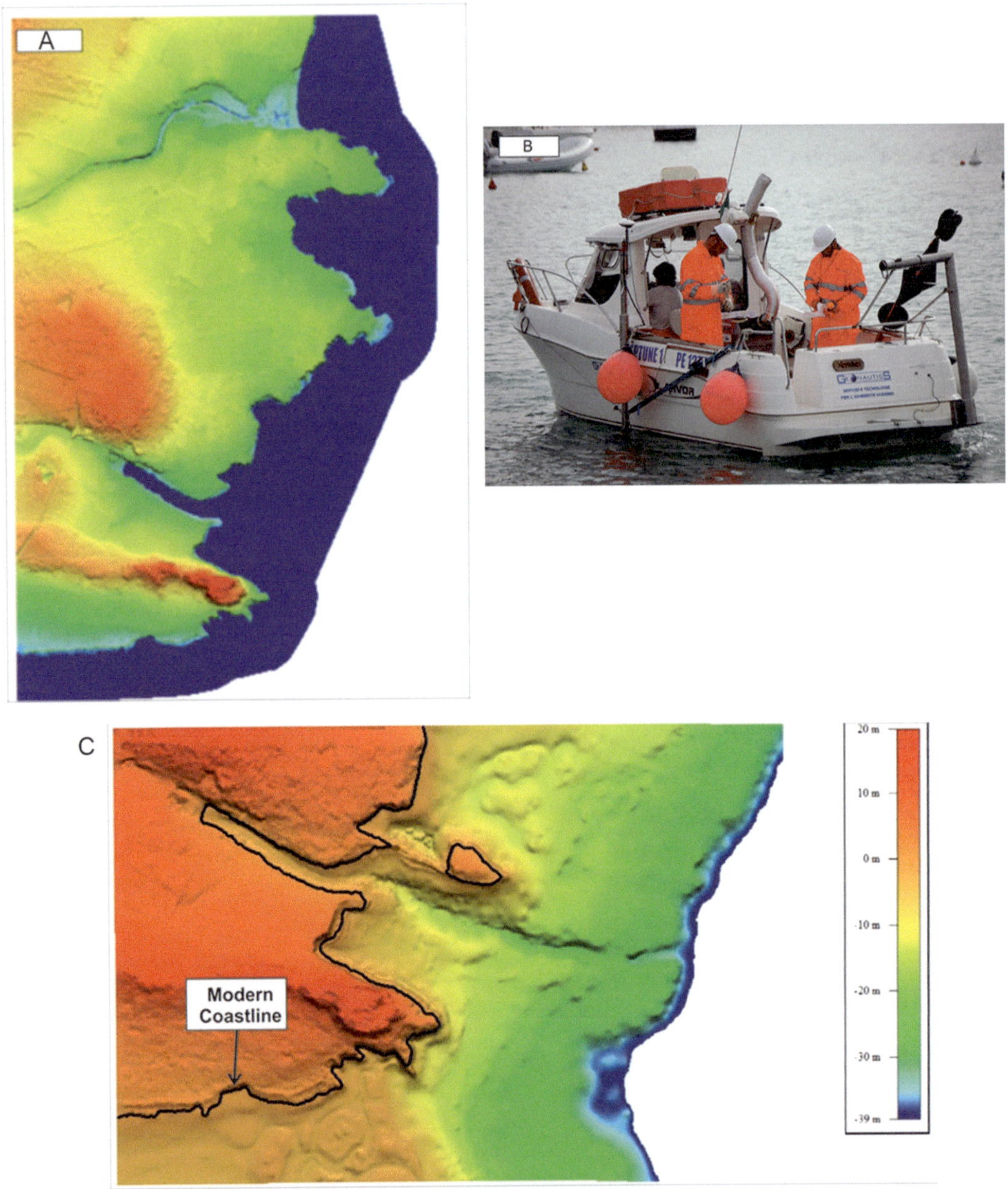


**Fig. 16.7** (**a**) DTM (2 × 2 m resolution) obtained from airborne LiDAR survey; (**b**) Morphobathymetrical survey operation; (**c**) DTM reconstructed for submerged and emerged area

topographical data. Morpho-bathymetric survey was conducted by using a shallow-water vessel and an Autonomous Underwater Vehicle (AUV). The AUV system has been adapted to cover the area where the vessel cannot move for safety reasons, and permitted us to make a better connection between terrestrial topography and marine bathymetry. Resulting data have been corrected for tides using the software WX Tide and gridded with Surfer to create a DTM with a cell size of 1 m (Fig. 16.7c).

### 16.4.2 Data Analysis

Due to the rapid sea level rise during the Holocene, palaeogeographic reconstruction of the coastal area in late prehistoric and historical time requires a very accurate topographical and bathymetrical dataset as well as reliable estimates of vertical rates of movement resulting from tectonic deformation and the effects of rock erosion. The coastal sector of south-eastern Sicily has been intensively studied in recent years with regard to late Quaternary and Holocene tectonic deformation. This is related to the richness and variety of palaeo-sea-level indicators. The occurrence of geomorphological markers such as marine terraces and notches, together with the presence of archaeological markers and geological markers such as speleothems or lagoon deposits has stimulated various authors to use the data from this area to test models of sea-level change.

Analyses of fossil assemblages of barnacle shells formed during sea level stands, recently discovered along the coast of the Ognina Cape, have allowed us to add further refinement to the palaeogeographic pattern. AMS (Accelerator Mass Spectrometry) $^{14}C$ analyses have been performed on the two best preserved balanids sampled from the upper shoreline located about 3–3.2 m above the present sea level and from the lower one located about 1–1.2 m above the present sea level (Fig. 16.6, Table 16.1). The samples have minimum radiocarbon ages indicating that their deposition has to be attributed to an earlier sea level high stand. Taking the models of sea level rise during the last 150 kyr (Waelbroeck et al. 2002), the most likely attribution is one of the sea-level stands in MIS 5 (Table 16.2).

We exclude MIS 5.5 and MIS 3, the first because it would imply very slow subsidence, and the second because it would imply very rapid uplift, both in total disagreement with existing data. For MIS 5.3 or 5.1, corresponding values of uplift of 0.24 and 0.27 mm/year, respectively, are in closer agreement with the results of Dutton et al. (2009), who estimated rates of 0.4 mm/year for the last 80 kyr. For these reasons we have selected a rate of vertical tectonic movement of 0.3 mm/year for the last 100 kyr, and applied this value to reconstruct the coastal paleogeography of the Ognina area.

A further adjustment is required to take account of progressive erosion of the exposed rocky surface, which is another important parameter to take into account in reconstructing the palaeo-landscape of coastal areas (Antonioli et al. 2014). According to estimates proposed by Furlani and Cucchi (2013) for carbonate rocks in climatic conditions similar to those of today, we use in our calculation an erosional rate of about 0.7 mm/year.

This rate has been applied to our scenario for the time spans during which the different coastlines have been emerged and exposed. Taking into account that since 6 ka the sea level rise noticeably decreased, it appears evident that the major contribution of took place during the Byzantine and Bronze Age periods, with results of 0.7 m and 0.4 m respectively.

Taking all these factors into account, we produce three distinct palaeogeographic reconstructions for the early Neolithic, the Bronze Age and the Byzantine period (Fig. 16.8), corresponding to the main phases of activity on Ognina.

In all the reconstructions, a river is present, mostly running along the present channel (Fig. 16.8). Bathymetrical data highlight the presence of a submerged channel perfectly aligned with the present emerged one. The bottom of the submerged channel appears to be strongly eroded showing a development similar to a river, the mouth of which has been identified at about 30 m depth and 800 m east of the present coastline. Direct field survey and remote sensing analyses of orthophotos revealed remains of a palaeo-river bed inland, corresponding to the present channel. The river was active until the beginning of the last century, when changes in the fluvial catchment probably diverted its natural course.

Analyses of old maps of the area support the presence of a river, as shown in the chart drawn by T. Spannocchi in 1578 (Fig. 16.9). During the Early Neolithic (8,000 ka cal BP) the present island was the head of a small promontory projecting from the Ognina Cape. In this period the river was probably a torrent characterized by strong currents able to carry pebbles, gravels and coarse sands responsible for the strong erosion of the bottom of the present submerged channel. During this period the southern

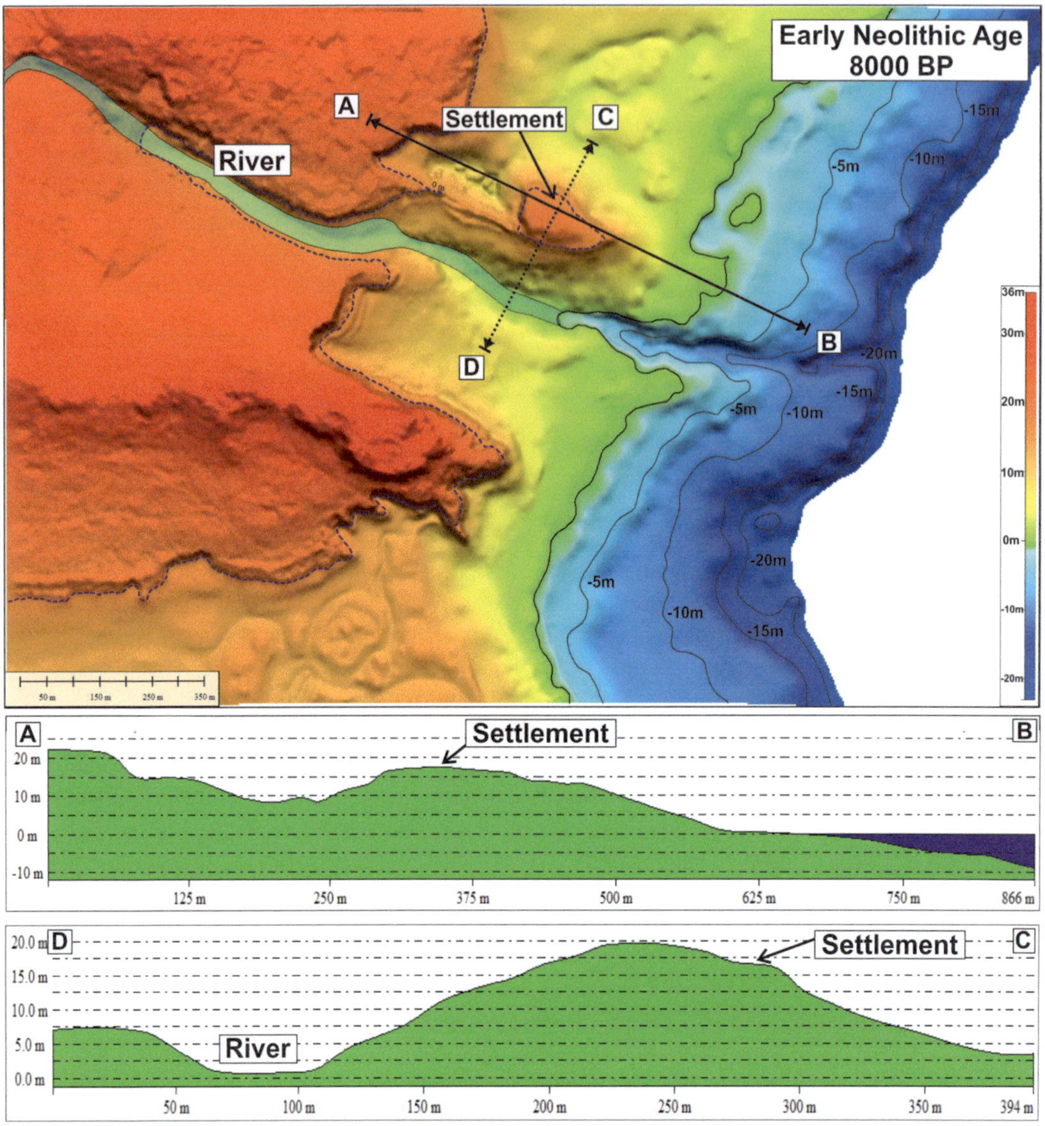


**Fig. 16.8** Palaeogeographical reconstruction for the site of Ognina during the Early Neolithic. The *dark continuous line* is the coastline at the time of reconstruction. The *blue dotted line* corresponds to the present coastline

shore of the river facing the Ognina Cape was probably similar to a small coastal alluvial plain characterized by flooding and fine sediment deposits. Between the Early Neolithic and Early–Middle Bronze Age a rapid sea level rise, reaching rates of more than 1 mm/year, strongly influenced the geography of the area. The northern promontory was partially submerged, remaining attached to the land by a narrow rocky isthmus, and the mouth of the river was significantly flooded. In this period, the southern sector of the Ognina Cape appeared as a dangerous shallow-water area, characterized by the presence of small, flat rocky islands rising just a few cm above sea level. During the Byzantine period, the isthmus underwent further submergence, leaving just a rocky causeway.

**Fig. 16.9** The area of Ognina as drawn by Tribucio Spannocchi (Spannocchi 1578)

## 16.5 Discussion and Conclusions

A survey of the archaeological sequence in the coastal area of Ognina, together with new geomorphological analyses of coastal change, have allowed us to interpret changes in the prehistoric and historical use of the area. Using published data and models of tectonic and sea level changes, supplemented by our own field investigations, we have reconstructed the changing coastal palaeo-landscape during 3 time slices: 8, 3.5 and 1.5 ka cal BP, and related these changes to the changing pattern of occupation.

During the Neolithic era, the island was connected to the mainland (Fig. 16.10). It would have been the sole high point on a low-lying coastal plain, and would have stood out as an imposing feature in the wider landscape. Evidence of a defensive settlement, and of post-holes belonging to possible long-house structures, may imply that it was the focus of religious and ritual practices, a talisman, totem or Neolithic 'acropolis' imbued with special powers (cf. Castagnino Berlinghieri 2003). Sea level change, geomorphological processes, landscape modification and marine erosion mean that archaeological evidence providing definitive proof is difficult to come by. However, future investigations of the submerged landscape may shed light on the land-practices of Neolithic Ognina and the connection between the remains of the settlement on the present-day coast edge and what we suggest is a sacred 'acropolis' located on the present-day island.

By the Bronze Age, the sea level changes and geomorphological processes discussed here resulted in the island being connected to the mainland solely by a rocky causeway, with much of the Neolithic landscape submerged (Fig. 16.11). This change could have heightened the significance of the island

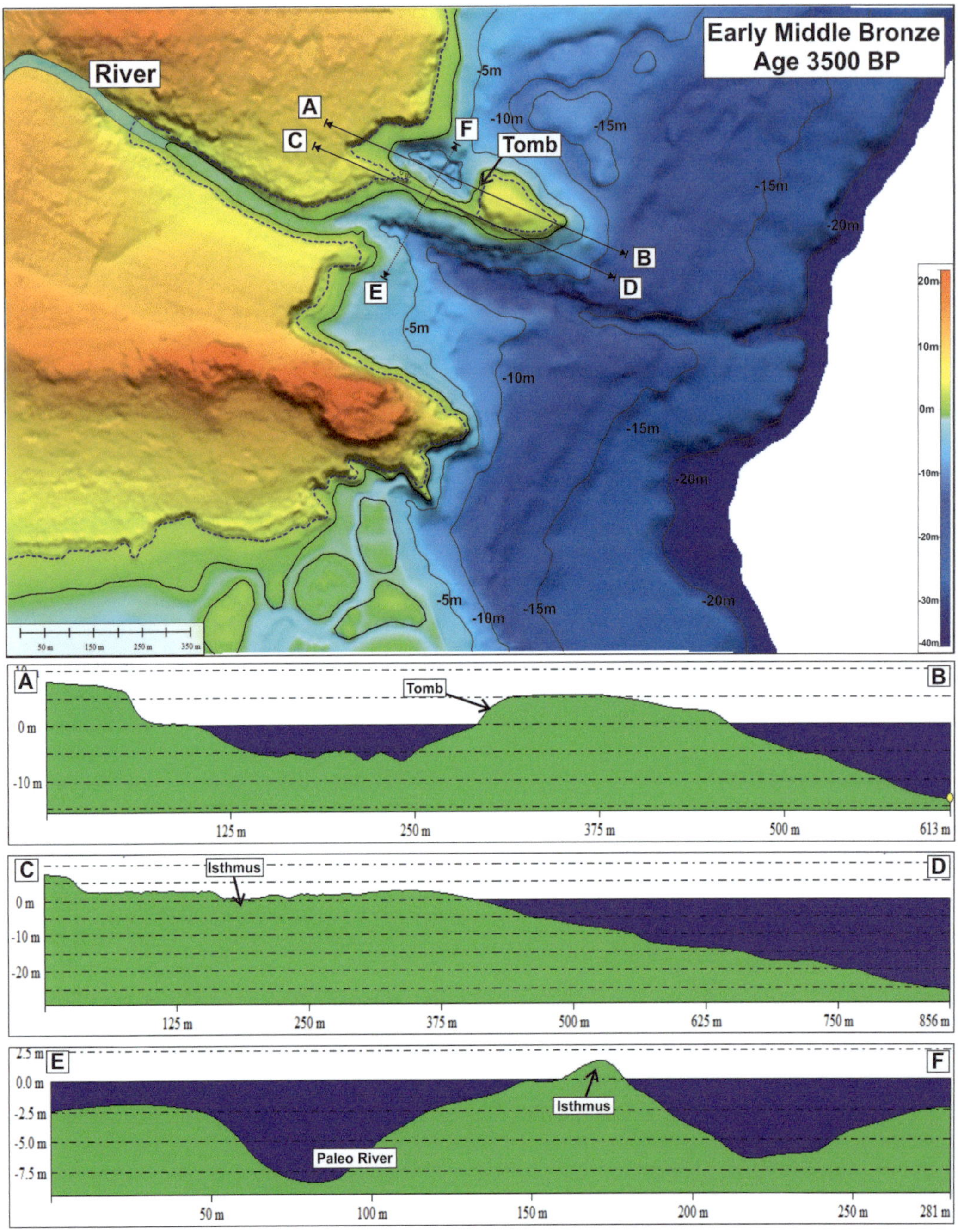


**Fig. 16.10** Palaeogeographical reconstruction of Ognina during the Early Middle Bronze Age. Conventions as in Fig. 16.8

as a special place with sacred significance. It is very important to stress that the occurrence of the Ognina tomb coupled with the assemblage of prestige artefacts from trench 'D' support a symbolic interpretation for all these finds, connected with the type of death-cult known from the Mycenaean era. This combination not only amplifies the special meaning played by this unique tomb—possibly

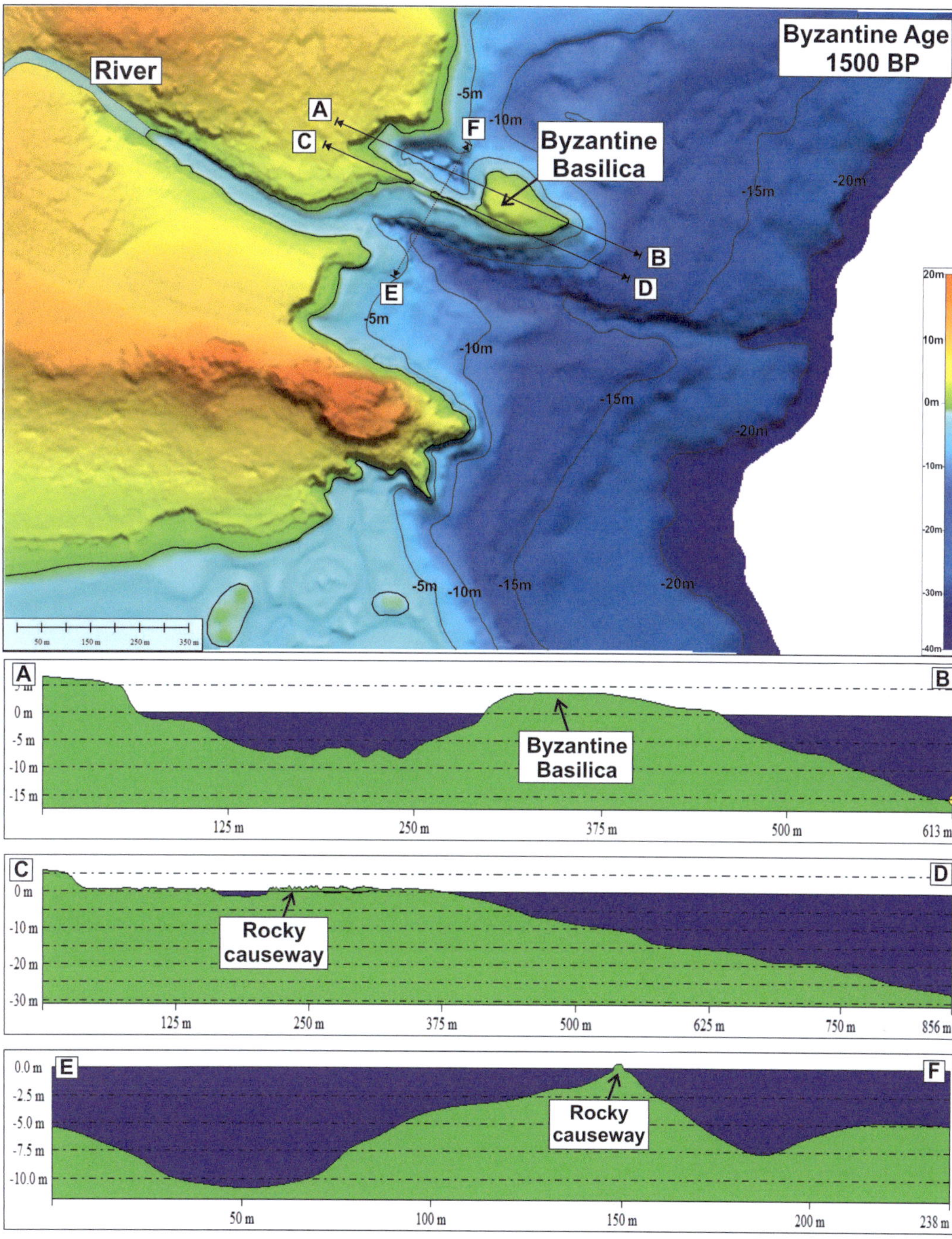


**Fig. 16.11** Palaeogeographical reconstruction for the site of Ognina during the Byzantine period. Conventions as in Fig. 16.8

acting as a 'sanctuary'—but also explains the ritual practices that we have archaeological evidence for in trench 'D'. If this area is a 'sacred landscape' that carried religious significance, the rock-cut tomb emphasizes its importance, which may be connected with the worship of the Earth and the powers of the underworld. The archaeological evidence is that Ognina capitalised on its religious and ritual power, whether through social or ritual means, becoming a significant centre for seafaring, trading contacts and mobility in the early Bronze Age of the western Mediterranean. The tomb would have functioned to legitimise power and authority through the uniqueness of a sacred space which functioned as a ritual 'sanctuary'. Thus formalised, ritual authorities could have dominated the religious, social and perhaps also the political landscape associated with this strategically located place of meeting and trade.

By the sixth century AD, the island was further isolated from the mainland, though not completely cut off from it (Fig. 16.11). Despite this—or perhaps because of it—the island was chosen as the site for a small triple-apse Byzantine basilica. There is no mention in any historical or liturgical records about the church—it is not even known which saint it was dedicated to or why it was built. This was perhaps the little church reached in Sicily by the North African Bishop Rufinianus with the purpose of practising asceticism while escaping Aryan persecutions from Byzacena (Eastman 2011). In this respect there are hints from the Vita Fulgentii (Rizzo 1991) where it is stated that in the early sixth century the religious Fulgentius, who later became Episcopi Ruspensis (507–533), went to meet Bishop Rufinianus in Sicily in a small boat that is described as a 'modica navicula', (Rizzo 2006, but see Sgarlata and Rizzone 2013). However, what is clear is that the religious authorities of the Byzantine Empire, like their Neolithic and Bronze Age predecessors, created yet another ritual space, perpetuating the long history of this unusual coastal feature as a sacred landscape.

In conclusion, our research confirms that the island of Ognina attracted human settlement from an early period, and that it was associated with a coastal landscape progressively modified by sea-level change, which may well have influenced the changing history of its human significance and in particular its role as a social and ritual centre. Initially, at the beginning of the Neolithic period, it was a prominent hill with a commanding control over a wider landscape and a nearby river that provided good access both to the hinterland and towards the sea. Subsequently, during the Bronze Age, it became progressively more isolated with changes in relative sea level, perhaps accentuating its significance as a sacred location set apart from everyday activities. By the time of the Byzantine period it had become virtually isolated as an island, though still within easy reach of the mainland, accentuating its virtues as a holy place of retreat and religious significance.

**Acknowledgements** All authors contributed extensively to the work presented in this paper and are jointly responsible for the results and the evaluation of their wider implications. EFCB and GS jointly conceived the study, designed and implemented the simulation model, and prepared the manuscript; EFCB is responsible for the introduction and the archaeological sections, and GS performed and supervised all the geophysical survey operations, produced live imaging experiments, generated the Digital Terrain Models and palaeogeographical reconstructions, and wrote the sections on geology, coastal geomorphology and sea-level change, with input from the other authors. EFCB and GS are jointly responsible for the discussion and conclusions with input from the other authors. FA and CM also gave technical support and conceptual advice. This study was motivated by discussions at the workshops of MOPP-MEDFLOOD (INQUA projects 1203 and 1603P).

## References

Antonioli F, Anzidei M, Lambeck K, Auriemma R, Gaddi D, Furlani S, Orrù P, Solinas E, Gaspari A, Karinja S, Kovačić V, Surace L (2007) Sea level change during the Holocene in Sardinia and in the north-eastern Adriatic (Central Mediterranean Sea) from archaeological and geomorphological data. Quat Sci Rev 26:2463–2486

Antonioli, F, Lo Presti, V, Gasparo Morticelli, M, Bonfiglio L, Mannino MA, Palombo MR, Sannino, G, Ferranti, L, Furlani, S, Lambeck K, Canese S, Catalano R, Chiocci FL, Mangano G, Scicchitano G, Tonielli R (2014) Timing of the emergence of the Europe-Sicily bridge (40–17 cal ka BP) and its implications for the spread of modern humans. In J. Harff, G. Bailey and F. Lueth (eds) Geology and archaeology: submerged landscapes of the continental shelf. Geological Society, London, Special Publications 411: 111–144. doi 10.1144/SP411.1

Anzidei M, Esposito A, Antonioli F, Benini A, Tertulliani A, Del Grande C (2006) I movimenti verticali nell'area di Briatico: evidenze da indicatori archeologici marittimi nell'area del terremoto del 1905. In: Guerra I, Bavaglio A

(eds) 8 settembre 1905 terremoto in Calabria. Deputazione di Storia patria per la Calabria. Università della Calabria, Regione Calabria
Anzidei M, Antonioli F, Lambeck K, Benini A, Soussi M (2011a) New insights on the relative sea level change during Holocene along the coasts of Tunisia and western Libya from archaeological and geomorphological markers. Quat Int 232:5–12
Anzidei M, Antonioli F, Benini A, Lambeck K, Sivan D, Serpelloni E, Stocchi P (2011b) Sea level change and vertical land movements since the last two millennia along the coasts of southwestern Turkey and Israel. Quat Int 232:13–20
Anzidei M, Antonioli F, Benini A, Gervasi A, Guerra I (2013) Evidence of vertical tectonic uplift at Briatico (Calabria, Italy) inferred from roman age maritime archaeological indicators. Quat Int 288:158–167
Auriemma R, Solinas E (2009) Archaeological remains as sea level change markers: a review. Quat Int 206:134–146
Barley N (1995) Dancing on the grave: encounters with death. Murray, London
Basile B, Di Stefano G, Lena G (1988) Landings, ports, coastal settlements and coastlines in southeastern Sicily from prehistory to late antiquity. In: Raban A (ed) Archaeology of coastal changes. Proceedings of the first international symposium "cities on the sea—past and present", British archaeological reports international series, vol 404. BAR, Oxford, pp 15–33
Baumann Z (1993) Mortality, immortality and other life strategies. Polity Press, Oxford
Benjamin J, Bonsall C, Pickard C, Fischer A (eds) (2011) Submerged prehistory. Oxbow, Oxford
Bernabò Brea L (1966) Abitato neolitico e insediamento maltese dell'età del Bronzo nell'isola di Ognina (Siracusa) e i rapporti fra la Sicilia e Malta dal XVI al XIII sec. a.c., Kokalos XII: 40–69
Bernabò Brea L (1976–1977) Eolie, Sicilia e Malta nell'età del Bronzo. Kokalos 22–23(1):33–110
Bianca M, Monaco C, Tortorici L, Cernobori L (1999) Quaternary normal faulting in southeastern Sicily (Italy): a seismic source for the 1693 large earthquake. Geophys J Int 139:370–394
Caputo M, Pieri L (1976) Eustatic variation in the last 2000 years in the Mediterranean. J Geophys Res 81:5787–5790
Castagnino Berlinghieri EF (1993–1995) Ognina (Siracusa): note preliminari all'indagine di topografia marittima. Klearchos 137–148, Taranto, 5–13
Castagnino Berlinghieri EF (2003) The Aeolian Islands: crossroads of Mediterranean maritime routes. A survey on their maritime archaeology and topography from the Prehistoric to the Roman periods. British Archaeological Reports, International Series 1181. Archaeopress, Oxford
Castagnino Berlinghieri EF (2011) Subsistence, mechanisms of interaction and human mobility in the Neolithic western Mediterranean: the nature of the occupation of Lipari (Aeolian Islands, Sicily). In Phoca-Cosmetatou P (ed) The first mediterranean islanders: initial occupation and survival strategies. Oxford University School of Archaeology, Monograph Series 73:109–139
Cazzella A, Recchia G (2013) Malta, Sicily, Aeolian Islands and southern Italy during the bronze age: the meaning of a changing relationship. In: Alberti MN, Sabatini S (eds) Exchange networks and local transformations. Oxbow, Oxford, pp 80–91
Cultraro M (2000) Considerazioni sull'architettura Funeraria in Sicilia durante l'età del Bronzo Antico. In: Melis MG (ed) L'Ipogeismo nel Mediterraneo: origini, sviluppo, quadri culturali: atti del Congresso internazionale, 23–28 maggio 1994, Sassari Oristano, Italia. Sassari, Università degli studi di Sassari, Facoltà di Letteree filosofia, Istituto di Antichità, arte e discipline etnodemologichee Dipartimento di Scienze umanistiche e dell'antichità. Vol 2, pp. 707–721
Di Grande A, Raimondo W (1982) Linee di costa pliopleistoceniche e schema litostratigrafico del Quaternario siracusano. Geol Romana 21:279–309
Dutton A, Scicchitano G, Monaco C, Desmarchelier JM, Antonioli F, Lambeck K, Esat TM, Fifield LK, Mcculloch MT, Mortimer G (2009) Uplift rates defined by U-series and 14C ages of serpulid-encrusted speleothems from submerged caves near Siracusa, Sicily (Italy). Quat Geochronol 4:2–10
Eastman, DE (2011) Paul the Martyr: the cult of the apostle in the Latin West. Writings from the Greco-Roman World Supplement 4. Society of Biblical Literature, Atlanta.
Farr H (2006) Seafaring as social action. J Marit Archaeol 1:85–99
Flemming NC (1968) Archaeological evidence for sea level changes in the Mediterranean. Underwater Association Report, 9–13
Flemming NC (1969) Archaeological evidence for eustatic changes of sea level and earth movements in the western Mediterranean in the last 2000 years. Geol Soc Am Spec Pap 109:1–125
Flemming NC, Webb CO (1986) Tectonic and eustatic coastal changes during the last 10,000 years derived from archaeological data. Zeitschrift fur Geomorphologie 62:1–29
Furlani S, Cucchi F (2013) Downwearing rates of vertical limestone surfaces in the intertidal zone (gulf of Trieste, Italy). Mar Geol 343:92–98
Grasso M, Lentini F (1982) Sedimentary and tectonic evolution of the eastern Hyblean plateau (southeast Sicily) during late cretaceous to quaternary times. Palaeogeogr Palaeoclimatol Palaeoecol 39:261–280

Hirn A, Nicolich R, Gallart J, Laigle M, Cernobori L (1997) Roots of Etna volcano in faults of great earthquakes. Earth Planet Sci Lett 148:171–191

Kapitaen G (1970) Perlustrazioni sottomarine sulla topografia originaria e la situazione portuale dell'abitato preistorico di Ognina (relazione preliminare 1969). Sicilia Archeologica 11:43–56

Kapitaen G (1991) Maritime aspects of early Neolithic coastal settlements in southeast Sicily. In: Lazarov M, Tatcheva M, Angelova, C, Georgiev M (eds) Quatrième symposium international ThraciaPontica IV: Sozopol, 6–12 octobre 1988: les thèmes, les agglomérations côtières de la Thrace avant la colonisation grecque, les sites submergés – méthodes des recherché. Izdatelska kŭshta "Galaktika", Sofia, pp 399–426

Lambeck K, Anzidei M, Antonioli F, Benini A, Esposito E (2004a) Sea level in roman time in the Central Mediterranean and implications for modern sea level rise. Earth Planet Sci Lett 224:563–575

Lambeck K, Antonioli F, Purcell A, Silenzi S (2004b) Sea level change along the Italian coast for the past 10,000 yrs. Quat Sci Rev 23:1567–1598

Lambeck K, Antonioli F, Anzidei M, Ferranti L, Leoni G, Silenzi S, Scicchitano G, Silenzi S (2011) Sea level change along the Italian coast during the Holocene and projections for the future. Quat Int 232:250–257

Lazzarini S, La Rosa G, Cappellani G (1965) Dieci anni di esplorazioni paleontologiche ed archeologiche nel siracusano. Archivio Storico Siracusano 11:136–146

Leighton R (1999) Sicily before history. Duckworth, London

Lo Presti V, Antonioli F, Auriemma R, Ronchitelli A, Scicchitano G, Spampinato CR, Anzidei M, Agizza S, Benini A, Ferranti L, Gasparo Morticelli M, Giarrusso C, Mastronuzzi G, Monaco C, Porqueddu A (2014) Millstone coastal quarries of the Mediterranean: a new class of sea level indicator. Quat Int 332:126–142

Orlandini P (1962) Il villaggio preistorico di Manfria presso Gela. Banco di Sicilia, Palermo

Palio O (2008) Ognina, Malta e l'Egeo. In: Bonanno A, Militello P (eds) Malta negli Iblei. Gli Iblei a Malta. Atti del Convegno, Catania 30 settembre –1 ottobre 2007 (KASA 2). Officina di Studi Medievali, Palermo, pp 71–80

Parker AJ (2001) Maritime landscapes. Landscapes 2(1):22–41

Pásztor E, Barna JP (2014) Neolithic longhouses and bronze age houses in Central Europe. In: Ruggles CLN (ed) Handbook of Archaeoastronomy and Ethnoastronomy. Springer, New York, pp 1307–1316

Procelli E (1995) Cultures and societies in Sicily between the Neolithic and the middle bronze age. In: Fisher Hansen T (ed) Ancient Sicily. Series Acta Hyperborea. Museum Tusculanum Press, Copenhagen, pp 13–22

Procelli EE (1991–1992) Considerazioni sul passaggio dall'antica alla media età del Bronzo nella Sicilia Orientale: Catania e Naxos. Rassegna Archeologica 10:561–568

Rizzo FP (2006) Sicilia cristiana. Dal I al V secolo (volume II). Supplementi a Kokalos, XVII. Bretschneider, Rome, pp 268–371

Rizzo EP (1991) Fulgenzio a Siracusa In: Studi di filologia classica in onore di Giusto Monaco. Palermo, Università di Palermo, Facoltà di lettere e filosofia, Istituto di filologia greca, Istituto di filologia Latina, Vol 4. Palermo, pp 1473–1482

Sartori R, Colalongo ML, Gabbianelli G, Bonazzi C, Carbone S, Curzi PV, Evangelisti D, Grasso M, Lentini F, Rossi S, Selli L (1991) Note stratigrafiche e tettoniche sul rise diMessina (Ionio Nord-occidentale). Giorn Geol 53:49–64

Scandone P, Patacca E, Radoicic R, Ryan WBF, Cita MB, Rawson M, Chezar H, Miller E, Mckenzie J, Rossi S (1981) Mesozoic and Cenozoic rocks from Malta escarpment (Central Mediterranean). Am Assoc Pet Geol Bull 65:1299–1319

Schmiedt G (1966) Antichi porti d'Italia. I porti delle colonie greche. L'Universo 46:253–296

Schmiedt G (1972) Linea di costa della Sicilia orientale compresa fra la foce del fiume Simeto ed il Capo Ognina. Il livello antico del Mar Tirreno, II: 201–221

Scicchitano G, Monaco C (2006) Grotte carsiche e linee di costa sommerse tra capo Santa Panagia e Ognina (Siracusa, Sicilia Sud-Orientale). Il Quaternario 19(2):187–194

Scicchitano G, Antonioli F, Castagnino Berlinghieri EF, Dutton A, Monaco C (2008) Submerged archaeological sites along the Ionian coast of south-eastern Sicily (Italy) and implications for the Holocene relative sea-level change. Quat Res 70:26–39

Scicchitano G, Costa B, Di Stefano A, Longhitano SG, Monaco C (2010) Tsunami and storm deposits preserved within a ria-type rocky coastal setting (Siracusa, SE Sicily). Z Geomorphol 54(Suppl. 3):51–77

Scicchitano G, Lo Presti V, Spampinato CR, Gasparo Morticelli M, Antonioli F, Auriemma R, Ferranti L, Monaco C (2011) Millstones as indicators of relative sea-level changes in northern Sicily and southern Calabria coastlines, Italy. Quat Int 232:92–104

Sgarlata MR, Rizzone V (2013) Vescovi e Committenza ecclesiastica nella Sicilia orientale: architettura e fonti, in Episcopus, Civitas, Territorium. In Cresci S, Lopez Quiroga J, Brandt O, Pappalardo C (eds) Atti del XV Congresso Internazionale di Archeologia Cristiana, Toledo, 8–12 settembre 2008. Studi di Antichità Cristiane 65: 789–801

Spampinato CR, Costa B, Di Stefano A, Monaco C, Scicchitano G (2011) The contribution of tectonics to relative sea-level change during the Holocene in coastal south-eastern Sicily: new data from boreholes. Quat Int 232:214–227

Spannocchi T (1578) Descripción de las Marinas de todo el Reio de Sicilia con otras importantes declaraciones notadas por el Cavallero Triburcio Spanoqui del Abito de San Juan Gentilhombre de la Casa de Su Magestad, dirigido al Principe Don Filipe Nuestro Senor en el Ano de MDCCLVIII. Biblioteca Nacional di Madrid, file-archive n. 788

Tanasi D (2014) Uniti e divisi dal mare, Mobilità di uomini e merci tra la Sicilia e l'arcipelago maltese nel II millennio a.C. In: Congiu M, Miccichè C, Amodeo S (eds) "Viaggio in Sicilia". Proceedings of the Atti del X Convegno di Studi di Sicilia Antica, 'Viaggio in Sicilia: Racconti, Segni e Città Ritrovate'. Caltanissetta, pp 99–116

Tusa S (1992) La Sicilia nella preistoria, Sellerio, second edn. Sellerio, Palermo

Waelbroeck C, Labeyrie L, Michel E, Duplessy JC, McManus JF, Lambeck K, Balbon E, Labracherie M (2002) Sea-level and deep water temperature changes derived from benthic foraminifera isotopic records. Quat Sci Rev 21:295–305

Westerdahl C (1992) The maritime cultural landscape. Int J Naut Archaeol 21:5–14

Westley K, Dix J (2006) Coastal environments and their role in prehistoric migrations. J Marit Archaeol 1(1):9–28

# Chapter 17
# Archaeological Potential of the Anchialine Caves in Croatia

**Irena Radić Rossi and Neven Cukrov**

**Abstract** Anchialine caves are a common phenomenon in the Mediterranean karst. Their entrances lie above sea level, and underground accumulations of fresh water in their interior float on top of the seawater, which communicates with the sea through porous carbonate rocks. The fresh water layer of such caves represents a significant resource of potable water for the ancient inhabitants of the eastern Adriatic coast. Due to the systematic work of a group of Croatian speleologists, more than 100 anchialine caves have been registered along the Croatian coast and its numerous islands. Submerged speleothems from two caves, situated on the islands of Krk and Lošinj, have contributed to the reconstruction of Late Pleistocene and Holocene sea level changes in the northern part of the eastern Adriatic. Archaeological finds have been observed in at least seven caves, but only one cave, Vodeni Rat on the Pakleni Islands near the island of Hvar, has been archaeologically researched. As with many other similar structures, there is evidence of human intervention on its rocky walls, and some Roman amphorae were recovered from the bottom of the cave. Another cave named Živa Voda, situated on the other extremity of the island of Hvar, revealed the presence of a great quantity of Bronze Age underwater finds, testifying to the intense use of fresh water from its interior, or maybe some other ancient function. Considering their distribution, accessibility, depths and position in relation to known prehistoric settlements, we can assume that the anchialine caves in Croatia have high archaeological potential that should be protected and researched.

## 17.1 Introduction

Anchialine caves are a specific speleological phenomenon around the world, located near the coast in tropical and moderately warm climatic zones (Fig. 17.1). They contain a freshwater layer at the top of the water body in their interior, which makes them extremely important sources of potable water in dry karstic areas.

A rocky carbonate coastline and a great number of islands and islets characterize the Croatian part of the Adriatic coast. The mainland and many of the larger islands possess water resources other than anchialine caves, but on the smaller islands they provide the only possibility of fresh water supply. In the past, rainwater was captured in cisterns, and the anchialine caves served as natural wells.

I.R. Rossi (✉)
Department of Archaeology, University of Zadar, Obala kralja Petra Krešimira IV. 2, HR-23000 Zadar, Croatia
e-mail: irradic@unizd.hr

N. Cukrov
Division for Marine and Environmental Research, Institute Ruđer Bošković,
Bijenička 54, HR-10000 Zagreb, Croatia
e-mail: ncukrov@irb.hr



G.N. Bailey et al. (eds.), *Under the Sea: Archaeology and Palaeolandscapes of the Continental Shelf*, Coastal Research Library 20, DOI 10.1007/978-3-319-53160-1_17

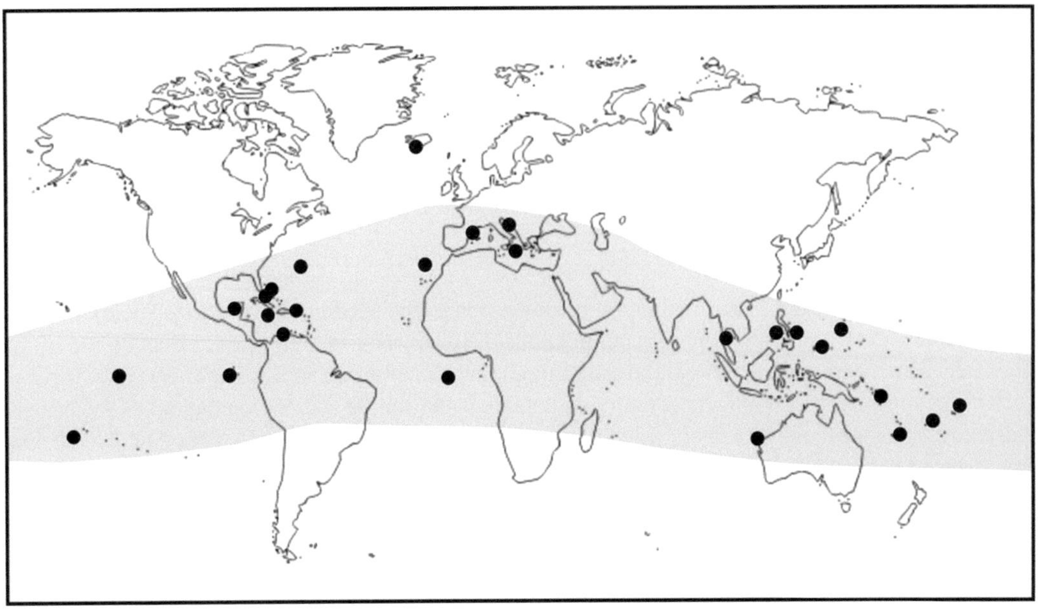

**Fig. 17.1** Position of Croatia in relation to the world distribution of anchialine caves (After Iliffe 2000)

Seafaring along the eastern Adriatic coast demanded safe anchorages and food and water supplies, whether for local needs or for more distant sea journeys, for example following the most suitable seafaring route linking the southern and northern Adriatic Basin. In that context, we can presume that the anchialine caves, especially when distant from well-organized ports, played an important role for renewing supplies.

In some anchialine caves, there are obvious traces of exploitation of fresh water since prehistoric times, which have been identified by speleologists and other researchers. The scarcity of well-documented archaeological evidence is the consequence of the lack of awareness of the archaeological potential of such places, which has already resulted in the disturbance and destruction of the archaeological context in some of the more frequently visited caves.

## 17.2 Definition and Distribution of Anchialine Caves

The anchialine (Gr. ἀγχίαλος = near the sea) environment was originally defined as a habitat consisting of 'pools with no surface connection with the sea, containing salt or brackish water, which fluctuates with the tides' (Holthuis 1973, p 3). Subsequent documentation of extensive, submerged cave systems by cave-diving explorers and scientists prompted Stock et al. (1986, p 91) to update the definition, as follows:

> Anchialine habitats consist of bodies of haline waters usually with a restricted exposure to open air always with more or less extensive subterranean connections to the sea and showing noticeable marine as well as terrestrial influences

Over the past three decades, research on the anchialine environment has expanded rapidly (Sket 1996; Iliffe 2000, 2005). However, anchialine ecosystems remain poorly understood. By definition, anchialine caves may exist anywhere between the coastal aquifer and the sea, but hardly any have been noted and investigated outside the tropical or warm temperate climatic zones (Fig. 17.1).

Geologically, they are located in two very different types of coastal environments: karstified carbonate rocks; and lava fields (Sket 2012).

## 17.3 Anchialine Caves in Croatia

The karst area covers more than 50% of the overall territory of Croatia, and has evidence of over 9000 caves. As far as we know today, more than a hundred, situated along the coast and on the numerous islands, belong to the category of anchialine caves (Surić et al. 2010), and are partially explored and/or described (Fig. 17.2). All of them are located in carbonate rocks, mostly limestone.

Anchialine caves are caves with lakes inside, which are stratified in terms of salinity, at least seasonally. Usually they have a fresh water layer at the surface, a brackish layer in the middle, and

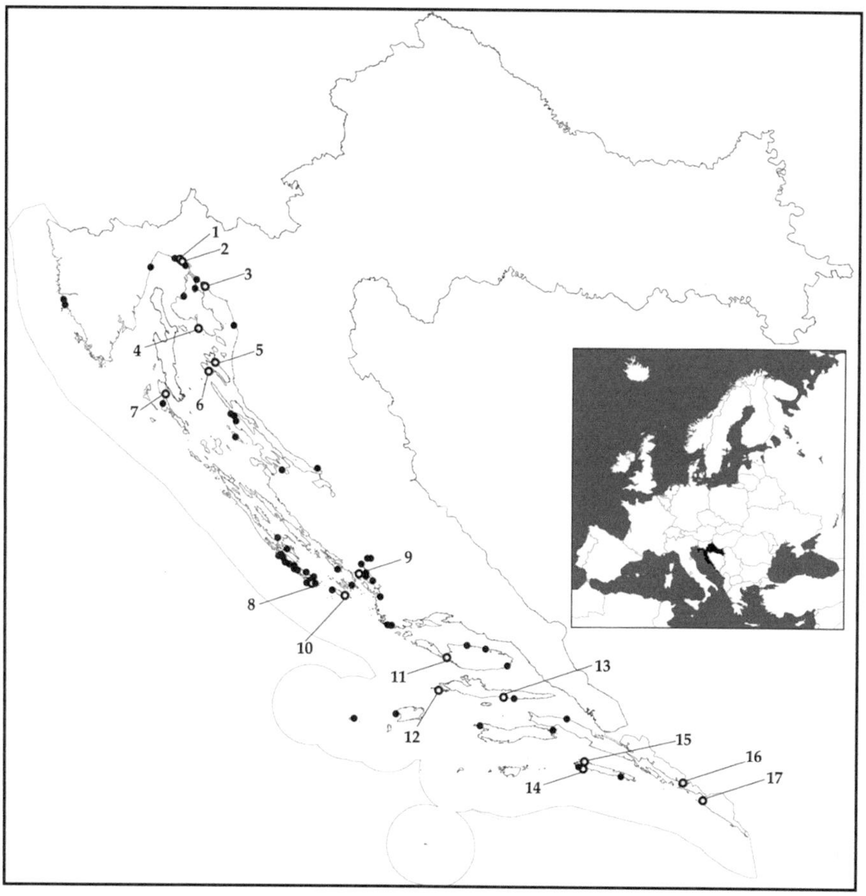


**Fig. 17.2** Distribution map of anchialine caves in Croatia showing the location of caves listed in the text (information from B. Jalžić, M. Cukrov, P. Kutleša, N. Cukrov): *1* Urinjska špilja, *2* Vrtare male, *3* Pijavica, *4* U vode, island of Krk, *5* Medova buža, island of Rab, *6* Mag, island of Rab, *7* Medvjeđa špilja, island of Lošinj, *8* Gravrnjača, island of Kurba Vela, Kornati Archipelago, *9* Vodena jama in Srima, *10* Gradina, island of Žirje, *11* Jama u Podstražišću, island of Brač, *12* Vodeni rat, island of Sv. Kliment/Hvar, *13* Živa voda, island of Hvar, *14* Bjejajka, island of Mljet, *15* Lenga, island of Mljet, *16* Sumporače, velika i mala, *17* Šipun

seawater at the bottom. The water has a more or less constant temperature, with very limited influence from outside conditions. Due to the lack of light, there is no photosynthesis, but bacteria-mediated chemosynthesis is present, with many endemic organisms.

Anchialine caves in Croatia are mostly small and not very attractive for speleological research. The longest one is Medvjeđa špilja on the island of Lošinj (Fig. 17.2, no. 7), with a length of 245 m, and the deepest one is Jama u Podstražišću (Fig. 17.2, no. 11) on the island of Brač, with a 45 m-deep dry part and a water column over 50 m deep. Most of the caves do not have a clear connection to the sea, but the tides exercise an influence to some degree in all of them. Just a few caves, like Medova buža on the island of Rab (Fig. 17.2, no. 5), have an open pathway to the sea. What attracts scientists is mostly their ecosystems, their deep-ocean-like water conditions, and the composition of the accumulated sediments. Recently, they have attracted some archaeological attention. Today, the anchialine caves in Croatia are recognized as unique habitats, and are under legal protection like all the other caves.

After the Last Glacial Maximum, the change in sea level of about 130 m (Surić 2009) transformed many anchialine caves of the time into marine ones (Iliffe 2012). At the same time, some caves with freshwater became anchialine.

### *17.3.1 History of Research*

In 1909 Stjepan Vuksan produced the first written report on an anchialine cave in Croatia. He accurately described the cave of Pijavica (Fig. 17.2, no. 3) near the town of Senj. In 1920, Croatian geologist Josip Poljak wrote a report about the anchialine cave of Urinjska špilja (Fig. 17.2, no. 1) near Rijeka, drew a topographical map, and reported on its brackish water. In addition, in 1963 Bruno Puharić realized in the same cave the first described snorkel dive (Cukrov et al. 2009).

The period of modern research started in the second half of the twentieth century, when in 1958 Boris Sket visited the anchialine cave of Šipun (Fig. 17.2, no. 17), located in the town of Cavtat south of Dubrovnik. Since then he has explored the fauna and ecology of more than 30 anchialine caves along the eastern Adriatic coast, mainly in the National Park of Kornati. His work has resulted in an ecological scheme for anchialine water bodies based on salinity stratification, local oxygen depletion, and faunal stratification by biotic exclusion (Sket 1986, 1996). During this period, many other speleologists and researchers visited anchialine caves and made early descriptions and sketches.

Branko Jalžić from the Croatian Biospeleological Society in Zagreb, Frane Kršinić from the Institute of Oceanography and Fisheries in Split, and a group of scientists from the Division for Marine and Environmental Research of the Ruđer Bošković Institute in Zagreb are the main promoters of modern systematic multidisciplinary research on the anchialine caves in Croatia. Besides speleological recognition and mapping, their projects focus on environmental conditions, habitat descriptions, and biogeochemistry of trace metals. As this systematic work started only recently, the anchialine caves along the Croatian Adriatic coast are still mostly unexplored.

### *17.3.2 Dating of Submerged Speleothems*

Surić et al. (2010) mention 235 partly or completely submerged caves on the Croatian coast and islands, of which 126 clearly demonstrate marine conditions. The rest of the group encompasses anchialine caves, springs and undetermined caves.

All of the explored formations proved to be of continental origin, and were submerged as a consequence of marine transgression. Speleothems are present, with a predominance of stalactites due to the better conditions for their preservation. They provide a good record of formation processes, and changes of palaeoenvironment and sea level. Up to now, the presence of speleothems has been attested in 140 caves, and the deepest recorded lies at −71 m (Garašić 2006).

Sampling and analysis of submerged speleothems in Croatia, promoted by Surić et al. (2005a, b, 2009), includes two anchialine caves: U vode (Fig. 17.2, no. 4) on the island of Krk, and Medvjeđa špilja (Fig. 17.2, no. 7) on the island of Lošinj (Surić et al. 2009; Surić and Juračić 2010). In the first case, samples were taken from two stalagmites at a depth of 14.5 m and 18.8 m, and in the second case from a fallen stalactite at a depth of 1.5 m, and a stalagmite at a depth of 10 m. The chronological determinations of these and 12 other speleothems from five different caves (taken up to a depth of 41.5 m) were realized with three different dating techniques. The results allowed the authors to reconstruct a partial sea-level curve for the Croatian part of the Eastern Adriatic coast for the last 220,000 years (Surić and Juračić 2010), and to suggest the dynamics associated with the palaeoenvironmental changes.

## 17.4 Palaeontological and Archaeological Finds

As previously stated, over one hundred anchialine caves are present in Croatia but are poorly documented and researched, especially from the archaeological point of view. On the other hand, existing examples of the anchialine cave finds are important indicators of their archaeological and paleontological potential.

In the present state of research archaeological finds are certainly present in at least six anchialine caves, and two caves contain palaeontological evidence. The most significant caves of archaeological interest are Vodeni rat (Fig. 17.2, no. 12) and Živa voda (Fig. 17.2, no. 13) in Central Dalmatia, while Vrtare male (Fig. 17.2, no. 2) on the northern Croatian littoral has attracted significant palaeontological attention.

The other caves in which speleologists and geologists have noted the presence of archaeological finds are a small cave in Mag Cove on the island of Rab (Fig. 17.2, no. 6), Gravrnjača Cave on the island of Kurba Vela (Fig. 17.2, no. 8) in the Kornati Archipelago, the cave of Vodena jama (Fig. 17.2, no. 9) in Srima near Šibenik, Gradina cave (Fig. 17.2, no. 10) on the island of Žirje (Barišić 1994; Jalžić 1994), and Šipun cave (Fig. 17.2, no. 17) in Cavtat (Falcon-Barker 1960). While the first and last caves only exhibit traces of human activity in the water body of the caves, in the cave on Kurba Vela there are a number of amphorae sherds (Gottstein Matočec and Jalžić 2003). Some of them are attributed to the Late Roman Republican period (Lamboglia Type 2), and the others to the Early Imperial period (Dressel Types 2–4). On the other hand, the attribution to the prehistoric period of a fragment of a globular vessel in Vodena jama in Srima is possible, but not yet confirmed.

As for palaeontological remains, besides the cave of Vrtare male described below, the remains of *Ursus spelaeus* (cave bear), which became extinct during the Last Glacial Maximum, were found in the already mentioned Medvjeđa špilja on the island of Lošinj (Malez et al. 1979).

### *17.4.1 Cave of Vrtare Male Near Crikvenica*

In 2005, Jalžić et al. (2005) summarized everything that was then known about the cave of Vrtare Male (Fig. 17.2, no. 2; Fig. 17.3). According to their report, it is located about 100 m from the seashore; the depth of the dry part measures 29 m, and the known depth of the water body is about 10.5 m. About 14 m below the present-day entrance there is a south-eastern channel which ends in the so called Big Lake, 10.5 m deep. The south-western channel leads to the so called Small Lake, whose

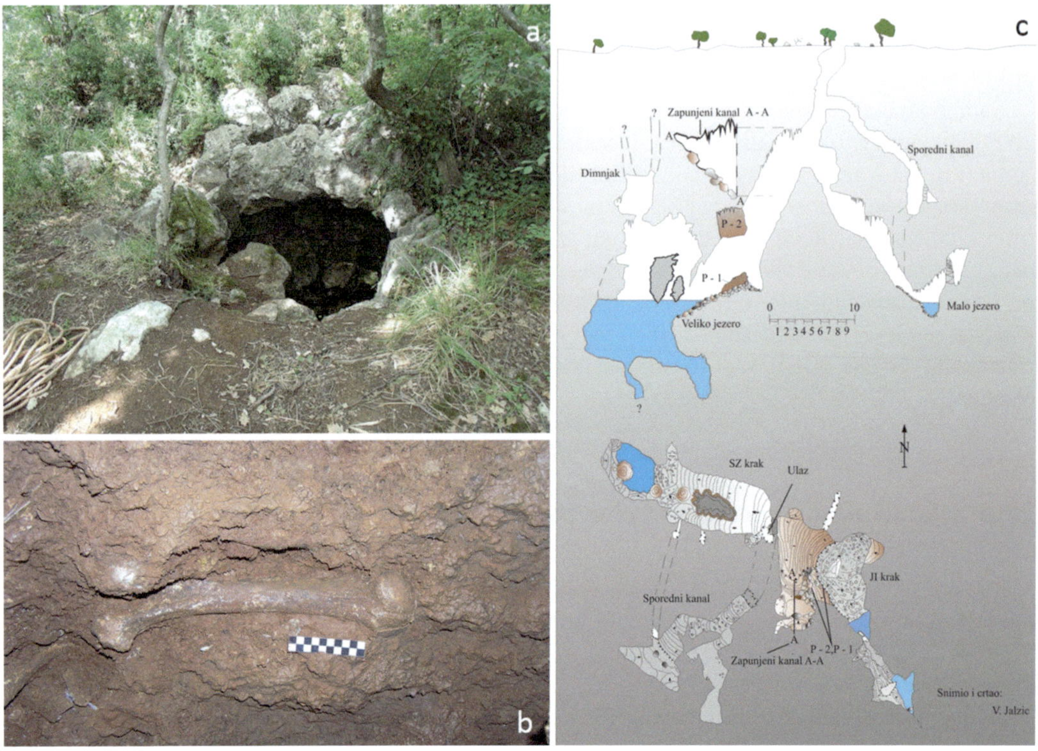


**Fig. 17.3** The Vrtare Male Cave near Crikvenica: (**a**) Entrance (Photo: K. Miculnić); (**b**) Palaeontological finds (Photo: K. Miculinić); (**c**) Plan and cross section (V. Jalžić)

depth does not exceed 1.5 m. The overall body of the cave is much more complex, and still not completely explored.

Since 1966, several expeditions to Vrtare Male have resulted in a detailed description of the geology, morphology, genesis and biology of the cave. They also led to the discovery and preliminary determination of the rich palaeontological material found in the dry sediment, mostly deposited in the south-eastern channel and in the near vicinity of the Big Lake. The underwater part of the cave is not yet palaeontologically explored.

According to speleologists, the variety of speleothems and their aspect suggest that the cave must be very old. The preliminary analysis of the animal remains attributed them to the families of Felidae (cats), Ursidae (bears), Canidae (dogs and dog-like mammals), Lagomorpha (hares, rabbits and pikas), Cervidae (deers), Proboscidae (trunked mammals) and Rhinoceratidae (rhinoceros). The lack of chronological data complicates determination of the age of the finds, although it is generally presumed that all the animal remains belong to the Pleistocene.

The remains of Proboscidae and Rhinoceratidae are quite rare in the Adriatic basin. The nearest analogy comes from the vicinity of the island of Rab, where one upper right molar and a left femur were recovered by a trawler fishing net from the seabed at a depth of 80 m. The finds were determined as belonging to *Mammuthus meridionalis adriacus* n. ssp., and dated to the lower part of the Middle Pleistocene (Malez and Lenardić-Fabić 1988). A recent publication emphasizes the need for a more detailed study of the finds (Mauch-Lenardić 2012).

### 17.4.2 Cave of Vodeni Rat on the Island of Sv. Kliment Near Hvar

Vodeni Rat is a small promontory with a cave in the form of a pit, situated on the southern coast of the island Sv. Kliment, the biggest island in the archipelago of Pakleni otoci, in front of the port of Hvar on Hvar Island (Fig. 17.2, no. 12; Fig. 17.4). The toponym Vodeni Rat literally means 'the water promontory', suggesting at first glance the presence of fresh water. The entrance to the cave lies about 30 m from the seashore, at a height of 13 m above sea level. A rock in the immediate vicinity of the entrance displays traces of marks made by ropes, obviously frequently used to raise water containers. The overall depth of the vertical channel of the cave is about 41 m, with a maximum width of about 6–8 m. The depth of the water body is 29 m. Speleothems are present on the walls and in the small caverns, down to a depth of 23 m, and at a depth of 24 m there is a small platform, measuring 1.5 × 2.0 m. It could have been used for fetching water at times of different water level, although there is still no evidence to support it. The bottom of the cave contains muddy sediment (Mesić 2006).

The first speleological mission was conducted in 1999, when speleologists from Split noticed the presence of archaeological finds, and informed the Heritage Museum of Hvar. A rescue archaeological operation took place in the same year, which resulted in the preliminary documentation and description of the cave. The research team raised the remains of five amphorae from the platform at the bottom of the cave, datable to the Late Roman Republican and the Late Roman period (Mesić 2006). As the scope of the action was merely protective, aiming to recover the ancient containers, the bottom of the cave was not accurately explored, and this work remains to be done.

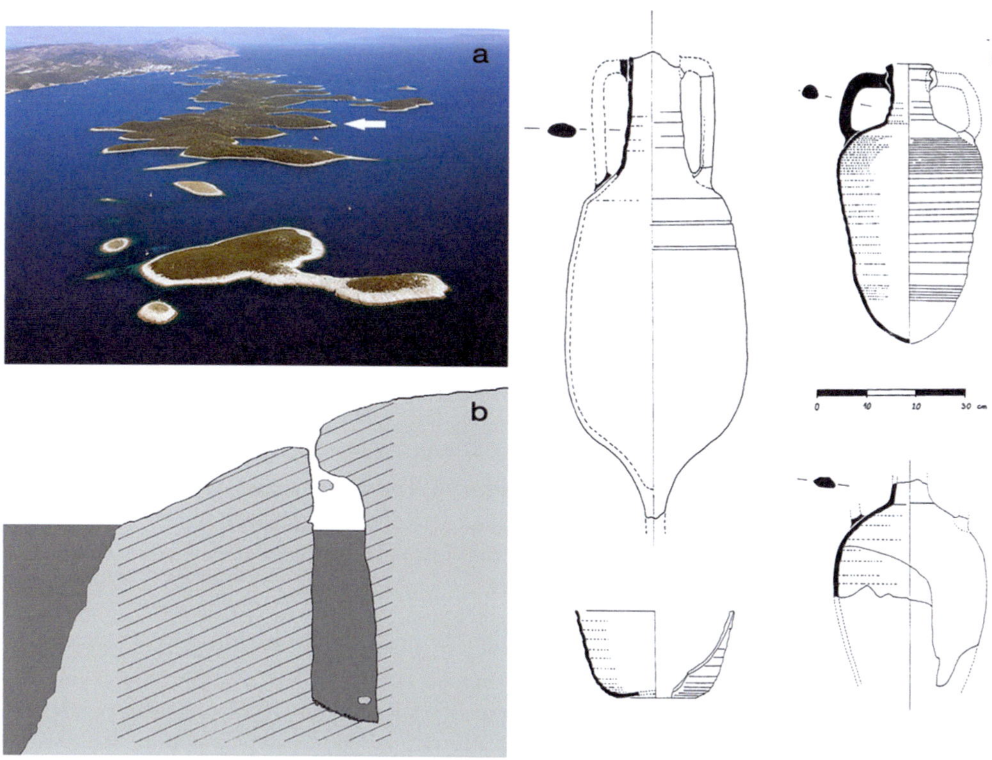


**Fig. 17.4** Vodeni Rat Cave on the island of Sv. Kliment near Hvar: (**a**) Position of promontory Vodeni Rat; (**b**) Cross section (After Mesić 2006); (**c**) Archaeological finds (After Mesić 2006)

**Fig. 17.5** Živa Voda Cave on the island of Hvar: **a**) and **b**) Plan and cross section (Drawn by B. Jalžić, A. Novosel, G. Polić, V. Jalžić); (**c**) Selection of finds from the Left channel (Photo: M. Petrić); (**d**) Archaeological finds in situ in the Right channel (Photo: V. Jalžić)

During the archaeological examination of the cave, traces of wooden structures came to light along the rocky walls, at depths of 6 and 9 m. These are, most probably, the remains of constructions built to facilitate access to fresh water. As no dating samples were taken, we currently have no chronological information concerning their creation and use.

### *17.4.3 Cave of Živa Voda in Kozja Cove on the Island of Hvar*

The cave of Živa Voda lies 55 m from the sea shore, in the Kozja Cove on the south coast of the island of Hvar, near the village of Bogomolje (Fig. 17.2, no. 13; Fig. 17.5). The entrance is 31 m above the sea level. After about 30 m the main channel splits into two. The western channel measures more than 60 m in length, with a water depth of about 26 m. The eastern channel measures about 80 m in length, with a water depth of about 38 m. The bottom of the cave contains muddy sediment (Ozimec and Jalžić 2003; Novosel et al. 2007; Cukrov et al. 2008).

Although the cave was never archaeologically researched, there is a record of steps carved in the rock (Vujnović 1990) and isolated finds consisting of a fragment of Neolithic pottery, a vessel from the Early Bronze Age and a fragment of Hellenistic Gnathia pottery of South Italian origin (Vujnović 2013). The speleological divers noticed a great quantity of ancient pottery distributed between the two channels. Apparently, the eastern channel bears more evidence of the exploitation of the cave in Classical Antiquity, while the western one attests to its importance in the Bronze Age, and maybe the Iron Age. A recently produced video testifies to the presence in the eastern channel of sherds of first century BC amphorae, typologically determined as Lamboglia 2.

Certain amounts of pottery fragments and well-preserved vessels recovered from the cave during the speleological missions are deposited in the Heritage Musum of Hvar, while others may be seen in private collections (Radić Rossi 2011). Geographically, the nearest analogies to particular pottery features come from the tumulus of Glava Maslinova near Bogomolje (Marović 1985). The pottery from this site is attributed to the regional cultures of the Early Bronze Age, such as the Cetina and Posušje cultures (cf. Marijanović 1981; Govedarica 1982; Marović 1999), but similar features seem to last throughout the whole Bronze Age (Forenbaher and Kaiser 2006).

According to speleological reports, there was a significant quantity of ceramic sherds and whole vessels in the western channel. The frequency of uncontrolled visits to the cave has probably resulted in loss or material and reduced the possibility of accurate chronological determination. Without additional archaeological research, it is difficult to establish whether the site represents the sunken evidence of water exploitation since the cave acquired the anchialine form as a consequence of the postglacial sea-level rise, or if it contains submerged evidence of some different activity in the past. In any case, the cave is the most notable example of the archaeological importance of the anchialine caves in the Adriatic discovered up to the present day.

## 17.5 Analysis of the Water Column and Sediments

Some anchialine caves like Bjejajka (Fig. 17.2, no. 14) and Lenga (Fig. 17.2, no. 15) on the island of Mljet have naturally elevated concentrations of trace metals in the water and sediment (Cuculić et al. 2011; Kwokal et al. 2014). Despite the absence of anthropogenic influence, quite high concentrations of ecotoxic metals were found: Hg up to 3.7 μg/L, Cd up to 0.3 μg/L, Cu up to 28 μg/L, Zn up to 9.6 μg/L (Cuculić et al. 2011; Kwokal et al. 2014). In contrast, the water column in the anchialine cave of Urinjska špilja (Fig. 17.2, no. 1), located in the industrial area, has extremely low concentrations of ecotoxic metal (Cuculić et al. 2012).

Naturally elevated concentrations of some very toxic metals, like mercury, indicate the importance of research in the zone between the sea and the coastal aquifers, in order to identify the main biogeochemical processes that control transport and influence the fate of trace metals and associated elements and compounds.

The two mentioned cases of natural contamination of water resources suggest the possibility of a similar situation in the past, when the naturally elevated concentrations of ecotoxic metals may have had a negative influence on the ancient inhabitants who were exploiting the cave water. This is an interesting issue to be kept in mind while analyzing the evidence and interpreting the cultural context, even if hard evidence of disturbance to the local community remains elusive.

## 17.6 Archaeological Potential

Anchialine caves are precious water resources along the Croatian coast, exploited through the ages by local populations. The typical coastal karst is characterized by a limited amount of surface water, forcing the inhabitants of the region to explore alternative solutions.

The quantity of the available fresh or brackish water depends upon the morphology of the cave, the quantity of incoming fresh water, and the intensity of its exploitation. Since prehistory, the anchialine caves were utilized as natural wells, acting as complementary water reserves to cisterns, ponds and other water sources.

In addition, the anchialine caves could have been used as significant water supplies for seafarers distant from well-organized ports, who probably partly planned their routes and stops according to the availability of water. The current literature has already addressed the need for drinking water during sea voyages of the past (e.g., Pryor 2001). There is no doubt about such needs in both local and long-distance seafaring among the Eastern Adriatic islands.

The presence of amphorae—the most frequent ancient containers for sea transport of liquid goods—in Vodeni rat, Živa voda and Gravrnjača Cave on Kurba Vela is clear evidence for the exploitation of cave water over a period of more than five centuries. The presence of prehistoric pottery in Živa voda could point to the exploitation of water resources during the Bronze Age, or even earlier. On the other hand, the diversity and quantity of the prehistoric pottery recalls another interesting characteristic of the caves, i.e., their sacred function (cf. Whitehouse 1992; Kleibrink 1998). Unfortunately, in the present state of research we cannot confirm any presumption about the rituals linked with the presence of fresh water in the anchialine caves, but it remains an interesting issue to be considered in the future.

The anchialine caves could also contain 'special' water, such as water rich in sulphur that may have serve for therapeutic purposes. Two examples in Croatia are the caves of Sumporača velika ('big') and Sumporača mala ('small') near Mokošica (Fig. 17.2 no. 16) on the river Ombla, north of Dubrovnik (Cukrov et al. 2012), which contain sulphur springs. Both caves belong to the group of hydrothermal sulphuric caves (after Gottstein 2010), and have been known since ancient times (Cukrov et al. 2012).

One of the rare systematically researched areas from the point of view of the anchialine caves in Croatia is the National Park of Kornati Archipelago. The presence of Roman pottery in the unnamed cave on the island of Kurba Vela has already been mentioned above, while other caves await systematic investigation to locate and categorize potential archaeological evidence. In any case, if we compare the map of the anchialine caves with the map of archaeological sites on land and underwater, we can immediately see the overlapping of the distribution areas. This confirms the logical conclusion that human settlements in the extremely dry environment of the islands depended also on the distribution of cave water resources.

## 17.7 Conclusion

Scarce but significant evidence confirms the past exploitation of the anchialine caves along the Eastern Adriatic coast. Whether utilized as simple water resources, cave sanctuaries, or for their medicinal properties, they certainly played an important role in the everyday life of the inhabitants of the Adriatic region. The presently available finds provide only a glimpse of what is available, and emphasize their archaeological potential and the need for future study.

The lack of initial archaeological involvement in early speleological missions resulted in poor documentation of the evidence for a long human relationship with the anchialine environments. Subsequently, some previously-known archaeological finds have disappeared from their original locations, without any documentation. This situation negatively impacts our perception of the past, and requires changes in the future. Promoting the archaeological potential and importance of these particular natural phenomena, and consequently fostering multidisciplinary research, will lead to a better understanding of the role of anchialine caves in the Adriatic past. Underwater exploration could also contribute to the research by identifying those caves that were anchialine during the last Ice Age, later becoming marine through the process of postglacial sea-level change. In addition, in the modern era of measurable climatic change, characterised by ongoing global warming and changes in precipitation patterns, the fresh water layer in anchialine caves may once more become an important resource.

**Acknowledgements** Promotion of interdisciplinary research on the archaeological potential of the anchialine caves is supported by the Croatian Science Foundation, in the framework of the project AdriaS (Archaeology of Adriatic Shipbuilding and Seafaring, IP-2014-09-8211). We also thank Branko Jalžić for sharing his extensive knowledge of Croatian anchialine caves with us, and making available to us all the available documentation.

## References

Barišić T (1994) Speleološka istraživanja otoka Žirja i Kornatskih otoka. In: Kale Mrvica J, Krnčević, Ž (eds) Žirajski libar, zbornik radova o otoku Žirju, Nakladno povjerenstvo Žirajskog libra and Muzej grada Šibenika, Šibenik, pp 183–193

Cuculić V, Cukrov N, Kwokal Ž, Mlakar M (2011) Distribution of trace metals in anchialine caves of Adriatic Sea, Croatia. Estuar Coast Shelf Sci 95(1):253–263

Cuculić V, Cukrov N, Omanović D, Jalžić B (2012) Preliminary study of trace metals distribution in the water column of Urinjska Špilja anchialine cave (Croatian Adriatic coast). Nat Croat 21:28–31

Cukrov N, Cukrov M, Jalžić B, Omanović D (2008) Koncentracije ekotoksičnih metala (Cd, Pb, Cu I Zn) u vodenom stupcu špilje Živa voda na otoku Hvaru. Subterranea Croat 10:28–32

Cukrov N, Jalžić B, Bilandžija H, Cukrov M (2009) Research history and anchialine caves characteristic in Croatia. In: Anchialine ecosystems: reflection and prospects. Symposium Abstracts. Palma de Mallorca, pp 19–20

Cukrov M, Ozimec R, Cukrov N, Bedek J, Jalžić B (2012) Sumporače – preliminary results from research into unique anchialine sulphur caves in Croatia. Nat Croat 21:36–42

Falcon-Barker T (1960) 1600 years under the sea. David McKay, New York

Forenbaher S, Kaiser T (2006) The pottery of Pupićina Cave. In: Miracle P, Forenbaher S (eds) Prehistoric herders of Northern Istria: the archaeology of Pupićina Cave. Arheološki muzej Istre, Pula, pp 163–224

Garašić M (2006) Pronađeni i snimljeni najdublji speleothemi u moru. Spelaeologia Croat 7:58

Gottstein Matočec S, Jalžić B (2003) Biospeleološka istraživanja vodene faune anhihalinih špilja I jama na području NP "Kornati", fieldwork report. Hrvatski prirodoslovni muzej, Zagreb, p 42

Gottstein Matočec S (2010) Priručnik za određivanje podzemnih staništa u Hrvatskoj prema direktivi o staništima EU. Državni zavod za zaštitu prirode, Zagreb (In Croatian)

Govedarica B (1982) Prilozi kulturnoj stratigrafiji paristorijskih gradinskih naselja u jugozapadnoj Bosni. Godišnjak Cent balkanološka istraživanja Akademije nauka umjetnosti Bosne Hercegovine 18:111–188

Holthuis LB (1973) Caridean shrimps found in land-locked saltwater pools at four Indo-West Pacific localities (Sinai Peninsula, Funafuti Atoll, Maui and Hawaii Islands), with the description of one new genus and four new species. Zool Verh 128:1–48

Iliffe TM (2000) Anchialine cave ecology. In: Wilkens H, Culver DC, Humphreys WF (eds) Ecosystems of the world, 30, subterranean ecosystems. Elsevier, Amsterdam, pp 59–76

Iliffe TM (2005) Biodiversity in anchialine caves. In: Culver DC, White WB (eds) Encyclopedia of caves. Elsevier, Amsterdam, pp 24–30

Iliffe TM (2012) Diving investigations of Bermuda's deep water caves. Nat Croat 21(S1):64–67

Jalžić B (1994) Jama Gradina na otoku Žirju. In: Kale Mrvica J, Krnčević Ž (eds) Žirajski libar; Zbornik radova o otoku Žirju. Nakladno povjerenstvo Žirajskog libra and Muzej grada Šibenika, Šibenik, pp 44–47

Jalžić B, Bedek J, Jalžić V, Lukić M, Miculinić K, Ozimec R (2005) Prirodoslovna istraživanja jame Vrtare Male, fieldwork report. Hrvatsko biospeleološko društvo, Zagreb, p 30

Kleibrink M (1998) The miniature votive pottery dedicated at the 'Laghetto del Monsignore', Campoverde. Palaeohistoria 39(40):441–512

Kwokal Ž, Cukrov N, Cuculić V (2014) Natural causes of changes in marine environment: mercury speciation and distribution in anchialine caves. Estuar Coast Shelf Sci 151:10–20

Malez M, Lenardić-Fabić J (1988) New subspecies of the southern elephant (*Mammuthus meridionalis adriacus* n. ssp.) from the bottom of the Adriatic Sea (Croatia, Yugoslavia). Palaeontol Jugoslav 37:1–36

Malez M, Sliepčević A, Srdoč D (1979) Određivanje starosti metodom radioaktivnog ugljika kvartarnim naslagama na nekim lokalitetima u Dinarskom kršu. Rad JAZU 383. Razred Prirodne Znanosti 18:227–271

Marijanović B (1981) Ravlića pećina (Peć Mlini). Glas Zemaljskog Muzeja Bosne Hercegovine Sarajevu 35(36):1–55

Marović I (1985) Iskopavanje kamenih gomila u Bogomolju na otoku Hvaru. Vjesn Arheol Hist Dalmatinsku 78:5–35

Marović I (1999) Jama u Podumcima. Vjesn Arheol Hist Dalmatinsku 90–91:9–41

Mauch-Lenardić J (2012) Miocene to Late Pleistocene proboscideans of Croatia. Quat Int 276–277:120–128. doi:10.1016/j.quaint.2011.06.037

Mesić J (2006) L'esplorazone archeologica a scopo protettivo della grotto di Vodeni Rat. In: Radić Rossi I (ed) Archeologia subacquea in Croazia. Studi e ricerche, Marsilio, Venezia, pp 91–100

Novosel M, Jalžić B, Pasarić M, Požar-Domac A, Radić I (2007) Ecology of an anchialine cave in the Adriatic Sea with special reference to its thermal regime. Mar Ecol 28(Suppl. 1):3–9

Ozimec R, Jalžić B (2003) Hvarska Živa voda. Meridijani 77:22–25

Pryor JH (2001) 'Water, water everywhere, Nor any drop to drink': water supplies for the fleet of the first crusade. In: Balard M, Kedar BZ, Riley-Smith J (eds) Dei gesta per Francos: études sur les croisades dédiés à Jean Richard. Ashgate, Aldershot, pp 21–28

Radić Rossi I (2011) Problematika prapovijesnih i antičkih nalazišta u hrvatskom podmorju/Problematics of prehistoric and Classical Antiquity underwater archaeological sites in the Croatian undersea. Unpublished dissertation, Zadar, University of Zadar
Sket B (1986) Ecology of the mixohaline hypogean fauna along the Yugoslav coast. Stygologia 2:317–338
Sket B (1996) The ecology of anchihaline caves. Trends Ecol Evol 11:221–225
Sket B (2012) Anchihaline (anchialine) caves and fauna. In: White WB, Culver DC (eds) Encyclopedia of caves. Elsevier, Amsterdam, pp 17–24
Stock JHT, Iliffe M, Williams D (1986) The concept "anchialine" reconsidered. Stygologia 2(1/2):90–92
Surić M (2009) Rekonstruiranje promjena morske razine na istočnoj obali Jadrana (Hrvatska) – pregled/Reconstructing sea-level changes on the Eastern Adriatic Sea (Croatia) – an overview. Geoadria 14(2):181–199
Surić M, Juračić M (2010) Late Pleistocene-Holocene environmental changes – records from submerged speleothems along the Eastern Adriatic coast (Croatia). Geol Croat 63(2):155–169
Surić M, Juračić M, Horvatinčić N, Krajcar Bronić I (2005a) Late Pleistocene-Holocene sea-level rise and the pattern of coastal karst inundation – records from submerged speleothems along the Eastern Adriatic Coast (Croatia). Mar Geol 214:163–175. doi:10.1016/j.margeo.2004.10.030
Surić M, Horvatinčić N, Suckow A, Juračić M, Barešić J (2005b) Isotope records in submarine speleothems from the Adriatic coast, Croatia. Bull Soc Géol France 176(4):363–373. doi:10.2113/176.4.363
Surić M, Richards DA, Hoffmann DL, Tibljaš D, Juračić M (2009) Sea-level change during MIS 5a based on submerged speleothems from the eastern Adriatic Sea (Croatia). Mar Geol 262:62–67
Surić M, Lončarić R, Lončar R (2010) Submerged caves of Croatia: distribution, classification and origin. Environ Earth Sci 62:1473–1480. doi:10.1007/s12665-010-0463-0
Vujnović N (1990) Prilozi arheološkoj karti otoka Hvara. Vjesn Arheol Hist Dalmatinsku 83:47–64
Vujnović N (2013) Kratki vodič kroz povijest općine Sućuraj. Janus Tours, Split
Whitehouse RD (1992) Underground religion: cult and culture in prehistoric Italy. Accordia Research Centre, University of London, London

# Part IV
# Landscapes of the Continental Shelf and Human Dispersals

# Chapter 18
# The Role of the Submerged Prehistoric Landscape in Ground-Truthing Models of Human Dispersal During the Last Half Million Years

**Nicholas C. Flemming**

**Abstract** Human genome analysis and research into fossil anthropogenic nuclear DNA and mitochondrial DNA are providing many new insights into hominin diffusion and migration over the past half million years. The beginning and end data on migration routes frequently imply that the migration involved crossing a present sea-channel or marginal basin, or migrating along the present continental shelf. However, there are very few attempts to correlate the models based on DNA with in situ archaeological and palaeoenvironmental data from the continental shelf or shelf marginal seas. Yet a significant number of sites are available for such correlation. Over 3000 submerged prehistoric archaeological sites on the continental shelf are known worldwide, varying in depth from the near-shore to about −100 m and ranging in age from 5000 years to >0.5 million years. Sites have been found off the coast of every continent except Antarctica. Most of the sites found so far are shallower than 10–20 m, with a few deeper than 40 m, and none are in the tropics. The submerged sites found so far exist in a very wide range of taphonomic conditions and climatic zones, confirming that sites could be found to provide empirical tests of the many different proposed migration routes. The principal exception is that no sites have yet been found between the Tropics of Cancer and Capricorn and the so-called Southern Route cannot yet be checked until the submerged landscape has been mapped in sufficient detail indicating where sites might survive and be identified. In all other geographic regions it is recommended that DNA models and seabed data are examined for consistency and mutual benefit. Further work is needed to identify submerged sites and landscapes in the tropics.

## 18.1 Introduction

Analysis of modern human DNA and fossil hominin DNA has recently provided an innovative basis for a number of plausible new models of global hominin dispersal from Africa (e.g., Barbujani and Betorelle 2001; Forster 2004; Soares et al. 2008; Oppenheimer 2009; Skogland et al. 2015; Fu et al. 2016). Recent advances in the analysis of ancient DNA show that valid genetic links can be established between deposits as old as 400 ka, allowing comparisons between anatomically modern humans (AMH), Neanderthals and Siberian Denisovans, revealing possibly new and previously unknown variants as well (Green et al. 2010; Fu et al. 2016; Meyer et al. 2016). However, earlier dispersals cannot yet be studied genetically by the analysis of fossils. Moreover, the DNA chronology derived from the

N.C. Flemming (✉)
National Oceanography Centre, European Way, Southampton SO14 3ZH, UK

Sheets Heath, Benwell Rd, Brookwood, Woking GU24 0EN, UK
e-mail: n.flemming@sheetsheath.co.uk



G.N. Bailey et al. (eds.), *Under the Sea: Archaeology and Palaeolandscapes of the Continental Shelf*, Coastal Research Library 20, DOI 10.1007/978-3-319-53160-1_18

analysis of modern variants and complex genetic indicators is also uncertain, and has been revised at times by significant factors (Scally and Durbin 2012). The genomic analysis of anthropogenic DNA plus a date for the originating material provides evidence that a variant of *Homo* arrived at a location by a certain date, with possible correlating evidence that it also existed at another location at a previous date. The two pieces of evidence indicate a movement of genetically similar peoples over multiple generations between the two locations, but it provides no data on the route taken, nor on the speed over different sectors of the possible routes, and dates that are open to dispute (Boivin et al. 2013). Average rates of transfer over long distances are typically 0.1 km per year, with extreme rates of 0.5–1.0 km per year, on average (Pope and Terrell 2008). The question also arises as to the earliest dates at which crossings or coastwise transport may have been undertaken by seacraft of some kind. For a global overview of the origins and development of prehistoric seafaring, which is closely interwoven with questions regarding occupation of the continental shelf, see Anderson et al. (2010) and for the eastern Mediterranean, Ammerman and Davis (2013–2014).

The slow speed of progress between end points implies that 'migrating' tribes were stationary for most of the interval between two known dated occurrences, or that the seasonal hunting-foraging zone was extending spatially in a consistent direction by a few hundred metres each year. Thus the terrain must have provided subsistence, either terrestrial or marine, on a generational timescale at all points on the route, although it is feasible that gaps of 10–20 km could be covered quite quickly on occasions.

In this chapter the word 'migration' will be used for a purposive relocation of a mass of people such that they move relatively rapidly in a single event from their previous area of occupation, and re-establish themselves in a new territory. Climate change, rapid rise of sea level, desertification, a previously under-exploited food source, or over-population might be causes of such events. The term 'migration' will also be used when discussing other publications or models in which the concept or the term is applied. The words 'dispersal' or 'diffusion' will be used when there is no teleological objective to reach a new destination, and a population extends its area of hunting or foraging in a way that gradually extends its range of territory.

The major submerged land areas and sea crossings relevant to global hominin dispersal are the Red Sea, the Mediterranean-Europe-North Sea, the Southern-Route—SE Asia–Australasia—and the Beringia-Aleutian Arc. For the first three of these there have been widely separated multiple dispersals, and in all cases except Australasia there are alternative potential dispersal pathways via land routes that do not depend on sea crossings or the availability of now-submerged coastal regions. Moreover, this terrestrial bias is reinforced by the still-widespread assumption that prehistoric archaeological remains on the continental shelf have been destroyed by marine transgression (e.g., Klein 1999, p. 566; Mithen 2003, p. 24; Apenzeller 2012, pp. 24–26). This assumption encourages the proposal of hypothetical shelf-wise migration routes, transit zones, or holding areas, for which it is assumed that archaeological ground-truthing or countervailing in situ cultural data will never be produced. For example, the uncritical promotion of the coastal rim of the Indian Ocean as a quick route to Australia (Mellars et al. 2013) assumes an unjustifiable uniformity and availability of a productive exposed shelf (cf. Groucutt et al. 2015). But the use of such models with broad sweeping arrows over long distances should not be acceptable without subsequent intermediate ground-truthing. The premise of this chapter is that the survival of known submerged archaeological material and features of the landscape globally on the continental shelf indicates that there are potentially sufficient data to verify or reject some hypotheses of postulated dispersal patterns based so far only on start and end points derived from DNA data.

Submerged prehistoric sites in primary or secondary condition have been found and studied in a wide range of coastal types off all inhabited continents (e.g., Stright 1990; Benjamin et al. 2011; Faught and Gusick 2011; Faught 2014), and in both the northern and southern hemispheres (Werz and Flemming 2001, Cartajena et al. 2011, Carabias et al. 2014). This paper summarises the taphonomic evidence for the correlation between offshore conditions and prehistoric site survival with examples, and relates discoveries to the regional questions regarding migration. A fuller analysis of taphonomic conditions for the European-Mediterranean region is in Flemming et al. (2017) in press.

## 18.2 Definition of the Problem

Submerged prehistoric sites have been discovered, mapped, excavated, and interpreted for over 100 years. Nevertheless, apart from the southern Baltic (e.g., Fischer 1995, 2007, Curry 2006, Harff et al. 2011, Holmlund et al. Chap. 4; Uldum et al. Chap. 5) and the Mediterranean coast of Israel (e.g., Galili et al. 2004, Chaps. 6 and 7), the number of seabed areas containing numerous researched sites forming a coherent, regional cultural assemblage is limited. Other chapters in this book provide the historical record of the many site discoveries and analyses in European seas. It is clear that the steady accumulation of data on submerged sites can build up a complex archaeological and palaeoanthropological picture that complements DNA-based models. The North Sea and English Channel are beginning to reveal data of sufficient density to consider such an approach (Cohen et al. 2014; Glorstad et al. Chap. 19; Gaffney et al. Chap. 20; Momber and Peters, Chap. 21). Both Gaffney et al. (2017) and Momber and Peeters (2017) are already using DNA analysis of sediments and their organic contents.

Progress in discovery and interpretation of submerged prehistoric sites is inevitably slow, although the technology is improving (e.g., Missiaen 2010; Grøn and Boldreel 2013; Grøn et al. 2013; Missiaen et al. Chap. 2). By comparison, research into palaeoanthropological DNA has been rapid and successful, with continuous technical improvements (e.g., Forster 2004; Green et al. 2010; Meyer et al. 2016). On the continental land masses, human genome research can be cross-calibrated with countless hundreds of archaeological records built up over nearly 150 years of professional research. For example, Skogland et al. (2015) use globally widespread genetic data to propose two founding populations crossing Beringia into the Americas. Another example is provided by Faught (2008), in which inward migration models to the Americas are compared with a wide range of archaeological sites including 63 stratigraphic locations where two or more reliable radiocarbon ages show human occupation earlier than 10,500 BP. A study of radiocarbon dates for over 5000 archaeological sites in South America but as yet without comprehensive DNA data (Goldberg et al. 2016) illustrates the potential in that area. By comparison, the offshore, archaeological data set is very sparse. There is thus a misfit between the two methodologies and the culture of the two different research communities, one with access to multiple datasets but without reference to the constraints or opportunities afforded by now-submerged landscapes subjected to variation with changes in sea level, the other focussing on the offshore record but constrained by a much more limited dataset. This is definitely not to suggest that one method or the other is actually misguided, or wrong, or produces misleading data. Research progresses by successive approximations to an idealised truth, and both approaches are generating progress. My intention is, rather, to suggest ways in which the two methodologies can best complement and reinforce each other, and benefit from the insights obtained in the other's work. Each methodology may suggest hypotheses that can be checked by the other, or be liable to weaknesses or gaps that can be resolved by the other.

## 18.3 Structure of the Argument

The use of anthropogenic DNA analysis in complex terrain of peninsulas and islands involving areas which are at present partially inundated by sea water has been progressively refined to suggest increasingly high-resolution models in time and space (e.g., Cavalli-Sforza and Minch 1997; Forster 2004; Hill et al. 2007; Soares et al. 2008; Henn et al. 2012; Fu et al. 2016). It should be noted that this selection from the literature on human genetics and migration models is only intended to give examples, and there is no attempt here to establish the validity of one thesis or another. Bailey (2009) has drawn attention to the fluidity and variability in models as different genetic indicators are used and techniques improved. As models are refined, and as they postulate more specific uses of the continental shelf, either as a route for rapid transit, to Australia for example (Oppenheimer 2009), or as an area of enduring residence such as Beringia (e.g., Tamm et al. 2007; Dixon and Monteleone 2014), or the Celtic-Sea and North Sea (McEvoy et al. 2004; Gaffney et al. 2017, Chap. 20), the need for more detailed bench-marks provided by archaeological evidence from the seabed becomes apparent. In some cases multiple alternative hypothetical models co-exist without apparent prospect of resolution in the near future. In most cases in the northern hemisphere, the alternative possibilities include diffusion through terrestrial landscapes and across land bridges rather than along the now submerged coast, so that the preference for terrestrial rather than coastal pathways of dispersal rests largely on a priori preferences for one or the other in default of any evidence from the submerged landscape that might provide confirmation or refutation.

If we propose as a global generalisation that models are to be checked by the study of occupied sites and tool assemblages on the submerged continental shelf, there must be some confidence that the data can be recovered, and that archaeological remains can in fact survive in a wide range of different geological and oceanographic conditions. It is not sufficient if submerged archaeological sites can only be found in one or two ideal ecological environments. Conversely, where the archaeological data from the continental shelf consist of one or two excavated sites in thousands of square kilometres devoid of other data, careful reference to models based on human genetics may greatly increase the value of the information and help to focus offshore research on the potentially most significant areas worth targeting in the search for new underwater finds.

It is therefore necessary to show that (1) submerged prehistoric sites can and do survive in different conditions, and (2) the conditions in which submerged sites have so far been found and studied are relevant to the conditions on key migration routes.

## 18.4 Survival of Submerged Prehistoric Sites in Different Environmental Conditions

The overall case for studying submerged prehistoric sites and the broad global assessment of the conditions of survival are provided by Bailey and Flemming (2008). During the work of the SPLASHCOS Action (TD0902), Working Group 2 concentrated on analysis of the conditions in different European regional seas which would favour or restrict the chances of survival of prehistoric archaeological deposits during and after inundation. The full results of this analysis are presented elsewhere (Flemming et al. in press), where the regional reports provided by a large number of researchers describe the geomorphological and oceanographic circumstances of different sea areas, and the taphonomic factors determining survival of underwater prehistoric sites on a wide variety of European coasts.

Viewed globally, the question is whether the range of coastal types shown to preserve submerged prehistoric data in European waters provides a basis for stating that all similar coastal types elsewhere

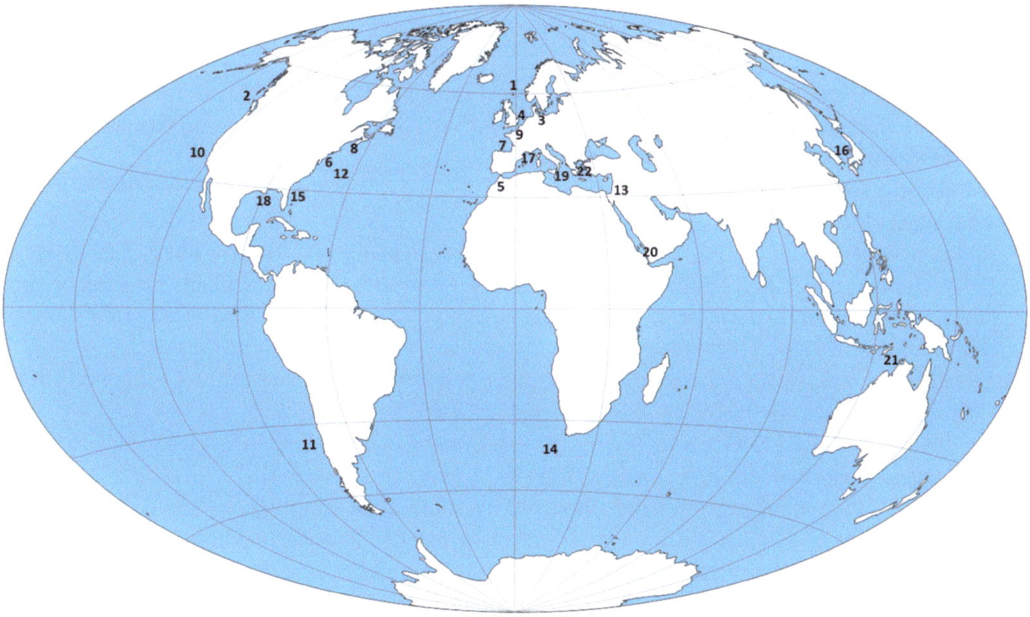


**Fig. 18.1** A selected sample of sites with submerged prehistoric materials, or submerged terrestrial landscapes that have been well mapped to locate sites: *1* Orkney coast and Viking Bank, *2* British Columbia, *3* South West Baltic (over 1500 sites), *4* North Sea (many tens of sites and thousands of paleontological retrievals), *5* Gorham's Cave, Gibraltar, *6* Chesapeake Bay, *7* Morbihan Bay, France, *8* Maine coast (Native American lithics offshore), *9* Fermanville, France, *10* California, *11* Quintero Bay, Chile, *12* Delaware offshore ('Cinmar' lithic), *13* Israeli Mediterranean coast (submerged Mesolithic villages), *14* Table Bay (Acheulean hand axes), *15* Cape Canaveral beach (Paleoindian site), *16* Tokonami River Bay (Jomon site), *17* Grotte Cosquer (cave paintings), *18* Florida coast, Gulf of Mexico, *19* Italy and Sicily submerged caves and Bronze Age tombs, *20* Farasan Islands, Red Sea, *21* Cootamundra Shoals, Australian shelf, *22* Aegean (Agios Petros and Pavlopetri)

can preserve sites. However, there are no known sites between the Tropic of Capricorn and the Tropic of Cancer (see Fig. 18.1), and no equivalent data from the European coast provides examples of comparable tropical coastal types. Also, it has so far been impossible to locate and excavate submerged prehistoric sites on coasts that are heavily coralline, populated with mangrove swamps or on massive tropical deltas. The task in this context is to identify sections of tropical coast that are free of modern reef coral growth. Hence, the evaluation of where sites may be preserved needs to take account of a wider range of coastal environments than those exhibited in Europe.

The following section is based on a selection of examples of known prehistoric sites from around the world, accompanied by a very brief summary of the conditions in which they have survived. Every site mentioned contains substantial prehistoric remains which have been published in the archaeological literature, except for the Farasan Islands (20) and Cootamundra Shoals (21), which have revealed numerous favourable seabed indicators but no submerged artefacts. Some of the general criteria for survival in mid-latitudes were summarised by Flemming (2004a). Here I will describe very briefly different coastal geomorphological and oceanographic conditions, starting at the Polar Regions, and working towards the equator, with examples from both the northern and southern hemispheres where available. The lists below are selective, not exhaustive and numbers in brackets refer to sites on Fig. 18.1.

### *18.4.1 Different Environments*

**Circum-Polar, High Temperate Latitudes** Orkney (1), in a sheltered loch with limited fetch, deep water exposed (Long et al. 1986), Deep water (50 m) in a sheltered archipelago, British Columbia (2) (Josenhans et al. 1997; Fedje and Josenhans 2000).

**Mid-Latitude Temperate Semi-Enclosed Seas** Baltic (3) archipelago, low tidal amplitude, limited wind fetch (Fischer 2007, 2011), North Sea (4) intermediate depth, protection from ocean swell, strong tides, artefacts embedded in surface sediments (Russell and Tizzard 2011; Tizzard et al. 2014; Hublin et al. 2009; Glimmerveen et al. 2004; Glorstad et al. 2017, Chap. 19; Momber and Peeters 2017, Chap. 21). Solent, narrow sea channel protected by large island, strong tidal currents (Momber et al. 2011).

**Mid-Latitudes, Ria, Natural Harbours, Inlets, Limited Fetch** Chesapeake Bay (6) (Blanton 1996). Golfe de Morbihan, er Lannic (7) (Gouezin and le Gall 1992).

**Mid-Latitude, Exposed to Oceanic Storms, Rocky Coast** Coast of Maine, USA (8) some protection from islands and boulders (Crock et al. 1993; Price and Spiess 2007; Kelley et al. 2013). English Channel (9) Fermanville, 20 m depth sheltered by submerged granite reef (Scuvée and Verague 1988; Cliquet et al. 2011). California (10) artefacts embedded in surface sediments or protected by kelp forest (Masters 1983). Exposed Pacific coast (11) Chile, 1 km offshore, low gradient (Cartajena et al. 2011).

**Mid-Latitude Deep Water Oceanic** Offshore Delaware (12) 80 m depth, open ocean, protected by depth now, protected by dune and lagoon features during inundation (Stanford et al. 2014, see also Long et al. 1986, for deep water preservation).

**Mid-Latitude Mediterranean/Florida Climate, Exposed Ocean Coast, Sandy Low Gradient** Aegean rocky coasts (Sakellariou and Galanidou 2017, Chap. 22). Israel (13) (Galili et al. 2004), and Table Bay (14) (Werz and Flemming 2001). The early date for site (14) shows survival through multiple glacial cycles of sea level change. Florida (15) (Cockrell and Murphy 1978). Sites on sandy beaches of very low gradient are protected by the constructive action of the rolling surf.

**Mid-Latitude, Sheltered Estuary, Semi-Enclosed Sea** Tokonami River Japan (16), 9000 year-old Jomon culture site at 25 m depth (Hayashida 1993), limited fetch, low tidal amplitude.

**Karstic, Submerged Caves** Gibraltar (5) Gorham's Cave on waterline. France (17), Grotte Cosquer (Clottes and Courtin 1994; Clottes et al. 2005). Florida sinkholes (18), Little Salt Springs (Clausen et al. 1979), Warm Mineral Springs (Cockrell 1975). Italy (19) (Antonioli et al. 1997).

**Tropics, Closed Sea Or Archipelago** Farasan Islands (20), complex archipelago restricts fetch, low tidal amplitude, low sedimentation, deep sink-holes and submerged drainage, salt karst (Bailey 2009; Bailey et al. 2012, 2015, 2017, Chap. 23, this volume; Lambeck et al. 2011). North west Australian coast (Ward and Veth 2017, Chap. 24).

The above list is by no means complete, and omits many sites where terrestrial Pleistocene fauna have been dredged up by fishermen, but which lack human indicators so far. Nevertheless it illustrates the great range of conditions, coastal types, and water depths which can preserve prehistoric deposits both during inundation and after complete submergence. In the next section I will examine whether these sites, or potential discovery of similar sites, can be correlated with human genetic models of migration.

## 18.5 Are the Seabed Conditions and Sites Researched so far Relevant to the Major Migration Routes?

### 18.5.1 Red Sea Crossing, Start of the Southern Route

Bailey et al. (2017, Chap. 23) describe recent work on the Farasan Islands, and summarise the literature on the southern Red Sea as a potential crossing zone. Both the Dahlak Islands on the west side, and the Farasan Islands on the eastern, provide complex indented coastlines where prehistoric sites could survive underwater. Discussion of various possible phases of migration or dispersal across the Red Sea is provided by Beyin (2011, 2013), while Petraglia (2011) considers possible routes across Arabia. The most probable Red Sea crossing point is just south of the Hannish Islands (Lambeck et al.: Figs. 19–21), but the political situation makes it difficult to work near these islands at present. It must be stressed that up to date (2016) there is not absolute proof that any crossings of any date occurred in the southern Red Sea, either at Bab el Mandab, or the Hannish Islands, or in the potentially large submerged living area further north surrounding the Farasan islands. Work is continuing in this area, and the latest research on the submerged terrestrial environment at the LGM is encouraging. To the east of Arabia itself the crossing of the Straits of Hormuz is probable at low sea levels and much of the Gulf would be dry land traversed by large rivers.

In summary, seabed research can provide definitively useful data for the southern Red Sea crossing itself, and strongly useful data for a postulated coastal route as far as the Straits of Hormuz and across those straits.

### 18.5.2 Europe and Crossings of the Mediterranean Marginal Basins

There have been multiple diffusions or migrations from Africa into and across Europe during the last 1 million years, with sites older than or approximating 1 million years in Georgia, northern Spain, and on the British North Sea coast (Parfitt et al. 2005). Seabed Neanderthal remains discussed by Hublin et al. (2009) and the submerged North Sea A240 site, with a date of 340 ka (Russell and Tizzard 2011; Tizzard et al. 2014), show that seabed archaeological data can correlate with migration phases before modern humans. The potential sea crossings where seabed data can provide bench-marking of migration models include the Strait of Gibraltar, the Sicily Channel (Antonioli et al. 2016), the Levant coast, crossing the Aegean (Runnels et al. 2014; Runnels and Howitt-Marshall 2017; Sakellariou and Galanidou, Chap. 22), the continental shelf of the Black Sea, crossing the central Adriatic, the Golfe de Lions, the Celtic Sea-English Channel, the North Sea (Flemming 2004a; Glorstad et al. Chap. 19; Gaffney et al. Chap. 20; Momber and Peters, Chap. 21) and the Baltic (Holmlund et al. Chap. 4; Larsson, Chap. 11; Hansson et al. Chap. 13). In every case, prehistoric sites have already been found offshore, or research has already started. There are strong reasons for the seabed archaeologists to assess DNA-based models and integrate these with their research. The SPLASHCOS database (http://splashcos.maris2.nl/) provides information which could be used in multidisciplinary projects of this kind.

### 18.5.3 Southern Route: Red Sea, Iran, India, SE Asia, Australia

In spite of considerable work on the coast of Australia (Flemming 1986; Nutley 2014; Ward and Veth, Chap. 24) no definitive sites containing prehistoric material older than 5000 years have so far been found offshore. The Tokonami River site (Hayashida 1993) illustrates the survival of AMH submerged prehistoric material off the coast of Japan, while the discovery by Chang et al. (2015) of an archaic hominin jaw bone in the Taiwan Straits at a depth of about 60 m is the first clear example of pre-AMH seabed prehistoric deposits in East Asia or SE Asia. The migration of modern humans into East Asia based on Y-chromosome analysis is reviewed by Su et al. (1999).

The possible crossing of the southern Red Sea has already been discussed above. The bulk of the so-called Southern Route, from the Straits of Hormuz to East Asia and Australia is where a sense of caution begins to raise serious questions, even before proposing that verification from offshore data is required (Erlandson and Braje 2015). The assumed route has been proposed and accepted so often, with so many broad coloured arrows on small maps, that it would be invidious to select one author or another to quote as illustrating its wide acceptance. We should also stress again that the word 'migration' is misleading with its modern connotation of goal-seeking purpose, since the rates of progression were of the order of 0.1–1 km/year, which shows that the process probably consisted of a gradual extension of hunting and foraging areas, rather than a purposeful intention of ambitious exploration.

The problems and uncertainties inherent in an over-simple acceptance of the Southern Route hypothesis are examined by Pope and Terrell (2008). A combination of phases of rapid rising and falling of sea level during MIS 4–3, the need to cross the deltas of some of the largest rivers in the world, variations of terrain caused by exposed coral reefs, mangrove forests, and deltaic swamps, all imply a complex and often inhospitable coastal zone, and this is likely to have been the case even if the conditions visible at the present-day coastline were different during periods of lowered sea level. For example, modern deltas are associated with complex braided channels and muddy islands with lateral extent of hundreds of km, and these may not have existed at intermediate sea levels. Pope and Terrell (2008) compare the rate of movement of the agricultural modern human frontier from the Middle East across Europe, which was about 0.3–0.4 km/year, with that necessary to account for human migration from the Red Sea to North Australia, which is more than 1 km/year. This apparent high rate of progress is even more curious when one considers that the coastline is by definition linear, with a narrow front of advance, while the European frontier was much broader. An obstacle encountered on land is likely to be by-passed through another line of advance. On the coast every obstacle must either be overcome on the spot, or a major diversion is needed. Any failure is total failure. In addition, the coast route requires lateral excursions around India, Sri-Lanka, Myanmar, and Malaysia, and is longer than the continental route across northern or central India.

The manner of calculating the rate of advance along the coast is not clear in the sense that defining the length of the coast is not a simple matter, since it is fractal. If the people are travelling on foot, even the smallest coast indentations, river mouths, deltas, or swamps would require a deviation, and often a choice between major deviations or crossing rivers by swimming or with boats. The real distance travelled may have been much more than that measured on easily available modern maps.

It is frequently suggested that the Southern Route migration achieved such relatively great speed because people used boats. This hypothesis is attractive, since it makes the first crossing of the southern Red Sea more plausible, and is in any case necessary to account for the final transit from the Sunda shelf to Australia by 48 ka BP (see also Bailey et al. Chap. 23). Again, hypotheses and models abound, with no suggested way to narrow the options and demonstrate the accuracy of one model as against another.

In 2004, the present author was invited to write a paper on the prehistoric archaeology of the Indian continental shelf (Flemming 2004b). In the circumstances the only thing to say was that it is an important area of research with potential to resolve many questions. In addition to the Southern Route itself, *sensu stricto*, there are many related problems and theories about migrations through SE Asia, and between SE Asia, China, Japan and the Philippines (e.g., Su et al. 1999). In every case, there is no reference so far to the potential for improving the accuracy of models with seabed archaeological data.

Although areas of active modern coral reef growth make the study of submerged Pleistocene landscapes impossible, there are many areas in the tropics where, for natural reasons, no modern reef building has taken place since the LGM. An extensive example is the northern shelf of Australia studied by Flemming (1986) in the Cootamundra Shoals project. Equivalent coral-free areas could be identified throughout the continental shelf areas proposed for the Southern Route. Analysis of data from oil and gas research off North Australia, and from the offshore mining work of BNP Billiton on the Sunda Shelf would help to focus ideas on submerged river valleys, buried coastlines, and other features where submerged archaeological sites could be found.

The previous discussion has considered only AMH. Earlier crossings of some channels are possible, and earlier occupation of the Indonesian and the Sunda Shelf is almost certain. The recovery of the archaic *Homo* jaw bone from the Taiwan Strait (Chang et al. 2015) illustrates the potential for E and SE Asia. Recent discoveries on the island of Flores in the Indonesian arc (Van den Berg et al. 2016; Brumm et al. 2016) show that the diminutive species known as *Homo floriensis* was already on the island, and already exhibiting its characteristic anatomy about 700 ka. This emphasises again that multiple species could have coexisted and may have occupied different climatic, geographical or environmental niches. Thus underwater research has the potential to provide data on previous migrations. Furthermore, the anomalous problems regarding the apparent rapid rate of progress to account for an AMH southern route to Australia do not mean that the south Asian coastline was not occupied by earlier peoples and cultures in a more static mode.

### 18.5.4 Beringia

The prehistoric transfer from Siberia to Alaska via the Bering land bridge has long been identified, but the mechanism, detailed structure and timing are difficult to define. Research shows that the process was rather complex, largely because southern Alaska and the Alaska Peninsula were heavily glaciated, while the exposed Beringia continental shelf and northern Alaska were not glaciated at the LGM (e.g., Hetherington et al. 2007). The exposed additional land area was 2.5 million $km^2$. In spite of the poor vegetation and cold climate of the exposed shelf, populations from Siberia diffused across to Alaska, although their passage south was then obstructed by the ice cap. Beringia thus served as a holding area or buffer zone from which people subsequently migrated in both directions as the climate improved and the sea rose.

The remoteness of the area and the severe weather and sea conditions have hampered seabed research and increased research costs. Dixon and Monteleone (2014) consider that both prehistoric terrestrial migration routes and coastal sea-craft routes are likely to have been used, and there is no evidence at present to exclude either model. The role of kayak canoes has been suggested to allow rapid progress along the southern margins of Beringia. Erlandson et al. (2015) have examined how changes in climate and sea level may have facilitated rapid diffusion southwards along the Pacific coast of North America, especially by peoples skilled in seafaring techniques. The coastal route southwards through the Canadian archipelago would have been viable by about 16,000 cal BP, that is several thousand years before a corridor between the inland ice caps (Dixon and Monteleone 2014, p 98).

The evidence for the exact position of the coastline, and the possible vegetation or fauna on the submerged lands, is now underwater, and the conditions for search, discovery, and examination of terrestrial landscape features and archaeological remains are adverse. Studies of local and regional relative sea level curves indicate the location of ancient shorelines, and have recovered a small number of mammoth tusk fragments and indicators of terrestrial fauna. Josenhans et al. (1997) and Fedje and Josenhans (2000) describe surveys on the British Columbia coast during which they recovered a retouched basalt blade-like flake from a depth of 53 m during mapping and sampling of the submerged landscape in Werner Bay.

Notwithstanding the steady increase of detail provided by human genetic data describing the earliest migration into the Americas, and the incidence of occurrence of archaeological sites extending southwards, there is still a great deal of uncertainty as to the routes taken, and the dates of different crossings. Seabed data has a great potential to provide answers to some of these questions, both in the Beringia area, and along the whole Pacific coast of the Americas.

## 18.6 What Next?

The simple linear model of human evolution and migration consisting of a central trunk with a few branches leading to extinct sub-species, and one main line leading distinctly and cleanly to modern humans began to crumble already in the 1980s, and has only been used as a reference benchmark against which to make exactly that negative point for the last few decades. Both archaeological excavations and fossil genome studies are revealing increasingly more co-existing variants and subspecies, and indicating others that may have existed but have not yet been found as fossils. There is increasing evidence for inter-variant breeding, and counter-current migrations. Given our perspective, where we look back at these events and tend to regard 5000 years as a short interval, it is scarcely surprising that the real picture is much more complex than we first expected. This being the case, the omission of the archaeological data from the continental shelf can no longer be regarded as a minor and marginal detail either figuratively or literally. If we are going to postulate events that may or may not have occurred on the continental shelf, the details matter. In order to establish these details, the researchers studying the human genome and its prehistoric variations need to work more frequently and collaboratively with the archaeologists studying and excavating seabed prehistoric sites.

The intensity of both chance discovery and funded seabed research on submerged prehistoric sites tends to correlate with the standard of living of the adjacent coastal state, population density on the coast, the development of technically advanced fishing gear and dredging, and the development of recreational scuba diving. Although the coastal states that border the so-called ‘Southern Route’ east of the Straits of Hormuz have high populations, all the other factors are at present missing. Unfortunately, many of the same positive factors are also absent from the Beringia area.

In summary, the study of the Mediterranean area and its sub-basins, and the study of the north-west European marginal seas are well advanced, and the evidence from the seabed is becoming more detailed. The southern Red Sea and the Gulf of Hormuz are being examined systematically, but the Southern Route is still devoid of offshore data, and is problematic for several reasons referred to above. Australia is being developed steadily, but there is a severe absence of offshore data in the area of Indonesia and the Philippines. Data are being gathered from Beringia, but conditions are extremely adverse, while offshore data from the eastern and western coasts of the Americas are increasing steadily, and there are new research investments, especially on the west coast.

## 18.7 Conclusion

The study of dispersal of AMH and pre-AMH hominins out of Africa and into all continents has been greatly facilitated by the analysis of human genetic data, both modern and fossil. The technique is essential if we are to make further progress. However, genetic data provides only a geographical start point and an end point for the movement of a genetic trait, and the dates are not always certain or accurate. Many models and hypotheses propose that migrating populations crossed sea channels on the continental shelf, or migrated along the shelf, or dwelt on the shelf zone as a buffer before subsequent movements, but provide no evidence in situ on the seabed.

The anthropogenic DNA data alone leaves uncertainty about the route or routes taken, the speed of movement on different segments of the route, and the duration of residence in buffer zones. Only archaeological data from the continental shelf can confirm the precise routes, provide correlation with dates and directions of movement, speed of movement, and even confirm that a particular route was used or not. Absence of evidence is not evidence of absence, but if in situ archaeological data are recovered routinely from some areas, and not from others, even after persistent searches, negative conclusions are eventually valid.

At present the discovery and analysis of seabed data has provided detailed information on continental shelf populations in the Baltic and on the coast of Israel with a time span of about 8–10 kyr. The volume and range of data from the North Sea and English Channel is beginning to provide sufficient information on the whole of the last rise of sea level since the LGM 20 ka, and some data on the whole of the last glacial cycle. The possibility of obtaining reliable DNA signals about the presence of plants and animals that lived on the submerged shelf is opening up a new range of research possibilities (Gaffney et al. Chap. 20), and the availability of rapid and cheaper analysis of genetic data from organic sediments and bones should be exploited at as many sites as possible in the North Sea and Channel.

The number of submerged sites found so far in the wider Mediterranean basins is limited because of the predominantly steep and rocky shores, the low sediment input, and the absence of research on the North African Coast. The Adriatic is an exception to this generalisation. Nevertheless, sites are being found, and will eventually help to define the crossing routes. The southern Red Sea crossing zone is exceptionally important because of its potential effect on the subsequent eastward migration routes to Asia. Research to date has revealed fascinating details of the submerged Pleistocene landscape around the Farasan Islands, which was habitable with large drainage channels and lakes. This research should continue, and be correlated with genetic data in East Africa and Arabia or southern Asia.

Very little if any research has been done offshore at any point between Hormuz and north Australia. From the experience in Australia there should be extensive zones along the southern route that are not overgrown with active modern coral reefs. Mapping of the submerged landscapes using acoustic technology would reveal the most probable areas where the Pleistocene landscape is exposed most completely. This is an essential precursor to any search for prehistoric artefacts offshore.

Active seabed research is growing in Australia, East Asia, and Beringia, though progress has been slow for a variety of reasons. The biggest total gap with the greatest interest is that of the Indonesian continental shelf. Survey data have been available from commercial activities for many decades, and these should be examined to recreate the Pleistocene landscape with as much detail as possible.

Joint meetings or discussion groups linking human genome experts with continental shelf prehistorians might identify specific projects or correlations that would be most helpful in advancing this research topic.

## References

Ammerman AJ, Davis T (eds) (2013–2014) Island archaeology and the origins of seafaring in the eastern Mediterranean. Proceedings of the Wenner Gren workshop held at Reggio Calabria on October 19–21, 2012. Eurasian Prehistory 10–11

Anderson A, Barrett JH, Boyle KV (eds) (2010) The global origins and development of seafaring. McDonald Institute for Archaeological Research, Cambridge

Antonioli F, Ferranti L, Lo Schiavo F (1997) The submerged Neolithic burials of the Grotta Verde at Capo Caccia (Sardinia, Italy): implication for the Holocene sea-level rise. Mem Descrittive Serv Geol Nazionale 52:329–336

Antonioli F, Lo Presti V, Morticelli MG, Bonfiglio L, Mannino MA, Palombo MR, Sannino G, Ferranti L, Furlani S, Lambeck K, Canese S, Catalano R, Chiocci FL, Mangano G, Scicchitano G, Tonielli R (2016) Timing of the emer-

gence of the Europe–Sicily bridge (40–17 cal ka BP) and its implications for the spread of modern humans. In: Harff J, Bailey G, Lueth F (eds) Geology and archaeology: submerged landscapes of the continental shelf, Special Publications, 411. Geological Society, London, pp 111–144. First published online September 11, 2014, http://doi.org/10.1144/SP411.1
Apenzeller T (2012) Eastern Odyssey. Nature 485(7536):24–26
Bailey G (2009) The Red sea, coastal landscapes and hominin dispersals. In: Petraglia M, Rose J (eds) The evolution of human populations in Arabia. Springer, Dordrecht, pp 15–37
Bailey GN, Flemming NC (2008) Archaeology of the continental shelf: marine resources, submerged landscapes and underwater archaeology. Quat Sci Rev 27(23–24):2153–2165
Bailey GN, King GCP, Devès M, Hausmann N, Inglis R, Laurie E, Meredith-Williams M, Momber G, Winder I, Alsharekh A, Sakellariou D (2012) DISPERSE: dynamic landscapes, coastal environments and human dispersals. Antiquity 86 (334). http://antiquity.ac.uk/projgall/bailey334/
Bailey GN, Devès MH, Inglis RH, Meredith-Williams MG, Momber G, Sakellariou D, Sinclair AGM, Rousakis G, Al Ghamdi S, Alsharekh AM (2015) Blue Arabia: palaeolithic and underwater survey in SW Saudi Arabia and the role of coasts in Pleistocene dispersals. Quat Int 382:42–57. http://dx.doi.org/10.1016/j.quaint.2015.01.002
Barbujani G, Betorelle G (2001) Genetics and the population history of Europe. Proc Natl Acad Sci U S A 98(1):22–25
Benjamin J, Bonsall C, Pickard C, Fischer A (eds) (2011) Submerged prehistory. Oxbow, Oxford
Beyin A (2011) Upper Pleistocene human dispersals out of Africa: a review of the current state of the debate. Int J Evol Biol. Volume 2011. Article ID 615094.http://dx.doi.org/10.4061/2011/615094
Beyin A (2013) A surface Middle Stone Age assemblage from the Red Sea coast of Eritrea: implications for Upper Pleistocene dispersals out of Africa. Quat Int 300:195–212
Blanton DB (1996) Accounting for submerged mid-Holocene archaeological sites in the Southeast: a case study from the Chesapeake Bay Estuary, Virginia. In: Sassaman KE, Anderson DG (eds) Archaeology of the Mid-Holocene Southeast. University Press of Florida, Gainesville, pp 200–217
Boivin N, Fuller DQ, Dennell R, Allaby R, Petraglia MD (2013) Human dispersal across diverse environments of Asia during the Upper Pleistocene. Quat Int 300:32–47
Brumm A et al (2016) Age and context of the oldest known hominin fossils from Flores. Nature 534:249–253
Carabias D, Cartajena I, Simonetti R, López P, Morales C, Ortega C (2014) Submerged paleolandscapes: site GNL Quintero 1 (GNLQ1) and the first evidence from the Pacific coast of South America. In: Evans AM, Flatman JC, Flemming NC (eds) Prehistoric archaeology on the continental shelf: a global review. Springer, New York, pp 131–150
Cartajena I, Lopez P, Carabias D, Morales C, Vargas G (2011) Arqueologia subacuatica y tafonomía: recientes avances en el studio de sitios Finipleistocénicos sumergidos en la costa Pacífica de Chile central. Antipoda 13:201–225
Cavalli-Sforza LL, Minch E (1997) Paleolithic and Neolithic lineages in the European mitochondrial gene pool. Am J Hum Genet 61(1):247–254
Chang C-H, Kaifu Y, Takai M, Kono RT, Grün R, Matsu'ura S, Kinsley L, Lin L-K (2015) The first archaic Homo from Taiwan. Nat Commun 6:6037. doi:10.1038/ncomms7037
Clausen CJ, Cohen AD, Emiliani C, Holman JA, Stipp JJ (1979) Little Salt Spring Florida: a unique underwater site. Science 203:609–614
Cliquet D, Coutard S, Clet M, Allix J, Tessier B, Lelong F, Baltzer A, Mear Y, Poizot E, Auguste P, Alix P, Olive J, Guesnon J (2011) The Middle Palaeolithic underwater site of La Mondrée, Normandy, France. In: Benjamin J, Bonsall C, Pickard C, Fischer A (eds) Submerged prehistory. Oxbow, Oxford, pp 111–128
Clottes J, Courtin J (1994) The cave beneath the sea. Harry N. Abrams, New York
Clottes J, Courtin J, Vanrell L (2005) Cosquer redécouvert. Seuil, Paris
Cockrell WA (1975) Warm Mineral Springs 1975: a multidisciplinary approach to a 10,000 B.P. archaeological site. Abstract. Fla Sci 38(Supplement 1):19
Cockrell WA, Murphy L (1978) Inundated terrestrial sites: 8 SL 17: methodological approaches to a dual component marine site on the Florida Atlantic coast. In: Barto-Arnold III J (ed) Beneath the waters of time. The proceedings of the ninth conference on underwater archaeology, Texas Antiquities Committee Publication No. 6. Austin, pp 175–182
Cohen KM, Gibbard PL, Weerts HJT (2014) North Sea palaeogeographical reconstructions for the last 1 Ma. Neth J Geosci 93:7–29. doi:10.1017/njg.2014.12
Crock JG, Petersen JP, Andersen R (1993) Scalloping for artifacts: a biface and plummet from eastern Blue Hill Bay, Maine. Archaeol East N Am 21:179–192
Curry A (2006) A Stone Age world beneath the Baltic Sea. Science 314:1533–1535
Dixon JE, Monteleone K (2014) Gateway to the Americas: underwater archaeological survey in Beringia and the North Pacific. In: Evans AM, Flatman JC, Flemming NC (eds) Prehistoric archaeology on the continental shelf: a global review. Springer, New York, pp 95–114

Erlandson JM, Braje TJ (2015) Coasting out of Africa: the potential of mangrove forests and marine habitats to facilitate human coastal expansion via the southern dispersal route. Quat Int 382:31–41. doi:10.1016/j.quaint.2015.03.046
Erlandson JM, Braje TJ, Gill KM, Graham MH (2015) Ecology of the kelp highway: did marine resources facilitate human dispersal from Northeast Asia to the Americas? J Island Coast Archaeol 10(3):392–411. doi:10.1080/15564894.2014.1001923
Faught MK (2008) Archaeological roots of human diversity in the New World: a compilation of accurate and precise radiocarbon ages from earliest sites. Am Antiq 73(4):670–698
Faught MK (2014) Remote sensing, target identification and testing for submerged prehistoric sites in Florida: process and protocol in underwater CRM projects. In: Evans AM, Flatman JC, Flemming NC (eds) Prehistoric archaeology on the continental shelf: a global review. Springer, New York, pp 37–52
Faught MK, Gusick AE (2011) Submerged prehistory in the Americas. In: Benjamin J, Bonsall C, Pickard C, Fischer A (eds) Submerged prehistory. Oxbow, Oxford, pp 145–157
Fedje DW, Josenhans H (2000) Drowned forest sand archaeology on the continental shelf of British Columbia, Canada. Geology 28(2):99–102
Fischer A (ed) (1995) Man and sea in the Mesolithic: coastal settlement above and below present sea level. Oxbow, Oxford
Fischer A (2007) Coastal fishing in Stone Age Denmark – evidence from below and above the present sea level and from the bones of human beings. In: Milner N, Bailey GN, Craig OE (eds) Shell middens in Atlantic Europe. Oxbow, Oxford, pp 54–69
Fischer A (2011) Stone Age on the continental shelf: an eroding resource. In: Benjamin J, Bonsall C, Pickard C, Fischer A (eds) Submerged prehistory. Oxbow, Oxford, pp 289–310
Flemming NC (1986) A survey of the late quaternary landscape of the Cootamundra Shoals, North Australia: a preliminary report. In: Flemming NC, Marchetti F, Stefanon A (eds) Proceedings of the 7th International Diving Science Symposium of CMAS 1983. Padova, pp 149–180
Flemming NC (ed) (2004a) Submarine prehistoric archaeology of the North Sea. Research priorities and collaboration with industry, CBA Research Report 141. Council for British Archaeology, York
Flemming NC (2004b) Submarine prehistoric archaeology of the Indian continental shelf: a potential resource. Curr Sci Indian Acad Sci 89(9):1225–1230
Flemming NC, Harff J, Moura D, Burgess A, Bailey G (eds) (2017) (In preparation) Submerged landscapes of the continental shelf. Quaternary palaeoenvironments. Chichester: Wiley-Blackwell.
Forster P (2004) Ice ages and the mitochondrial DNA chronology of human dispersals. Philos Trans R Soc Lond Ser B 358:255–264
Fu Q et al (2016) The genetic history of Ice Age Europe. Nature 534:200–205
Galili E, Lernau O, Zohar I (2004) Fishing and coastal adaptation at Atlit Yam, a submerged Neolithic village off the Carmel coast, Israel. Atiquot 48:1–34
Glimmerveen J, Mol D, Post K, Reumer JWF, van der Plicht H, De Vos J, van Geel B, van Reenen G, Pals JP (2004) The North Sea Project: the first palaeontological, palynological and archaeological results. In: Flemming NC. (ed) 2004aSubmarine prehistoric archaeology of the North Sea. Research priorities and collaboration with industry, CBA Research Report 141. Council for British Archaeology, York, pp 43–52
Goldberg A, Mychajliw AM, Hadly EA (2016) Post-invasion demography of prehistoric humans in South America. Nature 532:232–235
Gouezin P, Le Gall E (1992) Le site mégalithique d'Er Lannic. Association Archéo Douar Mor. Recherches archéologiques terrestres et sous-marines. Nouvelle topographie du site. Bilan scientifique DRASSM 1992
Green RE, Pääbo S et al (2010) A draft sequence for the Neandertal Genome. Science 238:710–722
Grøn O, Boldreel LO (2013) Sub-bottom profiling for large scale maritime archaeological survey. OCEANS – Bergen, 2013 MTS/IEEE, pp 1–8. doi: 10.1109/OCEANS-Bergen.2013.6608027
Grøn O, Dell'Anno A, Hermand J-P (2013) Investigations of deep submerged Stone Age settlements covered by seafloor sediments: preliminary methodological considerations. OCEANS – Bergen, 2013 MTS/IEEE, pp 1–4. doi: 10.1109/OCEANS-Bergen.2013.6608026
Groucutt HS, Petraglia MD, Bailey GN, Parton A, Clark-Balzan L, Jennings RP, Lewis L, Blinkhorn J, Scerri EML, Drake NA, Breeze P, Boivin N, Inglis RH, Devès MH, Meredith-Williams MG, Thomas MG, Scally A (2015) Rethinking the dispersal of *Homo sapiens* out of Africa. Evol Anthropol 24:149–164. doi:10.1002/evan.21455
Harff J, Bjorck S, Hoth P (eds) (2011) The Baltic Sea basin. Springer, Berlin
Hayashida K (1993) The archaeological materials from the Takashima seabed (Part II). 1993. Takashima Town Cultural Property Research Report. Educational Committee of Takashima Town, Nagasaki Prefecture (In Japanese)
Henn BM, Cavalli-Sforza LL, Feldman MW (2012) The great human expansion. Proc Natl Acad Sci U S A 109(44):17758–17764. doi:10.1073/pnas.1212380109
Hetherington R, Weaver AJ, Montenegro A (2007) Climate and the migration of early people into the Americas. In: Harff J, Hay WW, Tetzlaff FF (eds) Coastline changes: interrelation of climate and geological processes. Geological Society of America Special Paper 426

Hill C, Soares P, Mormina M, Macaulay V, Clarke D, Blumbach PB, Vizuete-Forster M, Forster P, Bulbeck D, Oppenheimer S, Rihards M (2007) A mitochondrial stratigraphy for island southeast Asia. Am J Hum Genet 80(1):29–43
Hublin J-J, Weston D, Gunz P, Richards M, Roebroeks W, Glimmerveen J, Anthonis L (2009) Out of the North Sea: the Zeeland ridges Neandertal. J Hum Evol 57:777–785
Josenhans H, Fedje D, Pienitz R, Southon J (1997) Early humans and rapidly changing Holocene sea levels in the Queen Charlotte Islands, Hecate Strait, British Columbia, Canada. Science 277:71–74
Kelley JT, Belknap DF, Kelley AR, Claesson SH (2013) A model for drowned terrestrial habitats with associated archaeological remains in the northwestern Gulf of Maine, USA. Mar Geol 338:1–16
Klein RG (1999) The human career. University of Chicago Press, Chicago
Lambeck K, Purcell A, Flemming N, Vita-Finzi C, Alsharekh A, Bailey GN (2011) Sea level and shoreline reconstructions for the Red Sea: isostatic and tectonic considerations and implications for hominin migration out of Africa. Quat Sci Rev 30(25–26):3542–3574
Long D, Wickham-Jones CR, Ruckley NA (1986) A flint artefact from the northern North Sea. In: Roe DA (ed) Studies in the Upper Palaeolithic of Britain and Northwest Europe, British Archaeological Reports International Series 296. BAR, Oxford, pp 55–62
Masters PM (1983) Detection and assessment of prehistoric artefact sites off the coast of southern California. In: Masters PM, Flemming NC (eds) Quaternary coastlines and marine archaeology. Academic, London. pp
McEvoy B, Richards M, Forster P, Bradley DG (2004) The Longue Durée of genetic ancestry: multiple genetic marker systems and Celtic origins on the Atlantic façade of Europe. Am J Hum Genet 75(4):693–702
Mellars P, Gori KC, Carr M, Soares PA, Richards MB (2013) Genetic and archaeological perspectives on the initial modern human colonization of southern Asia. Proc Natl Acad Sci U S A 110 (26): 10699–10704. www.pnas.org/cgi/doi/10.1073/pnas.1306043110
Meyer M, Pääbo S et al (2016) Nuclear DNA sequences from the Middle Pleistocene Sima de los Huesos hominins. Nature 531:504–507
Missiaen T (2010) The potential of seismic imaging in marine archaeological site investigations. Relicta 6:219–236
Mithen S (2003) After the ice. Weidenfeld and Nicolson, London
Momber G, Tomalin D, Scaife R, Satchell J, Gillespie J (eds) (2011) Mesolithic occupation at Bouldnor Cliff and the submerged prehistoric landscapes of the Solent, CBA Research Report 164. Council for British Archaeology, York
Nutley D (2014) Inundated site studies in Australia. In: Evans AM, Flatman JC, Flemming NC (eds) Prehistoric archaeology on the continental shelf: a global review. Springer, New York, pp 255–273
Oppenheimer S (2009) The great arc of dispersal of modern humans: Africa to Australia. Quat Int 202(1–2):2–13
Parfitt SA, Barendregt RW, Breda M, Candy I, Collins MJ, Coope GR, Durbidge P, Field MH, Lee JR, Lister AM, Mutch R, Penkman KEH, Preece RC, Rose J, Stringer CB, Symmons R, Whittaker JE, Wymer JJ, Stuart AJ (2005) The earliest record of activity in northern Europe. Nature 438:1008–1012
Petraglia MD (2011) Trailblazers across Arabia. Nature 470:50–51
Pope KO, Terrell JE (2008) Environmental setting of human migrations in the circum-Pacific region. J Biogeogr 35(1):1–21
Price F, Spiess AE (2007) A submerged prehistoric site and other fishermen's reports near Mount Desert Island. Maine Archaeol Soc Bull 47(2):21–35
Runnels C, DiGregorio C, Wegman KW, Gallen SF, Strassser TF, Panagopoulou E (2014) Lower Palaeolithic artifacts from Plakias, Crete: implications for hominin dispersals. In: Ammerman AJ, Davis T (eds) Island archaeology and the origins of seafaring in the eastern Mediterranean. Proceedings of the Wenner Gren workshop held at Reggio Calabria on October 19–21, 2012. Eurasian Prehistory 11: 129–152
Russell J, Tizzard L (2011) Seabed prehistory: site evaluation techniques (area 240) synthesis. OASIS/ADS: Wessex Archaeology, 70754.04. doi:10.5284/1008287
Scally A, Durbin R (2012) Revising the human mutation rate: implications for understanding human evolution. Nat Rev Genet 13:745–753
Scuvée F, Verague J (1988) Le gisement sous-marine du Paléolithique Moyen de l'anse de la Mondrée à Fermanville (Manche). Ministère des Affaires Culturelles, Autorisation No. 001740, 1971, CEHP- Littus, BP. 306, Cherbourg 50104
Skogland P, Mallick S, Bartolini MC, Chennagiri N, Hunemeier T, Perzel-Erle ML, Salzano FM, Patterson N, Reich D (2015) Genetic evidence for two founding populations of the Americas. Nature 525:104–107
Soares P, Trejaut J, Hill C, Mormina M, Macaulay V, Bulbeck D, Loo J-H, Lin M, Oppenheimer S, Richards MA (2008) Climate change and post-glacial human dispersals in Southeast Asia. Mol Biol Evol 25(6):1209–1218
Stanford D, Lowery D, Jodry M, Bradley B, Kay M, Stafford TW Jr, Speakman RJ (2014) new evidence for a possible paleolithic occupation of the eastern North American continental shelf at the Last Glacial Maximum. In: Evans A, Flatman J, Flemming NC (eds) Prehistoric archaeology of the continental shelf. Springer, New York

Stright MJ (1990) Archaeological sites on the North American continental shelf. Geol Soc Am 4:439–465Centennial Special Volume
Su B, Xiao J, Underhill P, Jin L et al (1999) Y-chromosome evidence for a northward migration of modern humans into eastern Asia during the Last Ice Age. Am J Hum Genet 65(6):1718–1724
Tamm E, Kivisild T, Reidia M, Malhi RS et al (2007) Beringian standstill and spread of Native American founders. PLoS ONE 2(9):e829. doi:10.1371/journal.pone.0000829
Tizzard L, Bicket AR, Benjamin J, De Loecker D (2014) A Middle Palaeolithic site in the southern North Sea: investigating the archaeology and palaeogeography of Area 240. J Quat Sci 29(7):698–710
Van den Berg GD, Kaifu Y, Kurniawan I, Kono RT, Brumm A, Setiyabudi E, Aziz F, Morwood MJ (2016) Homo floresiensis-like fossils from the Middle Pleistocene of Flores. Nature 534:245–248
Werz B, Flemming NC (2001) Discovery in Table Bay of the oldest handaxes yet found underwater demonstrates preservation of hominid artefacts on the continental shelf. S Afr J Sci 97:183–185

# Chapter 19
# The Northern Coasts of Doggerland and the Colonisation of Norway at the End of the Ice Age

**Håkon Glørstad, Jostein Gundersen, and Frode Kvalø**

**Abstract** Due to the need for a national strategy for offshore wind farms in Norway, a small and preliminary joint archaeological and geological research project was conducted. The aim of the investigation was to determine the possibilities for human occupation in the southernmost parts of the Norwegian sector of the North Sea during the Late Glacial and Preboreal periods. The research was in the shallowest area of the Norwegian sector of the continental shelf, south of the Norwegian Trench. By analysing 3D seismic reflection data, several layers with traces of drainage systems and other landforms could be identified in the sediments. These features were, however, mainly situated deep down below the surface of the seabed, and covered by more than 100 m of younger sediments. Their age is therefore uncertain, even if the depth alone indicates that they are much older than the Last Glacial Maximum (LGM). A sediment core through the shallow sea-bed sediments (0–13 m below present sea floor) was analysed for environmental and chronological data. The youngest observed transition, from glacio-lacustrine to marine sediments, was dated to approximately 14,000 cal. BC or the Older Dryas. This indicates that this part of Doggerland was not dry land after the LGM, but was first covered by an ice-dammed lake and later inundated by saltwater from the Atlantic Ocean. The analysis also indicates that the distance from Doggerland to Norway, in the periods when people began to inhabit southern Scandinavia, was far too long to be crossed by boats or on ice. This makes the western coast of Sweden and Bohuslän the most likely bridgehead in the colonisation of Norway. Humans arrived at the Norwegian shores first when a safe and sheltered passage was created between Bohuslän and the Oslo Fjord area, at approximately 9300 cal. BC.

## 19.1 Introduction

Doggerland—the prehistoric land bridge between present-day Britain and Denmark—has been a topic of fascination and debate in Norwegian archaeology since Clement Reid (1913) first published on the submerged forests of the North Sea. In 1923 Anders Nummedal (1923, pp 127–128) referred to Reid when mentioning the sunken landscape of the Dogger Bank, but did not conclude that it had

H. Glørstad (✉)
Museum of Cultural History, University of Oslo, Box 6762 St. Olavs plass,, NO-0130 Oslo, Norway
e-mail: hakon.glorstad@khm.uio.no

J. Gundersen
Directorate for Cultural Heritage, Dronningensgate 13, Box 8196 Dep, NO-0034 Oslo, Norway
e-mail: jog@ra.no

F. Kvalø
Norwegian Maritime Museum, Bygdøynesveien 37, NO-0286 Oslo, Norway
e-mail: frode.kvalo@marmuseum.no



G.N. Bailey et al. (eds.), *Under the Sea: Archaeology and Palaeolandscapes of the Continental Shelf*, Coastal Research Library 20, DOI 10.1007/978-3-319-53160-1_19

relevance to the coastal Stone Age sites along the west coast of Norway. Alfred Rust (1942) was the first to suggest that Doggerland or the former dry plains of the North Sea were an important habitation area for humans in the Late Glacial and Early Preboreal periods (see also Gaffney et al. Chap. 20 and Sturt et al. Chap. 28, for the history of the term 'Doggerland'). Rust also suggested that the plains of Doggerland connected the prehistoric people who are known to have occupied coastal regions surrounding the North Sea Basin. A few decades later, the hypothesis that Doggerland was crucial for early Holocene communication gained support from several Norwegian researchers (Odner 1966; Hagen 1967; Rolfsen 1972; Indrelid 1975, 1989; Mikkelsen 1978). The line of argument was, and still is, that a much lower sea-level at the Last Glacial Maximum (LGM) and in the first phases of deglaciation made the distance from the coast of the European continent—Doggerland—to Norway much shorter than the present distance from Norway to Denmark (minimal distance c. 110 km). People could therefore relatively easily cross this channel of open water, either by boat or on foot across winter ice. Such arguments have repeatedly been presented over the last two decades (Fuglestvedt 1989, 2001, 2007, 2009, 2012; Rokoengen and Johansen 1996; Skar 1995; Anundsen 1996; Bang-Andersen 1996, 2003; Johannessen 2009; Selsing 2012).

It would not be unfair to say that Norwegian archaeological research on human occupation of Doggerland has been speculative. There is very little, if any, relevant evidence on the continental shelf in the Norwegian part of the North Sea or the Skagerrak (Glørstad and Kvalø 2012). The speculative nature of the research has by no means been irrelevant or indeed unproductive. Speculations and the will to test new hypotheses are important parts of a healthy research strategy. However, it is nonetheless vital to produce some evidence in favour or disfavour of these hypotheses.

In this chapter, we present new data from the seabed of the Norwegian part of the North Sea pertinent to questions regarding Doggerland. Our new observations raise new speculations and hypotheses about early human migration to the Scandinavian Peninsula, and the location of the northern coasts of Doggerland at the transition from the Pleistocene to the Holocene.

## 19.2 Background, Scope and Research Area

Several Norwegian scholars have relatively recently claimed that Doggerland was the homeland of the first modern inhabitants of the Scandinavian Peninsula (e.g., Indrelid 1989; Johansen and Rokoengen 1996; Fuglestvedt 2009; Selsing 2012). The relevance of the Doggerland area for early human migration to Norway was accentuated by the very inspiring works of Per Blystad (1989) and Bryony Coles (1998). In Coles' fascinating, yet coarse-grained, reconstructions of the extension of the Doggerland area from the LGM and onwards, the distance from Norway to dry land on the Continent appears to be surprisingly short. At the LGM, only a narrow trench or fjord no more than 50 km broad divided Norway from the Doggerland plains. This is the Norwegian Trench, which today is the deepest part of Norway's continental shelf in the North Sea (Fig. 19.1). According to Coles' map, the distance between the Continent and Norway was still very short as late as the Allerød oscillation (approx. 11,800–10,600 cal. BC), with an archipelago of drowning islands situated between Scotland and western Norway (Fig. 19.1). In this scenario the people of Doggerland might have seen the Scandinavian glacier. If navigating along the coast, they might even have seen the highest peaks as dry land protruding through the ice (nunataks) at the LGM, and maybe in the Allerød as well. Even with the rise in sea level after the LGM, which made the distance a bit longer, it would still be possible to cross the Norwegian trench on the sea ice or by boats— that is, if people were present at these northern shores of Doggerland at the time.

This is a fascinating model, but what kind of evidence is it based on? How much do we know about the northern extent of the Doggerland area? At the end of the 1980s, the geologist Per Blystad (1989, p 64) reviewed the then-current state of knowledge. His main message was that there was very little

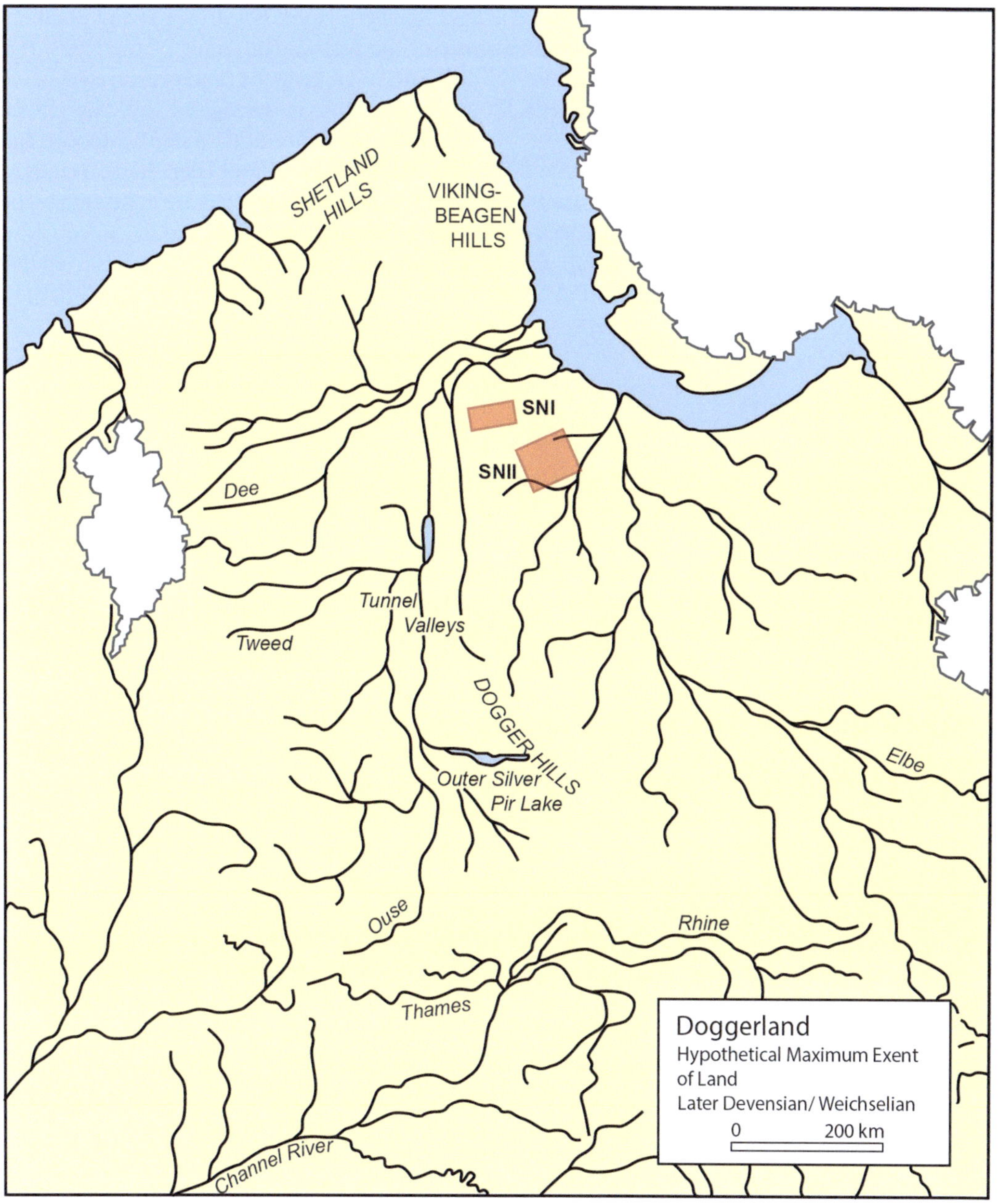


**Fig. 19.1** Hypothetical maximum extent of Doggerland at the Last Glacial Maximum. The designated areas for wind farms called Southern North Sea I (*SN I*) and Southern North Sea II (*SN II*) are postulated to be on dry land (Modified from Coles (1998))

firm evidence, and that much more research needed to be done in order to collect proxy data. Arguably, 25 years later, his conclusions are still valid.

In 2008 the Norwegian Parliament (Stortinget) decided to develop a national strategy for offshore production of renewable energy (Inst. S. nr. 145 2007–2008). As part of the strategy, the Norwegian Water Resources and Energy Directorate (NVE) made a strategic impact assessment (SIA) to evaluate selected areas for the potential for offshore wind farms along the Norwegian coast and in territorial waters (NVE 2012). Two of the selected areas, named Southern North Sea I (SN I) and Southern North Sea II (SN II), are located in the southern part of the Norwegian sector of the North Sea (Fig. 19.2). Located more than 140 km offshore, the areas are still among the shallowest waters in the Norwegian sector. Water depths vary between 50 and 80 m, with an average of less than 65 m. Consequently one would expect these areas to have been dry land during the late Pleistocene and maybe also at the beginning of the Holocene due to the low global sea level (Fig. 19.2). The SIA, financed by the Norwegian Ministry of Oil and Energy, gave a welcome opportunity to determine the nature of the palaeoenvironments in the selected areas, the history of inundation and the potential for earlier human occupation. A research group of archaeologists, geologists and geophysicists from five institutions handled the project (Hafeez et al. 2012; Glørstad and Kvalø 2012).

## 19.3 Methodology

Although the Directorate for Nature Management had asked for an archaeological evaluation of the selected areas, the available resources for the SIA did not provide finance for new surveys of the seabed. The whole project was therefore based on analysing already existing data—gathered by previous research projects, the petroleum industry, commercial fishing and other maritime activity.

The aim was rather straightforward. By reconstructing the palaeolandscape we should be able to determine the areas most suitable for human occupation. A more precise location of the Late Glacial estuary of the Elbe was considered as a particularly useful starting point for tracing potentially suitable places for human settlement sites (see Hepp et al. Chap. 14). Planning the analyses, we were particularly inspired by The North Sea Palaeolandscapes Project (Gaffney et al. 2007, Chap. 20). Their method of reconstructing the ancient landscape of the south-western part of Doggerland by using 3D seismic data seemed to be the best way to proceed. Other data sets, such as bathymetric data, 2D seismic and cores from boreholes, were also considered useful for our work.

SN I and SN II together cover an area of almost 4,000 km$^2$. However, the entire chosen study area is more than 50,000 km$^2$ in extent, making it considerably larger than, for instance, Denmark. Finding and getting access to available data from the southern part of the Norwegian North Sea was a challenging process. Although extensive petroleum-industrial activity has been present in the central North Sea since the early 1970s, only limited high-resolution geophysical and geological data is available (Hafeez et al. 2012). The bathymetric data is limited and of poor quality, and not sufficient to make a detailed seabed reconstruction (Hafeez et al. 2012, p 12). One petroleum exploration well exists in each of the two target areas, drilled in the years 1976 (Well 8/9-1 in SN I) and 2000 (Well 3/6-1 in SN II) (Hafeez et al. 2012, pp. 15–16). Fortunately, from the latter well both a seabed sediment core (3/6-1) and a small 2D high-resolution site survey were made available for the project by ENI.

In addition, Petroleum Geo-Services (PGS) granted access to conventional 3D seismic reflection data from those parts of the study area covered by their Central North Sea Mega Survey (PGS 2012) (see grey-shaded areas in Fig. 19.3). The strategy was, then, to reconstruct the prehistoric landscape with 3D seismic data according to the methodology developed in The North Sea Palaeolandscapes Project (Gaffney et al. 2007).

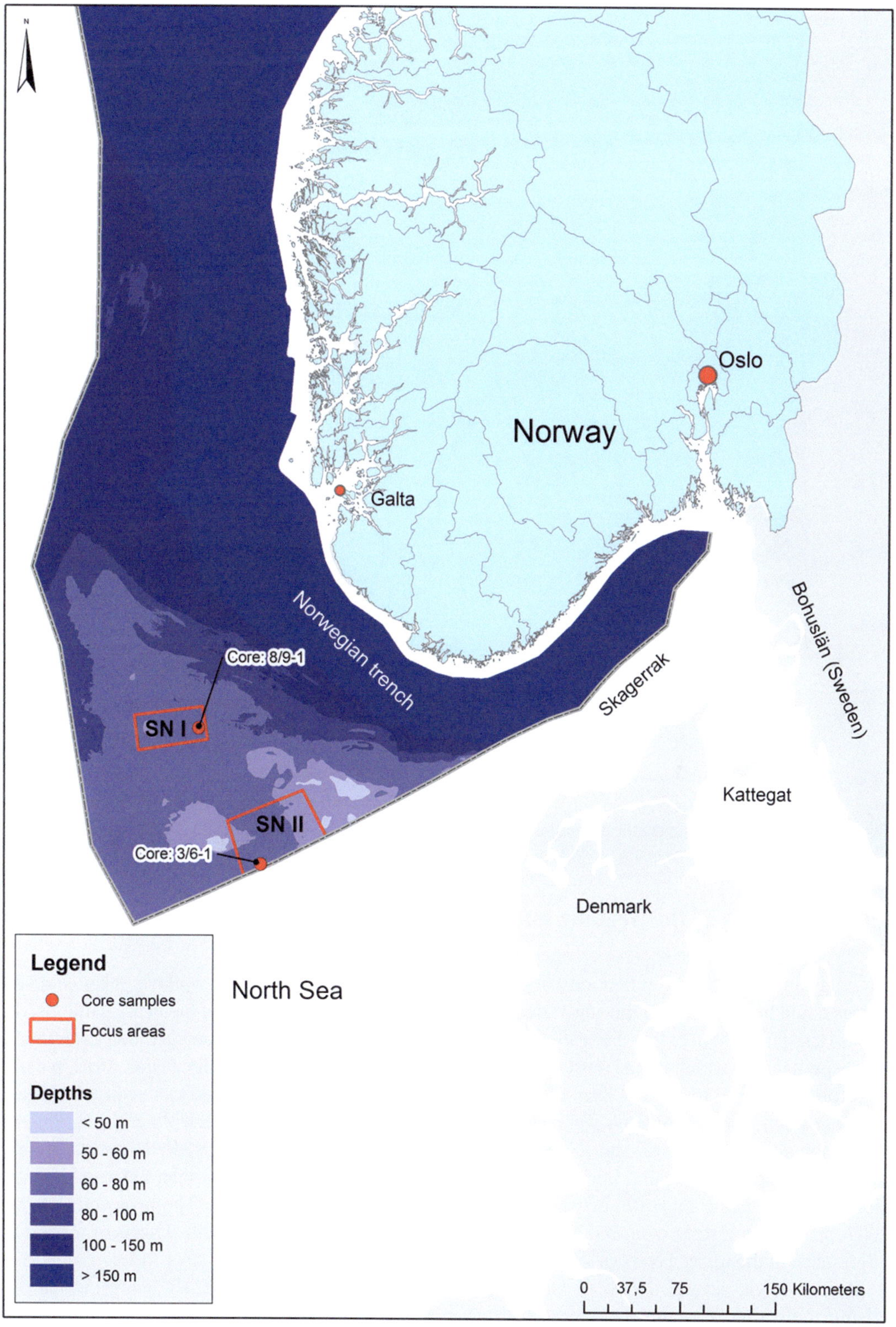


**Fig. 19.2** The focus areas for the present study. SN I covers an area of 1,375 km$^2$. The water depth varies between 50 and 70 m, with an average of 64 m. Distance to the closest point on the Norwegian coast today is 149 km. SN II covers an area of 2,591 km$^2$.The water depth varies between 53 and 70 m, with an average of 60 m. Distance to the closest point on the Norwegian coast today is 140 km

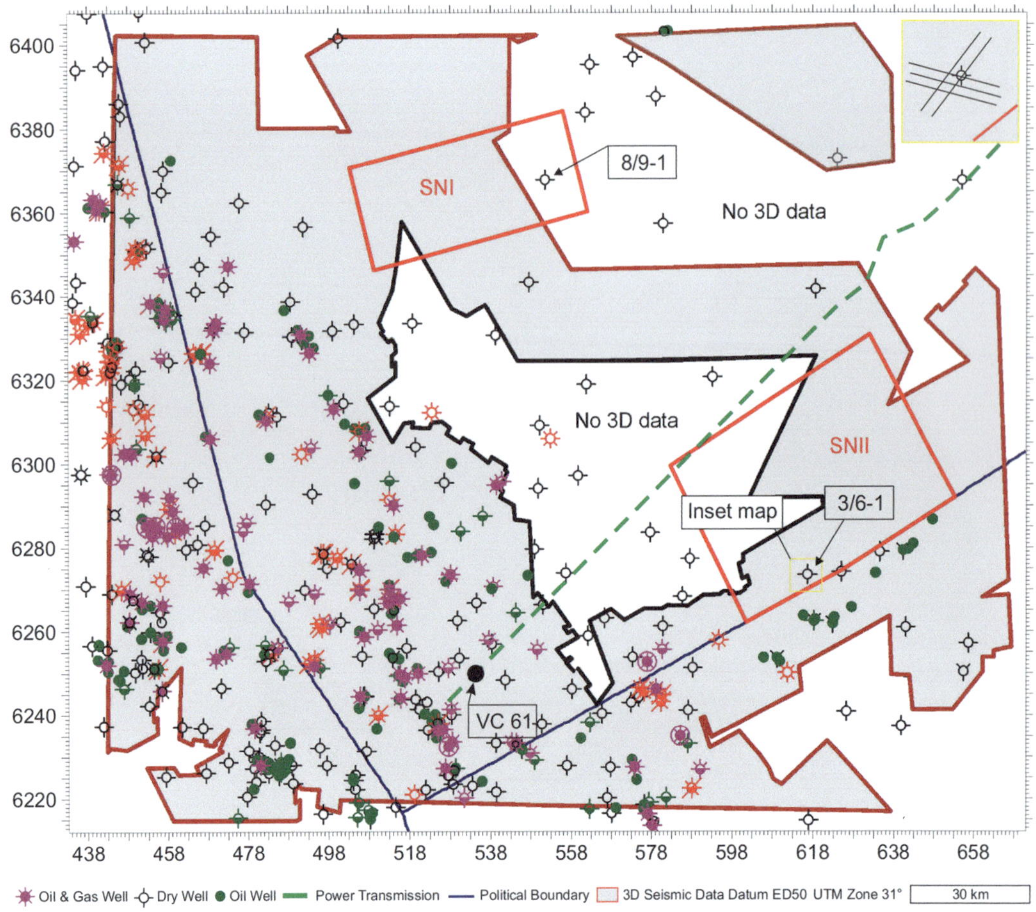


**Fig. 19.3** Location map showing the 3D seismic data coverage, boreholes, and high-resolution data (From Hafeez et al. (2012))

## 19.4 Analysis of the Seismic and Geological Data Sets

By analysing 3D seismic reflection data, distinct landforms such as rivers, valleys, lakes and salt domes could be identified in the study area. However, the seismic resolution of the near-seafloor sediments (less than 100 m below the seafloor) is poor. This is because the acquisition geometry is focused on deeper targets, which are of most relevance to the petroleum industry (Hafeez et al. 2012, p 4). It therefore turned out to be very difficult to study the upper 50–100 m of the seafloor sediments based on these data. These problems were also recognised in The North Sea Palaeolandscapes Project (Thomson and Gaffney 2007, pp. 23–31). They decided to rely on analysing these data sets anyway, as they provide better general horizontal resolution in determining landforms than 2D seismic data. However, 2D seismic data offer better vertical resolution than the 3D data sets (Thomson and Gaffney 2007, pp. 27–30; Hafeez et al. 2012, pp. 16–19; Glørstad and Kvalø 2012, p 29). This is of particular importance in the upper layers of the sediments.

Deeper down, several details of ancient landscapes were distinguishable. By analysing the 3D seismic reflection data, several layers with river systems, valleys and other landforms could be detected (Fig. 19.4). Some of the channels showed stability throughout the different layers, indicating relatively stability through time. The dating of the features, however, could not be determined—we

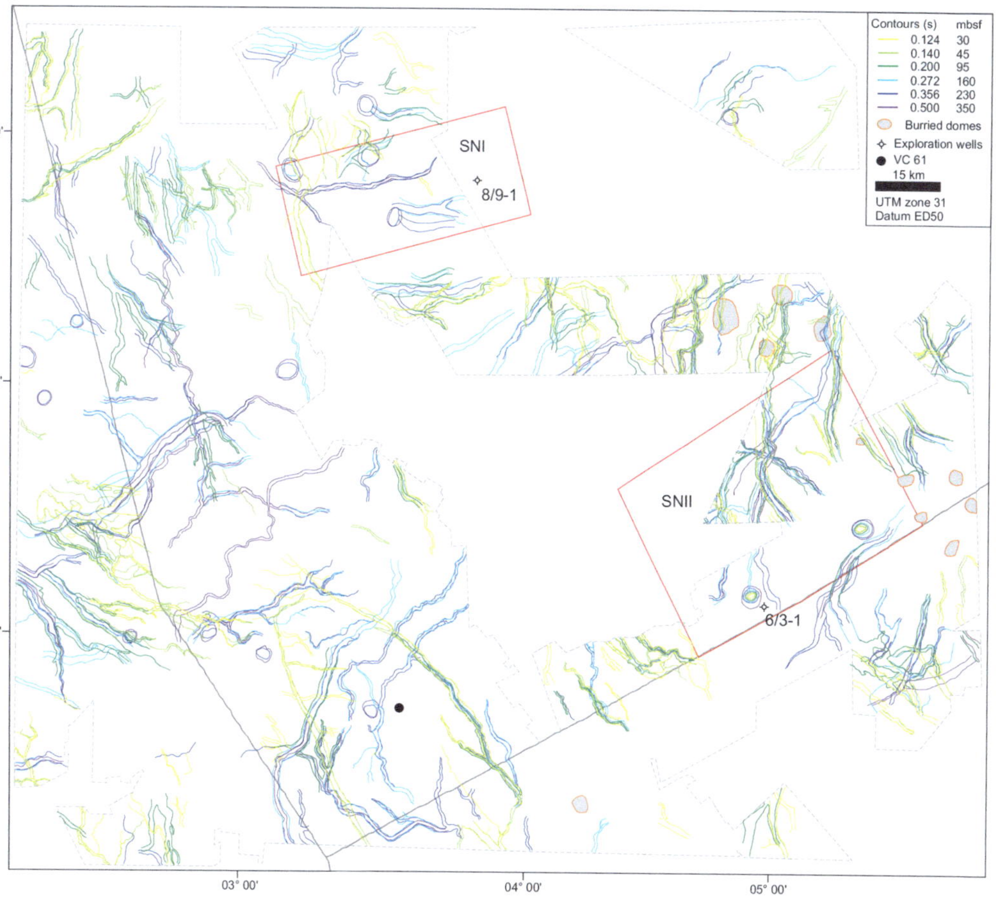


**Fig. 19.4** Compilation of interpreted tunnel valleys, salt dome, and structural domes (buried domes) at different depths. At least five major channels with varying geometries are present in SN II. These channels are typically 1.5–2.7 km wide and have a depth of more than 270 m (From Hafeez et al. (2012))

only know that they were covered with at least 50–100 m of younger sediments. Consequently, the former landscape we were looking at was situated more than 100 m below the present sea level, and therefore was possibly not the same dry landscape which existed during the LGM and the early part of the deglaciation (Fig. 19.4).

The most reliable way of dating the different landscape slices is to analyse and date sediment cores going through the same layers. Because most of the boreholes on the Norwegian continental shelf are made for petroleum prospecting, the shallowest parts of the cores are normally not stored. Blystad was alerted to this problem as early as the 1980s and recommended a change in legislation so that the whole core would be kept for future analyses (Blystad 1989, p 64). His recommendations have not been taken into serious consideration, and commercial companies are still only legally obliged to keep the deep parts of the cores of commercial relevance. In light of this we were therefore fortunate to find that samples from the shallow part of the core from 3/6-1 were still preserved. This core was drilled for geotechnical purposes into the upper seafloor sediments, and seven samples taken at 1–2 m intervals from the upper 13 m of the core had been kept (Hafeez et al. 2012, p 17, 20). The core had been taken at a water depth of 64 m in the southern part of SN II. The physical analysis of the material from the core gave very interesting results. The top layers showed continuous marine sedimentation in warm-temperate conditions, while the deeper layers showed a succession of clay sediments from a glacio-lacustrine environment, with a basal layer deposited in an arctic marine environment.

We were also able to radiocarbon-date the upper layers from the core (Fig. 19.5), which were quite old compared to the main focus of our study. Dating of the arctic marine sediments from 10 m below the seafloor gave a date of 38,400 uncal BP, placing them early in the Weichselian period. The youngest transitional phase, from glacio-lacustrine to marine sediments was found at 3 m below the seafloor and dated to approximately 13,500 uncal BP, that is, the Older Dryas or late Weichselian period (Fig. 19.5, Hafeez et al. 2012, p 24).

The analysis of the core demonstrated that the different layers identified by physical examination could also be identified in a few lines of high-resolution 2D seismic data. The 2D data are also distinct in the upper parts of the sediment (Fig. 19.5) and the 2D seismics and the layers identified in the core correspond quite neatly. Interestingly, the dating of the layers confirmed our suspicion from the analyses of the 3D seismic reflection data that the landscape reconstructed from them was much older than the final stages of the Last Ice Age and had little to do with the landscapes of the Final Palaeolithic or Early Mesolithic.

It is notable that the analysed core did not reveal any evidence for dry land in the Allerød, Younger Dryas or following phases, but rather a succession from lacustrine to marine conditions. It is, of course, a difficult task to prove the presence of dry land from layers in a sediment core, since sedimentation takes place for the most part under water, not on dry land. However, intervals of dry land may be indicated by evidence of erosion, oxidation or traces of terrestrial organisms. No such features were detected in this case, although this is by no means unequivocal proof for the absence of dry land in this period (compare Schmitt 1994, pp. 252–254). Source critical factors, such as the spacing of core samples (every 1.2 m), could also mean that evidence for dry land was missed (see discussion in Hammer et al. 2016.)

Despite these uncertainties concerning the presence of dry land, it seems that the inundation of this northern part of Doggerland, with water depths around 60 m today, could have happened at a very early stage—approximately 13,500 uncal BP or 14,450–13,210 cal BC (Fig. 19.6, Hafeez et al. 2012, p 27). This is earlier than the oldest evidence for modern human occupation in southern and northern Scandinavia (Grimm and Weber 2008; Petersen 2009). It is also earlier than the predicted inundation in existing models of changing sea-levels in the late Pleistocene and early Holocene (e.g., Blystad 1989; Lambeck 1995; Gyllencreutz et al. 2006; Lambeck et al. 2010, appendix).

Based on this analysis, the reconstruction of Doggerland in the LGM and Younger Dryas is quite different from the models usually applied by Scandinavian archaeologists. In these models Doggerland is displayed as a dry and ice-free land immediately south of the Norwegian Trench at a much later stage than our results indicate (e.g., Rolfsen 1972; Mikkelsen 1978; Blystad 1989, Bang-Andersen 1996; Schmitt et al. 2006; Fuglestvedt 2009, 2012). Based on the evidence from our analysis, other models may be more appropriate.

The geologist Bjørn Andersen (2000) offers an alternative reconstruction of northern Europe (Fig. 19.7) that is perhaps a more appropriate interpretation of the palaeotopographic situation around the LGM. In his reconstruction, the Scandinavian Glacier and the British Glacier are connected, sealing off the Doggerland area from the Atlantic and creating a large ice-dammed lake in the present North Sea. Andersen (2000) claims that natural drainage to the north was blocked by the glacier. Our observations correspond to this claim. A large ice-dammed lake in the same area is also indicated in the recently published work of C.D. Clark and his research team (Clark et al. 2012).

Our results seem to fit these recently presented geological models showing an ice-dammed lake created during the LGM, which was later refilled with seawater when the ice barrier blocking off the Atlantic Ocean melted towards the end of the Weichselian period. The consequence of this interpretation is that the passage from Doggerland to the Norwegian mainland would not have been a narrow fjord or bay when humans arrived in this part of Europe at the end of the Ice Age. Rather, the distance between Doggerland and the Scandinavian Peninsula at that time appears much larger than the present distance between Norway and Denmark (110 km).

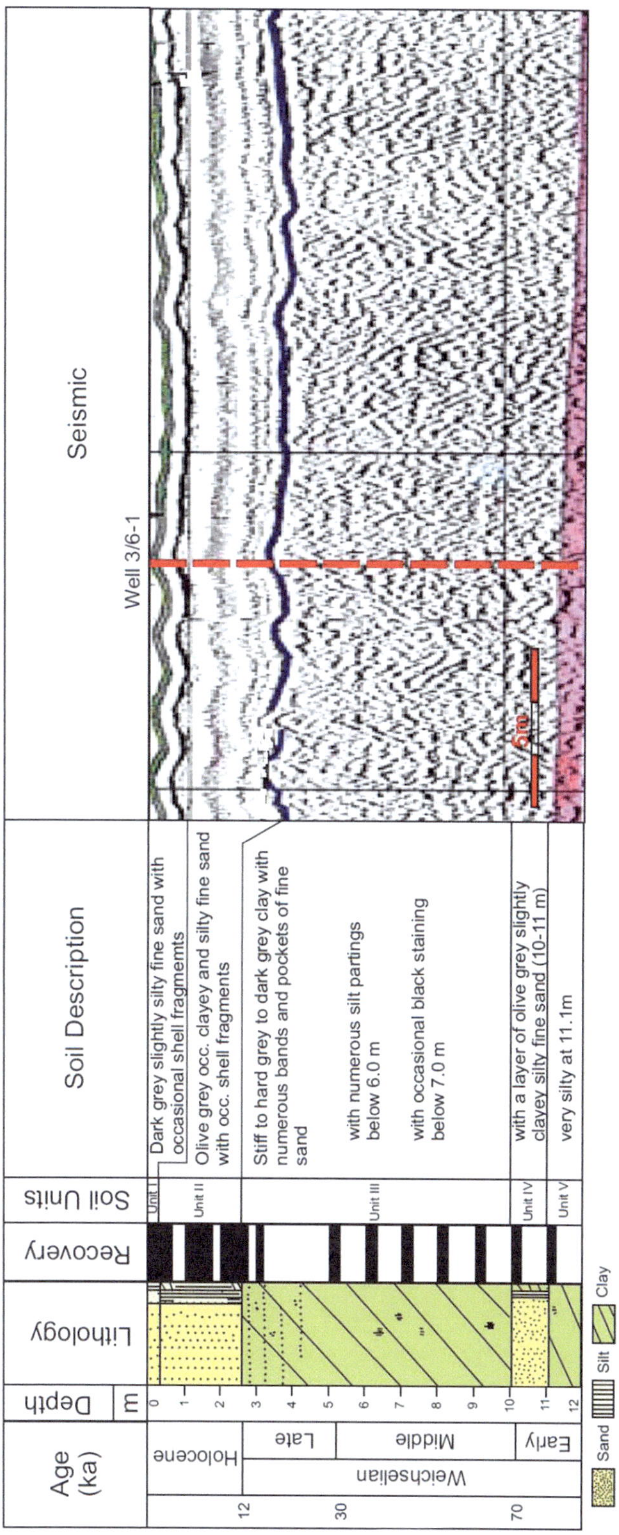


**Fig. 19.5** Borehole-seismic correlation for well 3/6–1. A high-amplitude reflection is clearly identified in the deep tow sparker data at about 2.5 m below the seafloor, separating a transparent unit from underlying scattered and semi-continuous dipping events. The boundary is the base of the Holocene, separating marine and glacial sediments (From Hafeez et al. (2012))

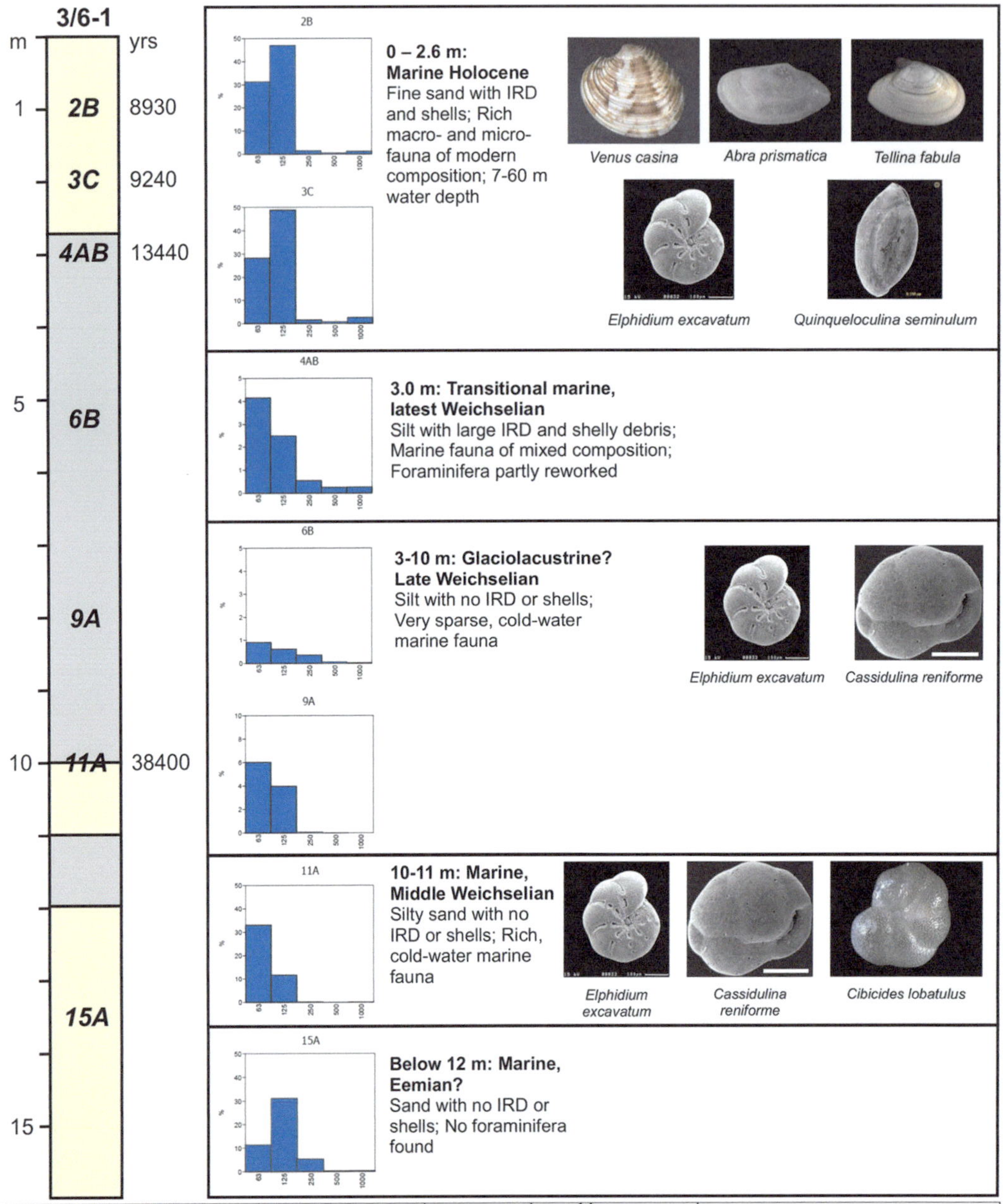


| Sample | Depth (m) | Material | Weight | $^{14}$C age | Calibrated age, 95% conf. |
|---|---|---|---|---|---|
| 2B | 1.00 – 1.75 | *A. prismatica* | 30 mg | 8930 ± 55 BP | 7810 – 7520 BC |
| 3C | 2.00 – 2.54 | Tellinacea | 50 mg | 9240 ± 55 | 8250 – 7935 |
| 4AB | 3.03 – 3.15 | Bivalves | 56 mg | 13440 ± 80 | 14450 - 13210 |
| 11A | 10.00 – 10.14 | Coal | 4 mg | 38400 ± 1500 | 42900 - 38000 |

**Fig. 19.6** Summary of core 3/6–1 analyses and interpreted environment with calibrated radiocarbon dates. The calibrations are calculated with Calib 6.0 and the Marine09 calibration curve, which includes a globally averaged marine reservoir correction. A local correction (Delta R) was not used because the modern value cannot be applied to the completely different hydrographic conditions in the Weichselian to early Holocene of the central North Sea. This implies an additional uncertainty of 200–300 years. The calibrated ages are given in BC, not BP. IRD is an abbreviation of ice rafted debris (From Hafeez et al. (2012))

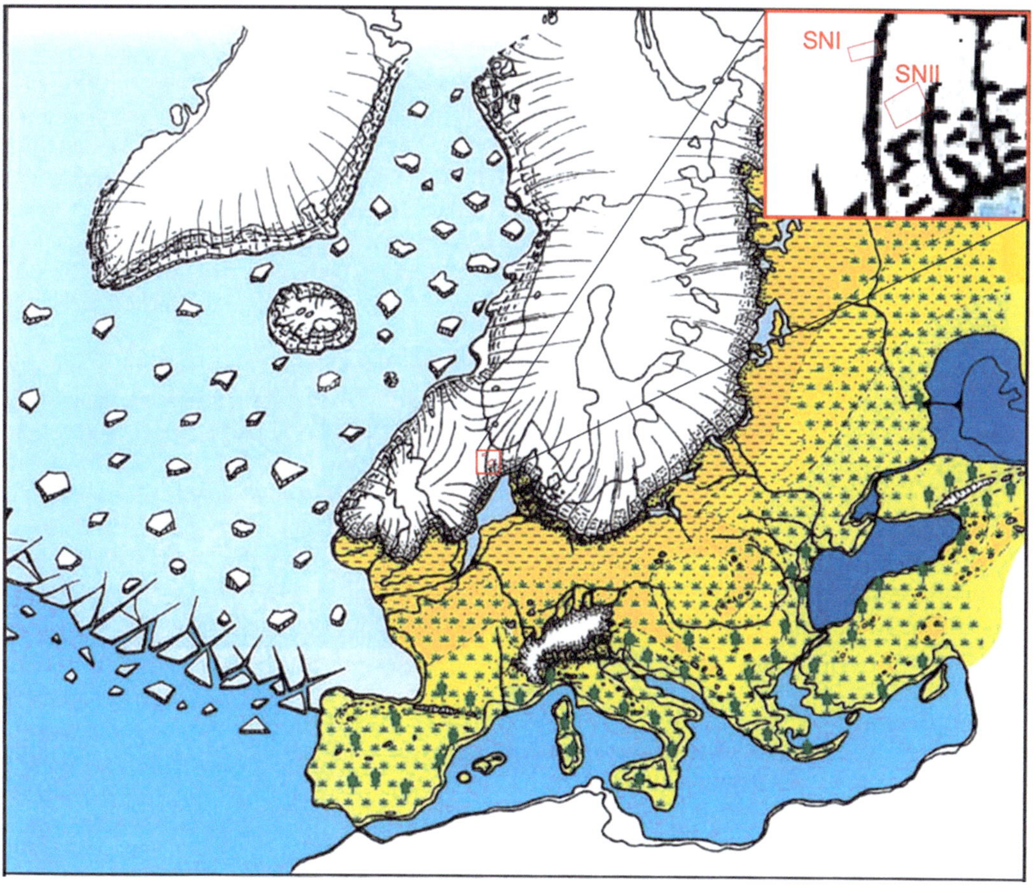


**Fig. 19.7** Sketch showing northern Europe at the LGM about 25.5–23.5 cal BP. Note ice- dammed lake in the central North Sea close to SN I and SN II (Modified from Andersen (2000))

## 19.5 Discussion: Doggerland and the Colonisation of Norway at the End of the Ice Age

When discussing the colonisation of the Norwegian mainland, most scholars seem to accept that the first immigrants came from the south (see overview in Bjerck 2008a). The origin of the oldest archaeological complex in Norway, the Fosna-Komsa (hereafter Fosna), is sought in the North European Ahrensburgian technocomplex of the Younger Dryas (Fuglestvedt 2007; Bjerck 2008a, and references therein). Several typological and technological connections between the early Fosna and the Ahrensburgian complex have been established, and the focus on reindeer hunting is seen as a shared feature as well (e.g. Fuglestvedt 2009, 2012). Although the Fosna and Ahrensburgian share reindeer hunting as a subsistence strategy, the Fosna complex seems to be characterised by a coastal orientation and a marine way of life. A large majority of the known sites are located on the ancient shorelines and relatively exposed islands off the Norwegian and western Swedish coasts. Boats and maritime knowledge must therefore have been part of the cultural package (Bjerck 2008a, b).

As initially discussed, it has been suggested that the first inhabitants of Holocene Norway were Ahrensburgian hunters who crossed the Norwegian Trench either by boat or on foot in the winter season (Rolfsen 1972, p 147; Bang-Andersen 1996, p 222; Fuglestvedt 2007). This is an interesting theory, but archaeologically speaking there is no convincing evidence for direct crossing of the

Skagerrak (the body of water between the west coast of Sweden, Jutland in Denmark and the south-east coast of Norway) until the Late Neolithic. All preceding traffic and communication seem to have followed the coastlines (e.g. Østmo 2005, 2011; Glørstad 2011). This is, for instance, demonstrated by the distribution of artefacts made in southern Scandinavia, and by the distribution of artefacts made in the same design tradition that dominated in Denmark and Scania. Such objects show a pattern of a more or less regular fall-off with increasing distance from the production centres in the south as one moves progressively further north and west around the coastlines of the Skagerrak from western Sweden via the Oslo Fjord area to the coastline of south-eastern Norway (Fig. 19.2). Consequently, such objects are much rarer on the Norwegian west coast than in the Oslo Fjord area (e.g. Glørstad 2009). This is quite different from the pattern one might expect if immigrants had moved across the Norwegian Trench.

Against this background, several scholars have concluded that only the Late Neolithic represents the first commencement of real overseas voyages in Scandinavia (Prescott and Walderhaug 1995; Østmo 2005, 2008, 2011; Kvalø 2007; Prescott 2011; Prescott and Glørstad 2011), a step change connected to the invention of plank boats made with metal tools (Østmo 2005) and supported by experiments that have demonstrated that such boats are solid, seagoing vessels, capable of crossing the Skagerak Sea (Crumlin-Pedersen and Trakadas 2003).

The developments taking place in the Late Neolithic are in fact relevant for our understanding of the initial colonisation of Norway after the end of the Ice Age. The longest overseas journeys that are documented in Scandinavia prior to the Late Neolithic are the Neolithic voyages between Gotland and Öland in the Baltic Sea, and voyages across the Kattegat, between Jutland and Bohuslän (Österholm 1988, 1989; Malmer 2002). These are distances of approximately 50 km. From the preceding Mesolithic, documented journeys are much shorter, with the longest approximately 20–36 km (e.g., Bjerck 1990; Fischer 2002). Mesolithic sea journeys appear to have been coast-bound, with land never out of sight (see discussion in Glørstad 2010). The closest distance between present-day Norway and Denmark is approximately 110 km. This makes the crossing of the Skagerrak from Denmark to Norway something rather outstanding and exceptional compared to known examples of Mesolithic sea crossings, approximately four times longer than any overseas journey so far documented from this period in Scandinavia.

If our data from the southern North Sea are representative for the rest of the northern coastal rim of Doggerland, prehistoric people must have crossed at least equally long distances of open water, or sea-ice, at the end of the Ice Age in order to reach the Norwegian coast from Doggerland. This also means that the Norwegian mainland was never visible from Doggerland (or vice versa). One may theoretically see over a distance of about 110 km of open sea from a height of 1,000 m. Yet, such high mountains were not present in the Doggerland area, or on the Norwegian coastal mainland closest to it.

Is it likely that long journeys across open water or ice, without sight of land in any direction for a long time, were part of the Early Mesolithic/Late Palaeolithic communication patterns? A late Palaeolithic boat-building tradition and technology capable of handling such waters has not yet been demonstrated in northern Europe (see discussion in Glørstad 2013). It must also be noted that in some recent publications about the first colonisation of the Scandinavian Peninsula, the importance of Doggerland as the bridgehead is reduced (Bjerck 2008b; Bang-Andersen 2012). Instead, it is the immigration route along the Swedish west coast that is considered to be the main communication line into Norway. This however, does not entirely exclude the theory of crossing from Doggerland to Norway over the sea-ice during winter.

Analysis of the sediment core form SN II revealed the presence of warm seawater in the Skagerrak Sea from the Older Dryas onwards. This would contribute to weaker sea-ice than a situation dominated by fresh melt-water. Today, two opposing tidal waves make the tidal differences in the Skagerrak basin quite small (Gjevik 2009). However, when the English Channel and a part of the Doggerland plains were dry land in the Late Glacial and Early Holocene periods, the tidal differences were likely quite different (Schmitt et al. 2006). A stronger tidal current must have challenged the stability of the sea ice during winter. The theory of crossing the sea-ice must therefore be considered to be weakened by the increasing distance, warmer seawater and stronger tidal currents—and again the lack of any plausible supporting evidence.

The colonisation of the Norwegian coast has been dated by $^{14}$C-samples from a number of sites (Fig. 19.8). The Fosna sites from where the samples are taken are also marked in Fig. 19.8. As indicated by the sum of the dates, colonisation can be dated back to approximately 9300–9200 cal. BC. According to these dates, the initial occupation of Norway therefore started quite late in the Preboreal period, when the climate had stabilised towards boreal conditions (e.g., Bos et al. 2007).

The radiocarbon dates from the Norwegian sites are younger than the dates of the final Ahrensburgian and seem to be contemporaneous with the oldest part of the Maglemose complex in Scania and Denmark (Glørstad 2013). From these dates it seems more parsimonious to see the Fosna complex as a continuation in space and time of the Ahrensburgian complex rather than a parallel technological or cultural group. The Fosna site of Galta from Rogaland has been dated by several scholars to the transition from the Younger Dryas to the Preboreal period (Prøsch-Danielsen and Høgestøl 1995; Fuglestvedt 2007; Bang-Andersen 2012) and could therefore be considered as contemporaneous with the final Ahrensburgian and several hundred years older than the radiocarbon-dated sites in Norway. The location of the Galta site strongly indicates a coastal orientation, and the date is primarily based on the shoreline displacement curve in the area (Prøsch-Danielsen 1993) Such a date is, however, affected by several source critical factors and an alternative shoreline dating of the Galta site could be 9300–8600 cal. BC (see Glørstad 2014 for the full argument). Consequently there are so far no Fosna sites that can be convincingly dated prior to the second half of the Preboreal period.

Why then, did colonisation of the coast of the Scandinavian Peninsula start so late? From the Allerød oscillation onwards, substantial parts of the Norwegian coast were ice-free and in principle habitable by humans (Bjerck 1994). In our opinion, the most reliable explanation for this is that Doggerland was of no direct relevance or importance for the colonisation process. This is in line with Hein Bjerck's evaluation of the current state of knowledge in the 1990s (Bjerck 1994, 1995). Recent re-examinations of Norwegian artefacts assumed to be of late glacial origin, from the North Sea (T 25726, B 13835), Kolevik VI (T 11701), Ingerdalen (B 13502), Blomvåg (B 14969), Snik (S 7780), Breiviksklubben (S 11678) and Utvik (S 10170) also confirm Bjerck's analysis. None of these artefacts are Late Glacial. They all belong to the younger Mesolithic Fosna tradition (Bjerck 1994; Eigeland 2012; Fischer 2012, see also Eigeland and Solheim 2012). In sum, whilst there could have been people in the northern parts of Doggerland in the final stages of the Pleistocene, they probably did not reach Norway.

If Doggerland is not a plausible candidate as a bridgehead in the colonisation of southern Norway after the Ice Age, it is more likely that it was from the Swedish west coast or Bohuslän that people first colonised Norway. From the Allerød oscillation to the Preboreal period, the Scandinavian glacier front was situated approximately along the so-called Limes Norrlandicus running roughly from the outer Oslo fjord in the west to Stockholm in the east (Andersen 2000). Due to the drainage of the Baltic Ice Lake through the Mälardalen area, Bohuslän was an archipelago with a very productive ecosystem. Lou Schmitt in several articles has made strong arguments for quite a large population of marine-oriented foragers in this area from the end of the Younger Dryas onwards (Schmitt 1994; Schmitt et al. 2006, 2009). However, the entrance to the Norwegian coast from this area was still blocked by a large ice barrier at that time. The Scandinavian glacier consisted of a >100 km-long ice lobe in the Oslo Fjord. This made it very difficult—potentially impossible and certainly unattractive—for humans to reach the southern and western ice-free parts of the Norwegian coast.

Taking a closer look at deglaciation and land uplift at the bridgehead connecting the coast of Norway with the Swedish west coast, it is quite interesting to note that around 9300–9100 cal. BC the glacier had retreated from the Oslo Fjord (Fig. 19.9, Bargel 2005). A sheltered archipelago between Sweden and Norway was created, and it is from this time on that we can trace and convincingly date human settlements along the Norwegian coast. In our opinion this demonstrates that foragers first

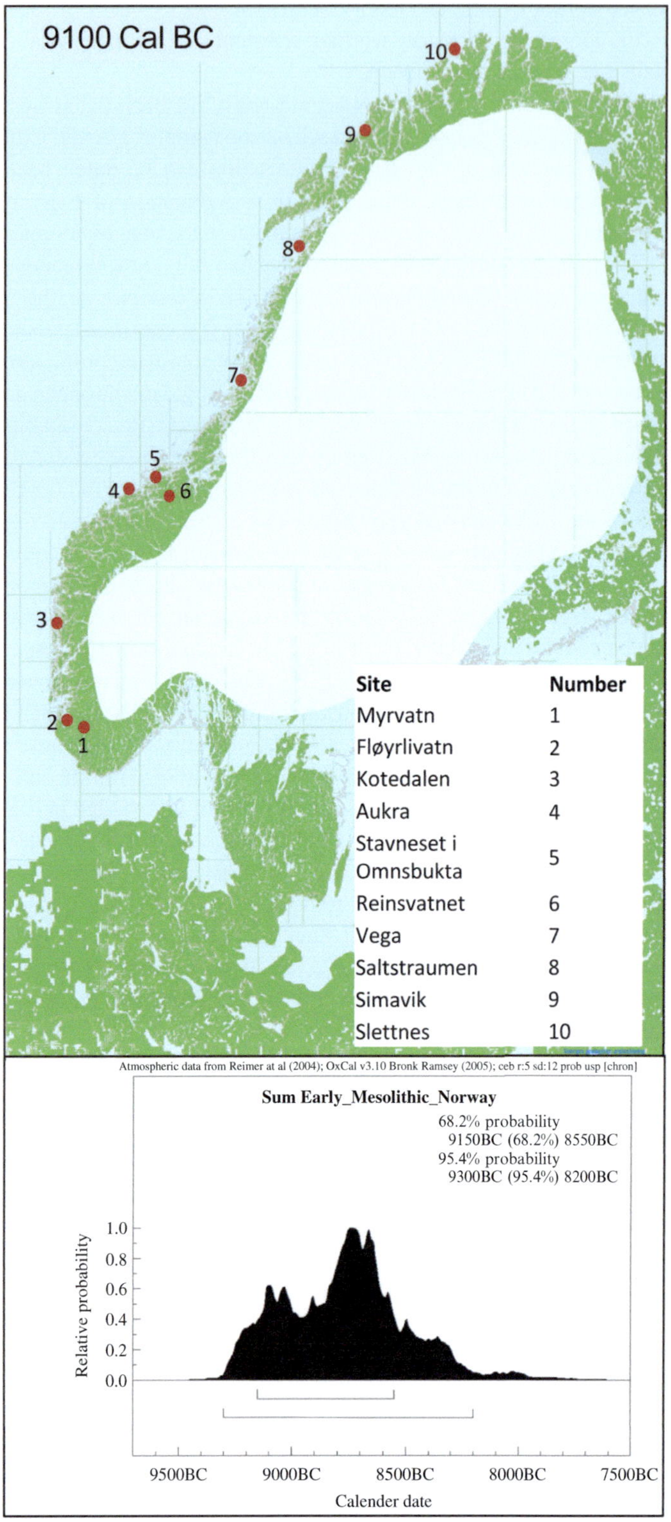


**Fig. 19.8** Radicoarbon dates and locations of earliest Mesolithic sites in Norway (After Glørstad (2013))

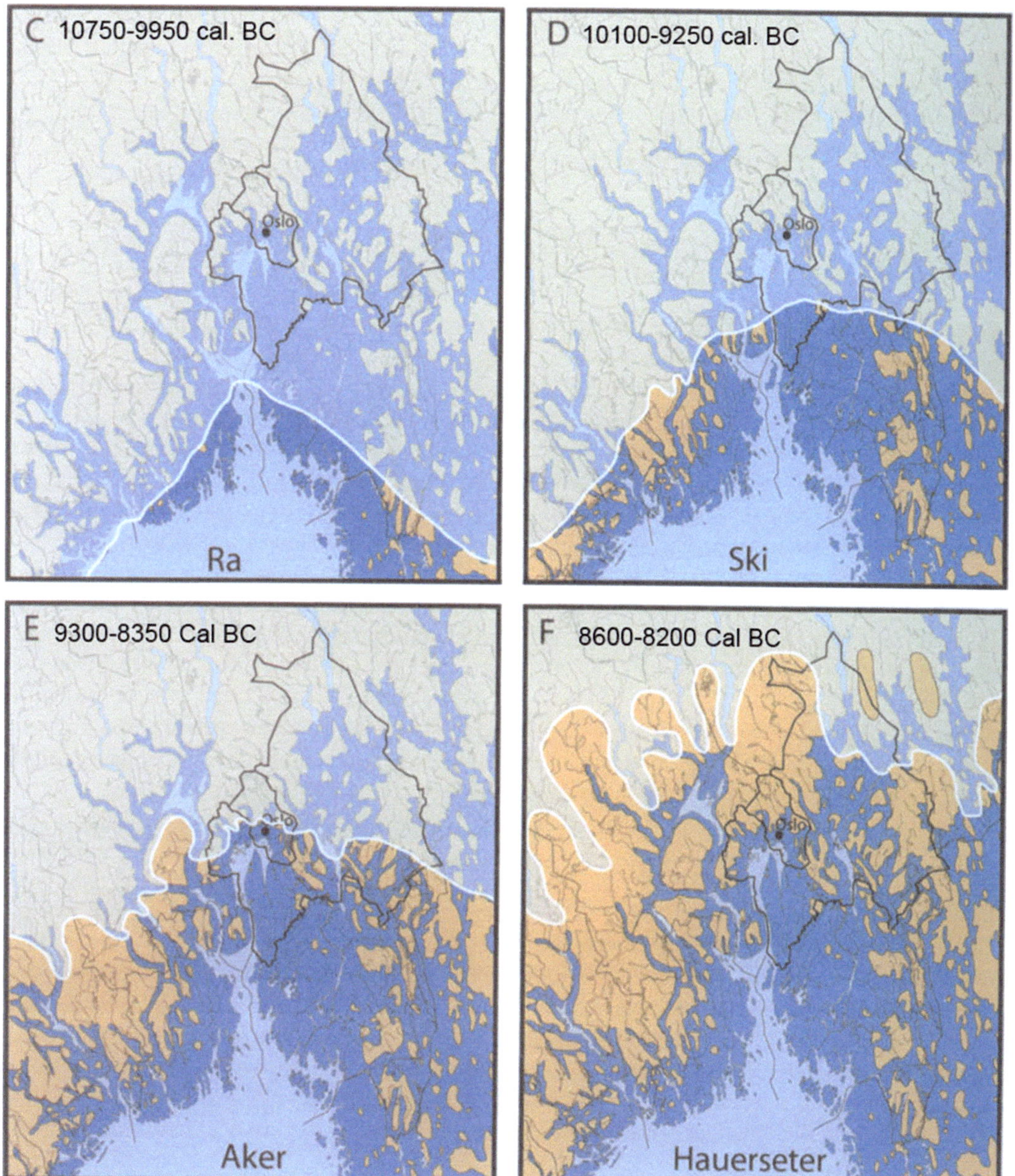


**Fig. 19.9** The different stages in the deglaciation and land rise of the Oslo fjord. Note that a sheltered archipelago is created at approximately 9300–9100 cal BC (Aker stage). The main moraines covering the whole fjord-area are, in spatial and chronological order, the Ra, Ski, Aker and Hauerseter stages. Radiocarbon dates for the moraines are displayed for each stage. The city of Oslo and Akershus County surrounding the city are marked with *bold black lines* to help orientation. Other present-day administrative borders are marked with *thinner black lines*. Present sea is coloured in *light blue*. Present land covered with sea during and after glaciation is coloured in *dark blue*. Dry land at the different stages of deglaciation is coloured in *brown*. The ice-front is marked with a white line and ice cover in transparent grey (Reworked after Bargel (2005))

entered the south Norwegian coast from the Bohuslän area, via the Oslo Fjord corridor. This colonisation process was dependent on sheltered passages and a relatively predictable environment (Glørstad 2014). This enabled safe risk-limited transport and reliable subsistence possibilities. The Early Mesolithic transport and communication system must therefore be considered as not so very different from the rest of the Mesolithic (Glørstad 2013; Åstveit 2014).

## 19.6 Conclusion

In the wake of a national strategy for offshore wind farms in Norway, a small and preliminary joint archaeological and geological research project has been conducted. The aim of this research was to determine the possibilities for human occupation in the southernmost parts of the Norwegian sector of the North Sea during the Late Glacial and Preboreal periods. One target area for the research was to determine the position of the northern shores of Doggerland in the Late Glacial and Preboreal periods as a starting point for finding traces of prehistoric settlement. The area where the research was conducted was located in the shallowest area of the Norwegian part of the continental shelf, just south of the Norwegian Trench. The chances of finding evidence for dry land in the periods in question should in principle be fairly good. However, the results of our work were quite surprising. Analysis of 2D and 3D seismic reflection data did show evidence of dry land forms such as rivers and lakes. The observed features are, however, covered by up to 100 m of sediments. Their age remains somewhat uncertain, but their depth and date suggest that they predate the LGM. An analysis of a sediment core taken from a borehole in SN II was used to reconstruct the environment associated with sedimentation and the age of the shallow sea-bed sediments (0–13 m below the present sea floor). The youngest observed transition, from glacio-lacustrine to marine sediments, was dated to approximately 14,000 cal. BC, the Older Dryas, indicating that this part of Doggerland was in fact not dry land after the LGM, but actually covered by an ice dammed lake and later inundated by salt water from the Atlantic Ocean. This scenario fits well with geological models recently presented by Norwegian and British researchers. In such models a large ice-dammed lake was created by a continuous glacier front from Britain to Norway that blocked drainage of meltwater north- and westwards into the Atlantic Ocean.

Our investigations have yielded no evidence of dry land after the LGM in the study areas. There are several source critical factors that must be taken into consideration before a decisive conclusion can be made. Yet, despite these caveats, our analysis strongly indicates that the distance from Doggerland to Norway, after the LGM and throughout the Late Glacial, was far too long to be crossed by boats or on sea-ice. Our knowledge of Mesolithic voyages in this area unambiguously points towards coast-bound navigation at sea. Distances of open water comparable to the size of the Norwegian Trench dividing Doggerland from Norway do not seem to have been crossed before the end of the Stone Age, 2400 cal. BC. This in turn makes the Swedish west coast and the region of Bohuslän the most likely bridgehead for the colonisation of Norway. People first arrived on the Norwegian shores when a safe and sheltered passage was created between Bohuslän and the Oslo Fjord area at approximately 9300 cal. BC.

## References

Andersen B (2000) Istider i Norge: landskap formet av istidens breer. Universitetsforlaget, Oslo

Anundsen K (1996) The physical conditions for earliest settlement during the last deglaciation in Norway. In: Larsson L (ed) The earliest settlement of Scandinavia. Almquist & Wiksell International, Stockholm, pp 207–217

Åstveit LI (2014) Noen synspunkt på den tidligmesolittiske bosetningen i Sør-Norge. Primitive Tider 16:87–104

Bang-Andersen S (1996) The colonization of southwest Norway: an ecological approach. In: Larsson L (ed) The earliest settlement of Scandinavia and its relationship with neighbouring areas. Almqvist & Wiksell International, Stockholm, pp 219–234

Bang-Andersen S (2003) Southwest Norway at the Pleistocene/Holocene transition: landscape development, colonization, settlement types, settlement patterns. Nor Archaeol Rev 36(1):5–25

Bang-Andersen S (2012) Colonizing contrasting landscapes: the pioneer coast settlement and inland utilization in southern Norway 10,000–9500 years before present. Oxford J Archaeol 31:103–120

Bargel T (2005) Spor etter istiden i Oslo og Akerhus. Gråsteinen 10. Norges Geologiske Undersøkelser, Trondheim

Bjerck H (1990) Mesolithic site types and settlement patterns at Vega, northern Norway. Acta Archaeol 60:1–32

Bjerck H (1994) Nordsjøfastlandet og pionerbosetningen i Norge. Viking LVII:25–58

Bjerck H (1995) The north Sea Continent and the pioneer settlement of Norway. In: Fischer A (ed) Man and sea in the Mesolithic: coastal settlement above and below present sea level. Oxbow Books, Oxford, pp 131–144

Bjerck H (2008a) Norwegian Mesolithic trends: a review. In: Bailey G, Spikins P (eds) Mesolithic Europe. Cambridge University Press, Cambridge, pp 60–106

Bjerck H (2008b) Tidligmesolittisk tid og fosnatradisjon 9500–800 BC. In: Bjerck H, Åstveit LI, Meling T, Gundersen J, Jørgensen G, Normann S (eds) NTNU Vitenskapsmuseets arkeologiske undersøkelser Ormen Lange Nyhamna. Tapir akademisk forlag, Trondheim, pp 552–570

Blystad P (1989) Nordsjøen i seinkvartær tid AmS–Rapport 1. Arkeologisk museum, Stavanger

Bos JAA, van Geel B, van der Plicht J, Bohncke SJP (2007) Preboreal climate oscillations in Europe: wiggle–match dating and synthesis of Dutch high-resolution multi-proxy records. Quat Sci Rev 26:1927–1950

Clark CD, Hughes ALC, Greenwood SL, Jordan C, Sejrup HP (2012) Pattern and timing of retreat of the last British-Irish Ice Sheet. Quat Sci Rev 44:112–146

Coles BJ (1998) Doggerland: a speculative survey. Proc Prehist Soc 64:45–81

Crumlin-Pedersen O, Trakadas A (eds) (2003) Hjortspring: a pre-Roman iron-age warship in context, Ships and Boats of the North, vol 5. Viking Ship Museum, Roskilde

Eigeland L (2012) En teknologisk vurdering av mulige senglasiale funn i Norge. In: Glørstad H, Kvalø F (eds) Havvind – Paleogeografi og arkeologi. Norwegian Maritime Museum, report 2012:12. Norwegian Maritime Museum, Oslo, pp 57–69

Eigeland L, Solheim S (2012) Blomvågfunnet – veid og funnet for let. Viking LXXV:7–26

Fischer A (2002) Food for feasting? In: Fischer A, Kristiansen K (eds) The Neolithisation of Denmark: 150 years of debate. J R Collins Publications, Sheffield, pp 341–394

Fischer A (2012) Vurdering af Vestnorske fund, som kandiderer til palæolitisk datering. In: Glørstad H, Kvalø F (eds) Havvind – Paleogeografi og arkeologi. Norwegian Maritime Museum, report 2012:12. Norwegian Maritime Museum, Oslo, pp 50–56

Fuglestvedt I (1989) Norges landnåm? Nicolay 51(1):17–46

Fuglestvedt I (2001) Pionerbosetningens fenomenologi Sørvest-Norge og Nord-Europa 10200/10000–9500 BP. University of Bergen, Doctoral thesis

Fuglestvedt I (2007) The Ahrensburgian Galta 3 site in SW Norway. Dating, technology and cultural affinity. Acta Archaeol 78(2):87–110

Fuglestvedt I (2009) Phenomenology and the pioneer settlement on the western Scandinavian Peninsula. Bricoleur Press, Göteborg

Fuglestvedt I (2012) The pioneer condition on the Scandinavian Peninsula: the last frontier of a 'Palaeolithic Way' in Europe. Nor Archaeol Rev 45(1):1–29

Gaffney V, Thomson K, Fitch S (eds) (2007) Mapping Doggerland. The Mesolithic landscapes of the southern North Sea. Archaeopress, Oxford

Gjevik B (2009) Flo og fjære langs kysten av Norge og Svalbard. Farleia forlag, Oslo

Glørstad H (2009) The Northern Province? The Neolithisation of southern Norway. In: Glørstad H, Prescott C (eds) Neolithisation as if history mattered. Processes of Neolithisation in North-Western Europe. Bricoleur Press, Gothenburg, pp 135–168

Glørstad H (2010) The structure and history of the Late Mesolithic Societies in the Oslo Fjord Area 6300–3800 BC. Bricoleur Press, Gothenburg

Glørstad H (2011) Historical ideal types and the transition to the Late Neolithic in South Norway. In: Prescott C, Glørstad H (eds) Becoming European: the transformation of third millennium Europe – and the trajectory into the second millennium BC. Oxbow Books, Oxford, pp 82–99

Glørstad H (2013) Where are the missing boats? The pioneer settlement of Norway as long-term history. Norwegian Archaeological Review 46(1):57–80

Glørstad H (2014) Deglaciation, sea-level changes and the Holocene colonisation of Norway In: Harff J, Bailey G, Lüth F (eds) Geology and archaeology: submerged landscapes of the continental shelf. Geological Society, London, Special Publications, 411: 9–25

Glørstad H, Kvalø F (2012) Havvind – Paleogeografi og arkeologi Norsk maritimt museum – Arkeologisk rapport 2012:12. Norsk Maritimt Museum, Oslo

Grimm SB, Weber M-J (2008) The chronological framework of he Hamburgian in the light of old and new 14C-dates. Quartär 55:17–40

Gyllencreutz R, Backman J, Jakobsson M, Kissel C, Arnold E (2006) Postglacial palaeoceanography in the Skagerrak. Holocene 16:975–985

Hafeez A, Planke S, Hammer Ø, Faleide JI (2012) Post-Glacial paleogeography of the southern Norwegian North Sea. Volcanic Basin Petroleum Research, Oslo

Hagen A (1967) Norges oldtid. JW Cappelens forlag, Oslo

Hammer Ø, Planke S, Hafeez A, Hjelstuen BO, Faleide JI, Kvalø F (2016) Agderia – a postglacial lost land in the southern Norwegian North Sea. Nor J Geol 96(1):43–59

Indrelid S (1975) Problems relating to the Early Mesolithic settlement of Southern Norway. Nor Archaeol Rev 8(1):1–18

Indrelid S (1989) Dei fyrste menneskja på. Vestlandet Naturen 113(1):25–35

Inst. S nr. 145 (2007–2008) Innstilling til Stortinget frå energy- og miljøkomiteen St.meld. nr. 34 (2006–2007) om Norskklimapolitikk

Johansen AB, Rokoengen K (1996) Possibilities for early settlement on the Norwegian continental Shelf. Norsk Geol Tidsskr 76(2):121–125

Johannessen L (2009) Ahrenburgkulturens lokalitetsplassering. En redegjørelse av forholdet mellom kyst og inland. Master thesis, University of Oslo

Kvalø F (2007) Oversjøiske reiser fra Sørvest-Norge til Nordvest-Jylland I elder bronsealder. In: Hedeager L (ed) Sjøreiser og stedsidentitet. Unipub forlag, Oslo, pp 11–134

Lambeck K (1995) Late Devensian and Holocene shorelines of the British Isles and North Sea from models of glacio-hydro-isostatic rebound. J Geol Soc Lond 152:437–448

Lambeck K, Purcell A, Zhao J, Svensson NO (2010) The Scandinavian ice sheet: from MIS 4 to the end of the last glacial maximum. Boreas 39 (2): 410–435. Full data set published at http://www.peoplersesanueduau/lambeck_k/indexphp?p=research

Malmer MP (2002) The Neolithic of south Sweden: TRB, GRK, and STR. The Royal Swedish Academy of Letters, History and Antiquities, Stockholm

Mikkelsen E (1978) De første «nordmenn» på bunnen av Nordsjøen. Naturen 3:99–105

Nummedal A (1923) Om flintpladsene. Norskgeologisk Tidsskrift VII:89–141

NVE (Norges vassdrags- og energidirektorat) (2012) Havvind Strategisk konsekvensutredning

Odner K (1966) Komsakulturen i Nesseby og Sør-Varanger, Tromsø museums skrifter, vol XII. Universitetsforlaget, Tromsø

Österholm S (1988) I utriggarkanot over Östersjön. Populär Arkeologi 6(1):26–30

Österholm I (1989) Bosättningsmönstret på Gotland under stenåldern: en analys av fysisk miljö, ekonomi och social struktur. Stockholms universitet, Institutionen för arkeologi, Stockholm

Østmo E (2005) Over Skagerak i steinalderen. Noen refleksjoner om oppfinnelsen av havgående fartøyer i Norden. Viking LXVIII:55–82

Østmo E (2008) Some notes on the development of shipbuilding and overseas connections in Scandinavian prehistory. In: Chilidis K (ed) Facets of archaeology. Essays in honour of Lotte Hedeager on her 60th birthday. Oslo Academic Press, Oslo, pp 265–274

Østmo E (2011) Late Neolithic expansion to Norway. The beginning of a 4000 year-old shipbuilding tradition. In: Prescott C, Glørstad H (eds) Becoming European: the transformation of third millennium Europe – and the trajectory into the second millennium BC. Oxbow Books, Oxford, pp 63–69

Petersen EB (2009) Late glacial and early holocene climatic oscillations and human waves of immigration into Scandinavia. IOP Conf Ser: Earth Environ Sci 6(7):1–13

Petroleum Geo–Services (PGS) (2012) Central North Sea electronic overview of datasets. Available online http://pgscom/en/Data_Library/Northwest–Europe/Central–North–Sea/

Prescott C (2011) Third millennium transformations in Norway: modeling an interpretative platform. In: Prescott C, Glørstad H (eds) Becoming European: the transformation of third millennium Europe – and the trajectory into the second millennium BC. Oxbow Books, Oxford, pp 115–127

Prescott C, Glørstad H (eds) (2011) Becoming European: the transformation of third millennium Europe – and the trajectory into the second millennium BC. Oxbow Books, Oxford

Prescott C, Walderhaug E (1995) The last frontier? Processes of Indo-Europeanization in Northern Europe: the Norwegian case. J Indo-Eur Stud 23(3–4):257–280

Prøsch-Danielsen L (1993) Naturhistoriske undersøkelser i Rennesøy og Finnøy kommuner, Rogaland, Sørvest-Norge. Arkeologisk museum, Stavanger

Prøsch-Danielsen L, Høgestøl M (1995) A coastal Ahrensburgian site found at Galta, Rennesøy, Southwest Norway. In: Fischer A (ed) Man and sea in the Mesolithic: coastal settlement above and below present sea level. Oxbow Books, Oxford, pp 123–130

Reid C (1913) Submerged forests. Cambridge University Press, London

Rokoengen K, Johansen AB (1996) Possibilities for early settlement on the Norwegian continental Shelf Norsk. Geologisk Tidsskrift 76(2):121–125

Rolfsen P (1972) Kvartærgeologiske og botaniske betingelser for mennesker I Sør-Norge I seinglasial og tidlig postglasial tid. Viking XXXVI:131–153

Rust A (1942) Über die endglaciale Kulturentwicklung im rechtselbischen Nordwesteuropa unter Berücksichtung der geologischen und siedlungsarchäologischen Verhältnissen. Offa 6(7):52–75

Schmitt L (1994) The Hensbacka: a subsistence strategy of continental hunter–gatherers, or an adaptation at the Pleistocene-Holocene Boundary? Oxford J Archaeol 13(3):245–263

Schmitt L, Larsson S, Schrum C, Alekseeva I, Tomczak M, Svedhage K (2006) 'Why they came': the colonization of the coast of Western Sweden and its environmental context at the end of the last glaciation. Oxford J Archaeol 25(1):1–28

Schmitt L, Larsson S, Burdukiewicz J, Ziker J, Svedhage K, Zamon J, Steffen H (2009) Chronological insights, cultural change, and resource exploitation on the west coast of Sweden during the Late Palaeolithic/Early Mesolithic transition. Oxford J Archaeol 28:1–27

Selsing L (2012) The early settlement of South Norway after the last deglaciation: a diasporic perspective. Nor Archaeol Rev 45(2):177–205

Skar B (1995) Lokalisering av undersjøiske steinalderboplasser i Sør-Norge. En foreløpig strategi. In: Marinarkeologi Kunnskapsbehov Rapport fra seminar 22–25 september 1993. Karshavn ved Lindesnes Norges forskningsråd program for forskning om kulturminnevern, Oslo, pp 51–62

Thomson K, Gaffney V (2007) 3D seismic reflection data, associated technologies and the development of the project methodology. In: Gaffney V, Thomson K, Fitch S (eds) Mapping Doggerland. The Mesolithic landscapes of the southern North Sea. Archaeopress, Oxford, pp 23–31

# Chapter 20
# Doggerland and the Lost Frontiers Project (2015–2020)

**Vince Gaffney, Robin Allaby, Richard Bates, Martin Bates, Eugene Ch'ng, Simon Fitch, Paul Garwood, Garry Momber, Philip Murgatroyd, Mark Pallen, Eleanor Ramsey, David Smith, and Oliver Smith**

**Abstract** As this volume, the final monograph of the SPLASHCOS network, was being finalised, the European Research Council agreed to fund a major new project relating to the marine palaeolandscapes of the southern North Sea. Emerging from the earlier work of the North Sea Palaeolandscapes Project (NSPP), the Lost Frontiers project seeks to go beyond the maps generated by that ground-breaking research. Led by researchers in the fields of archaeogeophysics, molecular biology and computer simulation, the project seeks to develop a new paradigm for the study of past environments, ecological change and the transition between hunter gathering societies and farming in North West Europe. Following from earlier work, the project will seek to release the full potential of the available seismic reflectance data sets to generate topographical maps of the whole of early Holocene Doggerland that are as accurate and complete as possible. Using these data, the study will then reconstruct and simulate the emerging palaeoenvironments of Doggerland using conventional palaeoenvironmental data, as well as ancient DNA extracted directly from sediment cores along the routes of two submerged river valleys. Using this base data, the project aims to transform our understanding of the colonisation

V. Gaffney (✉) • S. Fitch • P. Murgatroyd • E. Ramsey
School of Archaeological Sciences, University of Bradford, Bradford, West Yorkshire BD17 1DP, UK
e-mail: v.gaffney@bradford.ac.uk; S.Fitch@bradford.ac.uk; P.Murgatroyd@bradford.ac.uk; ellie.ramsey@gmail.com

R. Allaby • M. Pallen • O. Smith
School of Life Sciences, Gibbet Hill Campus, University of Warwick, Coventry CV4 7AL, UK
e-mail: R.G.Allaby@warwick.ac.uk; m.pallen@warwick.ac.uk; O.Smith@warwick.ac.uk

R. Bates
School of Geography and Geosciences, Earth Sciences, Irvine Building, St Andrews, UK
e-mail: crb@st-andrews.ac.uk

M. Bates
School of Archaeology, History and Anthropology, University of Wales Trinity Saint David, Lampeter Campus, Ceredigion SA48 7ED, UK
e-mail: m.bates@tsd.ac.uk

E. Ch'ng
School of Computer Science, University of Nottingham, Ningbo, China
e-mail: eugene.chng@nottingham.edu.cn

P. Garwood • D. Smith
Department of Classics, Ancient History and Archaeology, University of Birmingham, B15 2TT, Birmingham, UK
e-mail: p.j.garwood@bham.ac.uk; d.n.smith@bham.ac.uk

G. Momber
Maritime Archaeology Trust, National Oceanography Centre, Southampton SO14 3ZH, UK
e-mail: garry.momber@maritimearchaeologytrust.org



G.N. Bailey et al. (eds.), *Under the Sea: Archaeology and Palaeolandscapes of the Continental Shelf*, Coastal Research Library 20, DOI 10.1007/978-3-319-53160-1_20

and development of floral, faunal and human life, to explore the Mesolithic landscapes and to identify incipient Neolithic signals indicating early contact and development within the region of Doggerland.

## 20.1 Introduction

In the penultimate conference of the SPLASHCOS network held in Esbjerg, Denmark in April 2013, the lead author of this chapter provided an overview of how extensive analysis of the submarine landscapes of the southern North Sea might proceed within the context of larger industrial collaboration. In the period following that conference, and also following an immense amount of new work within the region, it remains true that many of the issues identified at that time have persisted as major research concerns. It had been clear for some considerable time that extensive, habitable landscapes had preserved vital evidence beneath water-borne deposits and that these were probably comparatively untouched by subsequent development (Coles 1998). However, aside from sporadic archaeological finds much of the area had subsequently remained 'terra incognita' until quite recently, when work by a number of the authors of this paper rendered the landscape amenable to extensive landscape mapping using a combination of remote sensing technologies (Gaffney et al. 2007, 2009, also Sturt et al. 2013). What has changed in recent times, perhaps, has been the manner in which these landscapes have been viewed by archaeologists. Despite its pivotal location as the gateway to the UK landmass and Scandinavia from continental Europe, the area of the North Sea has frequently been regarded as archaeologically marginal or even ignored as an inconvenient data gap or bridge (see also Sturt et al. Chap. 28). The work of SPLASHCOS and its partners has changed this perception and highlighted the importance of submerged sites in preserving and providing data that may challenge the current consensus regarding major shifts in lifestyle (Benjamin et al. 2011; Flemming 2004; Ransley et al. 2013; Flemming et al. 2014). The opening paragraph of the document produced by the North Sea Prehistory Research and Management Framework (NSPRMF) succinctly outlines the scientific importance of Doggerland and such landscapes:

> The North Sea Basin has undergone many changes driven by climate change and, at times, it may have been the largest wetland environment in Europe and a major focus of population. For these reasons, in relation to the terrestrial archaeological record, its archaeology may not just be 'more of the same', but perhaps qualitatively different from what we know already. Furthermore, the spectacular finds of the last decade have shown the excellent quality of the remains of former plant and animal communities (including early humans) within the sedimentary sequence; and several studies have demonstrated the extensive survival of entire submerged landscapes. The area has enormous potential for studying the relationships between early humans and their landscapes, and plainly this is of world-wide significance. (Peeters et al. 2009, 7)

Perceptions of marine palaeolandscapes have fundamentally changed, from a landscape that has remained frustratingly inaccessible to the interests of archaeologists, to one that now opens up unique opportunities to explore critical historical and cultural events centred on Doggerland and its pivotal role as a porous and moving frontier involving sea level changes, population movements and cultural shifts. These include the progress and nature of hunter-gatherer resettlement of the area following climatic amelioration, and the impact of land loss induced by climate change on human communities either on the coastal plain, on surviving islands or on the terrestrial periphery throughout a period that encompasses both the Late Palaeolithic and Mesolithic periods and, probably, the succeeding earlier Neolithic.

The aim of this chapter is to focus on the relationship between submerged landscapes of the North Sea or Doggerland in the final stages of sea-level rise and the transition from hunter-gatherer to farmer, to highlight the evidence that significant areas of now-drowned land were still present at the time of early Neolithic dispersal, to outline a new underwater project aimed at pursuing this theme through new mapping, coring and the analysis of ancient DNA signals embedded in underwater sediments, and to comment on the history and use of the term Doggerland.

## 20.2 Submerged Landscapes and the Transition to Farming

Recently, Flemming et al. (2014) have attempted to assess the discipline's perceived priorities (Table 20.1). Whilst the results exhibit an important emphasis on the understanding of earlier Holocene and Late Pleistocene topics, the relative lack of concern relating to the introduction of farming and domesticates into these environments is surprising and only achieves a rank of 1 (out of 10, where 1 ranks low). Although there has been passing academic interest in the final phases of inundation and the movement of people in relation to Neolithic developments (Coles 1999; Sturt and Van der Noort 2013), regional investigations of sea-level change have remained framed in terms of the transition to the current terrestrial context and the impact on the hunter-gatherer environment rather than the transition from hunter-gather to farmer societies. This situation exists despite a number of indicators that the final stages of sea-level rise were taking place even as Neolithic farmers were becoming established in some of the surrounding regions.

First, there is the likelihood that the potential offshore area available to farmers during the earliest Neolithic may have been considerable, although its extent remains uncertain and the territories largely unexplored. Secondly, the position of later prehistoric monuments at the marine interface such as Seahenge dated at c. 2049 BC (Brennand and Taylor 2003) clearly begs the question as to where the earlier Neolithic coastline actually stood and what the implications are in having an enlarged coastal strip during this period.

**Table 20.1** Ranking of research topics according to question 18 of a questionnaire distributed to member Departments and agencies of the European Archaeological Council

| Research theme | Score |
|---|---|
| Human response to rising/falling sea level during climate change | 10 |
| Origins of exploitation of marine resources and marine diet | 10 |
| Reconstruction of river channels and fresh-water drainage or karst on the submerged continental shelf | 9 |
| Demography and human response to climate change | 8 |
| Palaeo-environments and climate on the continental shelf at the Last Glacial Maximum | 7 |
| Earliest prehistoric occupation of islands presently separated from the mainland of Europe | 7 |
| Migration routes to and from the coast of your country | 7 |
| Origins of prehistoric seafaring | 6 |
| Reconstruction of vegetation and fauna of the continental shelf, providing an environment for hominins | 6 |
| Prehistoric non-lithic material culture which only survives in permanently waterlogged sediments | 6 |
| Palaeolithic re-population of recently deglaciated coastal zones | 5 |
| Study of population that has contributed to DNA of your region | 4 |
| Food, diet, population demographics, diseases, and life expectancy of Palaeolithic or Mesolithic populations | 4 |
| Changes in subsistence, such as the introduction of agriculture | 4 |
| Population centres as a refugium from nearby lands abandoned during glacial periods | 3 |
| Early hominin migrations and areas of occupation during previous glacial cycles | 3 |
| Domestication of animals and early farming and crops | 1 |
| Hominin and Hominin and human migration or diffusion pathways from Africa into Europe | 0 |

For further details of the questionnaire see Flemming et al. (2014, Annexe 5, pp. 155–160). The number of respondents was 15. The data in the table are from Flemming et al. (2014, Table 3.2, p. 40), Respondents were asked to indicate the main research objectives listed in the left-hand column which apply to the submerged prehistoric continental shelf of their country. The score for each question is shown in the right-hand column (After Flemming et al. (2014, Table 3.2, p. 4))

**Fig. 20.1** The North Sea and the Dogger Bank

Finally, occasional finds of artefacts on the seabed such as the early Neolithic Michelsberg axes from the Brown Banks (Gaffney et al. 2009; van der Noort 2013, p. 111) have usually been interpreted as votive deposits either at sea or at low tide as gifts to hunter-gatherer ancestors of the inundated plains. Events such as these might find a different interpretation if it could be proven that there had been a more extensive, earlier Neolithic landscape within the southern North Sea. Such a change of perception finds an earlier parallel in the archaeological historiography of the southern North Sea in the transition of the iconic 'Colinda harpoon', from being treated as an artefact dropped at sea to one derived from a terrestrial landscape following the Godwins' (1933) formative study of its associated peats (see also Sturt et al. Chap. 28).

The archaeological imperative to try to understand the nature of the later transitional period of human occupation of the North Sea has recently come to the fore with publication of the analysis of ancient traces of DNA within sediments (sedaDNA) from the submarine Mesolithic site at Bouldnor Cliff, off the Isle of Wight in the Western Solent (Figs. 20.1, 20.2, and 20.3, Momber et al. 2011; Logan et al. 2015; Smith et al. 2015a, b, Momber and Peeters Chap. 21).

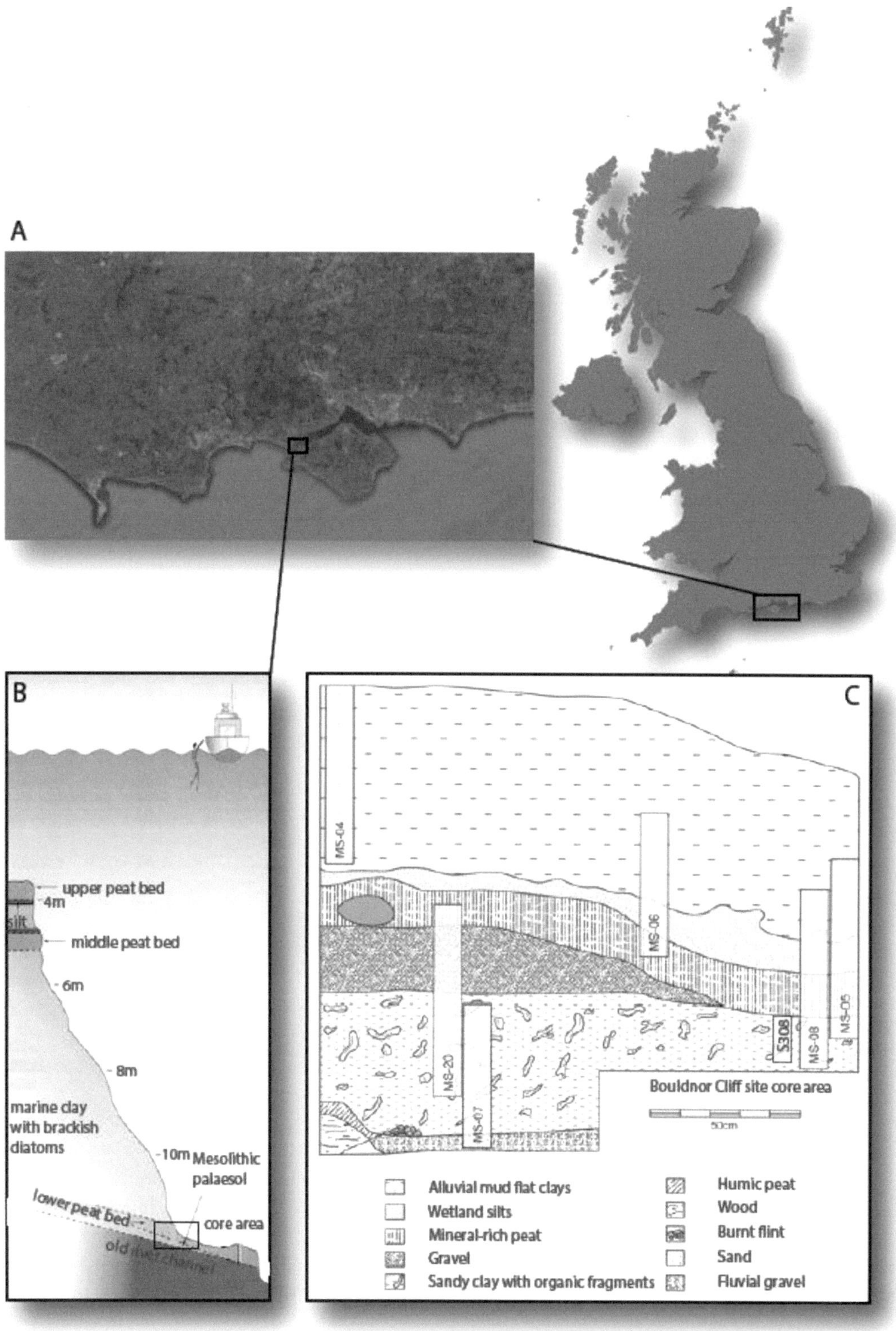


**Fig. 20.2** Sampling location of sedaDNA from Bouldnor Cliff. (**a**) Location of the Bouldnor Cliff site in the Solent on the coast of the Isle of Wight. (**b**) Large-scale stratigraphic profile of the site, indicating the depth and location of the Mesolithic palaeosol and the location of the area from which cores were taken. (**c**) Core area in detail, stratigraphic profile of the site indicating core sites (MS-04-8, and MS-20), and approximate location of the sediment sample taken for sedaDNA analysis

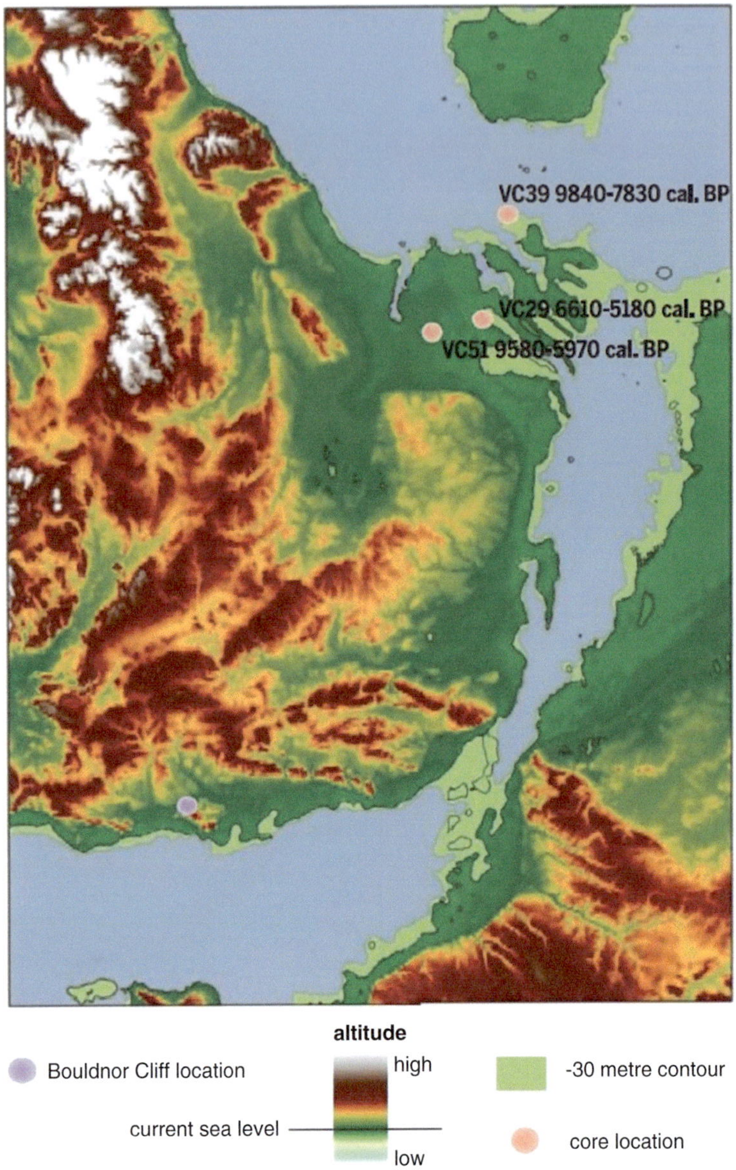


**Fig. 20.3** Generalized map of potential coastal extent around southern Britain 9840–7830 cal BP. Vibrocores through submerged old land surfaces off the coast of Britain at depths of 31.68 m (VC39), 24.01 m (VC51), and 23.9 0 m (VC29) were OSL-dated The map extrapolates the contour of the VC29 vibrocore site (Source: Humber Regional Environmental Characterisation Data from Tappin et al. (2011))

The site at Bouldnor has been dated to between 8030 and 7980 cal BP and therefore dates, conventionally, to the later Mesolithic in Britain. The results of study of the sedaDNA profile, unsurprisingly, revealed a wooded landscape that included oak, poplar, apple, and beech, with grasses and a few herbs. Oak and poplar were also detected in the pollen profile, whereas oak, apple, and alder have been reported in archaeological remains of worked wood at the site. SedaDNA analysis also revealed a faunal profile compatible with contemporaneous human activity, with an abundant presence of Canidae and Bovidae. Canidae may refer either to dog or wolf whilst material interpreted most likely as *Bos* appears to be supported by the find of an auroch bone at the site. The presence of deer, members of the grouse family, and rodents, all compatible with the contents of a Mesolithic diet shared by humans and dogs, essentially reinforces the impression of a later Mesolithic environment.

What is surprising is that the lower strata provide DNA evidence of Triticeae, that these finds relate to domesticated einkorn wheat, and that these indicators dominate the plant profile in the upper strata (81% of the signal for flowering plants). The pollen from this zone does not indicate the presence of larger cereals, which would be expected if they had been grown locally, Assuming that the evidence of DNA is not intrusive from later deposits, then the source of the *Triticum* signal must come from wheat imported from elsewhere.

The occurrence of domesticated wheat 8,000 years ago on the British continental shelf appears to be unexpectedly early, given that Neolithic assemblages are not established on the mainland of north-west Europe until 7500 BP in the central Rhineland, 7300 BP in the Rhine/Maas delta and adjacent areas, and 7400 BP in western France (Crombé and Vanmontfort 2007; Louwe Koojimans 2007; Marchand 2007; Robb 2013; Tresset and Vigne 2007).

Publication of these results has, not surprisingly, proven controversial and several responses have sought to counter the results largely on technical grounds. Despite this, the team involved in the study has not identified any reason to suggest that the results reflect intrusive genetic material nor any technical issue that might have led to a false positive identification of grain DNA within the sediment and during the late Mesolithic period assigned to the sediment layer by the excavators. Currently, the results of sedaDNA analysis at Bouldnor are interpreted as evidence for the presence of wheat, a domesticated plant associated with the Neolithic, at a site on the British continental shelf 2,000 years earlier than would be expected from the known archaeology of the British mainland and 400 years earlier than in proximate sites on the European continent.

Our current understanding of the marine data clearly suggests the existence of significant emerged landscapes along the east coast dating to the 9th–7th millennia BC. The presence of a large Dogger Island until at least the 8th millennium BC is also very likely and detailed modelling of coastlines during this period around south-east England (and similar coastlands on the French side of the Channel), is likely to demonstrate the former existence of extensive littoral plains and estuarine environments that were almost certainly occupied by large hunter-gatherer-fisher populations and probably later farmers. The fact remains that these landscapes are almost entirely unexplored and that recent interpretations of the distribution of late Mesolithic and early Neolithic populations in north-west Europe rely almost entirely on evidence from sites that would have been some way inland during the 9th–6th millennia BC. As a result of this absence of evidence, little account is taken of social landscapes in littoral, coastal-plain and estuarine areas where the majority of people may well have lived.

At this point, whilst the presence of wheat may indicate exchange via extensive social networks stretching across continental Europe, the most likely agency of movement for such material would have been by boat. It is clear that the dissemination of some Neolithic traits may have occurred very rapidly along the coasts in contrast to the terrestrial interior (Forenbaher and Kaiser 2011). If so, the evidence that coastal lansdcapes once existed but have now been drowned except in rare cases such as Bouldnor Cliff, and the lack of comparative studies using novel technologies including sedaDNA analysis and the investigation of submerged landscapes, suggests that ephemeral pioneer events relating to the spread of farming practice may never be detected using the currently available data sets and that further detailed research on the marine Mesolithic–Neolithic transition is urgently needed.

## 20.3 The Lost Frontiers Project

In 2014, as the Bouldnor Cliff sedaDNA data was being prepared for publication, the 'Lost Frontiers' project team prepared a European Research Council Advanced Research Grant application with the aim of investigating the following questions:

1. How did the Early Holocene Doggerland landscape develop in the face of the ameliorating climate and what was the impact of climate-related land loss on the plant, animal and, ultimately, human communities of the North Sea plain? (Scoring highly (10, 9, 8 and 6) as research priorities according to the recent assessment ((Table 20.1))
2. At what time did the Mesolithic people of the north-west plains make contact with Neolithic technologies and practices and what form did this contact take? (Scoring 7 and 4 (Table 20.1))
3. Has our view of the Mesolithic–Neolithic transition been drastically skewed by relying predominantly on land-based sites? If so, what changes need to be made to existing theories as a result of the new data?

In order to achieve these goals the project proposed the following primary objectives:

1. Produce a near complete topographic map of early Holocene Doggerland, primarily using seismic reflection data fully integrated with other data sources (e.g., sea-level curves, seabed cores).
2. Reconstruct the Early Holocene environments of Doggerland through conventional means and by using and developing the emerging methodologies for extracting plant and animal DNA directly from sediments cored from the sea-bed
3. Explore these data for evidence of the colonisation of plants and animals associated with climatic amelioration, and also for later markers associated with Neolithisation, including non-indigenous flora and fauna
4. Model possible dynamic scenarios for the geomorphological, ecological and, by inference, the human history of Doggerland using complex systems simulations.
5. Provide a robust, global framework for future research and management of these extraordinary scientific, heritage and educational resource associated with comparable landscapes around the world.

The proposal was submitted under the title 'Europe's Lost Frontiers: exploring climate change, settlement and colonisation of the submerged landscapes of the North Sea basin using ancient DNA, seismic mapping and complex systems modelling'. The project has since been accepted and was initiated in December 2015.

Whilst it is not the intention to reproduce the ERC application here, it is worth outlining some of the principles underpinning the project in order to disseminate awareness, encourage methodological debate and, hopefully, to stimulate collaborative research between the emerging research groups identified through the SPLASHCOS initiative and dotted around Europe, the Mediterranean and similar regions across the globe.

A key target within the project is to extend the mapping of Doggerland to achieve the maximum coverage possible and provide a suitable context for simulation studies of human movement and settlement. Whilst new data sources, including surveys associated with recent wind-farm development, have become available, the coverage of 3D seismic data previously used by the team remains of limited extent. Much of the North Sea is covered only by 2D seismic data, and the intensity of this coverage is variable (Fig. 20.4). The lack of detailed 3D survey throughout the area is problematic but the project team have demonstrated that 2D data may be used as a proxy when coverage is suitably dense (Fitch et al. 2011).

The interpreted data from earlier projects is, however, adequate to identify a series of palaeochannels that are known to possess sediments suitable to be cored for palaeoenvironmental data. Two

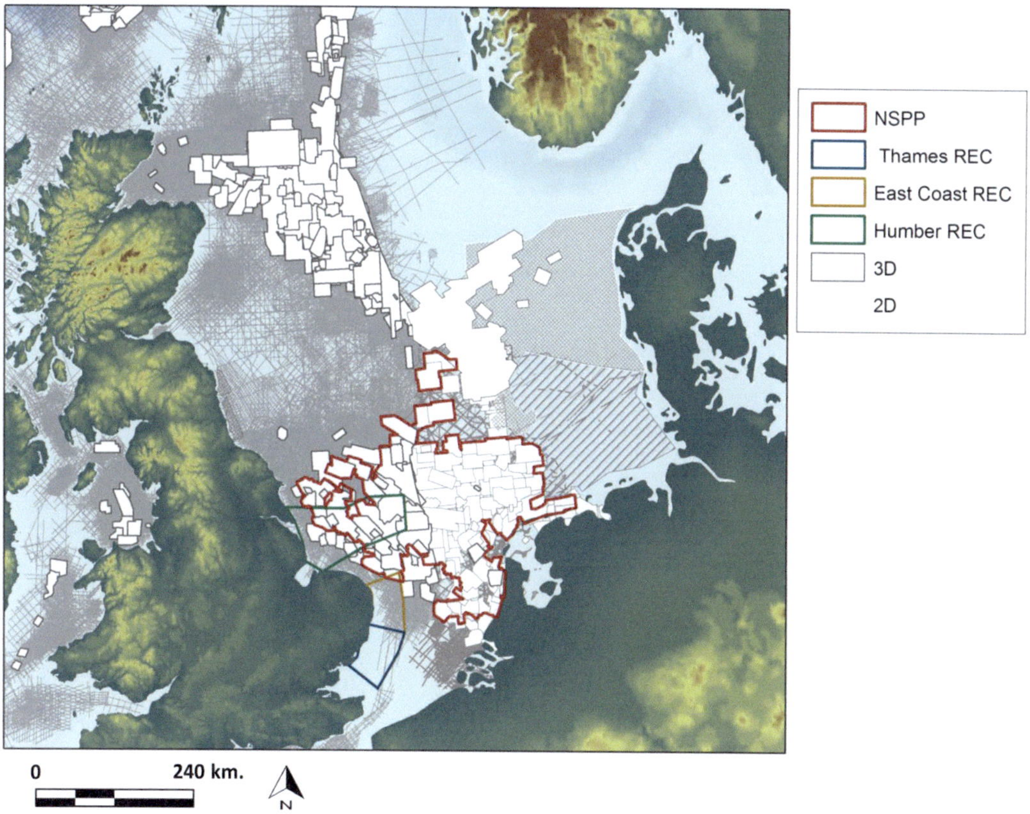


**Fig. 20.4** General map of selected available datasets for the southern North Sea and UK northern sector

valleys will act as sampling transects stretching from the mainland of Britain to the Outer Silver Pit Lake and from the Lake onto the area associated with the Dogger Island (Figs. 20.5 and 20.6). These transects seeks to cross the areas putatively associated with the Neolithic–Mesolithic transition and to establish the nature and rate of transgression in a consistent manner. A minimum of 100 cores will be assessed for the state of preservation of material suitable for 'conventional' palaeoenvironmental analysis (pollen, plant macrofossils, insect remains, ostracods/foraminifera and diatoms) and radiocarbon determination. Radiocarbon and OSL samples will be submitted initially for 'rangefinder' dates. Cores assessed to have sufficient potential will then be analysed in further detail for sedaDNA abstraction.

The data from improved, extensive seismic mapping, sedaDNA and palaeoenvironmental analysis will then be used to build dynamic models of the changing geomorphology and ecology of Doggerland, from the opening of the Holocene around 12,000 BP until its eventual total inundation around 7500 BP. In a departure from conventional approaches, the intention is to use a complexity systems modelling methodology such as Agent-Based Modelling (ABM) for the modelling of ecological processes and for simulating the dynamic interaction between the environment and the animals and plants which inhabit it. The scale of the modelling which will be attempted has little precedent, and developing the appropriate methodologies and software tools will form a significant part of the work programme over the following 5 years.

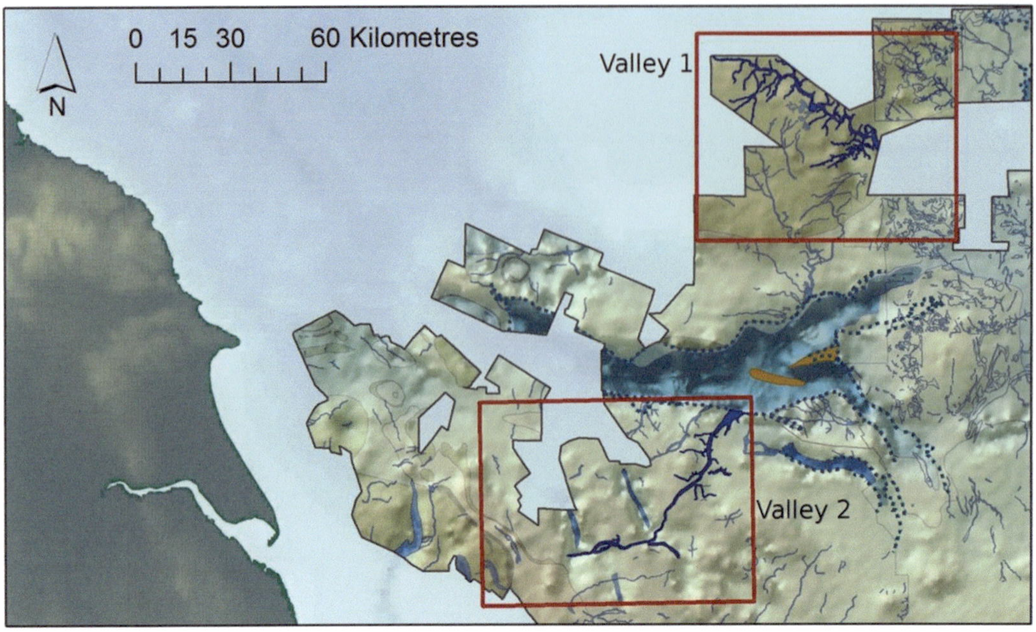


**Fig. 20.5** Map showing course of two submerged river valleys to be targeted for coring by the Lost Frontiers project team, overlaid on NSPP project base map (Gaffney et al. 2007)

## 20.4 Conclusion

Less than 3 months into the project such objectives are at this moment aspirational but the methodology described here should provide a number of advances in respect of traditional marine palaeolandscape investigation. Having developed methods of using 'legacy' 3D and 2D seismic reflection data to achieve maximal, near complete topographic mapping of Doggerland, we have the opportunity to develop and extend methodologies for the extraction and analysis of ancient plant and animal DNA directly from sediments. In doing so the extensive dating programme within the project can provide an insight into the poorly understood 'ending' of Doggerland and ascertain whether the inundated landscape, and its offshore islands, might possess a substantive, and essentially unrecorded, Neolithic history.

A dated sequence of palaeoenvironmental change within an enhanced digital landscape also provides significant opportunities for the development of new approaches to distributed simulation and visualisation methodologies. This, in many respects, is amongst the most challenging of research themes within the project. Whilst agent-based modelling has been undertaken by archaeologists for some time, it has yet to be fully accepted as a primary research methodology. In part this reflects the complexity of applying such technologies at extensive scales and when using inordinately large numbers of agents. However, the poor choice of archaeological projects in which to apply ABM modelling may be regarded, on occasions, as a key stumbling block. Not all archaeological processes are equally amenable to such modelling. In contrast, the modelling of plant and animal colonisation of Doggerland, along with the inference of human activity, is an attractive prospect given that the region, by definition, is largely inaccessible to traditional archaeological prospection. It is also true that, in the absence of other evidence, we cannot presume to fully understand the complexities of life on the great European plains simply on the basis of terrestrial finds. Consequently, the iterative simulation of climatic change, sea level rise and consequent landscape transformation, undertaken on the basis of a new geographical and temporal dataset, has much to offer archaeological enquiry within the North Sea and comparable landscapes elsewhere. In doing so the project's ultimate goal is to assess how human populations may have reacted to the evolving landscape and to generate novel research agendas for future archaeologists.

**Acknowledgements** We thank Lawrence Heyworth, Managing Director of Look and Learn Ltd., for his assistance in tracing Arthur Mee's work on the Dogger Banks in the extensive Look and Learn archives, and Dr. T.D. Holt-Wilson for drawing our attention to Beirne's interesting volume in a mail dated 15/04/2015. Nick Wilkinson and Susan Liberator at the Ohio State University library service kindly sought out the reference to the digital copies of Mee's compendium works held in Ohio, and Marion Watson excavated references to Ernest A. Bryant from the copy of Arthur Mee's 'The Children's Encyclopedia' given to her as a child. We also gratefully acknowledge the European Research Council's support for the Lost Frontiers Project through the provision of an Advanced Grant (Grant Agreement 670518 ERC-2014-ADG/ERC-2014-ADG).

## Appendix. The Naming of Parts: Doggerland, Agderia or Northsealand?

An increasing interest in the archaeology of the North Sea in recent years has been driven by significant amounts of new fieldwork, collaborations with offshore industry and, more recently, synthetic publications on the subject. All of this has stimulated debate in a fruitful manner. One area of discussion relates to the utility of the name Doggerland for the area under study. There have been proposals to replace this with the term Northsealand (Childe 1957; De Roest 2013, p. 13 note 2; Leary 2015, p. 113) whilst there has also been a suggestion that individual national sectors of the inundated land mass may also be separately named including the proposal to name the Norwegian sector 'Agderia' (Hammer et al. 2016). The Lost Frontiers project team has considered such proposals but remains of the opinion that to relinquish the term Doggerland seems unreasonable at this time and it is perhaps worth making some comment on this decision.

The underlying assumption of the proposed change, which as far as can be discerned is primarily mooted by archaeologists, is that somehow archaeology has imposed this name and perhaps that it is in some senses a 'loaded' term. There may also be a presumption that this is an English archaeological imposition given the general attribution of the name to Bryony Coles following her seminal paper in 1998. While we were writing this chapter, Professor Coles (2016) published a short paper on how she came to use the term Doggerland and the reasons for doing so. Nevertheless there is still a persistent view that the physical significance of the Dogger Bank as a geographical feature has been overstated and that this undermines its claim for toponymic extrapolation over the wider North Sea area.

At this point we suggest that at least some of the current discussion is confused. To begin with, the assertion regarding the nature of the banks is essentially a misunderstanding of the geomorphological evidence and literature. Whilst it is true that our own work has noted that the profile of the current banks largely reflects post-inundation deposition, this should not be understood as asserting that the area of the banks was topographically insignificant (Gaffney et al. 2009, p. 68). Indeed, the scarp, possibly a moraine feature, on the northern edge of the banks, must have been one of the more prominent features within the North Sea basin and this, separate from the modern Dogger Banks, was presumably associated with the area of the former Dogger Hills. The fact that the scarp was relatively prominent explains why even the earliest regional maps identified this as a significant upstanding coastal feature near the beginning of the Holocene. In geomorphological terms, it is significant, and was probably always significant to the geographical perceptions of the past peoples who lived on the North Sea plain.

The history of the term Doggerland is also intriguing. In respect of the nautical use of the term, Flemming (pers. comm. 2014) has noted that there is a real need for a formal study of the toponymy—the 'place names'—of the North Sea. The term Dogger does not appear to be English and may refer to a double-masted Dutch ship (from Dogge—trawler?). Its later transformation to a toponym representing a larger landscape is often credited to the archaeologist Bryony Coles. More recently, Leary (2015, p. 113) noted that Vere Gordon Childe referred to the area as Northsealand perhaps as early as 1957. Whilst this earlier claim to precedent may appear to give the discipline the authority to rename the area, archaeologists, in particular, should be cautious when faced with appeals to apparent authority.

In fact, the term Dogger Land (written as two words) seems to have been in use in respect of the inundated landscape as early as 1952 in Bryan P. Beirne's (1952) volume on the 'The Origin and History of the British Fauna', in which chapter VII is entitled 'The Cambrian Channel and Dogger Land Survivors'. The chapter title has echoes of Clement Reid's (1899) 'The Origin of the British Flora'. Given that Reid also brought the issue of the Dogger Bank to the attention of generations of geologists and archaeologists in his later work (Reid 1913), it is possible that there was a continuity of usage extending from Reid into later decades of the twentieth century. In following this line of thought, a Google Ngram search for the term 'Doggerland' revealed a number of references that occur from the 1950s but also a number of earlier phytogeographic references including the 1934 Acta Phytogeographioca Suecica (Samuelsson 1934, p. 302). More surprising is the fact that from 1919, only 6 years after the publication of Clement Reid's work, Arthur Mee, a noted children's writer, published or commissioned a number of articles on the archaeology and history of the Dogger Bank in the 'Children's Newspaper' (Bryant 1919[1]). In an edition of the paper published on the 21st of June 1919, an anonymous article, entitled 'Are the Welsh English?', states that:

> The English came chiefly from Doggerland, in the North Sea, and are of the same stock as the Scandinavians; but the Welsh came mainly from the West of France and the North-west of Spain; and we find that, as a rule, they are darker and shorter. (Anon 1919; Mee 1919, p. 95)

**Fig. 20.6** History of the term 'Doggerland' in Google Ngram viewer (1900–2008, unsmoothed) (https://books.google.com/ngrams/graph?content=Doggerland&case_insensitive=on&year_start=1910&year_end=2007&corpus=15&smoothing=0&share=&direct_url=t1%3B%2CDoggerland%3B%2CcO)

[1]The article is signed by "Our Natural Historian" and E.A.B. This was probably Ernest A. Bryant, a regular contributor to the paper on the subject of natural history (Holland 2006, 16).

This paper trail suggests that the term Doggerland may have been the product of the work of Arthur Mee and the specialists he commissioned to write for children and that the name has been in use for nearly a century. It is also a sobering thought that, given the very wide circulation of Mee's popular works (the Children's Newspaper sold more than 500,000 copies at its peak), the current presumption that knowledge of the archaeology of the North Sea was the preserve of the educated few may be the product of our discipline's ignorance of the popular literature of the early twentieth century. It is more than likely that nearly a century ago children were discussing the history of the Dogger Banks on the streets of Britain. It is equally clear that the immense digitisation projects of the past decade can now provide access to literature that we had never previously considered but which may hold relevant information on the history of public archaeology. What should certainly be asserted at this time is that the archaeological presumption that we can replace such historic toponyms at whim seems less than appropriate.[2]

The final issue relates to the proposed use of the name Northsealand. We appreciate that the term Doggerland may be challenged on the basis of its relatively limited regional extent in contrast to the North Sea, which approximates the entire area of concern.[3] However, this alone does not justify a change in terminology. There must be an argument that the term Doggerland at least has relevance to a significant geographical feature that would have been of consequence to the people who lived on the plain for the majority of its existence. Uniquely, the area would also have retained significance during the final stages of inundation. During this period, the Dogger Hills would have still existed as one of the last islands in the southern North Sea during a time when most of the surrounding landscape had been submerged and was, perhaps, being slowly forgotten or mythologised. In such circumstances it is not difficult to imagine that the island/hills/banks may have achieved considerable meaning to the peoples whose ancestors had lived on the surrounding plain. Consequently, if we deign to name the lands of those shadowy peoples, then the only longstanding topographic feature that must have retained some level of cultural significance throughout the late Pleistocene and earlier Holocene periods, and presumably was named, was almost certainly the area associated with the Dogger Bank. In contrast to this, the North Sea does not represent a comparable feature. No doubt, the developing seascape was itself significant to the occupants of the plain, although equally the rise of the sea occurred over a period of c.13,000 years, and for much of that time was of no consequence to peoples living far away from the coastline. The North Sea, then, is a later feature and simply overlies the landscape with which we are concerned. It surely has no real claim to precedence in contrast to the Dogger Hills and then the Dogger Island.

Consequently, we argue that archaeologists should retain the use of the term Doggerland in preference to Northsealand—or any other name. The toponym has a history of use which precedes the recent period of archaeological interest, and most likely has a rationale for the people who lived there in respect of the topography of the plain and the history of inundation which led to the later creation of the North Sea. Moreover, in a fundamental manner the North Sea was ultimately the destroyer of a named land and wiped clean millennia of traditions and cultural geography—it should have no further claim on the landscape it so effectively eradicated.

[2]Care should be taken with the use of Google Ngram data (Zhang 2015). Similar runs were made using the term 'Northsealand' with no positive results. Given the testified use of the term during the 1950s this suggests a need for caution but this may also indicate the relative frequency of use of the two terms within the general academic context with which we are concerned.

[3]It should also be noted that Leary (2015, p. 113, note 2) actually extends Northsealand beyond the North Sea to include the English Channel, whilst the existing toponym *Nordsjælland* refers specifically to the northern part of the Danish island of Zealand and the area north of Copenhagen. The specificity of competing terms would therefore appear to be more complex than some arguments have acknowledged.

## References


Anon (1919) Are the Welsh English? In: Mee A (ed) The children's newspaper, 21 June 1919, p 7

Beirne BP (1952) The origin and history of the British fauna. Methuen, London

Benjamin J, Bonsall C, Pickard C, Fischer A (eds) (2011) Submerged prehistory. Oxbow Books, Oxford

Brennand M, Taylor M (2003) The survey and excavation of a Bronze Age timber circle at Holme-next-the-Sea, Norfolk, 1998–9. Proc Prehist Soc 69:1–84

Bryant EAB (1919) Dogger Bank men: sleepers in the North Sea bed. Island of life the water swallowed up. The children's newspaper, 19th April, 1919

Childe VG (1957) The Dawn of European Prehistory. Published 1925, 6th edn revised, this repr. 1973. St Albans. Paladin (1957)

Coles BJ (1998) Doggerland: a speculative survey. Proc Prehist Soc 64:45–81

Coles BJ (1999) Doggerland's loss and the Neolithic. In: Coles B, Coles J, Schon Jorgensen M (eds) Bog bodies, sacred sites and wetland archaeology. WARP Occasional Paper 12, Exeter, pp 51–57

Coles BJ (2016). The Naming of Doggerland. PAST 83: 14–15. http://www.prehistoricsociety.org/files/PAST_83_for_web.pdf. Accessed 13 July 2016

Fitch S, Gaffney V, Gearey B, Ramsey E (2011) Between the lines: enhancing methodologies for the exploration of extensive, inundated palaeolandscapes. In: Cowley DC (ed) Remote sensing for archaeological heritage management. EAC Occasional Paper 5, Occasional Publication of the Aerial Archaeology Research Group No. 3. Europae Archaeologiae Consilium, Brussels, pp 173–204

Crombé P, Vanmontfort B (2007) The Neolithisation of the Scheldt basin in western Belgium. In: Whittle A, Cummings V (eds) Going over: the Mesolithic-Neolithic transition in north-west Europe. Proceedings of the British Academy 144, Oxford, pp 263–285

De Roest K (2013) How to cope with a drowning landscape? A research history into changing representations of Doggerland in relation to climate change. In: Raemaekers DCM (ed) A kaleidoscope of maritime perspectives: essays on the archaeology, art history and landscape history of the maritime world view. Groningen Institute of Archaeology and Barkhuis Publising, Groningen, pp 7–15

Flemming NC (ed) (2004) Submarine prehistoric archaeology of the north Sea: research priorities and collaboration with industry, CBA Research Report, vol 141. Council for British Archaeology, York

Flemming NC, Cagatay MN, Chiocci FL, Galandiou N, Lericolais G, Joens H, Missiaen T, Moore F, Rosentau A, Sakellariou D, Skar B, Stevenson A, Weerts H (2014) Land beneath the waves: submerged landscapes and sea level change. A joint geoscience-humanities strategy for European continental shelf prehistoric research. In: Chu NC, McDonough N (eds) Position paper 21 of the European Marine Board. European Marine Board, Ostend

Forenbaher S, Kaiser T (2011) Palagruža and the spread of farming in the Adriatic. In: Phoca-Cosmetatou N (ed) The first Mediterranean islanders: initial occupation and survival strategies. University of Oxford School of Archaeology, Oxford, pp 99–111

Gaffney V, Thompson K, Fitch S (eds) (2007) Mapping Doggerland: the Mesolithic landscapes of the southern north Sea. Archaeopress, Oxford

Gaffney V, Fitch S, Smith D (2009) Europe's lost world, the rediscovery of Doggerland CBA Research Reports 160. Council for British Archaeology, York

Godwin H, Godwin M (1933) British Maglemose harpoon sites. Antiquity 7:36–48

Hammer Ø, Planke S, Hafeez A, Hjelstuen BO, Faleide JI, Kvalø F (2016) Agderia – a postglacial lost land in the southern. Nor N Sea Nor J Geol 96(1):43–60

Holland S (2006) Look and learn: a history of the classic children's magazine. Look and Learn Ltd, London. PDF available at: http://www.lookandlearn.com/history/Look-and-Learn-History.pdf. Accessed 04 Mar 2016

Leary J (2015) The remembered land. surviving sea level rise after the last Ice Age. Bloomsbury, London

Logan K, Smith O, Ware R, Momber G, Bates R, Garwood P, Fitch S, Pallen M, Gaffney V, Allaby RG (2015) Thermal age, cytosine deamination and the veracity of 8,000 year old wheat DNA. BioRxiv preprint first posted online 17 Nov 2015. doi: http://dx.doi.org/10.1101/032060

Louwe Kooijmans LP (2007) The gradual transition to farming in the lower Rhine basin. In: Whittle A, Cummings V (eds) Going over: the Mesolithic-Neolithic transition in north-west Europe, vol 144. Proceedings of the British Academy, Oxford, pp 287–309

Marchand G (2007) Neolithic fragrances: Mesolithic-Neolithic interactions in western France. In: Whittle A, Cummings V (eds) Going over: the Mesolithic-Neolithic transition in north-west Europe, vol 144. Proceedings of the British Academy, Oxford, pp 225–242

Mee, A (1919) The children's newspaper and children's pictorial, Issues 1–52. Amalgamated Press, London. https://books.google.co.uk/books?id=xfVDAQAAMAAJ&q=%22doggerland%22&dq=%22doggerland%22&hl=en&sa=X&ved=0ahUKEwjdxIXryJ_LAhXF8Q4KHS0GAdUQ6AEIMjAE. Accessed 01 Mar 2016

Momber G, Tomalin T, Scaife R, Satchell J, Gillespie J (2011) Mesolithic occupation at Bouldnor Cliff and the submerged prehistoric landscapes of the Solent, CBA Research Report, vol 164. Council for British Archaeology, York
Peeters H, Murphy PL, Flemming N (2009) North Sea prehistory research and management framework. Rijksdienst Voor Cultureel Erfgoed, Amersfoort
Ransley J, Sturt F, Dix J, Adams J, Blue L (eds) (2013) People and the sea: a maritime archaeological research agenda for England, CBA Research Report, vol 171. Council for British Archaeology, York
Reid C (1899) The origin of the British flora. Dulau, London
Reid C (1913) Submerged forests. Cambridge University Press, London
Robb J (2013) Material culture, landscapes of action, and emergent causation: a new model for the origins of the European Neolithic. Curr Anthropol 54(6):657–683
Samuelsson G (1934) Die Verbreitung der höheren Wasserpflanzen in Nordeuropa (Fennoskandien und Dänemark), Acta Phytogeographica Suecica, Svenska Växtgeografiska Sällskapet, vol 6. Almqvist & Wiksells, Uppsala
Smith O, Momber G, Bates R, Garwood P, Fitch S, Pallen M, Gaffney V, Allaby RG (2015a) Sedimentary DNA from a submerged site provides evidence of wheat in the British Isles 8000 years before present. Science 347(6225):998–1001. doi:10.1126/science.1261278
Smith O, Momber G, Bates R, Garwood P, Fitch S, Pallen M, Gaffney V, Allaby RG (2015b) Response to comment on "Sedimentary DNA from a submerged site reveals wheat in the British Isles 8000 years ago". Science 349:247. doi:10.1126/science.aab2062
Sturt F, Garrow D, Bradley S (2013) New models of North West European holocene palaeogeography and inundation. J Archaeol Sci 40(11):3963–3976
Sturt F, Van de Noort R (2013) The Neolithic and Early Bronze Age. In: Ransley J, Sturt F, Dix J, Adams J, Blue L. (eds) 2013People and the sea: a maritime archaeological research agenda for England, CBA Research Report, vol 171. Council for British Archaeology, York, pp 50–74
Tappin DR, Pearse B, Fitch S, Dove D, Geary B, Hill J, Chambers C, Bates CR, Pinnion C, Green M, Gallyot J, Georgiou L, Brutto D, Marzialetti S, Hopla E, Ramsay E, Fielding H (2011) The Humber regional environmental characterisation. British Geological Survey Open Report OR/10/54, Keyworth
Tresset A, Vigne J-D (2007) Substitution of species, techniques and symbols at the mesolithic-neolithic transition in western Europe. In: Whittle A, Cummings V (eds) Going over: the mesolithic-neolithic transition in north-west Europe, Proceedings of the British Academy 144. Published for the British Academy by Oxford University Press, Oxford, pp 189–210
Van der Noort R (2013) Climate change archaeology: building resilience from research in the world's coastal wetlands. Oxford University Press, Oxford
Zhang S (2015) The pitfalls of using Google Ngram data to study language. Wired. (10/12/15, http://www.wired.com/2015/10/pitfalls-of-studying-language-with-google-ngram/)

# Chapter 21
# Postglacial Human Dispersal and Submerged Landscapes in North-West Europe

**Garry Momber and Hans Peeters**

**Abstract** This chapter examines the evidence of extensive human movements in the lands adjacent to the North Sea basin following the Last Glacial Maximum (LGM). We consider recent evidence from submerged sites in the southern North Sea and English Channel, and assess the potential for preservation of archaeological material under water by reviewing examples from coastal sites that have become exposed due to coastal change. We show how these site types hold organic sources of data that can be better preserved and survive in richer concentrations and greater quantities than material found on land. We place this evidence in geographical and temporal context to consider patterns of cultural dispersal and distribution from the late Pleistocene through to the Holocene. We demonstrate how the land would have been desirable and occupied, and how maritime pathways facilitated movement as sea level rose, resulting in wide-ranging transport networks for goods and people. The new discoveries of submerged archaeological material provide unique data that needs to be assessed if we are to gain a coherent understanding of human adaptation and dispersal across north-west Europe and particularly Britain following the LGM.

## 21.1 Introduction

Climate change and fluctuations in sea level during the Pleistocene enabled people to move repeatedly into north-west Europe exploiting extensive landscapes that are now underwater (Lambeck and Chappell 2001; Bailey 2004, 2011). The last recorded low stand was during the glacial maximum of the Devensian Ice Age. As this period drew to a close, large areas of the continent were accessible but frozen until the warming climate thawed the permafrost allowing vegetation to grow, animals to graze, and people to move north. As conditions became increasingly favourable, however, the melting ice caps led to a progressive rise in sea level, which, albeit with a time lag of several millennia, covered thousands of square kilometres of land to form the modern coastline.

This chapter will examine human dispersal into north-west Europe following the last glacial maximum. We will address the responses of cultural groups to the changing environments, look at evidence of mobility, the potential for occupation on the now-submerged North Sea basin and in particular the challenges faced by Mesolithic populations as their world became increasingly maritime. We will

G. Momber (✉)
Maritime Archaeology Trust, National Oceanography Centre, Empress Dock, Southampton SO14 3ZH, UK

Department of Archaeology, The King's Manor, University of York, Y01 7EP, York, UK
e-mail: garry.momber@maritimearchaeologytrust.org

H. Peeters
Groningen Institute of Archaeology (GIA), Poststraat 6, 9712 ER Groningen, The Netherlands
e-mail: j.h.m.peeters@rug.nl



G.N. Bailey et al. (eds.), *Under the Sea: Archaeology and Palaeolandscapes of the Continental Shelf*, Coastal Research Library 20, DOI 10.1007/978-3-319-53160-1_21

assess archaeological evidence from submerged landscapes and consider its significance when compared with terrestrial data with particular reference to human dispersal prior to the separation of Britain from mainland Europe. We will show how well-preserved material evidence from the submerged landscape has provided new insights into the technical ability, cultural links and dispersal patterns of the Mesolithic, and highlight the potential to provide a great deal more.

## 21.2 Human Movement Across the North Sea

Pioneering groups of Magdalenian hunter-gatherers moved towards northern France and into the southern North Sea (or 'Doggerland', see Gaffney et al., Chap. 20)[1] as climatic conditions allowed around 15–16,000 years ago (Miller 2012, p. 211). This was the first of numerous waves of migration to follow.

The Bølling/Alleröd interstadial or British Windermere climatic upturn from 12,700 cal BC showed an improvement in temperature and with it an expansion of human activity including the earliest evidence for re-occupation of southern Britain (Tolan-Smith 2008; Miller 2012). This warm spell was interrupted for a few hundred years between c.12,200 and 11,900 cal BC by the colder 'Older Dryas', during which northern latitudes were de-populated, but warmer conditions returned with the succeeding Windermere interstadial, and Magdalenian groups expanded into a region south of the European loess belt, extending into Belgium, north-eastern France and Britain (Jacobi and Higham 2011). The relatively rapid response to climatic amelioration indicates that populations had not retreated far, and were able to adapt easily to changing conditions. The warmer climate also saw the appearance of the Hamburgian, a tradition which is generally regarded as a subgroup of the Late Magdalenian, in the northern half of the Netherlands, northern Germany, Poland and extending as far west as Roberthill and Howburn in Scotland (Audouze and Enloe 1991; Rensink 1995; Street 1998; Ballin et al. 2010). By the end of the Windermere interstadial around 10,950 cal BC, additional groups included the Creswellian (Stapert 1985; Jacobi 1991; Barton et al. 2003), the Azilian (Bodu and Mevel 2008) and the Federmesser Gruppen (Fig. 21.1). All these groups would have had access to the North Sea Basin, which was dry land at the time.

The Creswellian forms a techno-complex that appears to be geographically centred in the drowned lands of the North Sea (Peeters and Momber 2014). The technology has been identified in the English Midlands and southern counties, in the Netherlands and possibly Belgium (Jacobi 1991; Barton et al. 2003). The Azilian, like the Creswellian, is believed to be a derivation of the Late Magdalenian. It originated in the Basque region but is also found in the foothills of the Alps, and in the Paris Basin where it persisted until the end of the Pleistocene.

The Federmesser Gruppen spanned the North European Plain, ranging from the Ukraine to Britain. They consumed a varied spectrum of foodstuffs and have been recorded at short-lived dwelling places located adjacent to rivers and lakes. The archaeological evidence indicates high mobility and seasonal resource exploitation (Deeben 1988; Street 1998; De Bie and Caspar 2000; Crombé et al. 2003, 2013; Baales 2004). At a regional level, the Federmesser sites of the Paris Basin and western Belgium hosted common technologies demonstrating cultural links despite distances of several hundred kilometres (De Bie and Caspar 2000; De Bie and Van Gils 2009; Miller 2012, p. 220). Furthermore, the recovery of marine shells from Bois Laiterie in Belgium provides evidence for interaction with the coastal zone that once lay to the west but is now lost beneath the southern North Sea. At the time, sea levels were around 50–60 m lower than today, putting the coastline hundreds of kilometres away and

[1] Since the area we refer to in this chapter includes the present-day land surrounding the North Sea as well as the now submerged area within it, we use North Sea as a generic short-hand reference, without prejudice to the arguments presented by Gaffney et al. (Chap. 20) in favour of the term 'Doggerland'.

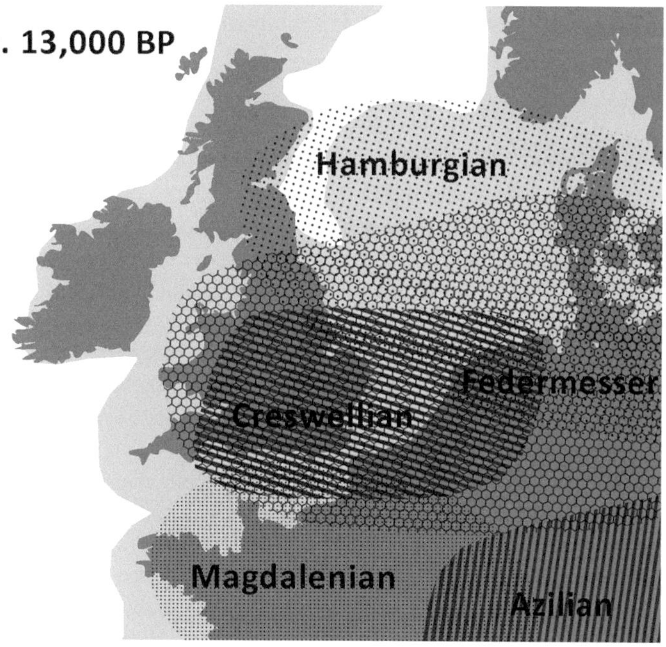


**Fig. 21.1** Map showing the approximate configuration of the coastline – areas shaded light grey would have been still exposed as land at lowered sea level – and a stylised pattern of different cultures across north-west Europe towards the end of the Windermere or Bølling/Alleröd interstadial. This indicates that large areas of the southern North Sea would have been accessible for human occupation

exposing large tracts of land that would have presented ideal lowland fluvial conditions for hunting and gathering. The archaeological assemblages with artefacts made of non-local raw materials show either large territories or extensive social networks.

The Windermere interval was terminated by the particularly cold Younger Dryas stadial. Forested areas in the north were replaced with tundra and human resource exploitation had to adapt to new challenges and opportunities. As the forests retreated and the land opened up, large herds of reindeer were able to move across the North European Plain, and with them people of the Ahrensburgian, showing common characteristics that ranged from north-east France, to northern Germany and across the North Sea into Britain, where it is known as the 'long-blade' tradition (Jöris and Thissen 1997; Barton 1998; Deeben et al. 2000). During more temperate phases of the stadial, the hunter-gatherers were able to exploit ranges further north in the warmer months, generally preferring the river valleys and lakes of mature river systems. These are patterns of subsistence that are comparable to the Early Mesolithic (Arts 1988).

Evidence of marine exploitation by hunter-gatherer groups at the termination of the Pleistocene and beginning of the Holocene is negligible, which is not surprising as shorelines of that date are now drowned along with any contemporaneous coastal archaeological sites. However, recent discoveries in north-west Scotland indicate that the sea was probably used as a means of movement. Excavations at Rubha Port an t-Seilich on the Isle of Islay, western Scotland have uncovered a lithic collection of worked tools associated with volcanic ash dating to the mid Younger Dryas at 10,300–10,000 cal BC. The finds are similar to Ahrensburgian assemblages (Mithen et al. 2015 pp. 404–405) and are particularly important as they provide stratified evidence of human activity on the west coast of Scotland during the Younger Dryas. They have an added significance because access to this area from the east would only have been possible around the north of the Scottish landmass at a time when the hinterland was covered by an ice sheet. This fact, along with the distribution of comparable sites in the north of Britain (Ballin and Saville 2003; Ballin et al. 2010), and the dearth of sites in the south, has prompted Mithen et al. (2015) to argue that the movement of 'Ahrensburgian' groups into the UK was not from south to north, but rather from east to west along the coastal fringes of the land that is now drowned by the North Sea.

## 21.3 The Diminishing Landscape of the Mesolithic

The Holocene began with a sharp rise in temperature around 9450 cal BC (Alley 2000). For a final time, the temperate ecosystems moved back north, enriching the landscape and altering the resource base. This enabled Early Mesolithic Maglemosian groups to spread through north-west Europe. Their presence has been recorded on both sides of the North Sea basin, from Scotland to Poland (Clark 1936, 1954; Leakey 1951; Rankine 1952; Reynier 2000; Bang-Anderson 2003; Conneller 2009; David 2009). The well-watered, low lying plains with nutrient rich alluvial soils would have provided a rich environment attractive to herding megafauna and hunter-gatherer groups. However these conditions were not to last as forests spread across the landscape, sea level rise turned fluvial systems into marine estuaries and low-lying pastures were drowned. These structural changes in palaeogeography and vegetation interrupted long-established migration routes, steadily leading to the replacement of reindeer by red deer, aurochs, wild boar and other woodland mammals. The Early Mesolithic toolkit shows similarities to the Ahrensburgian, but with an increased use of broad blade microliths that reflect the modified subsistence strategies. This change is evident in Early Mesolithic assemblages from Thatcham in southern England to Star Carr in the north-east (Wymer 1958; Conneller et al. 2012).

Around 8400 cal BC a new technology arrived on the margins of mainland Britain. The prevalent Early Mesolithic broad blade technologies were now being supplemented by the narrow blade tradition of the Late Mesolithic. The earliest sites are found on the fringes of the North Sea, along the north east coast of Britain. Many were accompanied by substantial structures. These have been found at Echline on the Firth of Forth, 8450–8240 cal BC (Waddington 2015), East Barnes, around 8000 cal BC (Suddaby 2007), Howick around 7800 cal BC (Waddington 2007a), Mount Sandel in Northern Ireland at 7700 cal BC (particularly complex construction measuring Woodman 2003) and Cass–ny–Hawin on the Isle of Man c. 8200–7950 cal BC (Oxford Archaeology 2016).

The Mesolithic settlement at Howick includes a round hut structure 6 m in diameter and was repeatedly occupied for over a century on the top of a low cliff near the Mesolithic coastline. The Cass ny Hawin structure is approximately 7 m in diameter with a ring of post holes around a sub-circular hollow and an internal redeposited gravel platform. Waddington (2007b) has suggested that substantial structures such as these may have been associated with sedentary or semi-sedentary settlements focused on a mixed economy of marine and terrestrial resources in a coastal ecotone, and that this in its turn may have been a response to loss of former extensive hunting territories resulting from sea-level rise. But, equally, similar structures and adaptations could have existed on the now submerged coastlines of the North Sea lowlands; if so, they still await discovery.

The large huts found on the British mainland during the early stages of the Mesolithic are not dissimilar to contemporaneous sites in continental Europe, particularly in Denmark in association with the Kongemose and the Ertebølle periods (Pedersen et al. 1997; Grøn 2003; Skaarup and Grøn 2004; Jenson 2009). The comparable structures, along with analogous lithic technologies, express a cultural link that extended across the North Sea basin. These links are clear in the archaeological record until soon after c.7000 cal BC when stone tool technologies in Britain and mainland Europe diverged. It should be noted that the archaeological record to date in Britain is skewed towards inland sites as coastal and estuarine sites that may have existed before the sea level stabilised around 3500 cal BC are now mostly drowned. Discovering such sites is problematic as they are not easy to locate, and assumptions that most would have been lost to erosion has stifled the motivation to look. However, a brief evaluation of known underwater archaeological evidence and coastal palaeo-landforms provides an insight into the potential of sites that could remain below water.

## 21.4 Insights from a Drowned Palaeolandcsape

South of Howick, at the mouth of the River Tees and on the beaches around Hartlepool, reamins of submerged forests are intermittently revealed in the intertidal zone. At Withensea, an exposed prehistoric landscape of tree stumps became known as Noah's Woods following its discovery in 1839, and an intertidal forest running for about 20 km between Grimsby and Skegness was recorded even earlier in the eighteenth century (Tann 2004; Hazell 2008). These prehistoric land surfaces remained hidden, protected and unknown until natural erosion exposed them. Traces of Mesolithic human activity can be found in such contexts and comprise not only worked flint, but also footprints (Jacobi 1976; Waughman et al. 2005; Bell 2007; Sheppard 1912). Other artefacts include a Late Palaeolithic barbed bone point and a Mesolithic barbed antler harpoon from below the low water mark at Hornsea, in the Tees estuary, and another found close to the low tide mark at Barmston (Brigham et al. 2008). In addition, there are midden deposits and a worked red deer antler from the Tees Estuary dated to 6750 ± 180 cal BC (BM-80) (Waughman et al. 2005, p. 8), and a wealth of Mesolithic artefacts in and adjacent to the low-lying east coast ria estuaries from the Humber to the Thames (Wymer and Robins 1994; Wilkinson and Murphy 1995, pp. 90–98, Bridges 1998, pp. 6–8; Brennand et al. 2003; Robertson et al. 2005). Indeed, wherever investigations are carried out beneath coastal peat deposits, either well preserved environmental or prehistoric archaeological material is found.

The sites dating back thousands of years demonstrate the length of time material can survive in stable environments. Preservation can be excellent in sheltered tidal inlets where rising sea levels force up the water table, protecting palaeoenvironmental deposits by forming anaerobic peat bogs, mires and sedimentary sinks. The process is incremental resulting in the deposition of a continually thickening blanket of sediment within the networks of channels. Many finds have come from flooded fluvial systems that followed courses into the North Sea basin. These buried palaeo-features are difficult to detect unless they become exposed on stretches of coastline where the balance has shifted from sedimentation to net erosion. Underwater, these sites are obviously even more difficult to find but the recovery of prehistoric artefacts and ecofacts from the North Sea during the last 100 years has demonstrated that comparable geomorphological conditions exist (Clark 1936; Coles 1998; Flemming 2004; Gaffney et al. 2009; Chap. 20, Sturt et al., Chap. 28). However, it has not been until recent geophysical surveys in the North Sea and the English Channel, that geo-archaeologists have been able to identify a complex of channels, plains, wetlands and estuaries associated with the rivers including the proto Thames, Solent and now obsolete Bytham River (Gaffney et al. 2007; Gupta et al. 2008) and in some cases associated archaeological material, such as the Lower/Middle Palaeolithic finds from the A240 site (Tizzard et al. 2014).

Similar palaeo-landscapes occur on the east side of the North Sea, notably within the Rhine/Meuse estuarine complex. Palaeolandscape modelling of in-filled channels in the adjacent Flevoland area has aided interpretation of the Mesolithic-Neolithic landscape (Peeters 2007). Evidence from the back-barrier, intertidal and coastal peats show how archaeological material can survive beneath land that is now covered by subsequent deposits, and in some cases, by the sea. A submerged example was found at the mouth of an estuary at the Maasvlakte-Europoort, the Netherlands, where geo-archaeological interpretation of the buried landscape led to the discovery of the Yangzte Harbour Mesolithic site in Rotterdam Harbour (Vos et al. 2010). Over 500 bone and antler implements, mainly harpoon points with parallels to Star Carr, Britain, and Hohen Viecheln, Germany, were collected from this area in the 1970–1980s (Glimmerveen et al. 2004; Verhart 2004). The recent investigations in the harbour during the development of the Yangtze extension zone between 2005 and 2014 demonstrated the presence of a Mesolithic occupation site on a sand dune with an intact Late Glacial to Mid Holocene sequence in 22–17 m of water dated to 7500–5800 BC (Vos et al. 2010; Moree and Sier 2015). In 2011, excavation recovered 46,067 plant remains and artefacts, indicating the exploitation of tubers, nuts, fish, birds and mammals (Peeters et al. 2015). The presence of Wommersom quartzite and amber indicate movement or trade links that reached at least 147 km to the east (Moree and Sier 2015 pp. 198–194). Water transport would have facilitated mobility and movement of goods through this estuarine and coastal seascape, and by the early fifth millennium BC widespread distribution of antler mattocks around the fringes of the North Sea indicates the development of extensive maritime networks (Elliot 2015).

Bouldnor Cliff is another key site that casts a unique light on questions of human dispersal in the Late Mesolithic (Momber et al. 2011; Momber 2014). The archaeological material lies 11 m below Ordnance Datum off the north-west coast of the Isle of Wight in the Solent (Figs. 21.2 and 21.3). Excavations here have been limited, yet the discoveries give tantalising glimpses of material culture and a lithic assemblage showing similarities and links with the continent. One retouched tool is an obliquely blunted blade similar to the Azilian category of 'une piece tronquée' recorded from sites in the Paris basin (Tomalin 2011, p. 152). The same type of blade is also found at the Powell site at Hengistbury Head (Barton 1992, p. 229). By contrast, a detached cutting tip of a bifacially prepared flint axe blade has been carefully formed with shallow skimming flakes. The regular blade edge has a weak S-shaped profile and the cross-section of the axe is a shallow ellipse. 'The care and symmetry displayed in this work is usually associated with Neolithic craftsmanship. The occurrence in a Mesolithic context is certainly unusual but perhaps not without Continental analogy (Tomalin 2011, p. 152). Axes and picks were used extensively in the region and many have been dragged up from the Solent during oyster fishing. Detached tranchet axe flakes were also found at Bouldnor Cliff. The abundance of finds in the Solent compares favourably with sites from northern France with similar tools, for example at the Mesolithic site of Acquiny (Eure) on the Lower Seine dated at 6510 + 170 cal BC and with the Middle Sauveterrian 'pics à crosse' at Grotte de Larchant south of the Seine (Hinout 1989a). After the beginning of the Atlantic Period around 5500 cal BC, picks are also present at the Final Tardenoisian site of Ferme de Chinchy at Villeneuve-sur-Fère (Hinout 1989b), at the Montmorencien site of Ile de la France, la Forêt de Montmorency (Guyot 1998), and at St Reine de Bretagne (Kayser 1989).

The organic material at Bouldnor Cliff includes some 80 pieces of worked wood. Some timbers have been fashioned by tangential splitting (Fig. 21.4) while others contain enigmatic workings and cuts for which there are no comparisons in the UK. Recent investigations have uncovered the edge of a platform or collapsed structure, made mainly from tangentially split timbers together with some radially split round-wood (Figs. 21.5 and 21.6). There is no other comparable structure in the Late Mesolithic record in the UK. Another piece measuring 0.94 m long and 0.41 m wide dated to 6240–6000 cal BC (Beta 249,735) represents a fragment of a much larger timber that was converted from the trunk of a tall slow-grown oak that would have been a couple of metres wide and several tens of metres high (Taylor 2011). Collectively, the artefacts from the site and their relationships indicate

**Fig. 21.2** Garry Momber, Maritime Archaeology Trust, with flints recovered from the sea floor after eroding from the submerged 8000 year-old Bouldnor Cliff site

**Fig. 21.3** A selection of flakes and bladelets recovered from the seabed at the Bouldnor Cliff site

**Fig. 21.4** Tangentially split round wood from the Bouldnor Cliff timber assemblage. Scale in cm

wood working, most probably for constructing a log boat. The wood working technology used to split the oak tangentially is of particular interest because this is something that is not seen again in the UK until the Neolithic, when it is used in the construction of the Haddenham long barrow at c. 3600 cal BC (Evans and Hodder 2006, pp. 185–87).

Pioneering extraction of sedimentary ancient DNA (sedaDNA) has revealed further information about activities on the site at Bouldnor Cliff (Smith et al. 2015). Samples were recovered from an

**Fig. 21.5** Edge of a platform or collapsed structure consisting of tangentially split timbers alongside radially split round-wood

**Fig. 21.6** Split and trimmed timber from the platform or collapsed structure at Bouldnor Cliff

archaeological horizon within a relict fluvial sand-dune context that was capped by peat deposits (see Gaffney et al., Chap. 20). The find demonstrates the presence of cultivated einkorn wheat at a UK site over 2000 years before agriculture is conventionally supposed to have reached Britain (Malone 2001). The distance of many hundreds of kilometres from any contemporaneous sites where einkorn was grown makes the discovery controversial and further evidence from sites that would help bridge the gap is needed. However, the presence of an outlier extending beyond the known frontier of Neolithic expansion is not impossible, given the pattern and rate of dispersal of einkorn in continental Europe, and the acknowledged mobility of Mesolithic groups as discussed above.

## 21.5 Discussion

The common cultural signatures seen on both sides of the present-day North Sea from the Magdalenian to the Late Mesolithic indicate extensive interconnections across the now-submerged landscape of the North Sea Basin. The range of Late Palaeolithic sites shows how the area was traversed by the Hamburgian culture while a Creswellian technocomplex was centred at its heart. The recolonisation was interrupted by climatic downturns but human populations were quick to reoccupy land once conditions improved. It would therefore appear that populations did not retreat far, and it is possible that the lowland areas of the North Sea Basin afforded a population refuge during periods when higher ground was climatically inaccessible, notably for the Federmesser and Ahrensburgian groups. The fluvial and alluvial systems made the plains mineral-rich and attractive for a range of resources including herds of reindeer or other large mammals. The extent to which these areas were exploited remains largely unknown as the landscape and coastlines have since become inundated. However, the presence of marine shells at the Federmesser site of Bois Laiterie in Belgium demonstrates interaction with the coastal zone through long distance movements or inter-group exchange (see also Larsson, Chap. 11 for similar indications in southern Scandinavia).

At the beginning of the Holocene, when the broad-blade Maglemosian extended to the north and west, it is probable that these groups also had the capability to move along coastlines and exploit marine resources. The isotope evidence from canine bones at Star Carr shows a diet containing seafood indicating connection with the coastline (Clutton-Brock and Noe-Nygaard 1990; Chatterton 2003; Fischer et al. 2007).

As the transgression advanced and the water rose, people evolved new strategies to make use of the changing conditions. The arrival of Late Mesolithic groups in northwest Britain in the middle of the ninth millennium BC was possibly a response to a loss of territory in the lowlands to the east. Watercraft are likely to have been used to facilitate the movement of people and resources although direct evidence is largely lacking at present. The archaeological evidence shows how networks expanded along coastal routes and boats would have allowed people to travel long distances to new territories relatively rapidly through an expanding system of coastlines and river estuaries. Waddington (2015) cites supporting evidence for a maritime dispersal in the pattern of dates and distribution of early narrow-blade technologies at the beginning of the eighth millennium BC along the north-east coast of England and lowland Scotland, around the north coast of Scotland, then south to Ireland, the Isle of Man and Caldey Island in the Bristol Channel off the coastline of south Wales.

A later southerly dispersal appears to have taken place in the south and east of the British mainland, showing connections with northern France and Belgium. These relatively late arrivals to the British mainland followed a more southerly route from the northern French region. It is uncertain whether the pathway into southern England was purely terrestrial or involved maritime travel along coastlines and rivers but most Late Mesolithic sites are associated with rivers or estuaries and recent discoveries on the Isles of Scilly show how developed maritime abilities appropriate to Atlantic waters were present by the beginning of the 6th millennium BC. The discovery of microlithic trapezes typologically linked to the Belgian region provides direct evidence for movement between the Islands and mainland Europe by boat around 6000 BC (Garrow and Sturt 2015), most likely via the south coast of England in the vicinity of the Solent and Bouldnor Cliff.

It is against this backdrop that the Bouldnor Cliff finds gain added significance, providing insights into human dispersal and links with mainland Europe at a time when the most likely pathways of movement were dominated by the progressive encroachment of estuarine and coastal conditions. Bouldnor Cliff contains a material culture that shows similarities with mainland Europe, wheat that must have been derived from Europe, a wood-working technology that is more akin to the Neolithic than the Mesolithic and microlithic trapezes similar to those in France and Belgium. It is therefore plausible to suggest that goods and cultural influences reached the Isle of Wight from western France or the lower Rhine Basin with the aid of watercraft via estuaries and coastlines at the western end of the 'proto-English channel' before Britain was finally separated from the European mainland.

## 21.6 Conclusion

The body of evidence in the archaeological record serves to demonstrate the value of the submerged landscapes to resolve questions about human occupation and dispersal across north-west Europe following the LGM. First, the high level of mobility, subsistence patterns and cultural links across the North Sea basin demonstrate the very strong likelihood that it was occupied and was the focal point of at least one techno-complex. As such, archaeological investigation within it could give information that would make our understanding of these cultures less fragmentary and indeed necessary if we wish for a coherent appreciation of post-LGM human occupation patterns.

Secondly, the evidence has a great deal to tell us about the use of coastal resources, adaptation to an increasingly marine environment and development of maritime skills as the sea level rose. Sea level rise was persistent throughout the Early Holocene and as such, most Mesolithic coastlines where this evidence is to be found now lie underwater.

Thirdly, it is the drowned landscapes that hold the data needed to provide insights into the cultural changes that followed the separation of Britain from mainland Europe. Bouldnor Cliff is an example of a site with high levels of technical skill that shows evidence of cultural links to the south-east and the west. It was occupied just before the network of rivers between Britain and the continental landmass was disrupted by the creation of the North Sea. This was a mature landscape rich in resources before it was drowned and the ensuing coastal squeeze forced its inhabitants to retreat upslope or migrate elsewhere.

Finally, the preservation potential for fine organic material and organic artefacts is invariably greater than in terrestrial deposits. It is a very rich archive that can provide high concentrations of archaeological material of a type that has rarely survived on sites in present-day terrestrial locations.

Despite the limited number of sites, the evidence for early human activity on the drowned lands of the continental shelf is significant and tangible. It remains an understudied archive of data that can contribute to our understanding of human dispersal, colonisation and behavioural variability. The sites have demonstrated that organic material and DNA- rich sediments can survive and remain well preserved for many millennia. These results are causing us to rethink the technical abilities of our Stone Age ancestors. The remains at Bouldnor Cliff and the targeted discovery in Yangtze Harbour indicate that there could be many more well preserved sites entombed within the submerged palaeo-landscape. The next question is where to find them. An understanding of geomorphological changes in response to rising sea levels can help target areas where the circumstances for preservation were greatest and where the environments remained stable (see in particular Hepp et al., Chap 14, Karle and Goldhammer, Chap 15, and Gaffney et al., Chap 20 for relevant work in the North Sea Basin. Many such examples are present along and offshore of the modern coastline and many of these landscapes survive underwater. Once we have sufficient data about these offshore landforms and we can use the evidence from currently known sites to relate human activity to the landscape, the resulting models should help us pinpoint sites with the highest potential.

**Acknowledgements** We thank the Department of Archaeology, University of York, and the DISPERSE Project (Grant No. 269586) funded by the European Research Council (ERC) under the EU Seventh Framework Programme, for providing intellectual and managerial support and research time for Garry Momber. We also thank Geoff Bailey and Clive Waddington for detailed comments that have helped to improve the text. We would also like to thank the Maritime Archaeology Trust under whose umbrella the work has continued for over eighteen years. This is DISPERSE contribution no. 34.

## References

Alley RB (2000) The younger dryas cold interval as viewed from Central Greenland. Quat Sci Rev 19:213–226

Alley RB, Marotzke J, Nordhaus WD, Overpeck JT, Peteet DM, Pielke RA, Pierrehumbert RT, Rhines PB, Stocker TF, Talley LD, Wallace JM (2003) Abrupt climate change. Science 299:2005–2010

Arts N (1988) A survey of final Palaeolithic archaeology in the southern Netherlands. In: Otte M (ed) De la Loire à l'Oder. Les civilisations du Paléolithique Final dans le nord-ouest Européen. Archaeopress, Oxford, pp 287–356

Audouze F, Enloe J (1991) Subsistence strategies and economy in the Magdalenian of the Paris Basin, France. In: Barton N, Roberts A, Roe D (eds) The Late Glacial in north-west Europe: human adaptation and environmental change at the end of the Pleistocene, CBA Research Report, vol 77. Council for British Archaeology, York, pp 63–71

Baales M (2004) Local and regional economic systems of the Central Rhineland Final Palaeolithic (Federmessergruppen). In: Crombé P (coord) Section C 7.1. Landscape-use during the Final Palaeolithic and Mesolithic in NW-Europe: the formation of extensive sites and site-complexes. In Crombé P, Vermeersch P (eds) Actes of the XIVth UISPP Congress, University of Liège, Belgium, 2–8 Sept. 2001, BAR International Series 1302. Archaeopress Oxford, pp 3–10

Bailey GN (2004) The wider significance of submerged archaeological sites and their relevance to world prehistory. In: Flemming NC (ed) Submarine prehistoric archaeology of the North Sea. Research priorities and collaboration with industry, CBA Research Report, vol 14. Council for British Archaeology, York, pp 3–10

Bailey G (2011) Continental shelf archaeology: where next? In: Benjamin J, Bonsall C, Pickard C, Fischer A (eds) Submerged prehistory. Oxbow Books, Oxford, pp 311–331

Ballin TB, Saville A (2003) An Ahrensburgian-type tanged point from Shieldaig, Wester Ross, Scotland, and its implications. Oxf J Archaeol 22:115–131

Ballin T, Saville A, Tipping R, Ward T (2010) An upper palaeolithic flint and chert assemblage from Howburn Farm, South Lanarkshire, Scotland: first results. Oxf J Archaeol 29:323–360

Bang-Anderson S (2003) Encircling the living space of early postglacial reindeer hunters in the interior of southern Norway. In: Larsson L, Kindgren H, Knutsson K, Loeffler D, Akerlung A (eds) Mesolithic on the move. Oxbow Books, Oxford, pp 193–204

Barton RNE (1992) Hengistbury head, Dorset, Vol 2, the late upper palaeolithic and early mesolithic sites. Oxford University Committee for Archaeology Monograph 34, Oxford

Barton R (1998) Long blade technology and the question of British Late Pleistocene/Early Holocene lithic assemblages. In: Ashton N, Healy F, Pettitt P (eds) Stone age archaeology. Oxbow Books, Oxford, pp 158–164

Barton R, Jacobi R, Stapert D, Street MJ (2003) The Lateglacial reoccupation of the British Isles and the Creswellian. J Quat Sci 18:631–643

Bell M (2007) Prehistoric coastal communities: the mesolithic in western Britain, CBA Research Report, vol 149. Council for British Archaeology, York

Bodu P, Mevel L (2008) Enquête autour des lames tranchantes de l'Azilien ancien. Le cas du niveau inférieur du Closeau (Rueil-Malmaison, Hauts-de-Seine, France). L'Anthropologie 112:509–543

Brennand M, Taylor M, Ashwin T, Bayliss A, Canti M, Chamberlain A (2003) The survey and excavation of a Bronze Age timber circle at Holme-next-the-Sea, Norfolk, 1998–9. Proc Prehist Soc 69:1–84

Bridges E (1998) Classic landforms of the north Norfolk coast. The Geographical Association, Sheffield

Brigham T, Buglass J, George R (2008) Rapid coastal zone assessment, Yorkshire and Lincolnshire. Bempton to Donna Nook. English Heritage Project 3729. Humber Archaeology Report 235. Humber Field Archaeology and English Heritage. https://archaeologydataservice.ac.uk/archiveDS/archiveDownload?t=arch-868-1/dissemination/pdf/RCZA_vol_1_Bempton_to_Donna_Nook/RCZA2a2breport.pdf

Chatterton R (2003) Star Carr reanalysed. In: Bevan L, Moore J (eds) Peopling the Mesolithic in a northern environment, BAR International Series, vol 1157. Archaeopress, Oxford, pp 69–80
Clark JGD (1936) The mesolithic settlement of north west Europe. Cambridge University Press, Cambridge
Clark JGD (1954) Excavations at Star Carr. Cambridge University Press, Cambridge
Clutton-Brock J, Noe-Nygaard N (1990) New osteological and C-isotope evidence on Mesolithic dogs: companions to hunters and fishers at Star Carr, Seamer Carr and Kongemose. J Archaeol Sci 17:643–653
Coles BJ (1998) Doggerland: a speculative survey. Proceedings of the Prehistoric Society 64:45–82
Conneller C (2009) Death. In: Conneller C, Warren G (eds) Mesolithic Britain and Ireland: new approaches. Tempus, Stroud, pp 139–164
Conneller C, Milner N, Taylor B, Taylor M (2012) Substantial settlement in the European Early Mesolithic: new research at Star Carr. Antiquity 86:1004–1020
Crombé PY, Perdaen J, Sergant J (2003) The site of Verrebroek 'Dok' (Flanders, Belgium): spatial organisation of an extensive early mesolithic settlement. In: Larsson L, Kindgren H, Knutsson K, Loeffler D, Akerlung A (eds) Mesolithic on the move. Oxbow Books, Oxford, pp 205–215
Crombé P, Smedt P, Davies N, Gelorini V, Zwertvaegher A, Langohr R, Van Damme D, Demiddele H, Van Strydonck M, Antrop M, Bourgeois J, De Maeyer P, De Reu J, Finke P, Van Meirvenne M, Verniers J (2013) Hunter-gatherer responses to the changing environment of the Moervaart palaeolake (NW Belgium) during the Late Glacial and Early Holocene. Quat Int 308:162–177
David E (2009) Show me how you make your hunting equipment and I will tell you where you come from: technical traditions, an efficient means of characterizing cultural identities. In: McCartan SB, Schulting RJ, Warren G, Woodman PC (eds) Mesolithic horizons. Oxbow Books, Oxford, pp 362–367
De Bie M, Caspar J-P (2000) Rekem. A Federmesser camp on the Meuse River Bank. Leuven University Press, Leuvan
De Bie M, Van Gils M (2009) Mesolithic settlement and land-use in the Campine region (Belgium). In: McCartan SB, Schulting RJ, Warren G, Woodman PC (eds) Mesolithic horizons. Oxbow Books, Oxford, pp 282–287
Deeben J (1988) The Geldrop sites and the Federmesser occupation of the southern Netherlands. In: Otte M (ed) De la Loire à l'Oder. Les civilisations du Paléolithique Final dans le Nord-Ouest Européen. Archaeopress, Oxford, pp 357–398
Deeben J, Dijkstra P, Van Gisbergen P (2000) Some new 14C dates from sites of the Ahrensburg culture in the southern Netherlands. Notae Praehistoricae 20:95–109
Elliot B (2015) Facing the chop: redefining British antler mattocks to consider larger-scale maritime networks in the early fifth millennium cal BC. Eur J Archaeol 18(2):222–244
Evans C, Hodder I (2006) A woodland archaeology: Neolithic sites at Haddenham. McDonald Institute for Archaeological Research, Cambridge
Fischer A, Olsen J, Richards M, Heinemeier J, Sveinbjörnsdóttir AE, Bennike P (2007) Coast-inland mobility and diet in the Danish Mesolithic and Neolithic: evidence from stable isotope values of humans and dogs. J Archaeol Sci 34:2125–2150
Flemming N (2004) The prehistory of the north sea floor in the context of continental shelf archaeology from the Mediterranean to the Nova Zemlya. In: Flemming NC (ed) Submarine prehistoric archaeology of the North Sea. Research priorities and collaboration with industry, CBA Research Report, vol 14. Council for British Archaeology, York, pp 11–20
Gaffney V, Thompson K, Fitch S (eds) (2007) Mapping Doggerland: the Mesolithic landscapes of the southern North Sea. Archaeopress, Oxford
Gaffney V, Fitch S, Smith D (2009) Europe's lost world, the rediscovery of Doggerland CBA Research Reports 160. Council for British Archaeology, York
Garrow D, Sturt F (2015) Continental connections: concluding discussion. In: Anderson-Whymark H, Garrow D, Sturt F (eds) Continental connections: exploring cross-channel relationships from the Lower Palaeolithic to the Iron Age. Oxbow Books, Oxford, pp 166–172
Glimmerveen J, Mol D, Post K, Reumer JWF, Plicht H, Vos J, Geel B, Reenen G, Pals JP (2004) The inundated landscapes of the western Solent. In: Flemming NC (ed) Submarine prehistoric archaeology of the North Sea. Research priorities and collaboration with industry, CBA Research Report, vol 14. Council for British Archaeology, York, pp 38–43
Grøn O (2003) Mesolithic dwelling places in South Scandinavia: their definition and social interpretation. Antiquity 77(298):685–708
Gupta S, Collier J, Palmer-Felgate A, Dickinson J, Bushe K, Humber S (2008) Submerged Palaeo-Arun and Solent Rivers: reconstruction of prehistoric landscapes. http://archaeologydataservice.ac.uk/archives/view/palaeoarun_eh_2007/
Guyot A (1998) Trouvailles en grès. Bulletin 3:1–6
Hazell Z (2008) Offshore and intertidal peat deposits, England – a resource assessment and development of a database. Environ Archaeol 13(2):101–110

Hinout J (1989a) Tardenoisien et facies Mauregny dans la Bassin Parisien. In: Mohen J-P (ed) Le temps de la préhistoire, tome 1. Le temps de la Préhistoire: actes du XXIIIe Congrès Préhistorique de France. Éditions Archéologia, Dijon. Societé Préhistorique Française, Paris, pp 346–347
Hinout J (1989b) Le Sauveterrien au sud de la Sein. In: Mohen J-P (ed) Le temps de la préhistoire, tome 1. Le temps de la Préhistoire: actes du XXIIIe Congrès Préhistorique de France. Éditions Archéologia, Dijon. Societé Préhistorique Française, Paris, pp 348–349
Jacobi R (1976) Britain inside and outside Mesolithic Europe. Proc Prehist Soc 42:67–84
Jacobi RM (1991) The creswellian, creswell and cheddar. In: Barton N, Roberts A, Roe D (eds) The Late Glacial in north-west Europe: human adaptation and environmental change at the end of the Pleistocene, CBA Research Report, vol 77. Council for British Archaeology, York, pp 128–140
Jacobi R, Higham T (2011) The later upper palaeolithic recolonisation of Britain: new results from AMS radiocarbon dating. In: Ashton N, Lewis SG, Stringer C (eds) The ancient human occupation of Britain, Developments in Quaternary Science, vol 14. Elsevier, London, pp 223–247
Jenson OL (2009) Dwellings and graves from the Late Mesolithic site of Nivâ 10, Eastern Denmark. In: McCartan SB, Schulting RJ, Warren G, Woodman PC (eds) Mesolithic horizons. Oxbow Books, Oxford, pp 465–472
Jöris O, Thissen J (1997) Microlithic tool assemblages associated with long blade technology. Ubach-Palenberg (Lower Rhine area) – a case study on a Late Palaeolithic site at the boundary Younger Dryas/Preboreal. In: Fagnart J-P, Thévenin A (eds) Le Tardiglaciaire en Europe du Nord-Ouest, Actes du colloque "Chronostratigraphie et environnement des occupations humaines du Tardiglaciaire et du début de l'Holocène en Europe du Nord-Ouest", 119ème Congrès national des sociétés historiques et scientifiques Amiens 26–30 octobre 1994. Editions du Comité des travaux historiques et scientifiques, Paris, pp 611–621
Kayser O (1989) L'Epipaléolithique et le Mésolithique en Bretagne. In: Mohen JP (ed) Le temps de la Préhistoire, tome 1. Société Préhistorique Française, Paris, pp 350–352
Lambeck K, Chappell J (2001) Sea level change through the last glacial cycle. Science 292:679–686
Leakey LSB (1951) Preliminary excavations of a mesolithic site at Abinger Common, Surrey. Res Pap Surrey Archaeol Soc 3:1–44
Malone C (2001) Neolithic Britain and Ireland. Tempus, Stroud, pp 11–45
Miller R (2012) Mapping the expansion of the Northwest Magdalenian. Quat Int 272–273:209–230
Mithen S, Wicks K, Pirie A, Riede F, Lane C, Banerjea R, Cullen V, Gittins M, Pankhurst N (2015) A Lateglacial archaeological site in the far north-west of Europe at Rubha Port an t-Seilich, Isle of Islay, western Scotland: Ahrensburgian-style artefacts, absolute dating and geoarchaeology. J Quat Sci 30(5):396–416. doi:10.1002/jqs.2781/pdf
Momber G (2011) Submerged landscape excavations in the Solent, southern Britain: climate change and cultural development. In: Benjamin J, Bonsall C, Pickard C, Fischer A (eds) Submerged prehistory. Oxbow Books, Oxford, pp 85–98
Momber G (2014) Submerged archaeology and cultural responses to climatic amelioration. In: Evans AM, Flatman JC, Flemming NC (eds) Prehistoric archaeology on the continental shelf: a global review. Springer, New York, pp 194–212
Momber G, Tomalin D, Scaife R, Satchell J, Gillespie J (eds) (2011) Bouldnor Cliff and the submerged Mesolithic landscape of the Solent, CBA Research Report, vol 164. Council for British Archaeology, York
Moree JM, Sier MM (eds) (2015) Interdisciplinary archaeological research programme Maasvlakte 2, Rotterdam. Bureau Oudheidkundig Onderzoek Rotterdam, Rotterdam
Oxford Archaeology (2016) A mesolithic pit house on the Isle of Man http://oxfordarchaeology.com/news/oanorth-news/423-a-mesolithic-pit-house-on-the-isle-of-man
Parfitt SA, Barendregt RW, Breda M, Candy I, Collins MJ, Coope GR, Durbidge P, Field MH, Lee RJ, Lister AM, Mutch R, Penkman KEH, Preece RC, Rose J, Stringer CB, Symmons R, Whittaker JE, Wymer JJ, Stuart AJ (2015) Happisburgh: the earliest record of human activity in Northern Europe. Nature 7072:1008–1012
Pedersen L, Fischer A, Aby B (1997) The Danish Storebælt since the ice age – man, sea and forest. A/S Storebælt Fixed Link, Copenhagen
Peeters JHM (2007) Hoge Vaart-A27 in context: towards a model of Mesolithic-Neolithic land use dynamics as a framework for archaeological heritage management. Rijksdienst voor Archeologie, Cultuurlandschap en Monumenten, Amersfoort
Peeters JHM, Momber G (2014) The southern North Sea and the human occupation of northwest Europe after the Last Glacial Maximum. Neth J Geosci 93:55–70. doi:10.1017/njg.2014.3
Peeters JHM, Brinkhuizen DC, Cohen KM, Kooistra LI, Kubiak-Martens L, Moree JM, Niekus MJLT, Schiltmans DEA, Verbaas A, Verbruggen F, Vos PC, Zeiler JT (2015) Synthesis. In: Moree JM, Sier MM (eds) Interdisciplinary archaeological research programme Maasvlakte 2, Rotterdam, Part 1 Twenty metres deep! The Mesolithic period at the Yangtze Harbour site – Rotterdam Maasvlakte, the Netherlands. Early Holocene landscape development and habitation. Bureau Oudheidkundig Onderzoek Rotterdam, Rotterdam, pp 287–317
Rankine WF (1952) A mesolithic chipping floor at the Warren Oakhanger, Selborne, Hampshire. Proc Prehist Soc 18:21–25

Rensink E (1995) On Magdalenian mobility and land use in north-west Europe. Some methodological considerations. Archaeol Dialogues 2:85–119

Reynier M (2000) Thatcham revisited: spatial and stratigraphic analysis of two sub-assemblages from site iii and its implications for early mesolithic typo-chronology in Britain. In: Young R (ed) Mesolithic lifeways: current research from Britain and Ireland, Leicester Archaeology Monographs, vol 7. Leicester University Press, Leicester, pp 33–46

Robertson D, Crawley P, Barker A, Whitmore S (2005) Norfolk rapid coastal zone archaeological survey: part one – assessment report. English Heritage Maritime Reports, Report Number 1045. Norfolk Archaeological Unit, Norfolk. https://historicengland.org.uk/images-books/publications/norfolk-rczas/

Sheppard T (1912) The lost towns of the Yorkshire coast. Brown and Sons, London

Skaarup J, Grøn O (2004) Møllegabet II. A submerged Mesolithic settlement in southern Denmark. BAR International Series 1328. Archaeopress, Oxford

Smith O, Momber G, Bates R, Garwood P, Fitch S, Pallen M, Gaffney V, Allaby RG (2015) Sedimentary DNA from a submerged site provides evidence of wheat in the British Isles 8000 years before present. Science 347(6225):998–1001. doi:10.1126/science.1261278

Stapert D (1985) A small Creswellian site at Emmerhout (province of Drenthe, the Netherlands). Palaeohistoria 27:1–65

Street M (1998) The archaeology of the pleistocene-holocene transition in the northern Rhineland, Germany. Quat Int 49–50:45–67

Suddaby I (2007) Downsizing in the mesolithic? The discovery of two associated post-circles at Silvercrest, Lesmurdie Road, Elgin, Scotland. In: Waddington C, Pedersen K (eds) Mesolithic studies in the North Sea Basin and beyond. Oxbow Books, Oxford, pp 60–68

Tann G (2004) Lincolnshire coastal grazing marsh. Archaeological and historical data collection, Lindsey Archaeology Services Report No. 770. Lindsey Archaeological Services, Lincoln

Taylor M (2011) Waterlogged wood. In: Momber G, Tomalin D, Scaife R, Satchell J, Gillespie J (eds) Bouldnor Cliff and the submerged Mesolithic landscape of the Solent, CBA Research Report, vol 164. Council for British Archaeology, York, pp 84–89

Tizzard L, Bicket AR, Benjamin J, De Loecker D (2014) A middle palaeolithic site in the southern North Sea: investigating the archaeology and palaeogeography of Area 240. J Quat Sci 29(7):698–710

Tolan-Smith C (2008) Mesolithic Britain. In: Bailey G, Spikins P (eds) Mesolithic Europe. Cambridge University Press, Cambridge, pp 132–157

Tomalin D (2011) The assemblage of lithic artefacts from the Mesolithic occupation features on the sea floor at Bouldnor Cliff. In: Momber G, Tomalin D, Scaife R, Satchell J, Gillespie J (eds) Bouldnor Cliff and the submerged Mesolithic landscape of the Solent, CBA Research Report, vol 164. Council for British Archaeology, York, pp 139–154

Verhart LBM (2004) The implications of prehistoric finds on and off the Dutch coast. In: Flemming NC (ed) Submarine prehistoric archaeology of the North Sea. Research priorities and collaboration with industry, CBA Research Report, vol 14. Council for British Archaeology, York, pp 57–61

Vos P, Van den Berg M, Bunnik F, Cremer H, De Keine M, Maljers D, Mesdag C (2010) Geoarcheologisch vooronderzoek ten behoeve van het Yangtzehavenproject: inventariserend veldonderzoek, fase 2. Deltares, Utrecht

Waddington C (2007a) Mesolithic settlement in the North Sea Basin. A case study from Howick, north-east England. Oxbow Books, Oxford

Waddington C (2007b) Rethinking mesolithic settlement and a case study from Howick. In: Waddington C, Pedersen K (eds) Mesolithic studies in the North Sea Basin and beyond. Oxbow Books, Oxford, pp 101–113

Waddington C (2015) A case for secondary Mesolithic colonisation of Britain following rapid inundation of the North Sea Plain. In: Ashton N, Harris C (eds) No stone unturned: papers in honour of Roger Jacobi, Lithic Studies Occasional Paper, vol 9. Lithic Studies Society, London, pp 221–232

Waughman M, Donaldson M, Innes J, Tooley M (eds) (2005) Archaeology and environment of submerged landscapes in Hartlepool Bay, England, Tees Archaeology Monograph Series No, vol 2. Tees Archaeology, Hartlepool

Wilkinson TJ, Murphy P (1995) The archaeology of the Essex Coast. Volume I. The Hullbridge survey. East Anglian Archaeol 71

Woodman P (2003) Colonising the edge of Europe: Ireland as a case study. In: Kindgren H, Knutsson K, Loeffler D, Åkerlund A (eds) Mesolithic on the move. Oxbow books, Oxford, pp 57–61

Wymer J (1958) Excavations on the Mesolithic site at Thatcham, Berks. Interim Report. ADS University of York. http://archaeologydataservice.ac.uk/archiveDS/archiveDownload?t=arch-787-1/dissemination/pdf/BAJ057_PDFs/BAJ057_A01_wymer.pdf

Wymer J, Robins P (1994) A long blade industry beneath Boreal peat at Titchwell, Norfolk. Norfolk Archaeol 42(1):13–37

# Chapter 22
# Aegean Pleistocene Landscapes Above and Below Sea-Level: Palaeogeographic Reconstruction and Hominin Dispersals

**Dimitris Sakellariou and Nena Galanidou**

**Abstract** The Aegean Region has remained marginal to research into human origins despite its key position in the multiple movements of animals between Europe and Asia. A possible explanation for this is that the Palaeolithic remains are invisible because they lie beneath the sea, whilst research in the field was hitherto developed on the mainland. In this chapter we make the submerged land, the coastal zones and the islands a unified research focus to examine the main, long-term and short-term geological and geotectonic processes which have controlled the development of Pleistocene landscapes in the Aegean Region above and below the fluctuating sea-level. We integrate evidence on the geology, tectonics, morphology and hydrogeology of the shallow coastal and shelf areas in order to reconstruct the palaeogeography. Given the variable tectonic evolution and geomorphological configuration of the coastal and shelf areas, we divide the Aegean into nine geographical units. Each unit has its own geotectonic and morphological history and offers a frame of reference to assess land-routes and the natural resources available to hominins at different times of the Pleistocene. We link this palaeogeographic reconstruction to the discussion of the early occupation of Europe. This allows the NE Mediterranean to become part of the discussion about hominin dispersals into Europe through a south-eastern route and gives a more complete view of the variations in Palaeolithic settlement.

## 22.1 Rationale, Aims and Methods

The Aegean Region, henceforth the Aegean, has remained marginal to research into human origins (Bar-Yosef and Belfer-Cohen 2001; Moncel 2010; Jöris 2014). Until recently, agreement about Lower Palaeolithic material involved only the Petralona Cave in Macedonia and the Kokkinopilos red beds in Epirus. Associated with a *Homo heidelbergensis* cranium and a handful of large cutting tools with bifacial retouch respectively, the two sites have remained controversial with respect to their stratigraphic context and date (see discussion in Darlas 2014; Galanidou 2016). Simultaneously, palaeontological work showed the central position of the Balkan Peninsula in the successive dispersal episodes of Early and Middle Pleistocene animal populations between west Eurasia and the east (Azzaroli

D. Sakellariou (✉)
Institute of Oceanography, Hellenic Centre for Marine Research,
46.7 km Athinon-Souniou Av, Anavyssos, P.O. Box 712, 19013 Athens, Greece
e-mail: sakell@hcmr.gr

N. Galanidou
Department of History and Archaeology, University of Crete, 74100 Rethymno, Greece
e-mail: galanidou@uoc.gr



G.N. Bailey et al. (eds.), *Under the Sea: Archaeology and Palaeolandscapes of the Continental Shelf*, Coastal Research Library 20, DOI 10.1007/978-3-319-53160-1_22

1983; Koufos 2001; Palombo et al. 2006; Rook and Martínez-Navarro 2010; Rodríguez et al. 2013; Khatib et al. 2014), thus emphasizing the potential of the region to yield material that would enhance the discussion on the early occupation of Europe. In recent years, excavations conducted by the University of Crete have unveiled two early sites: Rodafnidia an open-air site at Lisvori on Lesbos Island; and the Kythros Cave, a collapsed karstic complex on the rocky shore of the island of the same name in the Inner Ionian Archipelago. Both yield archaeology and radiometric dates that securely place hominin activity in the Middle Pleistocene (Galanidou 2013, in press; Galanidou et al. 2013, 2016b). The sites are situated on islands of different configuration and size, which at times of low sea levels—certainly during the periods of hominin presence—would have been joined to mainland Anatolia and Greece respectively. Their interpretation demands palaeogeographic reconstructions so as to understand the impact of changing sea levels on connections between the islands and their adjacent landmasses and the dynamic history of the drowned landscapes of the Aegean and Ionian shelves.

In this paper we make the submerged landscapes, the coastal zones and the islands of the Aegean a focus of research to review the main geological and tectonic processes which have controlled their evolution. We discuss the attractions that these three interconnected geographic entities may have offered to hominins either to settle or disperse into Europe following a north-western route from the Levantine corridor. Our discussion refers to the period of the Middle Pleistocene (<780–126 ka) and Upper Pleistocene (<126–12 ka). Our point of departure is the conviction that Palaeolithic research in this part of the world ought to move beyond a mainland-focused paradigm that regards continental Greece as the only area offering good prospects for finding early sites.

The initiative for studying the geological evolution of the submerged landscapes as a critical parameter has been developed within the framework of COST Action TD902 SPLASHCOS, (Bailey et al. 2012) and the writing of the European Marine Board's Position Paper 21 recommending archaeological and palaeoclimatic research on the drowned prehistoric landscapes of the European continental shelf (Flemming et al. 2014). In line with these developments, we view the submerged landscapes of the Aegean and its 'islandscapes' (see also Ward and Veth, Chap. 24) as areas worthy of systematic research, marking possible passageways and/or settlement areas for early hominins according to varying climatic and geographical opportunities.

The Aegean (Fig. 22.1) is an almost land-locked sea and coastal area in the NE Mediterranean that has repeatedly been transformed into an archipelago by sea level rise as in the present Holocene epoch, and an extended land mass with lake basins by falling sea level. During the low sea levels of the Pleistocene many of the present islands were connected to the Eurasian mainland, offering a much larger area of land for hominin occupation. Elsewhere we have argued that the archaeological record for the Pleistocene Aegean bears witness to a great geographical and temporal diversity, with both continuity and rupture (Sakellariou and Galanidou 2016). Although the Aegean as a geographic entity has been employed in larger scale reconstructions of the past as lying at the heart of Eurasia, in practice its Palaeolithic archaeology embraces the biogeographic influences of southern Europe in mainland Greece and the Ionian Sea, of western Asia along the islands of the East Aegean Sea and of mainland Greece, North Africa and possibly western Asia in Crete. In harmony with these multidirectional influences, we propose a palaeogeographic subdivision of the Aegean into nine units based on the particular geomorphological evolution of distinct regions, which stand out as individual tectonic blocks (or groups of blocks) and exhibit discrete geotectonic histories. We consider that the overall geodynamic outline and the morpho-tectonic structure of the Aegean act as a backbone for the discussion of the morphological evolution of the continental Eurasian shelf. We present a short description of the geological, tectonic and hydrogeological background against which the information on the natural resources available to hominins can be evaluated. Our palaeogeographical reconstructions of the shallow coastal and shelf areas in each unit are complemented by an overview of the Palaeolithic archaeology and palaeontology.

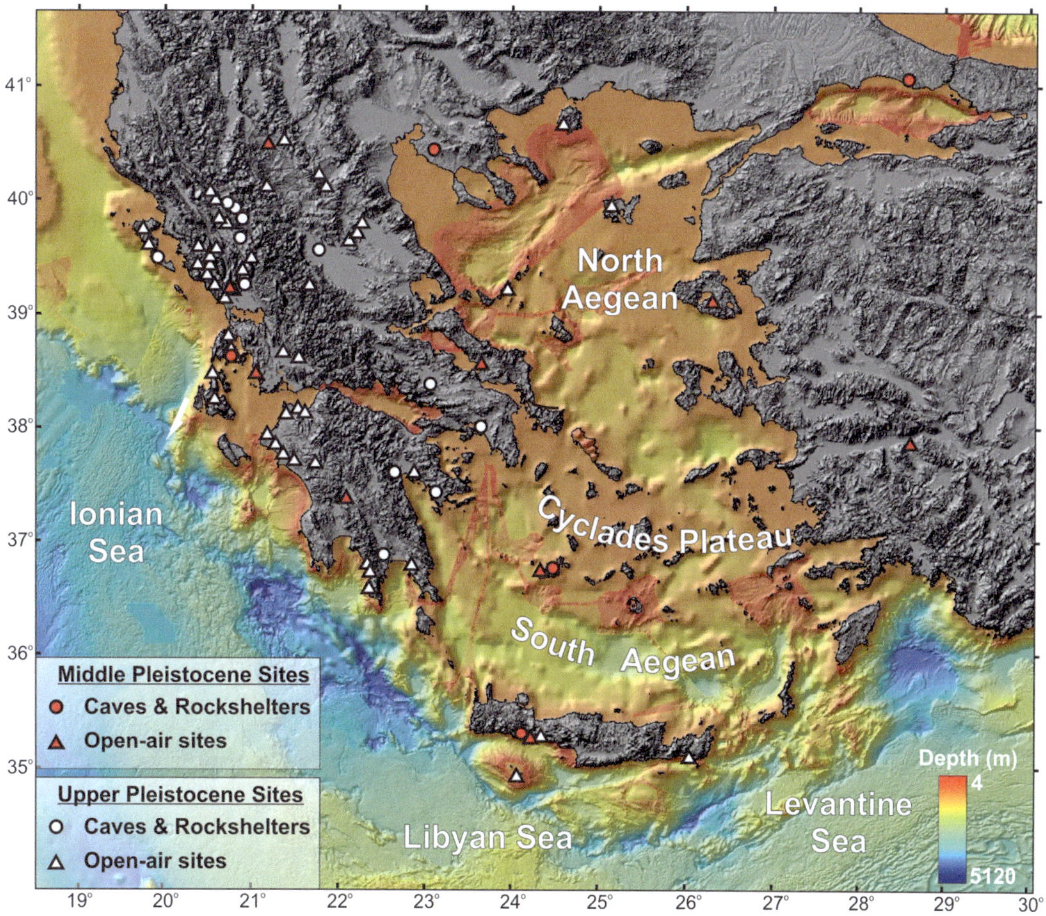


**Fig. 22.1** Shaded relief map of the Aegean region extracted from the CGMW/UNESCO Morpho-Bathymetry of the Mediterranean Sea (Brossolo et al. 2012) with location of Middle and Upper Pleistocene archaeological sites

## 22.2 Aegean Geodynamics, Geology and Hydrogeology

The Aegean is a unique environment in terms of active geodynamics, plate movements, on-going geological processes, formation of new relief and dynamically changing landscapes. The driving force is the convergence between the Eurasian and the African continental plates. The westward extrusion of the Anatolian continental block along the North Anatolian Fault, the NNE-ward subduction of the Eastern Mediterranean lithosphere beneath the Hellenic Arc, the subsequent SSW–NNE extension of the Aegean back-arc region, the collision of NW Greece with the Apulian block in the northern Ionian Sea north of the Kephallinia Fault and the incipient collision with the Libyan promontory south of Crete are the main, ongoing, geodynamic processes (McKenzie 1970, 1978; Dewey and Sengoer 1979; Le Pichon and Angelier 1979; Le Pichon 1982; Le Pichon et al. 1982, 1995; Angelier et al. 1982; Meulenkamp et al. 1988; Mascle and Martin 1990; Taymaz et al. 1991; Jolivet 2001; Armijo et al. 1999, 2004; Kreemer and Chamot-Rooke 2004).

The build-up of stresses along the boundaries and in the interior of the Aegean leads to brittle (and ductile) deformation in the upper crust expressed in normal, reverse or strike slip faulting. Long-term vertical tectonics associated with the activity on major tectonic elements and faults creates a puzzle of small-scale blocks each displaying a different history of uplift or subsidence.

The active, long-term, geodynamic processes, crustal movements and deformation of the Aegean also give rise to violent, short-term, geological events. Regional or local-scale phenomena such as onshore and offshore slope and coastal failures or earthquake-induced subsidence overprint the effect of long-term geodynamic movements and can temporarily amplify or moderate long-term trends of landscape evolution. Eustatic sea-level fluctuations are superimposed on these long-term geological processes. The result is a dynamically changing boundary at the land-to-sea interface, where vertical tectonic movements, sedimentation, erosion, coastal slope-failures and sea-level fluctuations force coastlines to move vertically and horizontally on a variety of time scales.

The geological structure of the Aegean (Fig. 22.2) resulted from the successive closure of different branches of the Mesozoic Tethys Ocean and evolved in the course of four orogenic cycles from Dogger to Miocene. The present arc-shaped configuration and rupturing derives from the overprinting of the Plio-Quaternary geodynamic regime on the Alpine structure. The basement geology of the mountainous regions of Northern Greece and North-western Turkey comprises Alpine and Pre-Alpine metamorphic rocks and sedimentary and volcanic rocks of Mesozoic and Cenozoic age. Extended Quaternary alluvial plains occur predominantly on the North Aegean coasts and are associated with major river-delta formations. The eastern parts of the Greek mainland, the Central Aegean Archipelago and the opposite coasts of Turkey are mostly built from Alpine sedimentary rocks and metamorphic rocks. Quaternary volcanic rocks and active volcanoes occur in the Central Aegean Island Bridge. The mountain chain of Western Greece, from Epirus in the north to the South Peloponnese in the south and through the islands of the Hellenic Arc to Southwest Turkey consists predominantly of folded and thrusted sedimentary rocks of Mesozoic to Cenozoic age. Late and Post-Alpine, Oligocene to Miocene and Pliocene sedimentary deposits occur at various regions in between the Mesozoic mountain chains, filling back-arc and molassic basins arranged parallel to the Alpine trend or neotectonic, Plio-Quaternary grabens cutting across the Alpine trend.

Mesozoic limestone and marble constitute a major component of the Aegean geological structure. Marble occurs in the metamorphic provinces of Northern Greece and NW Anatolia, the eastern part of Central Greece, the Central Aegean Islands (Cyclades) and opposite the Anatolian coast and along the Hellenic Arc (SE Peloponnese, Crete, Rhodes). Limestone forms the major part of the basement geology of the western part of mainland Greece, the Ionian and Aegean Islands and SW Anatolia. Clastic sedimentary rocks (Tertiary flysch, Triassic-Jurassic siltstones and early rift sandstones) form the second major component of the Alpine stratigraphy. Metamorphosed clastic sediments (mainly greenschists and phyllites) along with metamorphosed volcanics (greenschists, blueschists) complete the basement geology in the metamorphic provinces.

Folding and thrusting during the Alpine orogenic phases has led to the creation of complex geological structures and the deformation of the initial stratigraphic rock succession. Vertical and lateral transition between permeable limestone and marble and impermeable clastic sediments and metamorphic rocks results in suitable hydrogeological conditions for the formation of very significant aquifers in the karstified carbonate rocks. The interplay between hydrogeological structure and morphological relief defines the development of major karstic springs at the interface between marble/limestone and sedimentary/metamorphic rocks. Submarine, karstic, fresh or brackish water springs are known from very many places along the coastline or at shallow depths below the present sea level. Submarine springs constitute a major factor in the submerged landscape survey and are of paramount importance for the submerged Palaeolithic archaeology since fresh-water points have always been attractive to humans and animals.

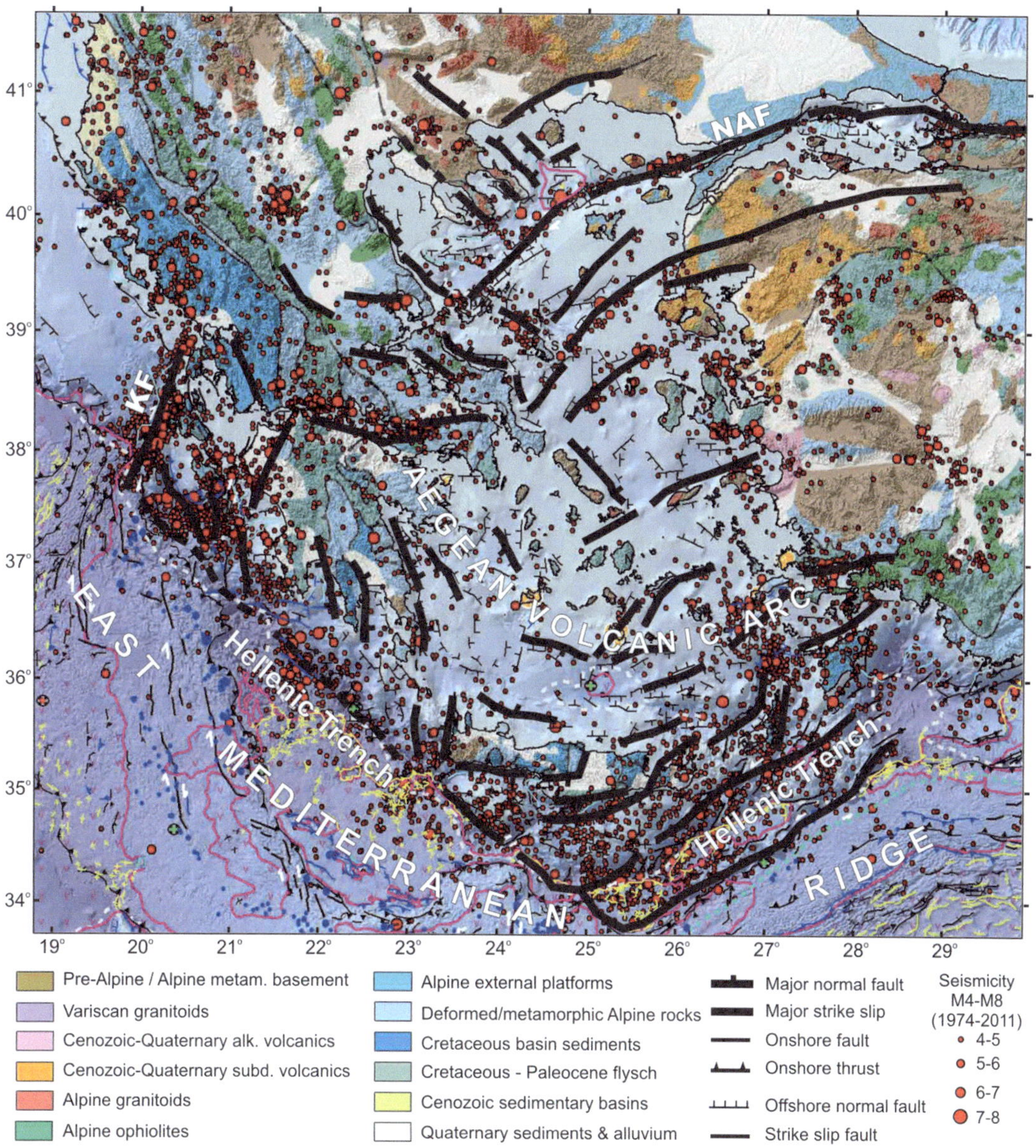


**Fig. 22.2** Geological and Morpho-Tectonic Map of the Aegean, extracted from the Geological and Morpho-Tectonic Map of the Mediterranean Domain (Mascle and Mascle 2012) and modified. *KF* Kephallinia Transform Fault, *NAF* North Anatolian Fault

## 22.3 Aegean Shelf Morphology and Palaeogeography

The North Aegean shelf, north of the North Aegean Trough, is the widest one in our study region. The largest rivers of the Southern Balkan Peninsula flow through mountainous Northern Greece and outflow here, feeding the shelf with huge quantities of new sediment. The Aegean Sea between the North Aegean Trough and the South Aegean Basin includes the vast majority of the numerous islands and islets which comprise the Aegean Archipelago. Many of them, like the Cyclades, are separated from each other by shallow shelves. Yet others again, like the East and Northeast Aegean Islands, the East

Dodecanese and the North Sporades Islands, are separated from each other and from the adjacent land masses of Greece or Turkey by shallow shelves. Two island bridges, a southern one in the Central Aegean and a northern one in the North Aegean Sea, connect the mainland of Greece with the landmass of Anatolia.

Fast but not uniform uplift prevails along the southern and eastern part of the Hellenic Arc, which is segmented by numerous faults. On the top of the uplifted blocks, the islands of the Hellenic Arc rise well above sea-level and are separated by narrow and deep, faulted trenches. This active environment has led to the formation of high relief and consequently does not favour the development of an extended shelf. The shelf off the mostly rocky coasts of the Hellenic Arc Islands is very narrow and nowhere exceeds a few kilometres in width.

The western part of the Hellenic Arc, from Corfu to South Peloponnese, displays a similarly narrow shelf. The Ionian margin hosts the deltaic plains of some of the largest rivers of Greece. Here, the shelf reaches maximum a width of up to 15–20 km near the river deltas but in most areas does not exceed 2–5 km.

## 22.4 Submerged and Coastal Landscape Analysis and Archaeology

Given the above geological and tectonic factors, we divide the Aegean into nine geographical units, each with its own morphological, geological, tectonic, hydrogeological and sedimentological characteristics (Fig. 22.3). The boundaries coincide closely with major fault zones separating crustal blocks with different tectonic characteristics (compare Figs. 22.2 and 22.3).

We present landscape reconstructions for each unit in turn, based on our understanding of long-term geological processes and likely patterns of landscape evolution since the Middle Pleistocene. We support the discussion with maps that show areas of the offshore shelf exposed during one or more low sea-level periods of the Middle and Upper Pleistocene. These offshore areas mainly comprise basement rocks with a thin sediment cover and show a close relationship with prevailing tectonic movements along major faults that define the boundary between areas of exposed land and areas subject to long-term subsidence and sediment accumulation. Finally, we integrate the available archaeological data on the distribution of Palaeolithic sites and highlight the important factors which need to be considered in future surveys of submerged landscapes and underwater Palaeolithic archaeology.

### *22.4.1 North Aegean Shelf*

The North Aegean Shelf belongs to the relatively 'non-deforming', rigid Eurasian continent (Fig. 22.4). Geological, sedimentological and tectonic processes and movements over longer periods have raised the mountainous regions of the Chalkidiki Peninsula and the Thasos and Samothraki Islands and formed sedimentary basins and lowlands comprising the present alluvial plains and submerged shelf area. Coastal subsidence varies between 0.05 and 1.2 mm/year along the coastline (Perissoratis and Conispoliatis 2003).

The shelf is up to 15–20 km wide. Four large rivers (Axios, Loudias, Aliakmon and Pinios) drain the eastern part of the Central North Greece mountainous area and outflow in the Thermaikos Gulf. The Strymon, Nestos, Filiouris, Kompsatos and Evros Rivers flow through the metamorphic basement of the Rhodope Mountains and outflow into the Strymonikos Gulf and Samothraki Plateau.

Systematic subbottom profiling (Perissoratis and Mitropoulos 1989; Piper and Perissoratis 1991) has identified the major riverbeds on the submerged landscape and connected them to the present rivers, while a number of ephemeral lakes and at least two permanent ones (Ierissos and Alexandroupoli)

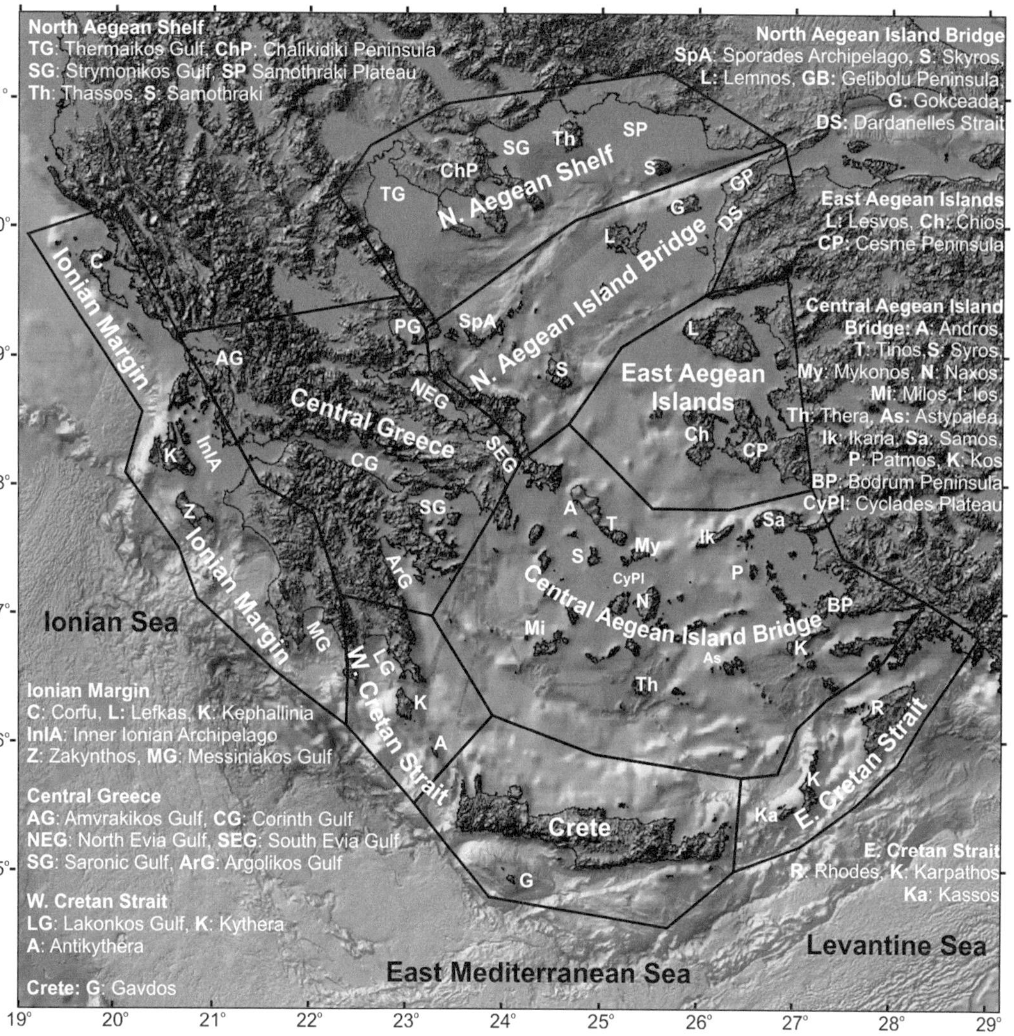


**Fig. 22.3** Bathy-morphological map of the Aegean (Aegean Sea and Hellenic Arc, IBCM/IOC 1981) divided into geographical units according to the nature of the active morpho-tectonic processes, evolution and configuration of the coastal and submerged landscape

existed in this area. The 14–16 ka shoreline has been mapped at about 120 m below present sea-level. The shelf area was exposed during the low sea-level stages of the Middle and Upper Pleistocene (Lykousis 2009), and the exposed land included the Thassos and Samothraki Islands, which would have been connected to the mainland to the North.

Hominins were present as early as the Middle Pleistocene according to the finds from the Petralona Cave (Hennig et al. 1982; Latham and Schwarcz 1992; Grün 1996). Upper Pleistocene evidence is represented by a small number of Middle Palaeolithic sites, namely the open-air site at Rachona near Pella (Darlas 2011), the Cave site by the Agitis Sources in Drama (Trantalidou 1989; Trantalidou and Darlas 1995), Therma near Strymon (Kourtessi-Philippakis et al. 1993) and further east in the Rhodope province a number of sites at Krovili (Ammerman et al. 1999) and the Petrota chert quarry (Fotiadis 2016). Upper Palaeolithic sites are conspicuously scarce, with the exception of the ochre quarry and associated lithic, bone and antler artefacts excavated at Limenaria on Thasos (Koukouli-Chrysanthaki and Weissgeber 1997; Koukouli-Chrysanthaki 2012).

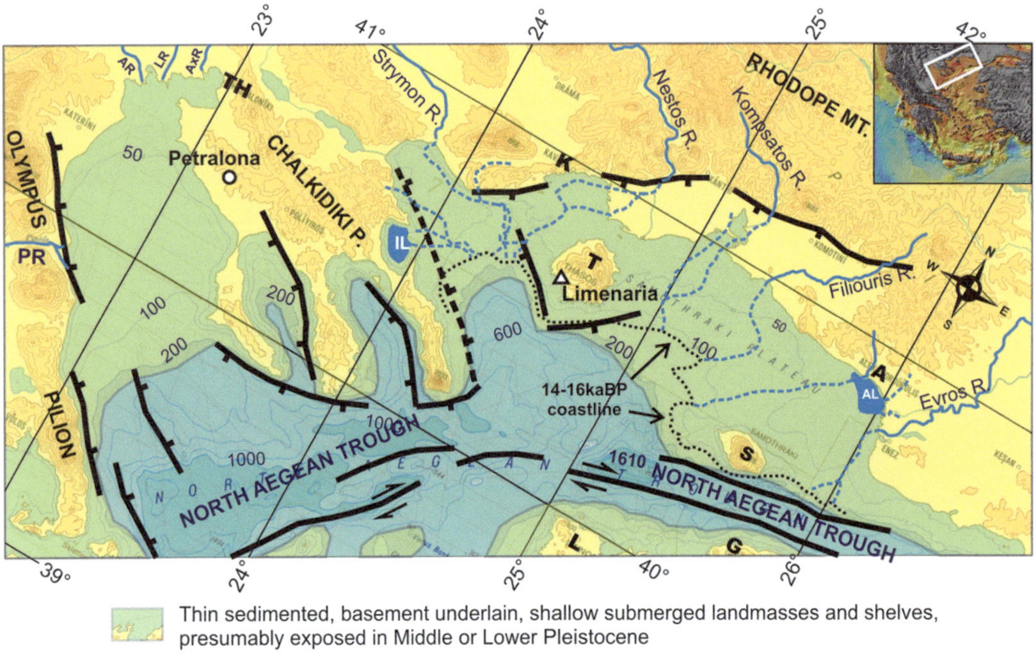


**Fig. 22.4** Bathymetry and morphology of the North Aegean Shelf (IBCM/IOC 1981) with major faults and rivers. *Green-shaded* areas indicate areas of the now-submerged shelf exposed during the low sea level stands of the Pleistocene. The dotted black line east of Chalkidiki Peninsula shows the location of the 14–16 ka BP coastline; the *dotted blue lines* are submerged river tracks; *IL* Ierissos Lake, *AL* Alexandroupolis Lake (From Perissoratis and Mitropoulos 1989). *PR* Pinios River, *AR* Aliakmon River, *LR* Loudias River, *AxR* Axios/Vardar River, *TH* Thessaloniki, *K* Kavala, *T* Thassos, *A* Alexandroupolis, *S* Samothraki, *L* Limnos, *G* Gokceada

## 22.4.2 North Aegean Island Bridge

Two clusters of islands, a western one with the North Sporades Archipelago and Skyros Island, and an eastern one with the Aghios Efstratios, Limnos, Gokceada Islands and other smaller islets constitute a series of stepping stones between Central Greece and Northwestern Anatolia (Fig. 22.5). During MIS (Marine Isotope Stage) 12, 10 and 8 the North Aegean Trough was isolated from the open sea and the North Aegean Island Bridge connected the mainland of Greece with that of north-western Anatolia. The Skopelos Basin was at that time an isolated lake located between the North Sporades Islands and North Evia. During MIS 6 the Sporades Archipelago remained connected to the Greek mainland to the west. A shallow sea separated the Sporades ridge from the exposed land attached to north-western Anatolia. The eastern islands remained connected to the Anatolian mainland throughout the LGM.

Active tectonics in the area are broadly associated with the prolongation of the North Anatolian Fault into the North Aegean Trough (Papanikolaou et al. 2002) and sub-parallel secondary branches, mostly strike slip and transtensional structures (Mascle and Martin 1990). Evidence of vertical tectonic activity is known from Skyros Island where Late Holocene submerged notches indicate subsidence (Evelpidou et al. 2012).

The Northern Sporades island cluster has a long-known Upper Pleistocene record at open-air locales on Alonissos (Panagopoulou et al. 2001) and its nearby islets Mikro Kokkinokastro (Theocharis 1970, 1971) and Agios Petros (Efstratiou 1985; Moundrea-Agrafioti 1992). The record comprises mainly Middle Palaeolithic finds recovered from low-density sites or smaller find spots and Late Upper Palaeolithic finds at Leptos Yialos on Alonissos (Panagopoulou et al. 2001, pp. 132–133). On

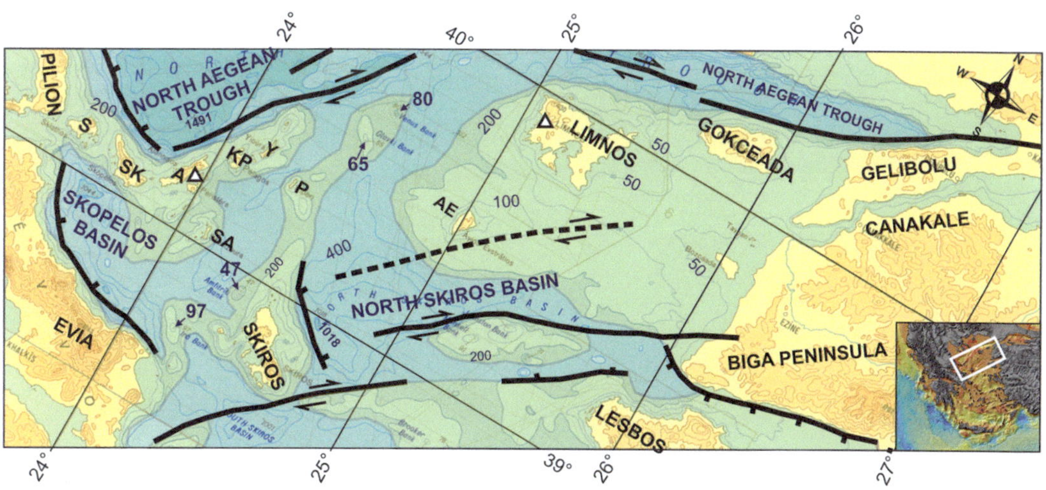


**Fig. 22.5** Bathymetry and morphology of the North Aegean Island Bridge (IBCM/IOC 1981) with major faults and location of Palaeolithic sites. *Numbers* and *arrows* indicate the depth of shallow ridges or banks. *Green-shaded* areas as in Fig. 22.4. *S* Skiathos, *SK* Skopelos, *A* Allonissos, *KP* Kyra Panagia, *Y* Youra, *P* Piperi, *AE* Aghios Efstratios

the shelf around the northern Sporades islands Middle Palaeolithic stone artefacts were found on the seabed. Some may originate from caves now at depths of −40 m off the shores of Kyra Panagia (Efstratiou 2001). The eastern counterpart of this unit has set its mark on the Palaeolithic map with the excavations at Ouriakos on Limnos (Efstratiou and Kyriakou 2011; Efstratiou et al. 2013; Efstratiou 2014). The closest comparanda to the Ouriakos archaeological material are found in the Upper Palaeolithic assemblage of Öküzini in SW Turkey, clearly suggesting an Anatolian origin.

### 22.4.3 *East Aegean Islands*

The two big islands, Lesbos and Chios, remained connected to the Anatolian land-mass during all major low-sea-level periods of the last 500 kyr. During the low sea-level MIS stages 12, 10 and 8, a large lake occupied the Central Aegean area. The lake comprises several deep basins developed along major tectonic lineaments and separates the exposed land of the East Aegean Island from the Greek land-mass to the west. NE–SW strike slip faults create a puzzle of elongated shallow ridges separated by deep trenches and basins (like the North and South Skyros Basins) forming a series of stepping stones between Lesbos and Skyros Islands (Fig. 22.6). After MIS 8, the Central Aegean and the North Aegean Trough remained connected to the open sea through shallow sea areas.

The Island of Lesbos is separated from the Asian coast by two sea straits, Muselim or Lamna to its north and Mytilene to its east. The Muselim Strait is an 8 km-wide, 300 m-deep strait developed in the hanging wall of the transtensional fault which runs along the southern edge of Biga Peninsula (Iler et al. 2008). The Mytilene Strait, between the eastern coast of Lesbos and the Anatolian shoreline, is 40–60 m deep and 6–8 km wide. Chios Island was part of the Çeşme Promontory during MIS 2–4 while Lesbos Island should have been attached to the area of Ayvalik. Several shallow banks occur in the sea-area west of Lesbos and north-west of Limnos. They represent basement ridges, bounded by extensional or transtensional faults, and rise from the surrounding 400–600 m-deep seafloor to heights as shallow as −30 m. Subsidence of the island, indicated by ancient harbour installations (Theodoulou

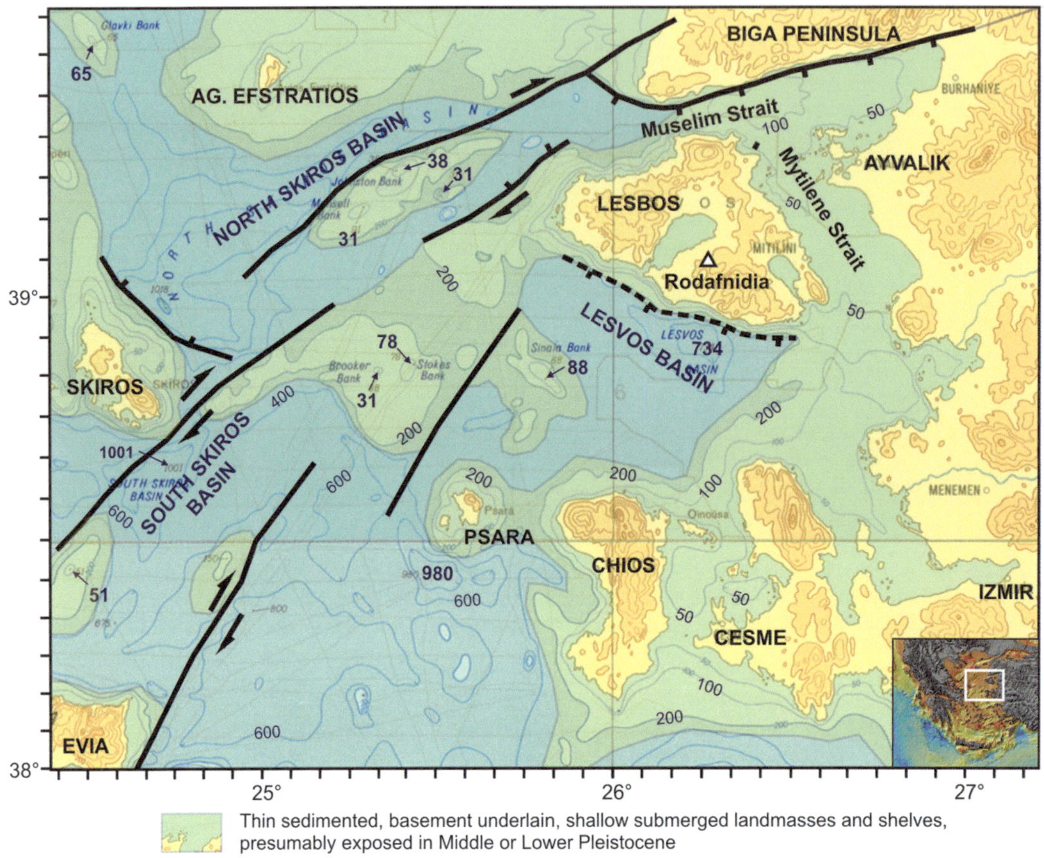


**Fig. 22.6** Bathymetry and morphology of the East Aegean Islands (IBCM/IOC 1981) with major faults and location of Palaeolithic sites. *Numbers* and *arrows* indicate the depth of shallow ridges or banks. *Green-shaded* areas as in Fig. 22.4

2008; Williams 2007), is mostly attributed to eustatic sea-level rise, although tectonically driven vertical movements cannot be ruled out.

Chios and Psara are so far devoid of any Palaeolithic sites, while Lesbos has only recently made an impressive entry to the Aegean Palaeolithic map with Rodafnidia, Lisvori, in the Kalloni Basin, securely placed by post-infrared infrared-stimulated luminescence (pIRIR) and relative dating based on artefact technology and typology to the Middle Pleistocene (Galanidou et al. 2013, 2016b). At Rodafnidia compelling evidence of hominin groups using early and late Acheulean industries is being excavated. Lower Palaeolithic activity took place in a fluvio-lacustrine environment near the shore of what, during Middle Pleistocene low sea level stands, was probably a palaeo-lake. The early industrial component shows affinities with other African and Asian Acheulean industries, and especially with the 'Large Flake Acheulean' (e.g., Goren-Inbar et al. 2000; Sharon 2007).

Rodafnidia is the first extensive Middle Pleistocene site in the Aegean with a corpus of archaeological evidence that supports a scenario of an eastern entrance to Europe by groups using Acheulean technology. The close industrial affinities are with material excavated at Gesher Benet Y'aaqov on the bank of the Jordan River near the paleo-lake Hula in northern Israel (Goren-Inbar et al. 2000) and the lower strata at Kaletepe Deresi 3 on a seasonal drainage in the Göllüdağ area near Cappadocia (Slimak et al. 2008). They suggest two possible but not mutually exclusive routes for hominin dispersal from the Asian mainland to Lesbos: a north–north-westward route from the Levantine corridor; or a westward route from central and eastern Anatolia. The presence of Middle Pleistocene hominins and Early Pleistocene fauna on Lesbos indicates that both humans and animals were reaching the island from the Asian mainland via land bridges opened by low sea level stands (Galanidou et al. 2013, 2016b). To judge by the archaeological variability observed at Rodafnidia, there must have been repeated hominin movements over long periods of time between Anatolia and mainland Greece via Lesbos from at least the early stages of the Middle Pleistocene. The palaeontological record of Lesbos suggests that animal crossings began much earlier. Taking into account that the nearby Anatolian record at Kocabaş (some 300 km in a direct line from the Kalloni Basin on Lesbos), with a very small number of artefacts and a *Homo erectus* skull found embedded in a travertine block, is securely dated to around 1.1 ma BP (Kappelman et al. 2008; Lebatard et al. 2014), we hypothesise that hominin presence on Lesbos may have begun during the Early Pleistocene.

### *22.4.4 Central Greece*

This is an area of Quaternary and ongoing extensional tectonics (Fig. 22.7). A series of WNW–ESE trending neotectonic grabens have developed as elongated marine gulfs cutting across the Alpine structure of the Hellenides mountain chain. The 900 m-deep Gulf of Corinth and the 450 m-deep North Evia Gulf are the largest and most active grabens in Central Greece. The Amvrakikos, South Evia, Pagasitikos and West Saronikos Gulfs are smaller, shallower and less active. All are connected to the open sea through narrow and shallow straits and were isolated lakes during the LGM (Perissoratis et al. 1993; Richter et al. 1993; Lykousis and Anagnostou 1994; Kapsimalis et al. 2005; Lykousis et al. 2007; Sakellariou et al. 2007a, b).

The first two gulfs display active, vertical tectonics, which modify significantly the effect of sea-level or lake-level fluctuations (Armijo et al. 1996; Leeder et al. 2005; Lykousis et al. 2007; Sakellariou et al. 2007b; Evelpidou et al. 2011a). Due to the ongoing opposite vertical movement of the two margins of the Gulf of Corinth, the shoreline of the LGM Lake can be found at −90 m along the northern margin and at −62 to 65 m along the southern margin. The lake level in the North Evia Gulf during the LGM was at about −90 m (Sakellariou et al. 2007b). The tectonic asymmetry of the basin forces the southern margin to subside (Evelpidou et al. 2011b) and the northern one to uplift (Sakellariou et al. 2007b).

The Argolikos Gulf is a deep neotectonic, asymmetric graben (Papanikolaou et al. 1994) and was always connected to the open sea of the South Aegean. Van Andel et al. (1990) have suggested a local sea-level history, while Van Andel and Lianos (1984) provided a reconstruction of the palaeoshorelines off Franchthi Cave. Preliminary results of a recent survey provide a detailed reconstruction of the submerged landscape (Sakellariou et al. 2015). Karstic, fresh or brackish water springs occur along the coasts of Central Greece or on the shelf at shallow depths below the present sea-level. The karstic spring of Anavalos is located at 6 m below the present sea-level on the western coast of the Argolikos Gulf.

Archaeological evidence dates to the Upper Pleistocene and is associated with both Neanderthal and AMH (anatomically modern human) populations. It is encountered in open air and in karstic settings, namely the Seidi Rockshelter in Boeotia (Stampfuss 1942), Kefalari Cave (Felsch 1973; Reisch 1976), Franchthi Cave (Farrand 2000; Perlès 1987, 1999; Douka et al. 2010) and Klissoura Cave 1 in the Argolid (Stiner et al. 2010). Middle Palaeolithic assemblages are commonly found in red-bed alluvial deposits and in the Argolid they have been uranium-series dated to around 50 ka (Pope et al. 1984). The Klissoura Cave 1 sequence spans the Middle and the Upper Palaeolithic.

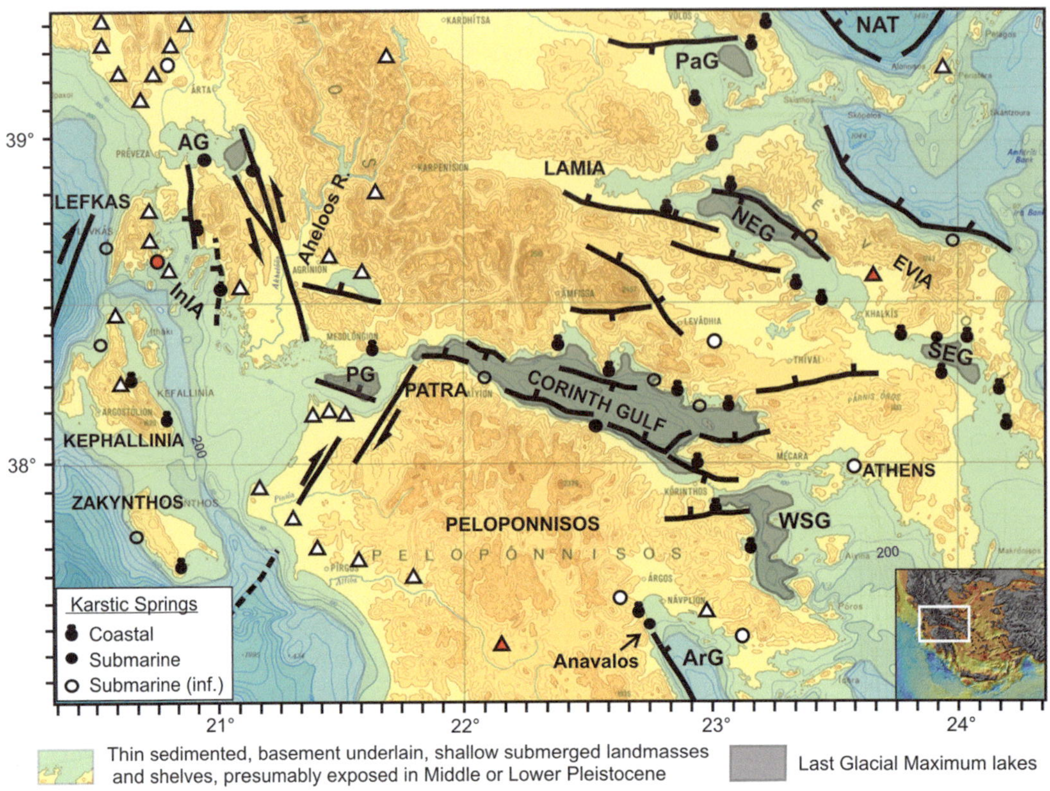


**Fig. 22.7** Bathymetry and morphology of Central Greece (IBCM/IOC 1981) with major faults, location of Palaeolithic sites and coastal and submarine karstic springs. *Grey* areas show Last Glacial Maximum lakes (see text). *AG* Amvrakikos Gulf, *InIA* Inner Ionian Archipelago, *PG* Patras Gulf, *NAT* North Aegean trough, *PaG* Pagasitikos Gulf, *NEG* North Evia Gulf, *SEG* South Evia Gulf, *WSG* West Saronic Gulf, *ArG* Argolikos Gulf

## *22.4.5 Central Aegean Island Bridge*

The Central Aegean Island Bridge comprises the Cyclades Archipelago and Plateau, the northern Dodekanese Archipelago and Ikaria and Samos Islands. The northern margin of the Central Aegean Island Bridge has been shaped by normal and strike-slip faults. The southern margin hosts the Hellenic Volcanic Arc. NW–SE and NE–SW normal and strike-slip faulting has created a puzzle of elongate basement ridges (Amorgos, Thera, Anafi-Astypalea) separated from each other by deep, elongate, basins. During the major low sea-level periods, an elongated land mass was exposed connecting Central Greece with Anatolia. Since MIS 6 the Central Aegean landmass has broken into two main parts. The western one (Cyclades Plateau) remained connected to Central Greece and became a large Island during the LGM (Kapsimalis et al. 2009). The eastern part remained attached to the Anatolian landmass. A narrow, 400–600 m-deep sea strait separates the Cyclades Archipelago to the west from the Dodecanese Islands to the east (Fig. 22.8).

The Central Aegean is characterized by relatively weak tectonic activity. The evolution of the coastline during the Late Pleistocene and Holocene is mostly controlled by eustatic sea-level fluctuations and to a lesser extent by isostatic movements. Even so, evidence of Late Holocene tectonic subsidence has been observed by comparison of submerged beach rocks (Desruelles et al. 2009).

Two sites on two islands of this region are noteworthy. The first is Triades Bay in west Melos where claims for Middle or possibly Lower Palaeolithic artefacts manufactured on rhyolite have been

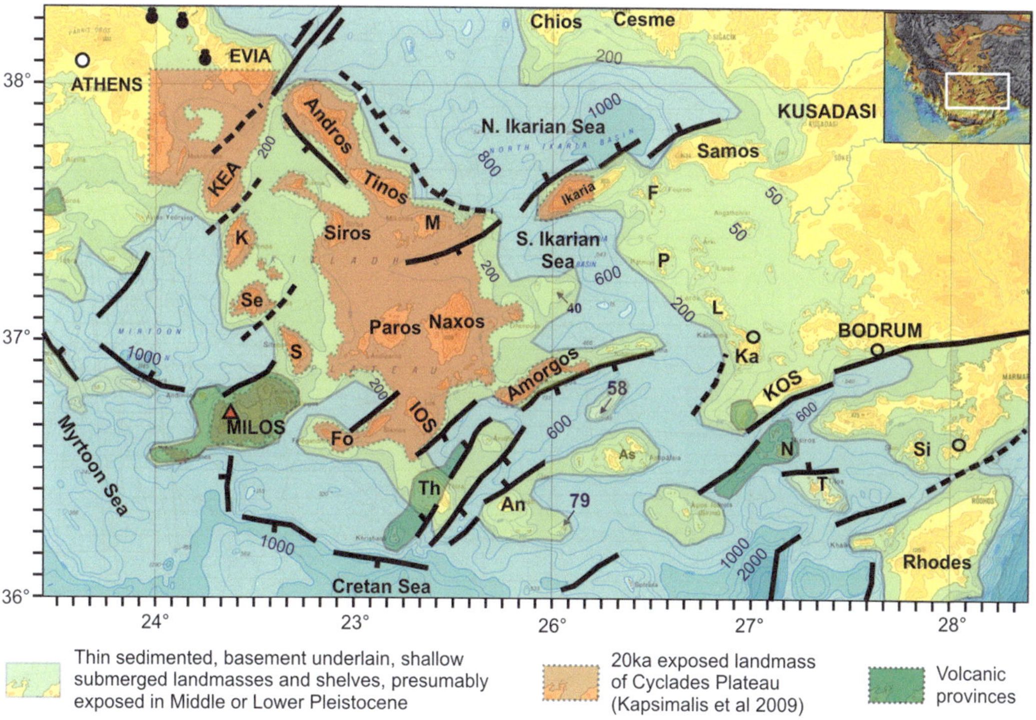


**Fig. 22.8** Bathymetry and morphology of Central Aegean Island Bridge (IBCM/IOC 1981) with major faults, location of Palaeolithic sites (symbols as in Fig. 22.1) and coastal and submarine karstic springs (symbols as in Fig. 22.7). *K* Kythnos, *Se* Serifos, *S* Sifnos, *Fo* Folegandros, *Th* Thera, *An* Anafi, *As* Astypalea, *M* Mykonos, *F* Fournoi, *P* Patmos, *L* Leros, *Ka* Kalimnos, *N* Nisyros, *T* Tilos

published in a preliminary report by Chelidonio (2001). The second is Stelida on Naxos where Carter et al. (2014) have reported the presence of an industry belonging to the Denticulate Mousterian facies of the Mediterranean coast. In the latter site ongoing research is expected to shed light on the context and date of deposition of the lithic finds, whereas in the former only systematic new research can answer questions of chronology, industrial affinity and depositional context.

## 22.4.6 *Ionian Margin*

The Ionian Margin belongs to the seismically most active area of the Aegean. Tectonic movements are associated with the thrusting of the overriding Aegean crustal block above the subducting Ionian lithosphere and with activity on the Kephallinia strike-slip fault (Fig. 22.2). Mesozoic evaporites, limestones with chert layers, radiolarites, Tertiary flysch and Late Alpine clastic deposits and Quaternary Post-Alpine sediments form the basement geology of the Ionian Margin area (Fig. 22.2).

Most of the Ionian Islands were connected to the Greek Mainland during the major low sea-level stands. Perissoratis and Conispoliatis (2003) suggest that Corfu Island was connected to the mainland with formation of a lake between them. The Inner Ionian Archipelago comprises over thirty islands, big and small (Fig. 22.9, Galanidou 2014a). Most of them were connected to each other and the mainland during low sea-level stages (Ferentinos et al. 2012). New marine geophysical surveys provide refined palaeogeography during the LGM and the MIS 6 low sea-level stands (Zavitsanou et al. 2015). Numerous caves, partly or totally submerged, are known along the coasts of the Ionian Islands. The

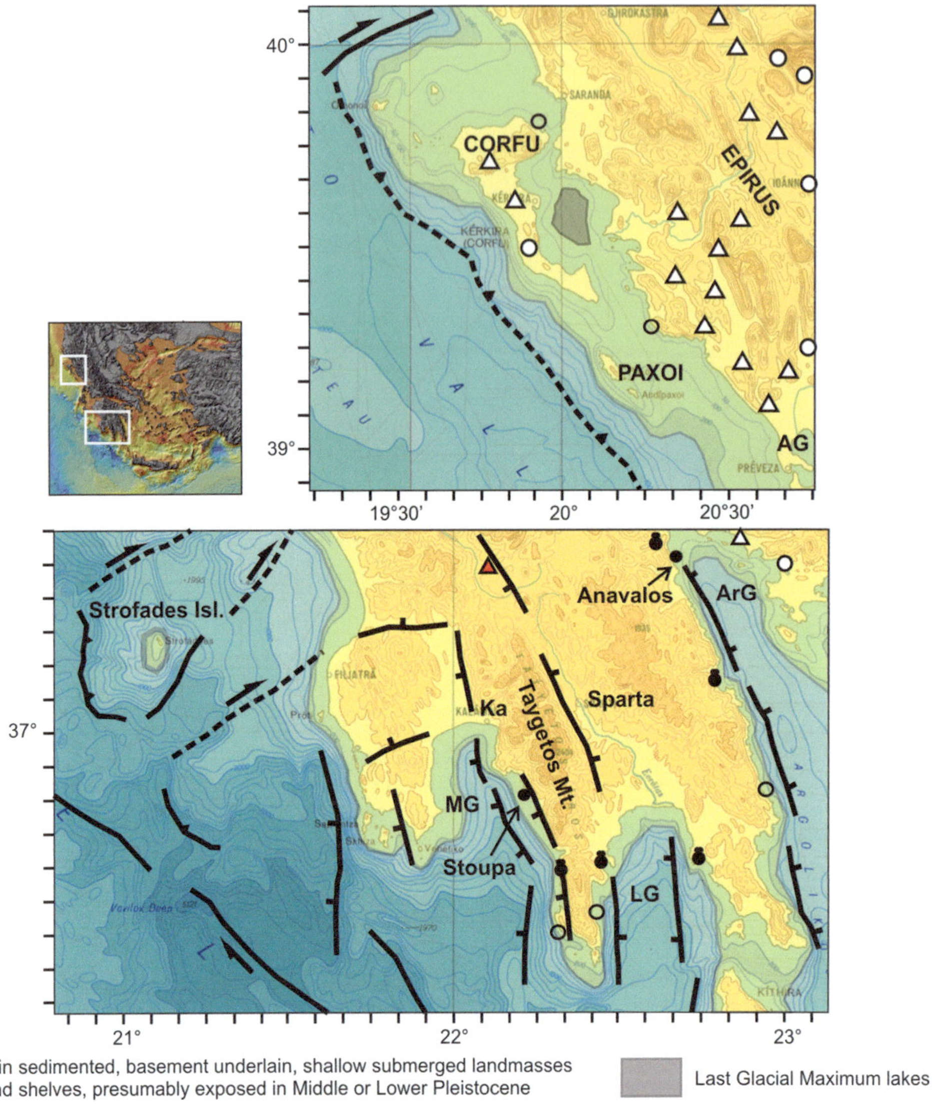


**Fig. 22.9** Bathymetry and morphology of the northern (*top*) and southern (*bottom*) Ionian Margin (IBCM/IOC 1981) with major faults, location of Palaeolithic sites and coastal and submarine karstic springs (symbols as in Fig. 22.7). *AG* Amvrakikos Gulf, *ArG* Argolikos Gulf, *LG* Lakonikos Gulf, *MG* Messiniakos Gulf

calcareous composition of the prevailing sedimentary rocks favours the formation of karstic caves in the fractured rocks. Further south, the shelf is fairly narrow and becomes negligible along the coasts of the western Peloponnese. Holocene sediment deposition on the narrow shelf is generally low. Strofades Islet is a structural high bounded by two NE–SW faults and has been separated from the land throughout the Pleistocene.

Two major gulfs, the Messiniakos and Lakonikos Gulf, represent long-term subsiding grabens formed on the hanging wall of N–S trending normal faults between the uplifting tectonic horsts of the three peninsulas of the southern Peloponnese. Shelves are fairly narrow and almost negligible along the coast of the Taygetos (Mani) Peninsula. The latter is built on Mesozoic limestone and marble with a well-developed karstic network which has favoured the formation of coastal and submarine springs and caves. Characteristic examples are the cave of Alepotrypa (Diros) and the submarine spring of Stoupa, the latter at −27 to 29 m (Rousakis et al. 2014).

The earliest evidence for hominin presence, as early as the Middle Pleistocene, comes from two key sites: Kokkinopilos in Epirus (Tourloukis et al. 2015); and the Kythros Cave in the Inner Ionian Archipelago (Galanidou in press). At Kokkinopilos in the Louros River valley a handful of impressive Large Cutting Tools with industrial affinities to the Lower and/or Middle Palaeolithic were recovered (Higgs 1964; Runnels and Van Andel 1993; Tourloukis 2009; Tourloukis and Karkanas 2012). Kythros Cave faces due NE on the rocky shore of Kythros, a pristine small island in the Inner Ionian Sea with no recent human activity. Its infill is a brecciated red clay deposit with calcitic matrix, containing limestone and chert clasts as well as fossil mammalian skeletal elements and flint artefacts with Middle Palaeolithic affinities. The site was once strategically placed to overlook a river valley between W Meganisi and E Lefkas. Systematic excavation on the site began in 2015 and two stratigraphic units separated by flowstone were identified. Two luminesence dates place the upper unit at the end of the Middle Pleistocene at around 200 ka BP and a full dating program is under way.

In the Upper Pleistocene a good number of sites containing Middle and Upper Palaeolithic archaeology are found all along the mainland coastal region from Thesprotia (Papoulia 2011; Ligkovanlis 2011; Galanidou et al. 2016a), to Preveza (Runnels and Van Andel 2003; Runnels et al. 2003), Aetoloakarnania (Sorrensen 2004; Darlas and Papaconstantinou 2004; Staikou in press), Achaia (Darlas 1985, 1989) and the western coast of Elis in the Peloponnese (Chavaillon et al. 1967). Likewise many of the Ionian Sea islands host Middle and Upper Palaeolithic sites: Corfu (Sordinas 1969, 1983; Darlas et al. 2007; Papadea and Georgiadou 2007), Lefkas, Meganisi and Kythros (Dousougli 1999; Zachos and Dousougli 2003; Galanidou 2014a, b, 2015, in press), Kephallenia (Foss 2002), and Zakynthos (Van Winjgaarden et al. 2006, 2008).

Given this wealth of archaeological sites on land across the Ionian margin, it is not by chance that the Ionian shelf has provided ample evidence for Palaeolithic sites. On the west coast of Corfu, Sordinas (1983, p. 340) reported heavily rolled Levalloiso-Mousterian tools and cores washed up from the sea and incorporated in modern beach material in several locations. Flemming and Kazianis (1987) reported underwater finds to the north and west as much as 2 km offshore and at various depths down to −30 m, and Flemming (pers. comm., 2014) observed a submerged cave with a collapsed roof and artefacts at a depth of −8 to −9 m. New marine-geoarchaeological and archaeological fieldwork explicitly focused on coastal sites may shed more light on this material. One such case study is the recent excavation on a naturally eroding section of land near the passage of the Korrisia lagoon into the Ionian Sea, which has verified the presence of a stratified site yielding part of a hippo jaw and Middle Palaeolithic finds (Darlas et al. 2007). The excavators stress in their report the intense erosional action that sea waves inflict on these Pleistocene deposits.

The Mani Peninsula coastline in the southern Peloponnese has numerous caves with Middle and Upper Palaeolithc material (Darlas and de Lumley 1998; Darlas and Psathi 2008; Panagopoulou et al. 2004). In addition, the only Neanderthal fossils from the Aegean were found in the same region, at the Kalamakia and Apidima Caves in the Oitylo Bay, and at Lakonis I near Gytheio (Harvati et al. 2003, 2010, 2013). All three cave sites are situated in the tidal zone, Kalamakia Cave and Apidima Cave on the west coast, and Lakonis I, part of which lies now submerged, on the east coast. The geography and topography of the areas around them as well as the bone and environmental remains suggest that the now-submerged lands immediately fronting the cave sites must have once comprised a varied mosaic of grasslands, parklands, woodlands, lagoons and marshes (Bassiakos 1993; Elefanti et al. 2008). The archaeological evidence is robust and shows that these littoral lowlands lying between the Pleistocene coastline and the cave sites were themselves important to the survival strategies of the Middle and the Upper Palaeolithic groups. Exploitation of marine and avian resources, as well as stalking and hunting, are believed to have taken place on what is now a submerged landscape (Lebreton et al. 2008; Roger and Darlas 2008a, b).

Glyfada Cave on the shore of Diro Bay offers a wealth of submerged Pleistocene faunal remains. The entrance of this large karstic complex lies at 0.5 m asl and a river runs through it. Of the 10.6 km explored, 1.8 km is underwater, the water drowning a palaeokarst at an average depth of −20 to

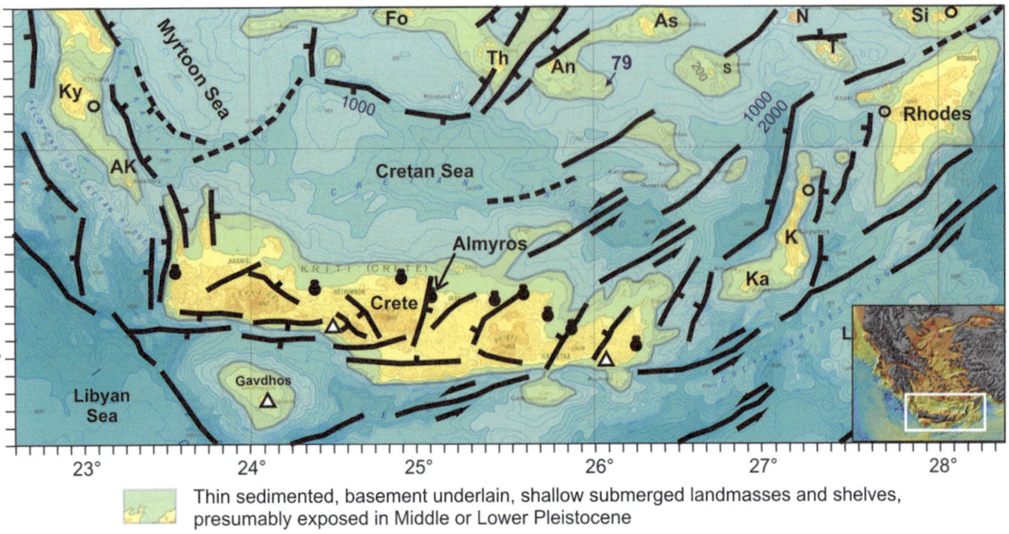


**Fig. 22.10** Bathymetry and morphology of Crete and the western and eastern Cretan Straits (IBCM/IOC 1981) with major faults, location of Palaeolithic sites and coastal and submarine karstic springs (symbols as in Fig. 22.7)

−25 m. Remains of Pleistocene land and sea mammals, such as deer, hyena, panther, lion, hippopotamus, rodents and seal as well as birds were collected from depths ranging between 0.5 and −2.5 m. They offer evidence for a littoral cave that was frequented by these animals. The largest assemblage of *Hippopotamus amphibius* in Greece, more than 350 whole bones and 900 fragments, comes from the submerged parts of Glyfada cave. The majority of the bones were collected from a stratigraphic horizon that post-dates 31,650 ± 550 BP (Giannopoulos 2000, p. 402). Forty thousand years ago, entrance to the cave was possible through a large opening.

## 22.4.7 *West Cretan Strait*

This includes the south-eastern part of the Peloponnese and the islands of Kythera and Antikythera, two stepping stones between the Peloponnese and Crete (Fig. 22.10). The area is known for its very high seismicity and long-term vertical tectonic movements. Uplifted Late Pleistocene marine terraces and Holocene shorelines (Pirazzoli et al. 1982) along the coastal area of SE Peloponnese indicate continuous tectonic uplift during the Quaternary.

The shelf off the coasts is fairly narrow and passes very speedily to steep submarine slopes leading to the deep basins of the Hellenic Trench to the south-west or the deep Myrtoon Sea to the east. A shallow, NW–SE running, shallow, basement ridge connects Cape Maleas in the Peloponnese with the Islands of Kythera and Antikythera. The shallowest parts of this ridge were exposed during the major low-sea-level periods. Kythera and Antikythera Islands were connected to the Peloponnese and the width of the sea-strait separating them from Crete may have been only 2–3 nautical miles (c. 3–5 km) (Fig. 22.10). Local tectonics and faulting is responsible for the formation of the deep basins and trenches and the rough seafloor morphology. The submerged prehistoric Bronze Age city at Pavlopetri in Vatika Bay close to Elaphonissos (Flemming 1968, 1978; Harding et al. 1969; Harding 1970; Henderson et al. 2011) is clear evidence of local subsidence within a region undergoing long-term uplift.

No Palaeolithic finds are known on Kythera and Antikythera, the two islands providing stepping stones between the Peloponnese and W Crete despite surface surveys conducted in recent years (Broodbank 1999; Bevan and Conolly 2013).

### 22.4.8 Crete

The geotectonic position of Crete is responsible for the very strong brittle deformation of the island. The largest earthquake in the Mediterranean during the historic era occurred on a NE-dipping fault-plane a few kilometres off SW Crete (Shaw et al. 2008). That earthquake uplifted western Crete abruptly by up to 8–9 m in the fourth century AD (Pirazzoli et al. 1982). Initial N–S extension followed by extension in an E–W direction formed the Cretan structural high and divided the island into several tectonic blocks (Van Hinsbergen and Meulenkamp 2006) with different tectonic movements. The high mountains of the island represent the faster uplifting blocks while the lower parts in-between them exhibit slower uplift or even relative subsidence (Pirazzoli 1988), as demonstrated by the Bronze Age sites now found submerged below the present sea level. The faulted, steep, southern flanks separate the >2000 m high mountains from the >2000 m deep basins south of the Island. Gavdos Island is the summit of a tectonic block south of Crete, separated from it by a 1000–3000 m-deep trench (Figs. 22.3 and 22.10). The linear morphology and rocky character of the steep west and east coasts of Crete is a good indication of their fault-related development. The sea-strait separating Crete from Kassos-Karpathos during Pleistocene low sea-level periods may have been only 2–4 (c. 3.5–7 km) nautical miles wide.

The geological structure of Crete has favoured the development of larger or smaller karstic springs, mainly along the northern coastline of the Island, like the Almiros coastal spring system. During low sea-level stands, the karstic spring was presumably discharging fresh water.

Throughout the Pleistocene, Crete was separated from any landmass of Africa, Asia or Europe by a considerable distance (Fig. 22.10). Its earliest prehistory goes hand in hand with the prehistory of seafaring in the Mediterranean. Systematic archaeological surveys conducted on Crete and on Gavdos Island, have brought to light artefacts and sites associated with activity dated to the early Holocene, the Upper Pleistocene and perhaps even stretching as far back as the Middle Pleistocene (Strasser et al. 2010, 2011; Kopaka and Matzanas 2009, 2011; Mortensen 2008).

On the north coast of west Crete, a large part of the Vamos Cave on the Drepano Cape is now submerged and offers compelling evidence for sea level fluctuation. Fossil bones of Pleistocene cervids and an elephant species endemic to Crete, *Elephas chaniensis*, were found at depths ranging between −1.5 and −4.5 m. The bones were deposited on the cave floor at periods when the sea level dropped by at least 10–20 m. The cave was on dry land during the Last Glacial Maximum some 20,000 years ago (Symeonides et al. 2001). Despite the good conditions of visibility and faunal preservation, there is no evidence of human activity from Vamos Cave in the form of artefacts or cut-marked bone recovered from its inundated deposits.

### 22.4.9 East Cretan Strait

The islands of Kassos, Karpathos and Rhodes and some smaller ones form an island chain between East Crete and the Anatolian landmass (Figs. 22.2 and 22.10). The islands of the East Cretan Strait represent the uplifted parts of the south-eastern segment of the Hellenic Arc, which is characterized by active strike-slip tectonics (Jongsma 1977; Mascle et al. 1982) and oblique thrusting (Stiros et al. 2010). Subsequent deformation and uplift of the Arc creates extensional tectonics of the upper plate, manifested by numerous normal and oblique faults (Ten Veen and Kleinspehn 2002). The East Cretan Strait is segmented into tectonic blocks, each with its own history of vertical movements. The island of Rhodes is characterised by a number of crustal blocks (Pirazzoli et al. 1989), each with a different tectonic history and with up to eight Late Holocene shorelines (Pirazzoli et al. 1982, 1989; Kontogianni et al. 2002). Despite the general uplift of the entire island there are still areas which exhibit submergence during the Late Holocene (e.g., the submerged Roman quarries at Koskinou) and which are separated from the uplifting blocks by normal faults.

The three main islands are surrounded by narrow shelves. The sea-strait between Rhodes and Turkey is shallow and it is possible that the island was connected to Anatolia during earlier low sea-level periods (Fig. 22.10). Karpathos and Kassos Islands were connected to each other during low sea-level stands but were always separated from Crete and Rhodes by deep sea straits. Distances between Kassos-Karpathos and Crete or between Karpathos and Rhodes may have been as narrow as 3–5 nautical miles (c. 5.5–9 km). No Palaeolithic remains have been reported from the islands of this unit. They do, however, have a rich animal fossil record that is essentially Asian, indicating that the islands were connected to the Asian mainland during low sea level stands.

## 22.5 Discussion

Our study shows that active tectonics in the Aegean is the major driving mechanism that has controlled and defined, first, the development of deep basins undergoing long-term subsidence and continuous sedimentation during the Quaternary, and, second, the crustal blocks which have escaped subsidence or exhibit long-term, continuous or episodic, relative uplift. The latter represent the areas which have been exposed during low sea-level periods and which have the potential to host Pleistocene landscapes and therefore traces of Palaeolithic archaeology. Understanding of the local and regional vertical tectonic movements is essential for the definition of the most promising areas to be surveyed for submerged landscapes by means of modern marine survey techniques. Interplay between fluctuating sea-level and long-term tectonic activity has led to the formation of extensive landmasses separated from each other by narrow sea-straits which may represent areas where early seafaring evolved.

The availability of fresh water resources is one of the main factors which make an area attractive for habitation. Water sources include rivers, lakes, streams and freshwater springs. Reconstruction of the course of the rivers and identification of possible lakes is a sensible first task in any survey of extensive shelves like the North Aegean Shelf, the eastern part of the North and Central Aegean Island Bridges, the East Aegean Islands and the Inner Ionian Archipelago. Late Pleistocene lakes have already been identified at several points on the North Aegean Shelf, in Central Greece and east of Corfu (in the northern part of the Ionian Margin). Fresh and brackish karstic springs are very common along the Aegean coastline. Most of the known underwater karstic springs in the Aegean are located on narrow shelves, off moderately or steeply sloping coasts. Large underwater karstic springs have been discovered on the eastern shelf of the Messiniakos Gulf, the Ionian Islands and off the Ionian coastline of Greece, the western shelf and coast of the Argolikos Gulf, off the steep southern and northern coasts of the Gulf of Corinth, the North and South Gulfs of Evia, the northern coast of Crete and many other areas.

Submerged karstic caves are sites of potential prehistoric occupation (see also Radić Rossi and Cukrov, Chap. 17). Like karstic springs, the caves are associated with calcareous rocks. Submerged caves have been discovered off the rocky coasts of the Ionian Islands, in the southern Peloponnese, Crete, the Sporadhes Islands and many other places. The two most important ones, Glyfada in the southern Peloponnese and Vamos in Crete, have yielded significant Pleistocene faunal remains, which exemplify the high potential of this type of site to future Palaeolithic research, despite the absence of Palaeolithic artefacts from these two particular sites.

A small number of prehistoric and palaeontological sites lie partly in shallow waters whereas less than a handful of sites containing Pleistocene remains are fully submerged at greater depths and distances from the shore. In addition, stray archaeological finds washed out from the shores of Greece and coastal sites, parts of which are currently underwater because of erosion, wave action or faulting, together build up the limited Pleistocene evidence now known from beneath the Greek seas. Submerged and coastal palaeontological and archaeological sites are testimonies to the much larger coastal mosaic of biotopes that were available to Palaeolithic groups during cold and arid periods. They are more eloquent about the potential of future coastal and underwater research aimed at discovering Palaeolithic sites rather than of an existing remarkable record.

During the twentieth century, Palaeolithic research in the NE Mediterranean operated with a mainland-focused epistemological paradigm that saw the NE Mediterranean Sea as peripheral to the Palaeolithic world and hence not worthy of systematic exploration (Galanidou 2014b). At the same time underwater explorations focused almost exclusively on the archaeology of shipwrecks. Now we emphasize that unless island, coastal and underwater geographies are incorporated into palaeogeographic research, many of the shortcomings of the Aegean Palaeolithic narrative are bound to remain. Our work is founded upon the premise that the coastal landscapes exposed at lowered sea levels provided relatively fertile and productive refugia for plants, land mammals and hominins, at a time during the cold, low sea-level periods when increased aridity would have reduced or deterred hinterland occupation (Bailey and Flemming 2008). During the last 20 years, the new sites discovered on the islands of Crete, Gavdos, Thasos, Limnos, Lesbos, Meganisi and Kythros lend credence to our claim that the 'islandscapes' and submerged seascapes of the Aegean are indeed highly promising targets for Palaeolithic investigation.

In the present study we have put together the evidence upon which a division of the Aegean into nine smaller, and rather coarse, units can be defined. We regard this as a first step to reconstructing the palaeogeography of the shallow and coastal shelf areas of the Aegean. In order to elaborate further on aspects of early human adaptation and dispersal, more work is needed, targeting individual units by means of underwater investigation of the shelf, coupled with archaeological survey on the coasts and islands (Bailey 2011; Benjamin et al. 2011; Flemming et al. 2014). The benefit of bringing the NE Mediterranean into focus and of linking these results to the discussion of 'the early occupation of Europe' is fundamental to progress. By changing the underlying paradigm of Palaeolithic research in the region, new possibilities for archaeological discovery will emerge, allowing the Aegean to take its rightful place in the study of human origins.

## References


Ammerman AJ, Efstratiou N, Adam E (1999) First evidence for the Palaeolithic in Aegean Thrace In: Bailey GN, Panagopoulou E, Perlès C, Zachos K (eds) The palaeolithic archaeology of Greece and adjacent areas: proceedings of the ICOPAG Conference, Ioannina, (September 1994) British School at Athens Studies 3, London, pp 211–214

Angelier J, Lyberis N, Le Pichon X, Barrier E, Huchon P (1982) The tectonic development of the Hellenic arc and the sea of Crete: a synthesis. Tectonophysics 86:159–196

Armijo R, Meyer B, King GCP, Rigo A, Papanastassiou D (1996) Quaternary evolution of the Corinth rift and its implications for the late cenozoic evolution of the Aegean. Geophys J Int 126:11–53

Armijo R, Meyer B, Hubert A, Barka A (1999) Westward propagation of the north Anatolian fault into the northern Aegean: timing and kinematics. Geology 27:267–270

Armijo R, Flerit F, King G, Meyer B (2004) Linear elastic fracture mechanics explains the past and present evolution of the Aegean. Earth Planet Sci Lett 217(1–2):85–95. doi:10.1016/S0012-821X(03)00590-9

Azzaroli A (1983) Quaternary mammals and the "end-Villafranchian" dispersal event – a turning point in the history of Eurasia. Palaeogeogr Palaeoclimatol Palaeoecol 44:117–139

Bailey G (2011) Continental shelf archaeology: where next? In: Benjamin J, Bonsall C, Pickard C, Fischer A (eds) Submerged prehistory. Oxbow Books, Oxford, pp 311–331

Bailey GN, Flemming NC (2008) Archaeology of the continental shelf: marine resources, submerged landscapes and underwater archaeology. Quat Sci Rev 2:2153–2165

Bailey G, Sakellariou D and members of the SPLASHCOS network (2012) SPLASHCOS: submerged prehistoric archaeology and landscapes of the continental shelf. Antiquity 86(334). http://antiquityacuk/projgall/sakellariou334/

Bar-Yosef O, Belfer-Cohen A (2001) From Africa to Eurasia – early dispersals. Quat Int 75:19–28

Bassiakos IE (1993) Dating of fossils from caves and speleothems: evidence from electron spin resonance (ESR) technique, the study of underground karst morphology and the relevant radiometric and geological conditions in speleoenvironments of Dyros, Mani. University of Athens, Unpublished PhD thesis (in Greek)
Benjamin J, Bonsall C, Pickard C, Fischer A (eds) (2011) Submerged prehistory. Oxbow Books, Oxford
Bevan A, Conolly J (2013) Mediterranean islands, fragile communities and persistent landscapes Antikythera in long-term perspective. Cambridge University Press, Cambridge
Broodbank C (1999) Kythera survey: preliminary report on the 1998 season. Ann Br Sch Athens 94:191–214
Brossolo L, Mascle J, Loubtrieu B (2012) Morpho-bathymetry of the Mediterranean Sea, Map 1/4000000, 1st edn. The Commission for the Geological Map of the World (CGMW) and UNESCO, Paris
Carter T, Contreras DA, Doyle S, Mihailovic DD, Moutsiou T, Skarpelis N (2014) The Stélida Naxos archaeological project: new data on the Mesolithic and Middle Palaeolithic Cyclades. Antiquity 88(341) http://journalantiquity-acuk/projgall/carter341
Chavaillon J, Chavaillon N, Hours F (1967) Industries paléolithiques de l' elide I – region d' Amalias. Bull Corresp Hell 91:151–201
Chelidonio G (2001) Manufatti liticisu ciottolo da Milos (isole cicladi)i (nota preliminare). Pegaso Riv Ann Cult Mediterr 1:117–144
Darlas A (1985) Palaeolithic finds from Kalamaki, Achaia. Archaeol Deltio (Studies) 40A:194–206
Darlas A (1989) The Aurignacian lithic industry from Elaeochori, Achaia. Archaeol Ephemer 128:137–159. (in Greek)
Darlas A (2011) Palaeolithic finds from Rachona, Pella. Archaeologiko ergo in Macedonia and Thrace 2. Archaeol Ergo Maced Thrace 21:109–114. (in Greek)
Darlas A (2014) In search of the identity of Petralona cave and its importance for the Greek and European prehistory. In: Stefani E, Merousis N, Dimoula A (eds) A century of research in prehistoric Macedonia, 1912–2012. International Conference Proceedings, Thessaloniki, pp 195–2044, (in Greek)
Darlas A, de Lumley H (1998) Fouilles franco-helléniques de la grotte de Kalamakia (Aréopolis; Péloponnèse). Bull Corresp Hell 122:655–661
Darlas A, Papaconstantinou E (2004) A new Middle Palaeolithic open-air site at Sardinia Amfilochia. In: Proceedings of the 2nd International Conference on the History and Archaeology of Aetoloakarnania, Agrinio 29–31 March 2002, Agrinio, pp 21–32 (in Greek)
Darlas A, Psathi E (2008) Le Paléolithique supérieur dans la péninsule du Mani (Péloponnèse, Grèce). In: Darlas A, Mihailović D (eds) The Palaeolithic of the Balkans, British archaeological reports international series 1819. Archaeopress, Oxford, pp 51–59
Darlas A, Karkanas P, Palli O, Papadea A (2007) Palaeolithic research in the Korission Lagoon area. In: Arvanitou-Metallinou G (ed) Prehistoric Corfu and its adjacent areas. Problems – perspectives. Proceedings of the conference dedicated to Augustus Sordinas, Corfu 17 December 2004, pp 77–84 (in Greek)
Desruelles S, Fouache E, Ciner A, Dalongeville R, Pavlopoulos K, Kosun E, Coquinot Y, Potdevin J-L (2009) Beachrocks and sea level changes since middle Holocene: comparison between the insular group of Mykonos-Delos-Rhenia (Cyclades, Greece) and the southern coast of Turkey. Glob Planet Chang 66:19–33
Dewey JF, Sengoer CAM (1979) Aegean and surrounding regions: complex multiplate and continuum tectonics in a convergent zone. Geol Soc Am Bull 90:84–92
Douka K, Perlès C, Valladas H, Vanhaeren M, Hedges REM (2010) Franchthi cave revisited: the age of the Aurignacian in South-Eastern Europe. Antiquity 85:1131–1150
Dousougli A (1999) Palaeolithic Leukas. In: Bailey GN, Adam E, Panagopoulou E, Perlès C, Zachos K (eds) The Palaeolithic archaeology of Greece and adjacent areas, proceedings of the ICOPAG conference, Ioannina, September 1994. British School at Athens Studies, 3, London, pp 288–292
Efstratiou N (1985) Agios Petros: a Neolithic site in the northern Sporades. Aegean relationships during the Neolithic of the 5th millennium BC, British archaeological reports international series 241. BAR, Oxford
Efstratiou N (2001) The Neolithic settlement of Agios Petros in Kyra-Panagia, Halonnesos and the island habitation patterns in the Aegean – a reconsideration. In: Sampson A (ed) Archaeology in the Northern Sporades Athens, pp 231–250 (In Greek)
Efstratiou N (2014) The final palaeolithic hunting camp of Ouriakos on the Island of Lemnos. Eurasian Prehist 11(1–2):75–96
Efstratiou N, Kyriakou D (2011) Following the traces of the last hunter-gatherers of East Mediterranean. Anaskamma 5:53–74. (in Greek)
Efstratiou N, Biagi P, Karkanas P, Starnini E (2013) A Late Palaeolithic site at Ouriakos (Limnos, Greece) in the north-eastern Aegean. Antiquity 87(335). http://antiquityacuk/projgall/efstratiou335/
Elefanti P, Panagopoulou E, Karkanas P (2008) The transition from the middle to the upper Palaeolithic in the southern Balkans: the evidence from the Lakonis I cave, Greece. Eurasian Prehistory 5:85–95
Evelpidou N, Pirazzoli PA, Saliege J-F, Vassilopoulos A (2011a) Submerged notches and doline sediments as evidence for Holocene subsidence. Cont Shelf Res 31:1273–1281

Evelpidou N, Pirazzoli PA, Vassilopoulos A, Tomasin A (2011b) Holocene submerged shorelines on Theologos area (Greece). Z Geomorphol 55(1):31–44
Evelpidou N, Vassilopoulos A, Pirazzoli PA (2012) Submerged notches on the coast of Skyros Island (Greece) as evidence for Holocene subsidence. Geomorphology 141–142:81–87
Farrand WR (2000) Depositional history of Franchthi cave: sediments, stratigraphy and chronology. In: Jacobsen TW (ed) Excavations at Franchthi cave, 12. Indiana University Press, Bloomington and Indianapolis
Felsch RCS (1973) Die Höhle von Kephalari eine jungpaläolithicsche Siedlung in der Argolis. Archaeol Analecta Athinon 6:13–27
Ferentinos G, Gkioni M, Geraga M, Papatheodorou G (2012) Early seafaring activity in the southern Ionian Islands, Mediterranean Sea. J Archaeol Sci 39:2167–2176. doi:10.1016/jjas201201032
Flemming NC (1968) Holocene earth movements and eustatic sea level change in the Peloponnese. Nature 217(5133):1031–1032
Flemming NC (1978) Holocene eustatic changes and coastal tectonics in the Northeast Mediterranean: implications for models of crustal consumption. Philos Trans R Soc Lond A 289(1362):405–458
Flemming NC, Kazianis D (1987) Preliminary report on the Otranto Straits project. Report submitted to the Department of Antiquities, Underwater Section, Greece
Flemming NC, Cagatay MN, Chiocci FL, Galandiou N, Lericolais G, Joens H, Missiaen T, Moore F, Rosentau A, Sakellariou D, Skar B, Stevenson A, Weerts H (2014) Land beneath the waves: submerged landscapes and sea level change. A joint geoscience-humanities strategy for European continental shelf prehistoric research. In: Chu NC, McDonough N (eds) Position paper 21 of the European Marine Board European Marine Board, Ostend
Foss P (2002) The lithics. In: Randsborg K (ed) Kephallénia: archaeology and history: the ancient Greek cities. Acta Archaeologica 73 (1–2): 77–147
Fotiadis M (2016) Leaf-points from Petrota (Greek Thrace) and the Palaeolithic chronology of the Vrachos chert quarry. Annu Br Sch Athens III(I):1–11
Galanidou N (2013) Looking for the earliest occupants of the Aegean – Palaeolithic excavations at Rodafnidia, Lisvori, Lesbos. In: Alvanou M (ed) Island identities. The contribution of the Secretariat General for the Aegean and Island Policy to the research and promotion of the culture of the Aegean Archipelago, Mytilene, pp 15–17
Galanidou N (2014a) Archaeological survey on the Inner Ionian Sea archipelago. In: Smith C (ed) Encyclopedia of global archaeology. Field archaeology. Springer, New York, pp 3882–3888
Galanidou N (2014b) Advances in the Palaeolithic and Mesolithic archaeology of Greece for the new millennium. Pharos 20(1):1–40
Galanidou N (2015) Seascape survey on the Inner Ionian Sea archipelago. In: Carver M, Gaydarska B, Monton-Subias S (eds) Field archaeology from around the world. Ideas and approaches. Springer, New York, pp 101–106
Galanidou N (2016) Before ancient Nericos what? Palaeolithic finds from Karyotes Lefkas. In: Papadatou-Yannopoulou C (ed) Nericos, Lefkas, Castro. The longest-living capital of Lefkas, Proceedings of a conference held at Lefkas in August 2010. Cultural Centre of the Lefkas Municipality, Lefkas. (In Greek), pp 83–111
Galanidou N (in press) Parting the waters: Palaeolithic archaeology and the Quaternary Ionian Sea. In: Driessen J, Garezou M-X, Iossif PP, Mendoni L, Tsingarida A (eds), Archaeology of the Sea. Proceedings of the Conference organized by the Greek Ministry of Culture and the Belgian School at Athens, Brussels, 21–22 March 2014, Athens
Galanidou N, Cole J, Iliopoulos G, McNabb J (2013) East meets West: the Middle Pleistocene site of Rodafnidia on Lesbos, Greece. Antiquity 87(336). http://antiquityacuk/projgall/galanidou336/
Galanidou N, Papoulia C, Ligkovanlis S (2016a) The middle Palaeolithic bifacial tools from Megalo Karvounari. In: Forsén B, Galanidou N, Tikkala E (eds) Thesprotia expedition III. Landscapes of nomadism and sedentism. Foundation of the Finish Institute at Athens, Helsinki, pp 29–58
Galanidou N, Athanassas C, Cole J, Iliopoulos G, Katerinopoulos A, Magganas A, McNabb J (2016b) The Acheulian site at Rodafnidia, Lisvori, on Lesbos, Greece: 2010–2012. In: Harvati K, Roksandic M (eds) Paleoanthropology of the Balkans and Anatolia. Human evolution and its context. Springer, Dordrecht, pp 119–138
Galanidou N, Iliopoulos G, Papoulia C (2016c) The Palaeolithic settlement of Lefkas. Archaeological evidence in a palaeogeographic context. J Greek Archaeol 1:1–32
Giannopoulos B (2000) Contribution to the study of modern and old environments of the most important Greek caves. University of Athens, Unpublished PhD thesis (In Greek)
Goren-Inbar N, Feibel CS, Versoub KL, Melamed Y, Kislev ME, Tchernov E, Saragusti I (2000) Pleistocene milestones on the out-of-Africa corridor at Gesher Ya'aqov. Israel Sci 289(5481):944–947
Grün R (1996) A re-analysis of electron spin resonance dating results associated with the Petralona hominid. J Hum Evol 30:227–241
Harding AF (1970) Pavlopetri. A Mycenaean town. Underw Archaeol 23:242–250
Harding AF, Cadogan G, Howell R (1969) Pavlopetri, an underwater bronze age town in Laconia. Ann Br Sch Athens 64:112–142
Harvati K, Panagopoulou E, Karkanas P (2003) First Neanderthal remains from Greece: the evidence from Lakonis. J Hum Evol 45:465–473

Harvati K, Stringer C, Karkanas P (2010) Multivariate analysis and classification of the apidima 2 cranium from Mani, southern Greece. J Hum Evol 6(2):246–250
Harvati K, Darlas A, Bailey SE, Rein TR, El Zaatari S, Fiorenza L, Kullmer O, Psathi E (2013) New Neanderthal remains from Mani peninsula, southern Greece: the Kalamakia middle Paleolithic cave site. J Hum Evol 64(6):1–14
Henderson JC, Gallou C, Flemming NC, Spondylis E (2011) The Pavlopetri underwater archaeology project: investigating an ancient submerged town. In: Benjamin J, Bonsall C, Pickard C, Fischer A (eds) Submerged prehistory. Oxbow Books, Oxford, pp 207–218
Hennig GJ, Herr W, Weber E, Xirotiris NI (1982) Petralona cave dating controversy. Nature 299:281–282
Higgs E (1964) A hand-axe from Greece. Antiquity 38:54–55
Iler EB, Aksu AE, Yaltýrak C, Hiscott RN (2008) Seismic stratigraphy and quaternary sedimentary history of the north-east Aegean Sea. Mar Geol 254:1–17
International Bathymetric Map of the Mediterranean (IBCM) 1st Edition (1981) Intergovernmental Oceanographic Commission
Jolivet L (2001) A comparison of geodetic and finite strain pattern in the Aegean, geodynamic implications. Earth Planet Sci Lett 187:95–104. doi:10.1016/S0012-821X(01)00277-1
Jongsma D (1977) Pliny and Strabo trenches, south of the Hellenic arc. Geol Soc Am Bull 88:797–805
Jöris O (2014) Early Palaeolithic Europe. In: Renfrew C, Bahn P (eds) The Cambridge world prehistory, vol III. Cambridge University Press, Cambridge, pp 1703–1746
Kappelman J, Alçiçek MC, Kazanci N, Schultz M, Özkul M, Sen S (2008) First Homo Erectus from Turkey and implications for migrations into temperate Eurasia. Am J Phys Anthropol 135(1):110–116
Kapsimalis V, Pavlakis P, Poulos SE, Alexandri S, Tziavos C, Sioulas A, Filippas D, Lykousis V (2005) Internal structure and evolution of the late quaternary sequence in a shallow embayment: the Amvrakikos gulf, NW Greece. Mar Geol 222–223:399–418
Kapsimalis V, Pavlopoulos K, Panagiotopoulos I, Drakopoulou P, Vandarakis D, Sakellariou D, Anagnostou C (2009) Geoarchaeological challenges in the Cyclades continental shelf (Aegean Sea). Z Geomorphol NF 53(1):169–190
Khatib S, Rochette P, Alçiçek MC, Lebatard A-E, Demory F, Saos T (2014) Études stratigraphique, sédimentologique et paléomagnetique des travertins de Kocabaş, Bassin de Denizli, Anatolie, Turquie, contenant des restes fossiles quaternaires. Anthropologie 118:16–33
Kontogianni V, Tsoulos N, Stiros S (2002) Coastal uplift, earthquakes and active faulting of Rhodes Island (Aegean arc): modelling based on geodetic inversion. Mar Geol 186:299–317
Kopaka K, Matzanas C (2009) Palaeolithic industries from the island of Gavdos, near neighbour to Crete in Greece. Antiquity 83 (321). http://antiquityacuk/projgall/kopaka321/
Kopaka K, Matzanas C (2011) Early marine travels in the Aegean and Crete. Reflections departing from the knapped-stone industries from Gavdos island. In: Andreadaki-Vlazaki M (ed) Proceedings of the I international cretological conference, Chania 1–8 October 2006, A1, pp 43–82 (in Greek)
Koufos G (2001) The Villafranchian mammalian faunas and biochronology of Greece. Boll Soc Paleontol Ital 40(2):217–223
Koukouli-Chrysanthaki H (2012) Prehistoric Thassos: assessment and perspectives of the archaeological research. In: Papadopoulos S, Malamidou D (eds) Ten years of excavation research of Limenaria prehistoric settlement Thassos (Conference Proceedings Thassos 11 July 2003) Thessaloniki, pp 9–36 (in Greek)
Koukouli-Chrysanthaki H, Weissgeber G (1997) Prehistoric ochre quarries in Thassos. Archaeol Ergo Maced Thrace 7:541–558. (in Greek)
Kourtessi-Philippakis G, Georgiou H, Dimadi L, Trantalidou K (1993) The exploration of the Palaeolithic settlement in East Macedonia: preliminary report 1989–1990. Archaeol Ergo Maced Thrace:553–559. (in Greek)
Kreemer C, Chamot-Rooke N (2004) Contemporary kinematics of the southern Aegean and the Mediterranean ridge. Geophys J Int 15:1377–1392. doi:10.1111/j1365-246X200402270x
Latham AG, Schwarcz H (1992) The Petralona hominid site: uranium-series re-analysis of "layer 10" calcite and associated palaeomagnetic analyses. Archaeometry 34:135–140
Le Pichon X (1982) Land-locked oceanic basins and continental collision: the eastern Mediterranean as a case example. In: Hsü K (ed) Mountain building processes. Academic, London, pp 201–211
Le Pichon X, Angelier J (1979) The Hellenic arc and trench system: a key to the neotectonic evolution of the eastern Mediterranean area. Tectonophysics 60:1–42
Le Pichon X, Angelier J, Sibuet J-C (1982) Plate boundaries and extensional tectonics. Tectonophysics 81:239–256. doi:10.1016/0040-1951(82)90131-7
Le Pichon X, Chamon-Rooke N, Lallemant S, Noomen R, Veis G (1995) Geodetic determination of the kinematics of Central Greece with respect to Europe: implications for eastern Mediterranean tectonics. J Geophys Res 100:12675–12690. doi:10.1029/95JB00317
Lebatard AE, Alçiçek MC, Rochette P, Khatib S, Vialet A, Boulbes N et al (2014) Dating the Homo Erectus bearing travertine from Kocabaş (Denizli, Turkey) at least 1.1 Ma. Earth Planet Sci Lett 390:8–18

Lebreton V, Psathi E, Darlas A (2008) Environnement vegetal des Neanderthaleniens de la grotte de Kalamakia (Areopolis, Grèce). In: Darlas A, Mihailovic D (eds) The Palaeolithic of the Balkans, British archaeological reports international series 1819. Archaeopress, Oxford, pp 61–68
Leeder MR, Portman C, Andrews JE, Collier RELI, Finch E, Gawthorpe RL, McNeil LC, Perez-Arlucea M, Rowe P (2005) Normal faulting and crustal deformation, Alkyonides gulf and Perachora peninsula, eastern gulf of Corinth rift, Greece. J Geol Soc 162(3):549–561
Ligkovanlis S (2011) Megalo Karvounari revisited. In: Forsén N, Tikkala E (eds) Thesprotia expedition II: environment and settlement patterns. Foundation of the Finish Institute at Athens, Helsinki, pp 159–180
Lykousis V (2009) Sea-level changes and shelf break prograding sequences during the last 400 ka in the Aegean margins: subsidence rates and palaeogeographic implications. Cont Shelf Res 29:2037–2044
Lykousis V, Anagnostou C (1994) Sedimentological and palaeogeographic evolution of the Saronic gulf during the late quaternary. Bull Geol Soc Greece 28(1):501–510. (in Greek)
Lykousis V, Sakellariou D, Moretti I, Kaberi H (2007) Late quaternary basin evolution of the Gulf of Corinth: sequence stratigraphy, sedimentation, fault-slip and subsidence rates. Tectonophysics 440:29–51
Mascle J, Martin L (1990) Shallow structure and recent evolution of the Aegean Sea: a synthesis based on continuous reflection profiles. Mar Geol 94:271–299
Mascle J, Mascle G (2012) Geological and morpho-tectonic map of the Mediterranean domain. Commission for the Geological Map of the World (CGMW) and UNESCO, Paris
Mascle J, Jongsma D, Campredon R, Dercourt J, Glacon G, Lecleach A, Lyberis N, Malod JA, Mitropoulos D (1982) The Hellenic margin from eastern Crete to Rhodes: preliminary results. Tectonophysics 86:133–147
McKenzie DP (1970) Plate tectonics of the Mediterranean region. Nature 226:239–243
McKenzie DP (1978) Active tectonics of the alpine-Himalayan belt: the Aegean Sea and surrounding regions. Geophys J Astron Soc 55:217–254
Meulenkamp JE, Wortel MJR, Van Wamel WA, Spakman W, Hoogerduyn Strating E (1988) On the Hellenic subduction zone and the geodynamical evolution of Crete since the late middle Miocene. Tectonophysics 146:203–215. doi:10.1016/0040-1951(88)90091-1
Moncel M-H (2010) Oldest human expansions in Eurasia: favouring and limiting factors. Quat Int 223–224:1–9
Mortensen P (2008) Lower to Middle Palaeolithic artefacts from Loutró on the south coast of Crete. Antiquity 83(321). http://wwwantiquityacuk/projgall/mortensen
Moundrea-Agrafioti A (1992) Agios Petros Kyra-Panagias. Elements of the knapped stone industry. In: Ancient Thessaly. Proceedings of the International Conference in Memory of Dimitris Theocharis. Archaeologico Deltio Publications 48:191–201 (in Greek)
Palombo MR, AFM V, Kostopoulos DS, Alberd MT, Spassov N, Vislobokova I (2006) Similarity relationships between the Pliocene to middle Pleistocene large mammal faunas of southern Europe from Spain to the Balkans and the north Pontic region. In: Kahlke R-D, Maul LC, Mazza PPA (eds) Late Neogene and quaternary biodiversity and evolution: regional developments and interregional correlations, Courier Forschungsinstitut Senckenberg 256, vol I. Schweizerbart'sche Verlagsbuchhandlung, Stuttgart, pp 329–347
Panagopoulou E, Kotzambopoulou E, Karkanas P (2001) Geoarchaeological research on Alonnisos: new data on the Aegean Palaeolithic and Mesolithic. In: Sampson A (ed) Archaeological research in northern Sporades, Community of Alonnisos, Alonnisos, pp 121–151. (in Greek)
Panagopoulou E, Karkanas P, Tsartsidou G, Kotzambopoulou E, Harvati K, Ntinou M (2004) Late Pleistocene archaeological and fossil human evidence from Lakonis cave, southern Greece. J Field Archaeol 29:323–349
Papadea A, Georgiadou G (2007) Prehistoric Othonoi: recent investigations. In: Arvanitou-Metallinou GG (ed) Prehistoric Corfu and its adjacent areas. Problems – perspectives. Proceedings of the conference dedicated to Augustus Sordinas, Corfu 17 December 2004, Corfu, pp 135–140 (in Greek)
Papanikolaou D, Chronis G, Metaxas C (1994) Neotectonic structure of the Argolikos gulf. Bull Geol Soc Greece 30(2):305–316
Papanikolaou D, Alexandri M, Nomikou P, Ballas D (2002) Morphotectonic structure of the western part of the north Aegean Basin based on swath bathymetry. Mar Geol 190:465–492
Papoulia C (2011) Mikro Karvounari in context: the new lithic collection and its implications for middle Palaeolithic hunting activities. In: Forsén N, Tikkala E (eds) Thesprotia expedition II: environment and settlement patterns. Foundation of the Finish Institute at Athens, Helsinki, pp 123–158
Perissoratis C, Conispoliatis N (2003) The impacts of sea-level changes during latest Pleistocene and Holocene times on the morphology of the Ionian and Aegean seas (SE alpine Europe). Mar Geol 196:145–156
Perissoratis C, Mitropoulos D (1989) Late quaternary evolution of the northern Aegean shelf. Quat Res 32:36–50
Perissoratis C, Piper DJW, Lykousis V (1993) Late quaternary sedimentation in the Gulf of Corinth: the effects of marine-lake fluctuations driven by eustatic sea level changes, Special publication dedicated to prof a Panagos. National Technical University of Athens, Athens, pp 693–744
Perlès C (1987) Les industries lithiques taillées de Franchthi (Argolide, Grèce) tome I: présentation générale et industries paléolithiques. In: Jacobsen TW (ed) Excavations at Franchthi cave, 3. Indiana University Press, Bloomington

Perlès C (1999) Long-term perspectives on the occupation of the Franchthi Cave: continuity and discontinuity. In: Bailey GN, Adam E, Panagopoulou E, Perlès C, Zachos K (eds) The Palaeolithic archaeology of Greece and adjacent areas, Proceedings of the ICOPAG Conference, Ioannina September 1994. British School at Athens Studies, 3, London, pp 311–318

Piper DJW, Perissoratis C (1991) Late quaternary sedimentation on the north Aegean continental margin, Greece. Am Assoc Pet Geol Bull 75:6–61

Pirazzoli PA (1988) Sea-level changes and crustal movements in the Hellenic arc (Greece). The contribution of archaeological and historical data. In: Raban A (ed) Archaeology of coastal changes, British archaeological reports international series 404. Archaeopress, Oxford, pp 157–184

Pirazzoli PA, Comment J, Thommer Y, Laborel J, Montaggioni LF (1982) Crustal block movements from Holocene shorelines: Crete and Antikythira (Greece). Tectonophysics 86:27–43

Pirazzoli PA, Montaggioni LF, Saliege JF, Segonzac G, Thommeret Y, Vergnaud-Grazzini C (1989) Crustal block movements from Holocene shorelines: Rhodes Island (Greece). Tectonophysics 170:89–114

Pope KO, Runnels CN, Ku TL (1984) Dating middle Palaeolithic red beds in southern Greece. Nature 312:264–266

Reisch L (1976) Beobachtungen an Vogelknochen aus dem Spätpleistozän der Höhle von Kephalari (Argolis Griechenland). Archäologisches Korrespondenzblatt 6:261–265

Richter DK, Anagnostou C, Lykousis V (1993) Aragonitishe whiting-Ablagerungen in Plio–/Pleistozaenen Mergelsequenzen bei Korinth (Griechenland). Zentralbl Geol Paläontol Teil 1(6):675–688

Rodríguez J, Martín-González JA, Goikoetxea I, Rodríguez-Gómez G, Mateos A (2013) Mammalian paleobiogeography and the distribution of homo in early Pleistocene Europe. Quat Int 295:48–58

Roger T, Darlas A (2008a) Upper Pleistocene bird remains from Kalamakia Cave (Greece). In: Darlas A, Mihailovic D (eds) The Palaeolithic of the Balkans. British Archaeological Reports International Series 1819, Oxford, pp 69–76

Roger T, Darlas A (2008b) Microvertébrés, paléo-environnement et paléoclimat de la grotte de Kalamakia (Péloponnèse, Grèce) In: Darlas A, Mihailovic D (eds) The Palaeolithic of the Balkans. British Archaeological Reports International Series 1819, Oxford, pp 77–84

Rook L, Martínez-Navarro B (2010) Villafranchian: the long story of a Plio-Pleistocene European large mammal biochronologic unit. Quat Int 219:134–144

Rousakis G, Karageorgis AP, Georgiou P (2014) Geological structure and seabed morphology of the Stoupa submarine groundwater discharge system, Messinia, Greece. Environ Earth Sci 71:5059–5069. doi:10.1007/s12665-013-2910-1

Runnels C, Van Andel TH (1993) A handaxe from Kokkinopilos, Epirus, and its implications for the Paleolithic of Greece. J Field Archaeol 20:191–203

Runnels C, Van Andel TH (2003) The early stone age of the nomos of Preveza: landscape and settlement. In: Wiseman J, Zachos K (eds) Landscape archaeology in southern Epirus, Greece *I. Hesperia* supplement 32. The American School of Classical Studies at Athens, Princeton, pp 47–134

Runnels C, Karimali E, Cullen B (2003) Early Upper Palaeolithic Spilaion: an artifact-rich surface site In: Wiseman J, Zachos K (eds) Landscape archaeology in Southern Epirus, Greece *I. Hesperia* Supplement 32, The American School of Classical Studies at Athens, pp 135–156

Sakellariou D, Galanidou N (2016) Pleistocene submerged landscapes and Palaeolithic archaeology in the tectonically active Aegean region. In: Harff J, Bailey G, Lüth, F (eds) Geology and archaeology: submerged landscapes of the continental shelf. Geological Society of London, Special Publications 411:145–178. doi:10.1144/SP411.9

Sakellariou D, Lykousis V, Alexandri S, Kaberi H, Rousakis G, Nomikou P, Georgiou P, Ballas D (2007a) Faulting, seismic-stratigraphic architecture and late quaternary evolution of the Gulf of Alkyonides basin – east gulf of Corinth, Central Greece. Basin Res 19(2):273–295

Sakellariou D, Rousakis G, Kaberi H, Kapsimalis V, Georgiou P, Kanellopoulos T, Lykousis V (2007b) Tectono-sedimentary structure and late quaternary evolution of the north Evia gulf basin, Central Greece: preliminary results. Bull Geol Soc Greece 37(1):451–462

Sakellariou D, Beck J, Rousakis G, Georgiou P, Panagiotopoulos I, Morfis I, Tsampouraki-Kraounaki K, Zavitsanou A (2015) Submerged prehistoric landscapes off Franchthi Cave, East Argolic Gulf: preliminary results. Proceedings of the 11th Panhellenic Symposium on Oceanography and Fisheries, Mytilene, Greece, pp 993–996

Sharon G (2007) Acheulean large flake industries: technology and significance, British archaeological reports international series 1701. Archaeopress, Oxford

Shaw B, Ambraseys NN, England PC, Floyd MA, Gorman GJ, Higham TFG, Jackson JA, Nocquet J-M, Pain CC, Piggott MD (2008) Eastern Mediterranean tectonics and tsunami hazard inferred from the AD 365 earthquake. Nat Geosci 1:268–276. doi:10.1038/ngeo151

Slimak L, Kuhn SL, Roche H, Mouralis D, Buitenhuis H, Balkan-Atlim N, Binder D, Kuzucuoðlu C, Guillou H (2008) Kaletepe Deresi 3 (Turkey): archaeological evidence for early human settlement in central Anatolia. J Hum Evol 54(1):99–111

Sordinas A (1969) Investigations of the prehistory of Corfu during 1964–1966. Balk Stud 10:393–424

Sordinas A (1983) Quaternary shorelines in the region of Corfu and adjacent islets, western Greece. In: Masters PM, Flemming NC (eds) Quaternary coastlines and marine archaeology: towards the prehistory of land bridges and continental shelves. Academic, London, pp 335–344
Sorrensen L (2004) The cave of Aghios Nikolaos near Kato Vasiliki in Aetolia: flint technological observations. In: Dietz S, Kolonas L, Houby-Nielsen S, Moschos I, Eiring J Greek-Danish excavations at Aetolian Chalkis 1999–2001. Third preliminary report. Proceedings of the Danish Institute AT aTHENS, Volume IV, Athens, pp 237–258
Staikou P (in press) Ακολουθώντας τους Νεάντερταλ στις όχθες του Εύηνου ποταμού. In: Archaeological work in NW Greece and the Ionian Islands, Ioannina 10–13 December 2014, Conference Proceedings Ioannina
Stampfuss R (1942) Die ersten altsteinzeitlichen Höhlenfunde in Girechenland. Mannus 34:132–147
Stiner MC, Kozlowski JK, Kuhn SL, Karkanas P, Koumouzelis M (2010) Klissoura Cve 1 and the upper Palaeolithic of southern Greece in cultural and ecological context. Eurasian Prehistory 7(2):309–321
Stiros S, Blackman D, Pirazzoli P (2010) Late Holocene seismic coastal uplift and subsidence in Rhodes Island, SE Aegean Arc, Greece: evidence from ancient shipsheds. IGCP 521 – INQUA 501 Sixth Plenary Meeting and Field Trip, 27 Sept–5 Oct 2010, Rhodes, Greece, Abstracts, 211
Strasser TF, Panagopoulou E, Runnels CN, Murray PM, Thompson N, Karkanas P, McCoy FW, Wegmann KW (2010) Stone age seafaring in the Mediterranean: evidence from the Plakias region for lower Palaeolithic and Mesolithic habitation of Crete. Hesperia 79:145–190
Strasser TF, Runnels C, Wegmann K, Panagopoulou E, McCoy F, DiGregorio C, Karkanas P, Thompson N (2011) Dating Palaeolithic sites in southwestern Crete, Greece. J Quat Sci 26:553–560
Symeonides NK, Theodorou GE, Giannopoulos VI (2001) New data on Elephas chaniensis (Vamos cave, Chania, Crete). In: Proceedings of the 1st International Congress 'The World of Elephants'. Rome, pp 510–513
Taymaz T, Jackson J, McKenzie D (1991) Active tectonics of the north and central Aegean Sea. Geophys J Int 106:433–490
Ten Veen JH, Kleinspehn KL (2002) Geodynamics along an increasingly curved convergent plate margin: late Miocene-Pleistocene Rhodes, Greece. Tectonics 21:3. doi:10.1029/2001TC001287
Theocharis D (1970) Palaeolithic finds on Alonnisos. Archaeologikon Deltion 25:276–279. (in Greek)
Theocharis D (1971) Alonnisos. Archaeol Deltion 26:296–300. (in Greek)
Theodoulou T (2008) The harbour network of ancient Lesbos. First step of an underwater approach. In: Tzalas H (ed) Tropis X, Proceedings of the 10th International Symposium on Ship Construction in Antiquity. Hellenic Institute for the Preservation of Nautical Tradition, Hydra. Athens, in press
Tourloukis V (2009) New bifaces from the Palaeolithic site of Kokkinopilos, Greece and their stratigraphic significance. Antiquity 83(320). http://antiquityacuk/projgall/tourloukis
Tourloukis V, Karkanas P (2012) The middle Pleistocene archaeological record of Greece and the role of the Aegean in hominin dispersals: new data and interpretations. Quat Sci Rev 43(12):1–15. doi:10.1016/jquascirev201204004
Tourloukis V, Karkanas P, Wallinga J (2015) Revisiting Kokkinopilos: middle Pleistocene radiometric dates for stratified archaeological remains in Greece. J Archaeol Sci 57:355–369
Trantalidou K (1989) The Agitis' sources (Maara) site in the Drama Basin. Data and perspectives. Archael Analekta Athinon 22:61–70. (in Greek)
Trantalidou E, Darlas A (1995) Research in the Drama prefecture caves, 1992. Archaeol Ergo Maced Thrace 6:587–603. (in Greek)
Van Andel T, Lianos N (1984) High resolution seismic reflection profiles for the reconstruction of post-glacial transgressive shorelines. An example from Greece. Quat Res 22:31–45
Van Andel TH, Zangger E, Perissoratis C (1990) Quaternary transgressive/regressive cycles in the Gulf of Argos, Greece. Quat Res 34:317–329
Van Hinsbergen DJJ, Meulenkamp JE (2006) Neogene supradetachment basin development on Crete (Greece) during exhumation of the south Aegean core complex. Basin Res 18:103–124. doi:10.1111/j1365-2117200500282x
Van Winjgaarden GJM, Sotiriou A, Pierers N, Tourloukis V (2006) The Zakynthos archaeology project 2006. Preliminary report of the first season. Pharos J Neth Inst Athens 14:29–46
Van Winjgaarden GJM, Sotiriou A, Horn Lopes J, Gkouma M, Koster E, Stoker A, Susan D, Tourloukis V (2008) The Zakynthos archaeology project. Preliminary report of the 2008 season. Pharos J Neth Inst Athens 16:63–86
Williams H (2007) The harbors of ancient lesbos. In: Betancourt P, Nelson MC, Williams H (eds) Krinoi Kai Limenes: studies in honor of Joseph and Maria Shaw. INSTAP Press, Pennsylvania, pp 107–116
Zachos E, Dousougli A (2003) Lefkas: an overview of its history and archaeology through the archaeological museum exhibits. Hellenic Ministry of Culture – IB' Ephorate of Prehistoric and Classical Antiquities, Athens (in Greek)
Zavitsanou A, Sakellariou D, Rousakis G, Georgiou P, Galanidou, N (2015) Paleogeographic reconstruction of the Inner Ionian Sea during Late Pleistocene low sea level stands: preliminary results. Proceedings of the 11th Panhellenic Symposium on Oceanography and Fisheries, Mytilene, Greece, pp 997–1000

# Chapter 23
# Africa-Arabia Connections and Geo-Archaeological Exploration in the Southern Red Sea: Preliminary Results and Wider Significance

**Geoffrey N. Bailey, Dimitris Sakellariou, Abdullah Alsharekh, Salem Al Nomani, Maud Devès, Panos Georgiou, Manolis Kallergis, Stefanos Kalogirou, Leonidas Manousakis, Prokopis Mantopoulos, Matt Meredith-Williams, Garry Momber, Ioannis Morfis, Ioannis Pampidis, Ioannis Panagiotopoulos, Panagiotis Renieris, Grigoris Rousakis, Vasilis Stasinos, and Spyros Stavrakakis**

**Abstract** We report on a preliminary exploration of the submerged landscapes in the Saudi Arabian sector of the southern Red Sea aboard the Hellenic Centre for Marine Research (HCMR) Research Vessel, AEGAEO, in May–June 2013. The survey sampled areas of the continental shelf down to the shelf margin at ~130 m depth in the vicinity of the Farasan Islands and combined high resolution acoustic techniques with sediment coring to reconstruct features of the now-submerged landscape of potential archaeological significance, including geological structure, topography, palaeoenvironment, and sea-level change. The region is currently of wide interest and significance: to archaeologists because it is currently regarded as one of the primary pathways of dispersal for early human populations expanding out of Africa during the Pleistocene, in which the extensive but now-submerged shelf region may have played a key role; and to marine geoscientists because the Red Sea offers unusual opportunities as a 'laboratory' for investigating Pleistocene sea-level change. Preliminary results

G.N. Bailey (✉) • M. Meredith-Williams • G. Momber
Department of Archaeology, University of York, the King's Manor, York YO1 7EP, UK
e-mail: geoff.bailey@york.ac.uk; m.meredith-williams@latrobe.edu.au; garry.momber@maritimearchaeologytrust.org

D. Sakellariou
Institute of Oceanography, Hellenic Centre for Marine Research, 46.7 km Athinon-Souniou Av, Anavyssos, P.O. Box 712, 19013 Athens, Greece
e-mail: sakell@hcmr.gr

P. Georgiou • M. Kallergis • S. Kalogirou • L. Manousakis • P. Mantopoulos • I. Morfis
I. Pampidis • I. Panagiotopoulos • P. Renieris • G. Rousakis • V. Stasinos • S. Stavrakakis
Hellenic Centre for Marine Research, 19013, Anavyssos, Greece
e-mail: pgeorg@hcmr.gr; manolis@hcmr.gr; skalogirou@hcmr.gr; sailor@her.hcmr.gr; prokopis@hcmr.gr; gianmor@hcmr.gr; gpamp@hcmr.gr; jpanagiot@hcmr.gr; prenie@hcmr.gr; rousakis@hcmr.gr; stasinos@hcmr.gr; stavrak@hcmr.gr

A. Alsharekh
Department of Archaeology, King Saud University, P.O. Box 2627, Riyadh 7524-12372, Kingdom of Saudi Arabia
e-mail: asharekh@hotmail.com

S. Al Nomani
Saudi Geological Survey, P.O. Box 54141, Jeddah 21514, Kingdom of Saudi Arabia
e-mail: nomani.sm@sgs.org.sa

M. Devès
Institut de Physique du Globe de Paris, 1 rue Jussieu, 75238 Paris Cedex 05, France
e-mail: deves@ipgp.fr



G.N. Bailey et al. (eds.), *Under the Sea: Archaeology and Palaeolandscapes of the Continental Shelf*, Coastal Research Library 20, DOI 10.1007/978-3-319-53160-1_23

indicate that the submerged landscape was characterised by a complex topography with fault-bounded valleys and deep basins, some of which may have hosted, at least intermittently, fresh water during periods of lowered sea level.

## 23.1 Introduction

The southern end of the Red Sea is currently of high interest in relation to the nature and direction of human expansion into Asia and Europe from a presumed African cradle of early human evolution. The current consensus is that there were at least two major episodes of expansion out of Africa: an earlier one involving the earliest members of the genus *Homo* (*H. ergaster* or *H. erectus*) at about 1.8 mya, and a later dispersal by the earliest members of anatomically modern *Homo sapiens* between about 150 and 60 kya (Grine et al. 2009; Groucutt et al. 2015). The conventional view is that the primary pathways of movement and the most attractive environments involved a land route via the Nile Valley and the Sinai Peninsula, and thence northwards via the Levant to Anatolia, and eastwards to Iran and the Indian Subcontinent. In this view, the Arabian Peninsula has typically been ignored as a predominantly arid and infertile cultural backwater, largely bypassed by the major flows of early populations. More recently, however, a number of new sources of evidence have focussed attention on the Arabian Peninsula as a key region for early human settlement and a gateway to onward expansion to Eurasia, especially through the 'southern corridor', referring to southern Arabia and its coastal regions, and especially in relation to the dispersal of *Homo sapiens*. These indications include growing evidence for an Arabian archaeological record extending back to the early Pleistocene, similarities in stone tool technology in East Africa and Arabia, palaeogenetic inference, evidence for periodic climatic 'greening' of Arabian deserts during the Pleistocene, evidence that for much of the glacial-interglacial cycle lowering of sea level reduced the southern Red Sea to a narrow and shallow sea channel dotted with islands that could easily have been crossed with minimal equipment or seafaring expertise, and growing evidence for a greater time depth in the Pleistocene for early developments in seafaring and exploitation of marine resources (Petraglia and Rose 2009; Lambeck et al. 2011; Bailey 2015; Bailey et al. 2015; Erlandson and Braje 2015; Petraglia et al. 2015).

Lowering of sea level would have exposed an extensive area of shelf, reaching a maximum width of about 100 km on either side of the southern Red Sea, with a relatively narrow and shallow sea-channel separating the two sides (Figs. 23.1 and 23.2). Both mapping of palaeoshorelines (Lambeck et al. 2011) and the stable isotope composition of the Red Sea as measured in deep-sea cores from the Red Sea (Siddall et al. 2003) demonstrate that a marine connection between the Red Sea and the Indian Ocean would have persisted throughout the lowest sea-level stand of the LGM, and most probably for earlier low stands, at least for the past 400,000 years. In other words, there would not have been a dry-land passage from Africa to Arabia at any time over that time span. However, a narrow channel dotted with small islands would have persisted for long periods during the glacial-interglacial cycle, affording relatively short and easy sea-crossings of no more than a few kilometres. For earlier periods of the Pleistocene, the topography of the southern channel is less clear because of uncertainties about the impact of plate motions and tectonics, but it is worth noting that ongoing separation of the Arabian and African plates is mostly accommodated in the Danakil depression rather than in the area of the Red Sea.

Over the past decade a series of investigations has been devoted to exploring both the onshore and offshore archaeological record of Southwest Saudi Arabia as a joint Saudi-British and international project (Bailey et al. 2007a, b, 2015; Bailey 2009; Alsharekh and Bailey 2014). Here we report on a joint archaeological and geoscientific exploration of the continental shelf around the Farasan Islands (Fig. 23.2), which was conducted in May–June 2013 aboard HCMR's Research Vessel AEGAEO with the aim of exploring systematically the submerged landscapes. This project developed as an international and interdisciplinary collaboration arising directly out of the SPLASHCOS COST Action

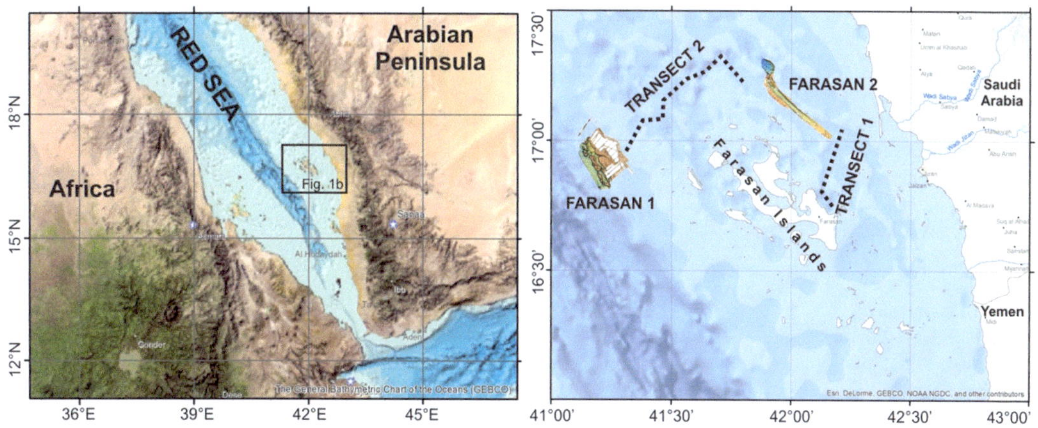


**Fig. 23.1** (**a**) (*left*) general relief map of the Southern Red Sea area with the location of the area of interest indicated in the *black box*; (**b**) (*right*): GEBCO bathymetry of the area of interest with the location of the survey areas Farasan 1, Farasan 2 and Transects 1 and 2

(www.splashcos.org; Bailey and Sakellariou 2012) and within the ERC-funded DISPERSE project (Bailey et al. 2012). This work builds on three strands of earlier investigation: one concerned with the impact of sea level change and submerged landscapes on the potential connections between Africa and Arabia and the dispersal of early humans expanding out of Africa during the Pleistocene (Bailey et al. 2007a, b; Lambeck et al. 2011); a second with the impact of active tectonics on the early landscapes of human evolution (King and Bailey 2006; Bailey and King 2011; Bailey et al. 2011); and a third with field investigation of the mid-Holocene shell mounds on the Farasan Islands and the search for earlier submerged shorelines and archaeological remains associated with them (Bailey et al. 2007a, b; Alsharekh and Bailey 2014).

In this chapter we concentrate on the deep-water survey of the offshore landscape. The main objectives of the cruise were to reconstruct the broad outlines of the now-submerged landscape on the shelf, to understand the underlying Quaternary geology and the role of active tectonics and sea-level change in shaping the changing topography, and to identify specific locations that might have preserved archaeological evidence of past human activity.

## 23.2 Materials and Methods

The marine survey comprised a wide variety of geological-geophysical techniques: (1) multi-beam bathymetry by using two hull-mounted systems (20 kHz and 180 kHz); (2) high-resolution sub-bottom profiling with a 3.5 kHz pinger; (3) deep-towed, 110/410 kHz, digital side scan sonar imaging; (4) deep penetrating airgun 10ci seismic profiles; (5) gravity coring, with cores 3–5 m long; (6) box coring, 40×40×60 cm; and (7) ROV dives to visually inspect sites selected from the bathymetric, acoustic and profiling data.

Two areas (FARASAN 1 and FARASAN 2) and two seismic transects (TRANSECT 1 and TRANSECT 2) were systematically surveyed (Figs. 23.3 and 23.4). The data set comprised about 500 km$^2$ of seabed morphology mapped with the multi-beam systems, 170 nautical miles (105 km) of airgun seismic profiles, 250 nautical miles (155 km) of 3.5 kHz sub-bottom profiles and 140 nautical miles (87 km) of side-scan sonar lines. Eighteen gravity cores and two box cores were recovered and five dives of the Max Rover ROV were accomplished.

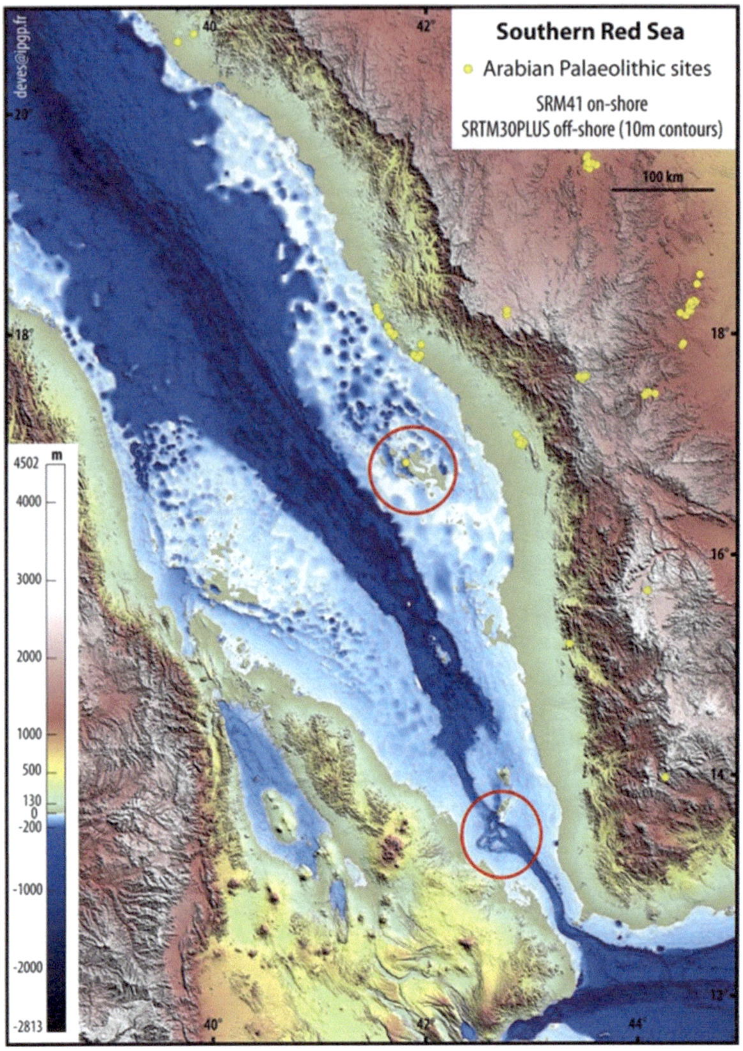


**Fig. 23.2** Enhanced satellite imagery of the southern Red Sea. *Light blue* area indicates the extent of the submerged landscape exposed at – 120 m. The upper *red circle* indicates the position of the Farasan Islands. The lower *red circle* indicates the group of islands that emerge in the region of the Hanish Sill at low sea level with sea crossings of less than 4 km. ASTER GDEM is a product of METI and NASA (created by Maud Devès)

## 23.3 Results

### *23.3.1 Survey Area Farasan 1*

Swath bathymetry mapping (Fig. 23.3) shows a relatively flat, 70–80 m-deep platform, which dips slightly towards the SE and occupies the outer part of the shelf. A shallower terrace, at approximately 40 m depth occurs at the north-eastern edge of the surveyed area. A deeper terrace is preserved locally at about 120 m depth on the steep slope off the shelf edge. Elongate or irregularly shaped flat-topped ridges are located at a short distance off the shelf edge. They tops of the ridges are at depths of about 80–90 m, very similar to the depth of the 80 m terrace on the shelf, and are separated from the shelf by deep and steep-sided troughs.

Airgun single-channel seismic profiles across the shelf and the ridges show that the steep, SW-facing slope off the shelf edge and the SW- and NE-facing slopes on both sides of the shallow ridges are controlled by normal faults running NW–SE, parallel to the rift axis of the Red Sea (Fig. 23.4). The latter is characterized by the deposition of massive, up to 2,500 m thick evaporites (salt deposits) in

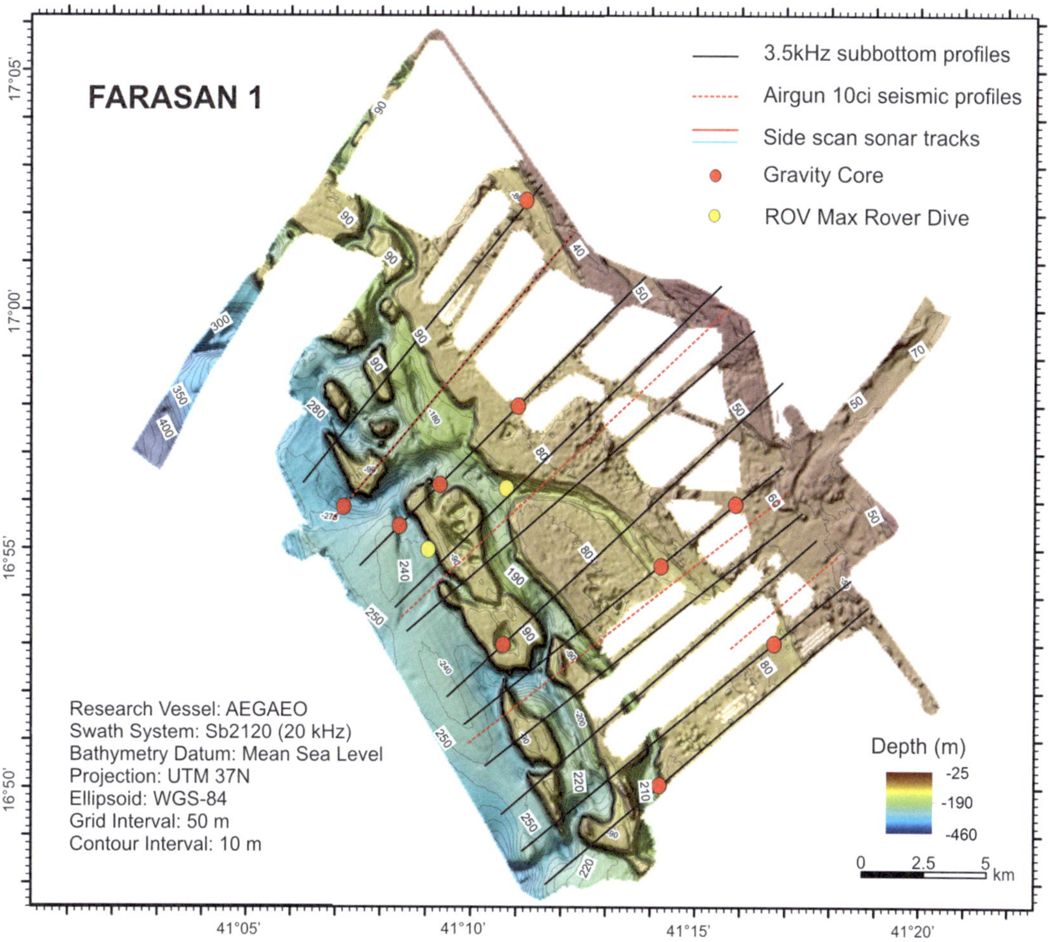


**Fig. 23.3** Swath bathymetry and location of seismic and subbottom profiles, side scan sonar track-lines, gravity cores and ROV dives in the Farasan 1 survey area

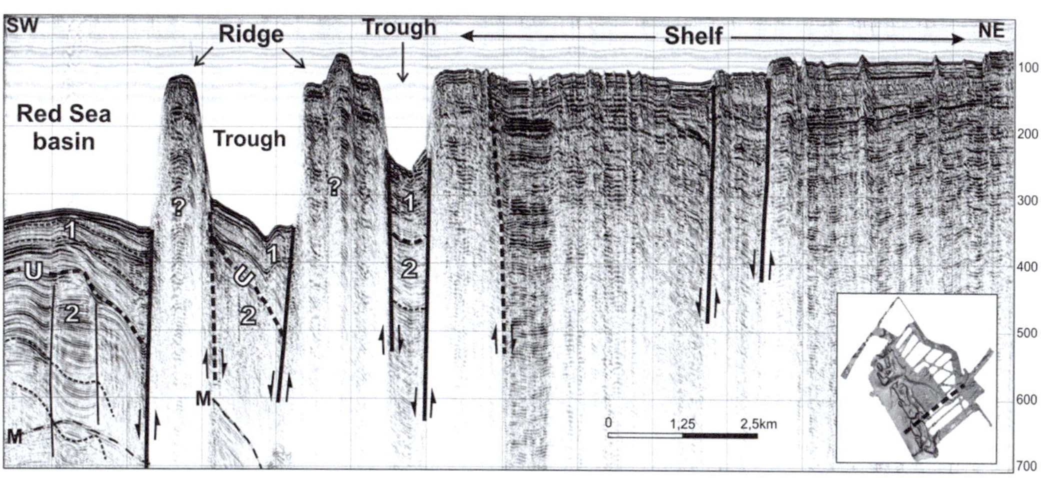


**Fig. 23.4** Airgun 10 cubic inches single channel seismic profile across the outer Farasan shelf, in Farasan 1 area

the Middle to Late Miocene (Bosence 1998). Shallow Pleistocene marine reef-limestone, underlain by Miocene salt and deformed by salt diapirism, forms the bedrock of the Farasan Islands (Bantan 1999). We expect that a similar stratigraphic succession prevails underneath the continental shelf around the Farasan Islands. Evidence for extensive salt flow towards the axis of the Red Sea has been provided by Augustin et al. (In press). On the basis of the observed stratigraphy on the Farasan Islands and the extensive salt flow observed on and underneath the seabed, we suggest that the flat-topped ridges, mapped off the shelf edge, are blocks dragged away from the shelf by salt flow that may occur in the deeper stratigraphic levels. This suggestion can explain both the steeply sloping sides of the troughs between the ridges and the similarity in height between the ridge tops and the continental shelf in the surrounding areas.

It is reasonable to assume that the ridges were exposed above sea-level during the last low sea-level stand during Marine Isotope Stage (MIS) 2 and probably during older, Pleistocene low sea-level periods, forming thus a series of flat islands, the 'prehistoric Farasan Archipelago', separated from the LGM-coastline by deep troughs. Normal faults dipping in a SW direction also crosscut the shelf and are responsible for the slight morphological anomalies, mainly depressions and incised valleys.

Holocene sediment deposition on the outer shelf is very limited and restricted to small depressions and valleys (Fig. 23.5). The sedimentological description of gravity cores indicates lacustrine-type sedimentation below the thin Holocene marine drape in the isolated depressions on the 80 m-deep terrace. The seismic stratigraphy of the seabed sub-surface displays horizontal sedimentary sequences in these depressions deposited presumably during the Pleistocene in relatively shallow waters, as shown by Bantan (1999) for the exposed sedimentary formations on the Farasan Islands.

### 23.3.2 Survey Area Farasan 2

This area comprises a 120 m-deep valley, bounded by NW–SE trending normal faults (Fig. 23.6). The valley is incised into the 70–75 m prominent terrace, which forms a prominent feature of the inner shelf and becomes narrower towards the north-west. A narrow gorge on the sea floor at the north-western tip of the valley connects it with a >200 m-deep, depression. The latter is presumably the result of the solution of a Miocene salt diapir and hosts a >250 m-thick sedimentary sequence of Quaternary age, possibly extending back into the Pliocene.

A terrace at about 112 m depth can be mapped along the flanks of the valley and is covered by seismically transparent Holocene deposits (Figs. 23.6 and 23.7). This terrace is interpreted as an indicator of the water-level in the valley during the last low sea-level stage. Gravity coring penetrated the

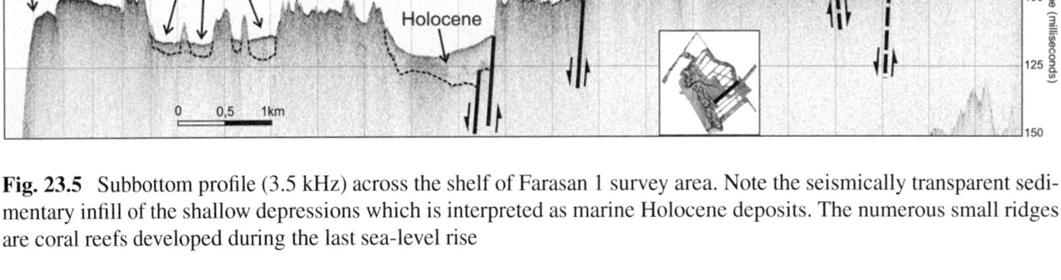


**Fig. 23.5** Subbottom profile (3.5 kHz) across the shelf of Farasan 1 survey area. Note the seismically transparent sedimentary infill of the shallow depressions which is interpreted as marine Holocene deposits. The numerous small ridges are coral reefs developed during the last sea-level rise

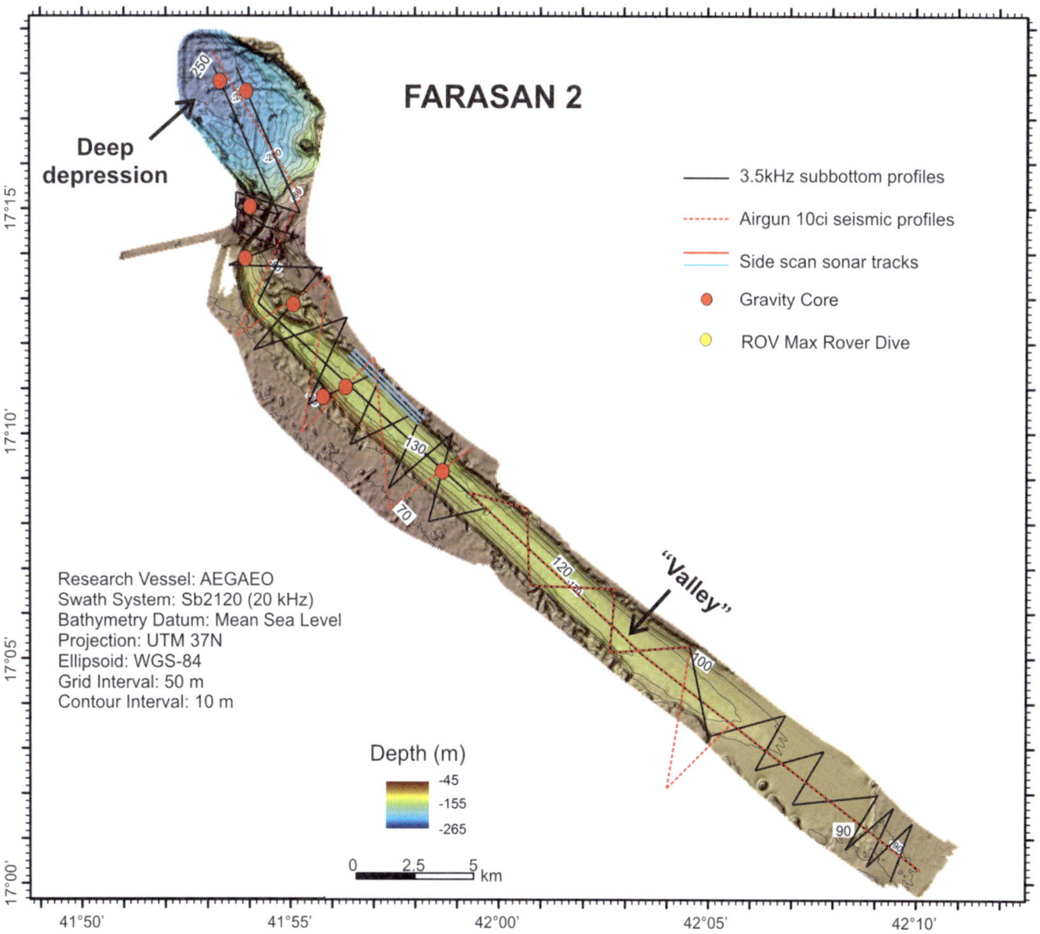


**Fig. 23.6** Swath bathymetry and location of seismic and subbottom profiles, side scan sonar track-lines, gravity cores and ROV dives in the Farasan 2 survey area

Holocene fine-grained, marine sediments, and recovered gypsum fragments presumably of Miocene age from the substrate at about 2–2.5 m below the seafloor.

Preliminary laboratory analyses on sediment cores from the deep depression reveal lacustrine-type sedimentation below the 1–2 m-thick Holocene marine silt deposits. Thus, both the 120 m deep valley and the >200 m deep depression in Farasan 2 area may have been lakes during the Last Glacial Maximum. One more core recovered from a small, 102 m deep depression located on the northern slope of the valley revealed evidence of lacustrine sedimentation (Fig. 23.8). Holocene marine silty sand prevails in the upper part of the core. Gradual coarsening of the grain size between 80 and 147 cm below the seafloor indicates a shallow marine to coastal environment. A sharp erosional contact at 147 cm separates the upper part of the core from the lower part, which is characterized by grayish to bluish-whitish colours. The sedimentological characteristics along with the increase in magnetic susceptibility below 147 cm indicate a significant change of the depositional environment from marine above to lacustrine below. Geochemical, micropalaeontological and geochronological analyses on the recovered marine and presumably lacustrine sediments of core FA13 and other cores are in progress, with the aim of shedding light on the environmental conditions (freshwater, saline or otherwise) prevailing in these lakes during the periods of lowered sea level.

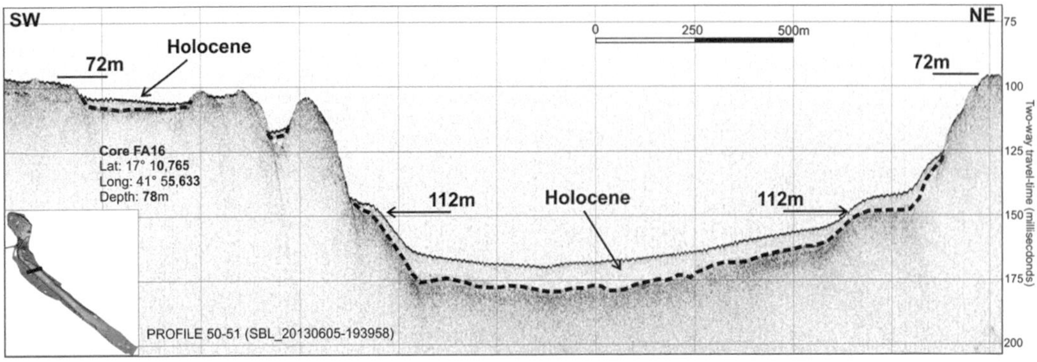


**Fig. 23.7** Subbottom profile (3.5 kHz) across the shallow, elongate basin in Farasan 2 survey area. Note the seismically transparent, sedimentary infill of the basin which is interpreted as marine Holocene deposits. The 112 m deep terrace on the flanks is shown on the profile

### *23.3.3 Combined Results*

The observations and preliminary results presented above can be summarised as follows:

(a) The numerous Plio-Pleistocene depressions of varying depth formed on the shelf around the Farasan Islands are due to the solution of Miocene evaporites and represent isolated water-bodies or lakes during the last low sea-level stand and possibly during older Pleistocene sea-level low stands (Fig. 23.9)
(b) The flat-topped ridges mapped off the shelf edge in the Farasan 1 survey area were disconnected from the exposed shelf, forming an island archipelago at a short distance from the LGM-shoreline
(c) The flat shelf observed in both survey areas was exposed during the Last Glacial Maximum and incised by river valleys, while shallow depressions were transformed into basins filled permanently or ephemerally with water, possibly as freshwater lakes
(d) This type of landscape might have served both as an attractor of human settlement and as a favourable location for the preservation of archaeological evidence

## 23.4 Discussion

The work reported above is still at an early stage, but there are several important outcomes that bring into focus three issues of wider relevance to the investigation of hominin dispersal and expansion. In the first place, it is clear that the landscape exposed at low sea-level stands would have afforded many of the characteristics that are known to be attractive to early human populations, in particular a complex topography comprising fault-bounded basins and topographic bottlenecks such as narrow connecting valleys that could have trapped freshwater and offered both extensive grazing territory for large mammals and also topographic opportunities for capturing them (King and Bailey 2006; Winder et al. 2013, 2015).

The availability of freshwater in this landscape is a key variable, and our results indicate the possibility, though not yet definitive evidence, that many of the basins would have contained, at least periodically, supplies of fresh water. It is well known to fishermen in this region, as elsewhere, that there are springs of fresh water emerging below modern sea level (see also Radić Rossi and Cukrov Chap. 17), and it has long been hypothesised that when sea level dropped, these springs would have

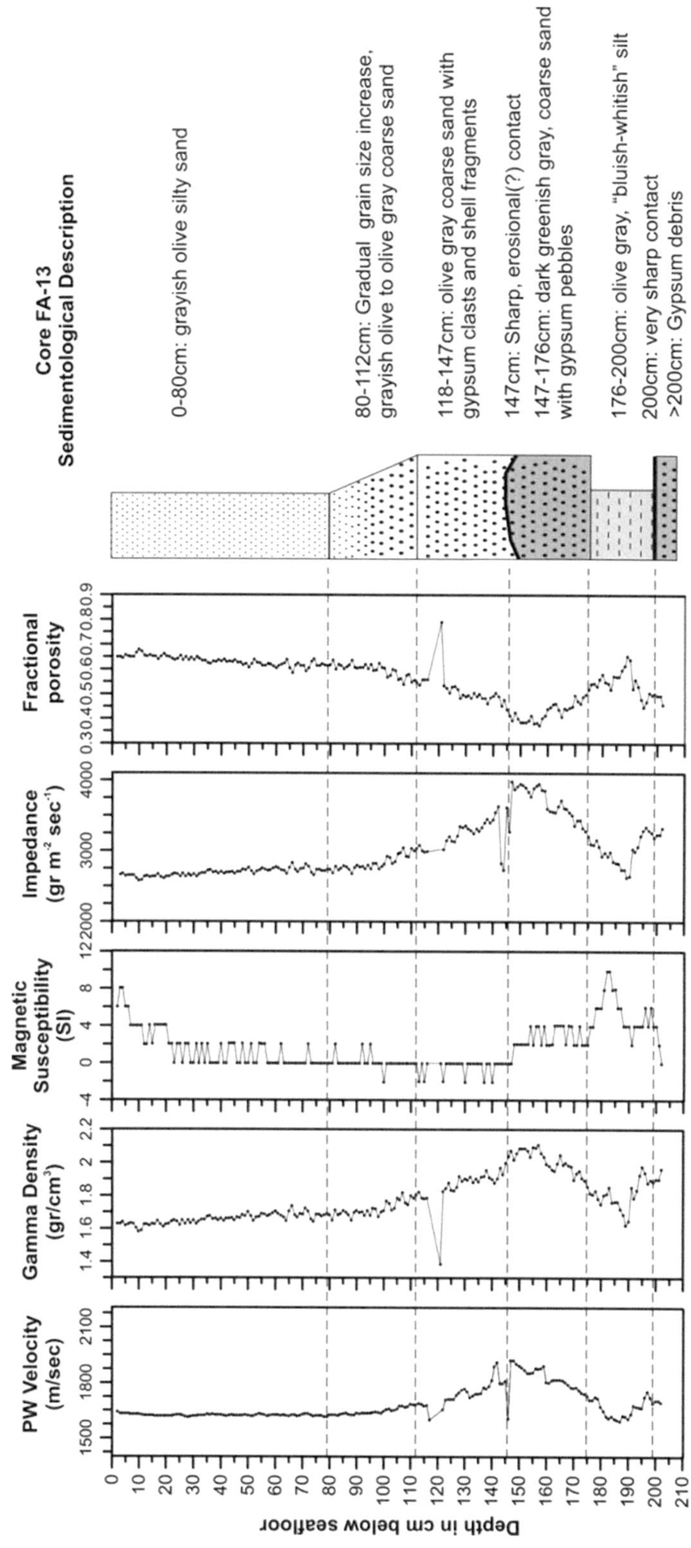


**Fig. 23.8** Sedimentological description and multisensor core logger measurements in core FA13. Note the remarkable increase in magnetic susceptibility at and below 147 cm, which corresponds with a sharp, erosional contact between the upper, marine section and the lower presumably lacustrine one. The location of the core is shown on the map in Fig. 23.6

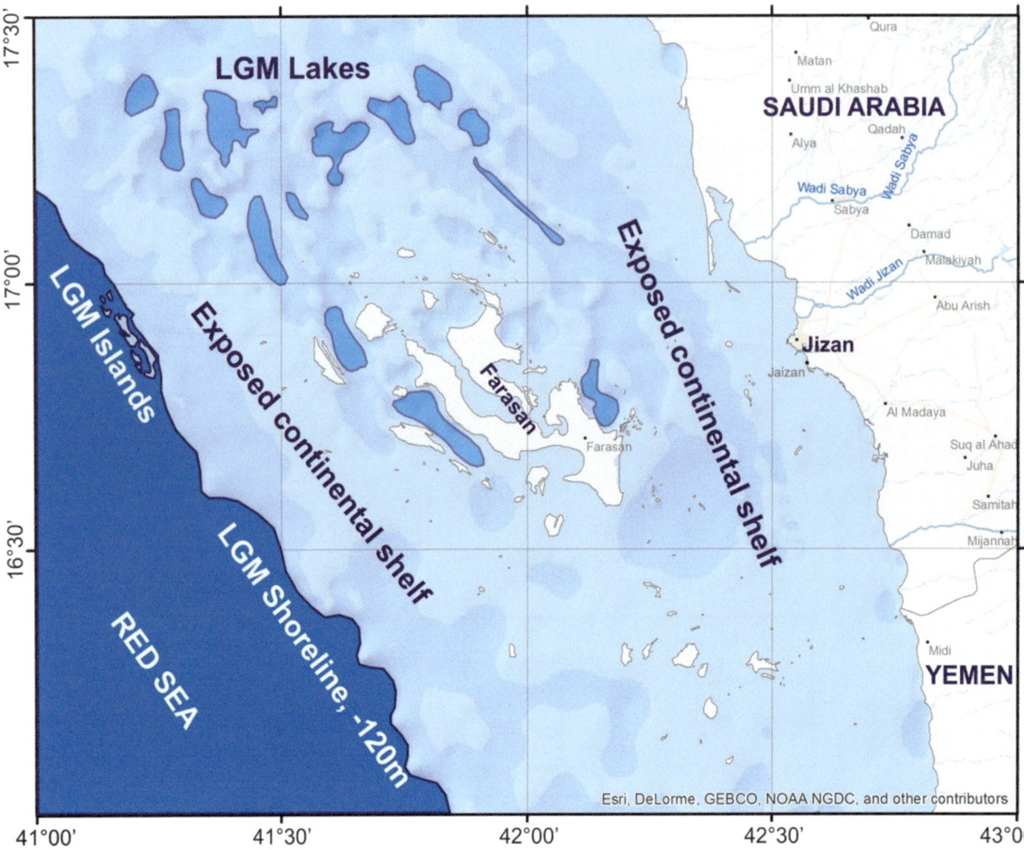


**Fig. 23.9** Palaeogeographic reconstruction of the Farasan continental shelf during the Last Glacial Maximum (LGM)

flowed more strongly, providing the possibility of a well-watered and relatively fertile landscape (Faure et al. 2002). The sediment cores that we have recovered from the Farasan survey provide the first opportunity to directly test this hypothesis, and sedimentological and foraminiferal analyses and dating of the sediments by optically stimulated luminescence (OSL) and radiocarbon are currently underway to pursue this.

Moreover, it is clear that these possibilities would have been available during periods of the glacial-interglacial cycle when the mainland of the Arabian Peninsula would have been exposed to conditions of maximum aridity and pressure on the survival of human populations in the region (Bailey 2015; Bailey et al. 2015, Breeze et al. 2016), providing a core area or refugium where populations displaced by the expansion of arid conditions in the interior of the Peninsula could have persisted.

We are still far from finding archaeological evidence of human habitation on the now-submerged landscape, and this brings into focus a second issue, which is the need to better understand the conditions under which archaeological material is likely to survive in underwater contexts, and where to look for it. To facilitate that understanding, our underwater investigations are being carried out in conjunction with archaeological survey on the adjacent mainland. This includes survey for Palaeolithic stone tools in the SW region of Saudi Arabia, demonstrating a time depth of human occupation extending back into the Middle Pleistocene if not earlier, and investigation of mid-Holocene coastal shell mounds, particularly on the Farasan Islands. Both types of investigations are directed towards researching archaeological sites in their wider geoarchaeological and landscape context, and to understanding the conditions under which material traces of human activity are variously accumulated, concentrated, preserved, eroded away, buried by subsequent sedimentation or exposed to discovery.

These investigations in their turn provide clues as to where archaeological materials may have been preserved or exposed on the submerged shelf and the likelihood that they will have survived destruction or burial by inundation during sea-level rise.

The combination of research under water and on dry land is an important feature of the wider research strategy with which the above underwater survey is associated. One reason for this is that the present-day coastline is an arbitrary boundary that cuts across what would have been a seamless territory extending from the mountain watershed of the western Arabian escarpment down to the contemporaneous shoreline as it existed at periods of low sea level, and this whole territory needs to be investigated as a single entity. Proceeding as if the landscape now offshore of the modern coastline did not exist must necessarily result in a highly biased and inaccurate evaluation of the archaeological record and the dynamics of human settlement and range expansion or contraction during the climatic fluctuations of the Quaternary.

Another reason is that the results from one sector can usefully inform on the research design and interpretation of results in the other, and on aspects of methodology that apply to both sectors. Understanding the factors that determine the location and visibility of archaeological sites on land clearly can provide clues as to where to search under water, and this is a well-practised strategy that has proved successful in other contexts (Faught 2004; Evans et al. 2014). However, this is not simply a one way relationship. Our underwater investigations are also influencing the way we think about research on land. In particular they focus on the all-important issue of 'landscape taphonomy' – namely the variables that determine the survival, preservation and visibility of archaeological material and the extent to which such variability confounds the interpretation of geographical patterning in the distribution of archaeological sites. This taphonomic factor is thrown into sharp relief when dealing with submerged landscapes because of the potentially destructive or obscuring influence of sea-level rise. But it is no less potent on dry land, where processes of erosion and sedimentation, and agricultural and industrial developments, may have equally powerful distorting effects on the visibility and recovery of archaeological remains. These are issues that are still at a very early stage of development and point to a new and emerging field of taphonomic investigation in need of future development.

Finally our results highlight the issue of the early development of simple methods for crossing sea barriers and the conditions under which this may have been favoured. We cannot be sure whether the narrow channel that existed at the southern end of the Red Sea during periods of low sea level would have been crossed. But we know from the evidence of early human expansion into Australia and New Guinea that sea crossings involving some form of water craft were being conducted on a regular basis at least 50,000 years ago and to Flores Island at 0.8 million years (Hiscock 2008; Ward and Veth Chap. 24). There is no reason to deny that possibility in other parts of the world. Moreover, it is likely that the earliest sea crossings would have been facilitated in archipelago environments with short travel distances, relatively sheltered conditions, land always in sight, proximity to a familiar mainland, and the prospect of attractive resources on the other side of the sea crossing. The presence of mid-channel islands at the southern end of the Red Sea during low sea level periods (Fig. 23.2) and the presence of islands offshore of the Farasan shelf at periods of lowest sea level (Figs. 23.3 and 23.8) are both examples of the conditions that may have encouraged early experiments in sea travel.

## 23.5 Conclusion

This is one of the first attempts to apply a suite of underwater techniques to the purposeful and systematic geoarchaeological exploration of a deeply submerged landscape (for comparable examples, see Dixon and Monteleone 2014; Pearson et al. 2014; Flemming Chap. 18; Gaffney et al. Chap. 20). It is clear that a landscape with interpretable features of geological structure, geomorphology, topography, and potential for human settlement lies now submerged on the extensive shelf region surrounding the Farasan Islands. This forms an important first step in archaeological survey and a promising basis for future investigations, but we emphasise that we are still at a very early stage in developing this research agenda, and that underwater archaeological material has yet to be discovered. Nevertheless, the research has clarified ways in which improvements in approach and the deployment of additional technologies can be applied in future work, and sharpened our focus on the issues that need to be addressed when attempting to incorporate submerged landscapes into the interpretation of human dispersal and range expansion during periods of low sea level.

**Acknowledgements** We thank the Saudi Commission for Tourism and National Heritage (SCTH) for permission to undertake fieldwork, and the President, HRH Prince Sultan bin Salman bin AbdulAziz al Saud, Vice-President, Professor Ali Al-Ghassan, and the Director General, Jamal al Omar, for their support. We also thank HRH Crown Prince Salman bin Abul Aziz Al Saud and the Department of General Survey of the Ministry of Defense for permission to undertake the cruise of R/V AEGAEO. We also thank Lt Fahad Al Shwish, Observer from the Hydrographic Department of the Saudi Ministry of Defense, for his support and valuable assistance in overcoming unexpected logistical difficulties during the cruise, and Captain Theodoros Kanakaris and the crew of R/V AEGAEO for their untiring efforts to ensure the smooth running of the scientific operation during the offshore survey work. The research is funded by the European Research Council through Advanced Grant 269586 DISPERSE under the 'Ideas' Specific Programme of the Seventh Framework Programme. This is DISPERSE contribution no. 35.

## References

AlSharekh AM, Bailey GN (eds) (2014) Coastal prehistory in southwest Arabia and the Farasan Islands: 2004–2009 field investigations. Saudi Commission for Tourism and Antiquities, Riyadh

Augustin N, Devey C, Van der Zwan F (In press) A modern view on the Red Sea rift: tectonics, volcanism and salt blankets. In: Rasul N, Stewart I, Nawab Z (eds) The Red Sea, vol Volume 2. Springer, Dordrecht

Bailey GN (2009) The Red Sea, coastal landscapes and hominin dispersals. In: Petraglia M, Rose J (eds) The evolution of human populations in Arabia. Springer, Dordrecht, pp 15–37

Bailey GN (2015) The evolution of the Red Sea as a human habitat in the quaternary period. In: Rasul N, Stewart CF (eds) The Red Sea: the formation, morphology, oceanography and environment of a young ocean basin. Springer, Heidelberg, pp 595–610

Bailey GN, King GCP (2011) Dynamic landscapes and human dispersal patterns: tectonics, coastlines and the reconstruction of human habitats. Quat Sci Rev 30:1533–1553

Bailey GN, Sakellariou D (2012) SPLASHCOS: submerged prehistoric archaeology and landscapes of the continental shelf. Antiquity 86:334. http://antiquity.ac.uk/projgall/sakellariou334/

Bailey GN, King GCP, Flemming NC, Lambeck K, Momber G, Moran LJ, Al-Sharekh AM, Vita-Finzi C (2007a) Coastlines, submerged landscapes and human evolution: the Red Sea basin and the Farasan Islands. J Island Coastal Archaeol 2:127–160

Bailey GN, AlSharekh A, Flemming NC, Lambeck K, Momber G, Sinclair A, Vita-Finzi C (2007b) Coastal prehistory in the southern Red Sea basin: underwater archaeology and the Farasan Islands. Proc Sem Arab Stud 37:1–16

Bailey GN, Reynolds SC, King GCP (2011) Landscapes of human evolution: models and methods of tectonic geomorphology and the reconstruction of hominin landscapes. J Hum Evol 60(3):257–280

Bailey GN, King GCP, Devès M, Hausmann N, Inglis R, Laurie E, Meredith-Williams M, Momber G, Winder I, Alsharekh A, Sakellariou D (2012) DISPERSE: dynamic landscapes, coastal environments and human dispersals. Antiquity 86:334. http://antiquity.ac.uk/projgall/bailey334/

Bailey GN, Devès MH, Inglis RH, Meredith-Williams MG, Momber G, Sakellariou D, Sinclair AGM, Rousakis G, Al Ghamdi S, Alsharekh A (2015) Blue Arabia: Palaeolithic and underwater survey in SW Saudi Arabia and the role of coasts in Pleistocene dispersal. Quat Int 382:42–57. http://dx.doi.org/10.1016/j.quaint.2015.01.002

Bantan RA, (1999, January) Geology and sedimentary environments of Farasan Bank (Saudi Arabia) southern Red Sea: a combined remote sensing and field study. Ph.D. Thesis, Department of Geology, Royal Holloway, University of London

Bosence DJ (1998) Stratigraphic and sedimentological models of rift basins. In: Purser BH, Bosence DWJ (eds) Sedimentation and tectonics in rift basins: Red Sea-gulf of Aden. Chapman & Hall, London, pp 9–26

Breeze P, Groucutt HS, Drake NA, White TS, Jennings RP, Petraglia MD (2016) Palaeohydrological corridors for homini dispersal in the Middle East ~250–70,000 years ago. Quternary Sci Rev 144:155–185. doi.org/10.1016/j.quascirev.2016.05.012

Dixon JE, Monteleone K (2014) Gateway to the Americas: underwater archeological survey in Beringia and the North Pacific. In: Evans AM, Flatman JC, Flemming NC. (eds) (2014)Prehistoric archaeology on the continental shelf: a global review. Springer, New York, pp 95–114
Erlandson JM, Braje TJ (2015) Coasting out of Africa: the potential of mangrove forests and marine habitats to facilitate human coastal expansion via the southern dispersal route. Quat Int 382:31–41
Evans AM, Flatman JC, Flemming NC (eds) (2014) Prehistoric archaeology on the continental shelf: a global review. Springer, New York
Faught MK (2004) The underwater archaeology of palaeolandscapes, Apalachee Bay, Florida. Am Antiq 69:235–249
Faure H, Walter RC, Grant DR (2002) The coastal oasis: ice age springs on emerged continental shelves. Glob Planet Chang 33:47–56
Grine FE, Fleagle JG, Leakey RE (eds) (2009) The first humans – origin and early evolution of the genus homo. Springer, Dordrecht
Groucutt HS, Petraglia MD, Bailey G, Scerri EML, Parton A, Clark-Balzan L, Jennings RP, Lewis L, Blinkhorn J, Drake NA, Breeze PS, Inglis RH, Devès MH, Meredith-Williams M, Boivin N, Thomas MG, Scally A (2015) Rethinking the dispersal of *Homo sapiens* out of Africa. Evol Anthropol 24:149–164. doi:10.1002/evan.21455
Hiscock P (2008) Archaeology of ancient Australia. Routledge, London
King GCP, Bailey GN (2006) Tectonics and human evolution. Antiquity 80:265–286
Lambeck K, Purcell A, Flemming N, Vita-Finzi C, Alsharekh A, Bailey GN (2011) Sea level and shoreline reconstructions for the Red Sea: isostatic and tectonic considerations and implications for hominin migration out of Africa. Quat Sci Rev 30:3542–3574
Pearson CE, Weinstein RA, Gagliano SM, Kelley DB (2014) Prehistoric site discovery on the outer continental shelf, Gulf of Mexico, United States of America. In: Evans AM, Flatman JC, Flemming NC. (eds) (2014)Prehistoric archaeology on the continental shelf: a global review. Springer, New York, pp 53–72
Petraglia M, Rose J (eds) (2009) The evolution of human populations in Arabia. Springer, Dordrecht
Petraglia MD, Groucutt H, Parton A, Alsharekh A (eds) (2015) Green Arabia: human prehistory at the cross-roads of continents. Quat Int 382:1–7
Siddall M, Rohling EJ, Almogi-Labin A, Hemleben C, Meischner D, Schmelzer I, Smeed DA (2003) Sea-level fluctuations during the last glacial cycle. Nature 423:853–858
Winder IC, King GCP, Devès MH, Bailey GN (2013) Complex topography and human evolution: the missing link. Antiquity 87:333–349
Winder IC, Devès MH, King GCP, Bailey GN, Inglis RH, Meredith-Williams M (2015) Dynamic landscapes and complex topography as agents in human evolution: the dispersals of the genus homo. J Hum Evol 87:48–65. doi.org/10.1016/j.jhevol.2015.07.002

# Chapter 24
# To the Islands: The Archaeology of the Archipelagos of NW Australia and its Implications for Drowned Cultural Landscapes

**Ingrid Ward and Peter Veth**

**Abstract** In this chapter we consider the archaeological records of islands from three archipelagos lying off NW Australia and their implications for the submerged landscapes of which they were once a part. We draw attention to unique human-landscape configurations from circa 42,000 cal BP to 7500 cal BP, for which there may not be analogues in modern cultural landscapes. A holistic understanding of the genesis of maritime cultures is currently based on a truncated record in which the most significant part (the drowned landscapes) is usually missing. Here we make the case for renewed investigation of the drowned landscapes of the NW Shelf, in order to better understand the role of islands, archipelagos and coastlines in human history.

## 24.1 Introduction

Since the earliest human occupation of Australia currently dated to circa 50,000 cal BP (Veth and O'Connor 2013) changing sea-levels have had dramatic effects on the Australian continental shelf, repeatedly exposing and drowning vast lowland areas. Prior to the most recent period of sea-level rise between 20,000 cal BP and 8000 cal BP (the postglacial transgression) the archipelagos and islands of NW Australia were part of an extensive lowland plain that extended in the south from NW Cape to as far north as the Bonaparte Gulf and westward well over 100 km from its current position (Jones 1973). This plain was dissected by riverine courses and scattered hills (Semeniuk et al. 1982; Semeniuk and Wurm 1987), some of which now make up the islands of the Dampier and Bonaparte Archipelagos and the Montebello and Barrow Island complexes (Fig. 24.1).

For many Indigenous Australians in the northwest, their 'country' is not conceptually restricted to the present landscape but includes the past, and correspondingly includes the landscapes drowned by postglacial sea-level rise. This 'Sea Country' or 'Saltwater Country' includes the mainland coastal, island and marine environments that together make up the traditional estates of maritime Indigenous groups in Australia (Smyth 1997). Like Hayward's (2012, p. 1) notion of the 'aquapelago', the Indigenous concept of Sea Country provides a framework for understanding the cultural continuum

I. Ward • P. Veth (✉)
School of Social Sciences, M257, Faculty of Arts, The University of Western, Crawley, WA 6009, Australia
e-mail: ingrid.ward@uwa.edu.au; peter.veth@uwa.edu.au



G.N. Bailey et al. (eds.), *Under the Sea: Archaeology and Palaeolandscapes of the Continental Shelf*, Coastal Research Library 20, DOI 10.1007/978-3-319-53160-1_24

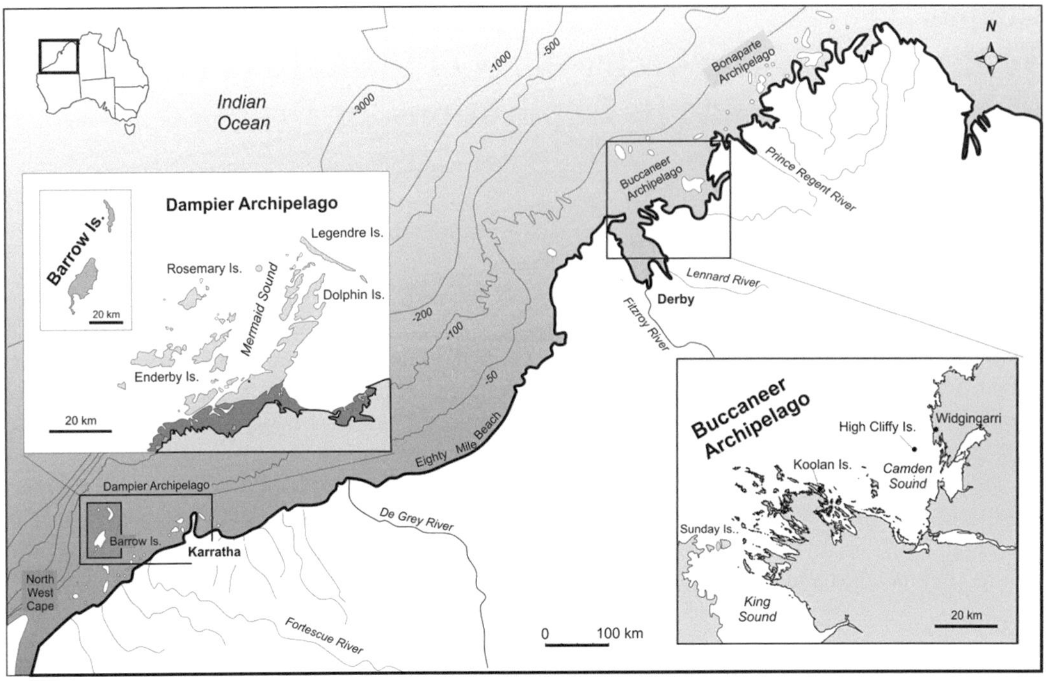


**Fig. 24.1** Archipelagos of NW Australia. (**a**) Dampier Archipelago, Barrow Island and Montebello Islands and (**b**) the Buccaneer Archipelago

between land and sea by emphasising 'the interrelation of marine and terrestrial spaces in areas of the planet in which small fragments of land are aggregated' (Hayward 2012, p. 1).

From an archaeological perspective, land and sea shed light on each other but, as argued by Dawson (2012, p. 19), we are more likely to find evidence of maritime cultural interactions on the islands than underwater, and this has certainly been the case for NW Australia so far. However, archaeologists in Australia are now beginning to explore these submerged landscapes (Ward et al. 2013, see also Flemming, Chap. 18) and a range of shelf edge sites (Veth et al. 2014; Ward et al. 2016). Indeed, we argue that the investigation of these former coastlines and submerged landscapes is critical to our understanding of maritime adaptations in this region (Ward et al. 2015, 2016, see also Benjamin et al. 2011).

Archaeological evidence demonstrates a continuity of marine resource exploitation in the north-west, from as early as ~42,000 cal BP (Veth et al. 2007, 2014, Morse 1988, 1993, Przywolnik 2002, 2005). At the same time, new understandings about the nature of human settlement of the plains and past coastlines, and reliance on maritime versus terrestrial resources of the now drowned NW Shelf are emerging from midden records and macrofaunal remains on Barrow and Montebello Island complex (Veth et al. 2007, 2014, 2016; McDonald and Veth 2009; Ward et al. 2016) and from some of the extensive engraved rock art (petroglyphs) in the Dampier Archipelago region (Mulvaney 2013, 2015). In this chapter, we summarise the evidence for maritime adaptations from the archipelagos of NW Australia and their implications for drowned cultural landscapes of northern Australia more generally. In contrast to the wealth (>3000) of submerged sites in the northern hemisphere (Benjamin et al. 2011), there are still no documented underwater sites from Australian waters that date to before mid-Holocene sea-level stabilisation.

## 24.2 Archaeology of the Archipelagos of NW Australia

### 24.2.1 Dampier Archipelago

The 42 islands that make up the Dampier Archipelago (now known as *Murujuga*) are located within the Pilbara Offshore marine bioregion on the inner part of the North West Shelf (NWS) (Fig. 24.1). Legendre Island and some of the small outer islets (e.g., Kendrew Island, Brigadier Island, Cohen Island) are remnants of consolidated limestone ridges which delineate a previous coastline (Semeniuk et al. 1982). In contrast, the other islands and the Burrup Peninsula (formally Dampier Island) are predominantly composed of granophyres, gabbro and basalt (Donaldson 2011), cut by dolerite dykes which extend offshore (Semeniuk et al. 1982; Dortch 2002).

The Dampier Archipelago formed when rising seas flooded the coastal plain between approximately 8500 and 7500 years ago. Older formations dating to the terminal Pleistocene have been identified including lithified dune, beach and offshore bar deposits located within and immediately beyond the Dampier Archipelago. These provide a physical clue to the changing sea-levels and former shoreline systems around which people may have foraged (Kojan 1994). Indeed shell midden sites from Burrup, Enderby and Rosemary Islands dating to between 10,000 and 7500 cal BP (Vinnicombe 1987; Bradshaw 1995; McDonald and Veth 2009; McDonald 2015) show an intriguing proximal association to some of these submerged shorelines sequences (see Ward et al. 2013), suggesting continuous use of the coastline before it was completely cut off from the mainland. This is the same pattern recently identified for the nearby Carnarvon bioregion, containing the Barrow and Montebello island groups, where over 40,000 years of continuous reliance on marine resources has been identified (Manne and Veth 2015).

Rising seas would have first started to encroach on the outer islands of *Murujuga* around 9000 years ago (Fig. 24.2). The use of coastline resources is evident from *Terebralia* (mud whelk) middens on Rosemary Island (Bradshaw 1995) and Enderby Island (Ward et al. 2013). Rising sea-levels progressively separated bedrock hills (e.g., Enderby Island, Rosemary Island, Malus Island and West Intercourse Island) from the mainland (Semeniuk et al. 1982; Bird and Hallam 2006; Ward et al. 2013) and may have even been abandoned (McDonald 2015). Around 7000 cal BP, Dolphin and Dampier Islands still formed a peninsula but by 6000 cal BP the Archipelago took its present form (see Fig. 24.1).

The wider use of the *Murujuga* landscape-seascape from the time before sea level rise through to islandisation is mirrored in the changing themes of the engraved rock art (petroglyphs) of the archipelago (McDonald and Veth 2009; Mulvaney 2013, 2015; McDonald 2015). A shift is seen through time from a preponderance of terrestrial species, including some not found in the region today such as the Emu, to increasing proportions of marine and wetland resources. Late Pleistocene (c. 47,000–11,700 cal BP) rock art, when Murujuga was just a small rangeland in the wider Abydos Plain, is focused around terrestrial fauna, anthropomorphic figures and simple geometric elements (Table 24.1). The early transgressive rock art phase (c. 11,700–8000 cal BP) has pecked birds and macropods, dynamic grouped humans and geometric designs but importantly sees the introduction of marine subjects and distinctive anthropomorphic types, such as the *Murujuga* 'rainbow-man' (Table 24.1, McDonald 2015).

The mid-Holocene (8000–6500 cal BP), when *Murujuga* truly forms an archipelago, is characterised by the dominance of fish and other marine fauna, and anthropomorphs with distinctive local stylistic traits. It is at this time that the outer islands (e.g., Rosemary, Enderby) are thought to have been abandoned. This phase of abandonment of distant outer islands is matched by the termination dates from both the Barrow and Montebello island groups further west on the NW Shelf. The final mid–late Holocene (6400 cal BP to Modern) rock art found on the inner islands accessible from the mainland is argued to represent a distinctive *Murujuga* style with distinctive maritime themes,

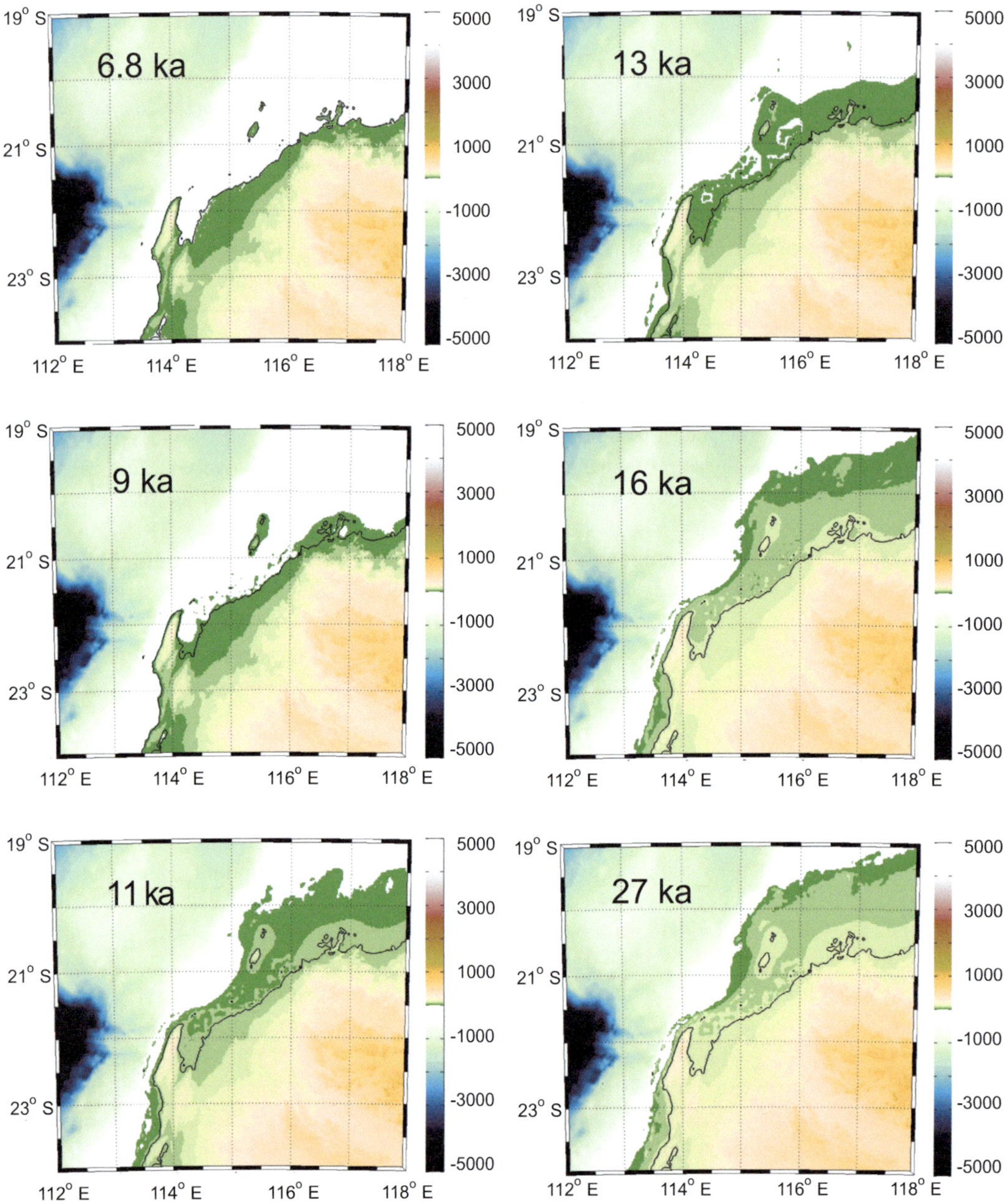


**Fig. 24.2** Modelled sea-level rise on the North West Shelf for key periods (Modelling after Jaqueline Austermann, Harvard University, Dec 5 2014, Fleming and Lambeck 2004; Peltier 2004; Mitrovica and Milne 2003; Kendall et al. 2005); D'Alpoim Guedes et al. 2016)

including large stylised water birds, more complex turtle compositions and crustaceans. Anthropomorphs show group behaviours and locally-distinctive headdresses and material culture combinations likely signalling the establishment of new territories after sea-level stabilisation (Table 24.1).

There is limited evidence that at some time during the most recent period (4000 cal BP to Modern) people from the coastal Pilbara region voyaged to some of the outer islands through the development

**Table 24.1** Landscape, sea level and climate, art phases and archaeological correlates

| Period age (ka BP) | Landscape and climatic context | Archaeological context | Mulvaney"s rock art phases |
|---|---|---|---|
| Late Pleistocene 47.0–22.0 | Abydos Plain: coast 110 km away Increasingly arid Weak monsoon | Early colonisation and establishment of regional broad-based economy. Art shows broad regional connections and long distance chains of stylistic connection | Phase 1: Regional graphics include archaic faces, elaborate geometric and anthropomorphic figures; unique forms disarticulated blob-heads |
| Late Pleistocene 22.0–18.0 | Abydos Plain: coast 160 km away Maximum aridity | Peak LGM occupational hiatus in many sites. Art demonstrates regional social connections between Dampier Range refugia and many major rock art complexes through the Pilbara and into the Western Desert | Phase 2: Regional style: Outline, large terrestrial faunal and anthropomorphic figures, simple geometric elements |
| Late Pleistocene 18.0–11.7 | Marine transgressive: Rapid sea-level rise – coast within 30 km, reintroduction of summer monsoon | Small population groups, high residential mobility; social pressure through territorial retraction. Art used to establish territoriality but distribution extends beyond the boundaries of the current Archipelago | Phase 3a: 1st distinct Murujuga style: Outline, solid, internal patterned terrestrial faunal and anthropomorphic figures, simple geometric elements |
| Pleistocene – Holocene transition 11.7–8.0 | Rapid sea-level rise – coast reaches outer 'islands'. Increasingly humid e return of monsoon | Larger population groups with decreasing residential mobility as territorial pressures increase and coastal resources are proximal and become more reliable: art and stone arrangements are used to assert territoriality. Art switches to marine focus | Phase 3b: Pecked intaglio stylised birds and macropods, dynamic grouped humans, simple and linear geometric designs |
| Mid Holocene 8.0–6.5 | Wetter: continuing sea-level rise results in formation of the Archipelago. Outer islands become separated; Great Mangrove forest is primary resource focus | Increasingly coastal focus, with decreasing residential mobility. Distinctive local signaling in the art repertoire from increased territorial pressure. Stone structures delineate space on larger habitation sites and modification of landscape is widespread. Outer islands may have become inaccessible | Phase 4: Murujuga style but with coastal connections: Outline and solid fauna and geometric elements; anthropomorphs have distinctive local stylistic traits, marine faunas begin to dominate |
| Mid-late Holocene 6.5–4.0 | Semi-arid with monsoonal influence. Sea-level highstand, results in reduction of mangal forests | Marine A – Predominant use of marine and intertidal (mangrove) resources: higher sea level creates an environmental change with apparent dire consequence for mangrove forests. Outer islands abandoned | Phase 5a: Distinctive Murujuga style: Outline and internal design marine and terrestrial fauna, anthropomorphs showgroup behaviours and distinctive headdresses and material culture |
| Late Holocene 4.0–0 | Modern island configuration. The last 1500 years represents a stable sea-level at current height | Marine B – Predominant use of marine resources with switch to sand flat and rocky resources after highstand. Increased intensity of site occupation (e.g. large shell mound building) and accelerated ritual and ceremonial cycles | Phase 5b: Murujuga style: Outline, internal and solid design marine-dominated fauna with increasing schematisation; human figures have exaggerated anatomical features and different ceremonial paraphernalia to preceding phases |

From McDonald (2015)

and use of log rafts, and an intimate knowledge of tides and currents (King 1827; Smyth 1997, 2007; Green 1998). Indeed, the different site patterning observed in the Holocene art repertoire across the Archipelago is argued to reflect the relative accessibility of the respective islands: by watercraft with respect to the outer and intermediate islands (e.g. East Lewis, Gidley and Malus); by swimming, to the intermediate islands; or by walking across low tidal flats, to Burrup Island and other proximal islands (e.g. Dolphin, West Intercourse).

Arguably *Murujuga* only represents Hayward's (2012) 'aquapelagic' society from the terminal Pleistocene when people began to make use of the coast and later as they navigated to the islands of the Archipelago, whilst remaining fundamentally connected to the mainland. However, there is increasing archaeological evidence that connection to Sea Country in the Pilbara was in place from as early as 42,000 cal BP (Veth et al. 2014), with artefacts fashioned from shellfish found up to 50 km inland. Regardless of whether they were utilitarian or dietary in nature, the inland presence of shellfish (including *Pinctada* sp., *Trochus* sp. and *Melo amphora*) indicates that people were both familiar with, and reliant on, coastal resources from a very early period (Veth et al. 2007, 2014, 2016, see also Ward et al. 2015, 2016).

### *24.2.2 Barrow Island: Montebello Island Complex*

Contrasting Dampier Archipelago's predominant volcanic geology, the offshore Barrow and Montebello Islands are comprised entirely of limestone (calcarenite), which has had a major influence on the preservation of certain landforms and zooarchaeological remains (Manne and Veth 2015; Ward et al. 2016). Barrow Island is the largest island on the inner shelf with nine smaller islands nearby (Fig. 24.3). Measuring some 20 km by 10 km in area, Barrow island is the second largest island off the entire Western Australian coastline. Lying 20 km north of Barrow Island, the Montebello Island group comprises 265 distinct, low lying islands and islets (Fig. 24.3).

Virtually all coastlines dating between about 130,000 and 15,000 cal BP now lie submerged and distant from the modern coast (Erlandson 2001), hence the record from Barrow and Montebello Island complex is particularly important because of its position on the continental shelf edge (Fig. 24.1). Moreover, the abandonment of these islands around 7500 cal BP, probably coinciding with the timing of insulation, provides a unique 'time capsule' of terminal Pleistocene behaviours without admixture from later Holocene occupation (Veth 1993, Veth et al. 2007, 2014, 2016). Indeed the high diversity of near-continental shelf rockshelters, islands and prograding shorelines in the wider region indicates at least 42,000 years of contact with the coast and a unique window into marine resource-use through long periods of glacio-eustatic and climate change (Veth et al. 2014, 2016).

Early occupation is registered at Noala Cave on the Montebello Island to 31,400 cal BP (based on an AMS date from *Melo amphora,* Veth et al. 2014). At this time, sea-level would have been ~80 m below present and the then limestone plateaux would have been a topographic highpoint in an an ecologically-rich and diverse coastal sand plain, as represented in the zooarchaeological record by burrowing bettong (*Bettongia lesueur*), spectacled hare wallaby (*Lagorchestes hirsutus*), western barred bandicoot (*Paremeles bougainville*), rock wallaby (*Petrogale* sp.), northern quoll (*Dasyurus hallucatus*), and Olive Python (*Liasis olivaceus*), a species that frequents permanent springs in rocky outcrops (Manne and Veth 2015).

After this early human presence, there is a hiatus until 14,500 cal BP, after which time episodic occupation is registered at Noala Cave up to 8700 cal BP and at nearby Haynes Cave until 8300 cal BP just prior to the insulation and inundation of the Mary Ann Passage (Veth et al. 2007; Manne and Veth 2015). Like the Dampier Archipelago, there is a clear shift in the representative faunal assemblages over the Pleistocene–Holocene transition, as distance to the coast decreased. Between ca. 11,500 and 8300 cal BP, there is a clear increase in exploitation of marine resources—particularly the

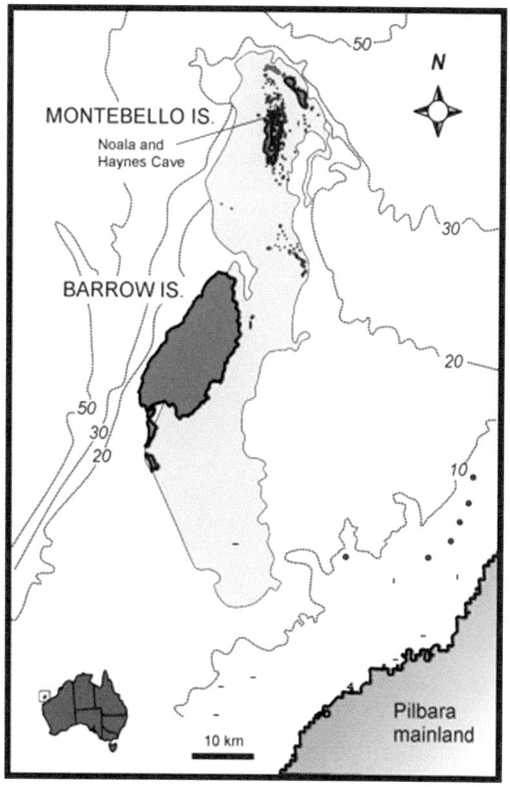


**Fig. 24.3** Map with bathymetric data for the Barrow-Montebello Islands, as well as the locations of Noala, Haynes and Boodie Caves (Sourced from Veth et al. 2007, 2014)

mangrove mudwhelk *Terebralia* sp.— in the Montebello Island record as the sea approached its current position (Veth et al. 2014; Manne and Veth 2015). At nearby Barrow Island, the archaeological record indicates that *Terebralia* sp. was consumed from as early as 17,000 cal BP—if not earlier (see Veth et al. 2016), with continuous exploitation of the transgressive coastal resources for both technological (*Melo* sp.) and economic purposes for a further 9000 years until it too was abandoned c. 7400 cal BP (Veth et al. 2014, 2016).

Not only do the archaeological records provide evidence of movement of people outward to the contemporaneous coastline, but the flaked and ground stone record also shows movement of people landward. Preliminary analysis of the lithic assemblages on Barrow Island indicates that artefacts are manufactured from both local limestone and also a significant proportion of exotic materials including quartz, volcanic and metasedimentary rocks (Veth et al. 2007; Basgall et al. 2014; Zeanah et al. 2014). The sources of these materials are found inland of the current Pilbara coastline and may thus have been traded over larger distances (Veth et al. 2007), and likely even from the Dampier Archipelago.

Lithic records show two distinct systems of lithic circulation and possibly two periods of occupation corresponding to a period when the island was part of the extensive NW Shelf, and a subsequent period when it was increasingly insulated. The model entertains two distinct stone transport systems (Fig. 24.4). A working hypothesis sees these distributions reflecting riverine sources that may have once run south and north of Barrow Island, with Ashburton and Cane Rivers in the south predominantly draining through metamorphic and sedimentary bedrock geology, and the Fortescue and Robe Rivers to the north draining through igneous bedrock geology. While there is no doubt that palaeoriver channels and coastal estuaries would have been a key resource focus for any past occupants, there is a clear need to research the past morphological extension of fluvial and estuarine systems across the shelf to be fully confident of any interpretation of past resource use (see Ward et al. 2015).

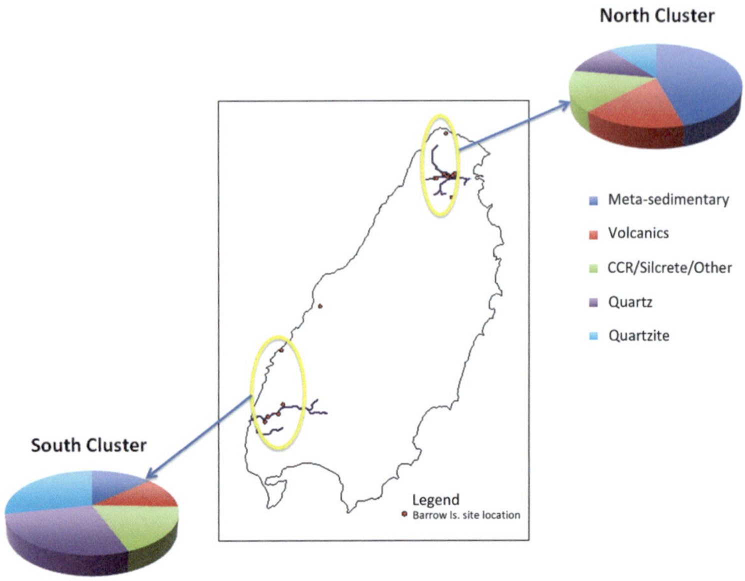


**Fig. 24.4** Pie charts showing different proportions of rock types within the lithic assemblages on different parts of Barrow Island, based on preliminary data (Reproduced with permission from Mark Basgall, David Zeanah and David Glover)

### *24.2.3 Buccaneer Archipelago*

The Buccaneer Archipelago occurs further north in the Kimberley Marine bioregion (KIM) in remote northwestern Australia (Fig. 24.1). The rocky sandstone islands of the Buccaneer Archipelago number over 800 and cover over 50 km$^2$ between King Sound and Camden Sound (Fig. 24.1). The largest of these, Koolan Island (22 km$^2$), is no more than 1 km from the mainland, with Round Island providing a stepping stone to the mainland.

Archaeological surveys on Koolan Island provide some of the earliest evidence for the use of marine shellfish in northern Australia, with dates from a mangrove dwelling bivalve (*Geloina coaxans*) of 27,300 ± 1100 cal BP (O'Connor 1999). This age estimate demonstrates the existence of developed mangrove communities, the shellfish of which were then transported considerable distances. At this time, sea level would have been about 125 m below present and over 150 km further northwest. However, King Sound has one of the largest tidal ranges (maximum tidal range 11.5 m) of any tide-dominated delta in the world (Brocx and Semeniuk 2011) and the tidal range may have been even larger during the LGM (Ward et al. 2013). Hence it is highly likely that complex intertidal facies were available for exploitation during the period leading into the LGM. Even now, many traditional marine activities in NW Australia revolve around the huge tidal range and gently sloping seabed, which results in vast areas of intertidal land and reef flats available for exploitation (Smyth 2007, p. 15). Thus coastlines with large tidal ranges are likely to have offered very attractive resources, and to have been intensively used throughout the Late Pleistocene and Holocene.

However, contrary to the implied expectation of continued coastal use, there is an apparent occupational hiatus between 24,600 cal BP and 10,900 cal BP in the SW Kimberley (O'Connor 1999). Whether this represents cultural abandonment or sedimentary stasis and differential preservation of archaeological sites along this tide-dominated shelf is not certain (O'Connor 1999, see also Ward and Larcombe 2003) although a change in sediment composition between the Late Pleistocene and mid-Holocene layer (shell midden) implies the latter. If people followed the coastline as it retreated and transgressed back again (see also Bowdler 1977) then a large part of the Pleistocene evidence for occupation may either be destroyed or lie buried and submerged on the drowned shelf. It should be noted that the date of 27,300 cal BP at Koolan shelter is not a basal date and depth-age curves indicate that it may have been in use 40,000 years ago, when the sea was again close to the site (O'Connor 1999).

King Sound (Fig. 24.1) itself formed when rising sea levels flooded the coastline at the end of the LGM around 12,000 years ago (Semeniuk 1980) and the sea reached the northern coastline of Koolan about 1000 years later, which is when the records of coastal occupation—and, importantly, net sediment accumulation—are re-established (O'Connor 1999). In the Pacific, many islands became inhabitable only when their coastal plains emerged sufficiently for them not to be inundated at high tide, something called the 'crossover point' (Nunn 2009). The attainment of the crossover point is coincident with the earliest-known date for the human settlement of many such islands (Dickinson 2003), and perhaps by inference, the re-establishment of coastal occupation along many parts of the mainland following the post-glacial transgression.

Other islands, including the outermost islands of the Archipelago, have not been systematically excavated. Hence there is both a temporal and spatial gap in the record for the Buccaneer Archipelago that, if filled, may provide new insights in regard to maritime adaptation associated with changing sea-level and changing tidal regimes (e.g. Fa 2008). Indeed a more precise understanding of past tidal regimes and past intertidal environments that may have been used by past occupants is necessary along all parts of this tide-driven continental shelf (see also Ward et al. 2013).

## 24.3 Continuity Versus Contiguity

Hayward's (2012) emphasis is on the archipelago as an entity constituted by human presence whereby the utilisation of the environment matters more than the islands as a geographical entity. Yet they are not mutually exclusive because archipelagos, to paraphrase Hayward (2012, p. 6, added emphasis) '*wax and wane as climate patterns* [and hence sea-level] *alter and as human socio-economic organisations and technologies, and/or the resources they rely on change and develop in these* [changing coastal] *contexts*'. The examples we have profiled briefly above illustrate that coastlines, coastal ecosystems and coastal occupation records are dynamic (see also Manne and Veth 2015; Ward et al. 2015, 2016).

This leads to the question of continuity and contiguity in the 'aquapelagical' record. Continuity relies heavily on the resolution of the available archaeological, environmental and chronological data (Stein 1993). Natural and cultural processes are necessarily coupled and operate irregularly over space and time, hence no single island or site can be expected to provide a continuous record of occupation. Rather these individual island records need to be linked with records from adjacent contexts both above and below water. It is this interdisciplinary 'contiguous landscape approach' between the terrestrial and marine, island and mainland, past and present that provides the best opportunity to resolve a significant missing element within Australia's ancient and rich archaeological heritage (McDonald and Veth 2009).

This *contiguous* approach can be well demonstrated for the Barrow Island and Montebello Island complex. Fig. 24.5 shows the pooled archaeological radiocarbon data for the wider northern Carnarvon bioregion lying to the west and south of the Pilbara island chains. It illustrates the break in radiocarbon dates from the island records precisely when they cut off from the mainland, with post-insulation dates being picked up immediately from the coastal mainland records at nearby Onslow and Cape Range. Here steep offshore shelves and prograding shoelines have preserved the earliest post-transgression records of marine resource use. The inference is that maritime people retreated to the hinterland as transgressive seas cut off the islands but maintained their reliance on coastal maritime resources. The presence of exotic materials in the lithic records indicates ongoing links with the mainland, either through long-distance movements of groups or exchange networks, even before the islands were abandoned (see also Veth et al. 2014). This is the 'coastal coalescence' described by Veth et al.

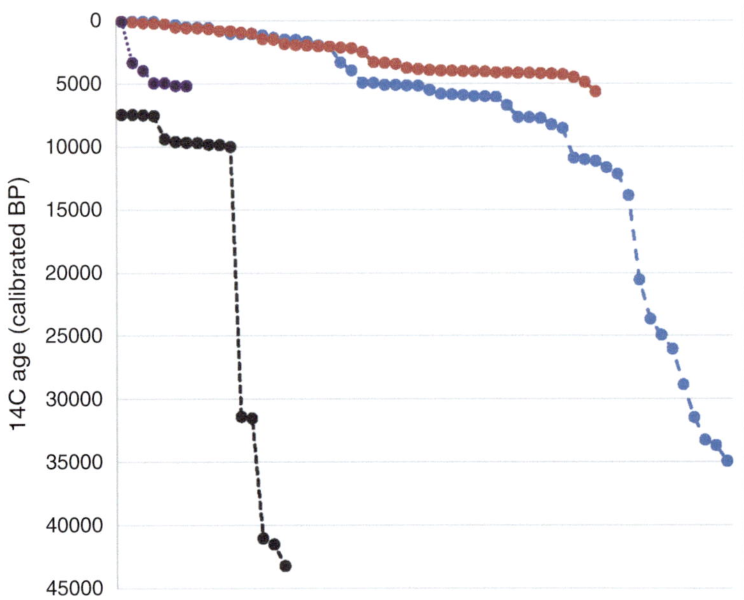


**Fig. 24.5** Pooled archaeological radiocarbon frequency data for the Carnarvon bioregion (current for 2014), which show a 'switching off' of the islands (*black line*) to Holocene mainland records (*purple and red lines*). NW Cape's steep offshore profile registers continuous occupation. Dates are calibrated (at 95%) using Oxcal v. 4.2, using the Marine Calibration, with a regional offset (delta R) of 54 ± 30 based on Squire et al. (2013)

(2014) in which Aboriginal people developed 'hybrid' maritime-desert societies focused both on the plentiful littoral zone and more dispersed arid interior resources.

Hence archipelagos and island complexes provide an insight into the spatio-temporal mosaic that is presented as the residue of extinct patterns and processes (McGlade and van der Leeuw 2013, p. 6). At the macro-scale, these are the effects of addition/loss of accessible lands such as LGM refugia and archipelagos or islands, and at the micro-scale, they are the effects of variable survival and exposure of occupation within any geomorphic unit. As such, any extrapolation between present-day terrestrial and maritime archaeological landscapes, particularly in shallow shelf areas, must be based on sound geological and geomorphological principles. Wherever possible, exhaustive and precise chronologies must be obtained across multiple sample points (Bates et al. 2007; Ward et al. 2015, 2016). In offering multiple sample points in the form of relict shelf islands (archipelagos), nearshore uplifted coral terraces with steeply shelving profiles and procumbent prograding shorelines, a unique window into past marine cultural and resource configurations has been uncovered from the Carnarvon bioregion and is also likely to be found in the Kimberley bioregion.

## 24.4 Conclusion: Drowned Cultural Landscapes and Human Dispersals

Hayward (2012) indicated that the study of cultural seascapes should not stop at its surface but should delve deeper. While he was mainly refering to aquaculture and fisheries in a modern context, we argue that any study of 'aquapelagic societies' would be incomplete without considering the submerged landscapes of which these archipelagos and island complexes were once a part (see also Erlandson and Fitzpatrick 2006). At present our understanding of the genesis of maritime cultures is inevitably based on a truncated record in which the most significant part (the drowned landscape) is usually missing (Ward et al. 2013, 2015). Whilst underwater exploration for traces of Indigenous societies on the now-submerged shelf is still nascent in Australia, it is unquestionably necessary to fully understand the role of oceans, islands and coastlines in human history.

The islands and archipelagos of NW Australia are geologically, ecologically and archaeologically unique, linked together by their common development through postglacial sea-level rise. There is no question that, at times of lower relative sea level, the extended coastal shelf of NW Australia was occupied by Aboriginal groups relying on both marine and land-based resources (McDonald 2011, p. 12; Veth et al. 2014, 2016). The archaeological records from the archipelagos of NW Australia are testament to this. To this extent islands are not necessarily entitities with 'defined geographical parameters' (Bowdler 1995, p. 945) but rather entities where our understanding of their past is limited by our lack of knowledge of their former geographic extent. The present and past coast and islands all in effect constitute Sea Country.

Yet until there is systematic study of drowned cultural landscapes, these islands and archipelagoes provide the only real link with the submerged context, and the only means of inferring, indirectly, the nature of that submerged landscape, for which we only have limited geological and chronological data, at present, and still no cultural records. What is missing is an archaeology of the *seabed* to match that of the land (following Dawson 2012), i.e., a systematic exploration of the now-submerged landscapes occupied by anatomically modern humans for the first 45,000 years of the occupation of Sahul. In short, the islands and archipelagos of NW Australia are not simply isolates of land surrounded by a mass of sea, but rather are all part of a complex tapestry with a long-standing connection to both the mainland and the seas that connect them.

The archaeological record for near-continuous use of maritime resources by peoples of NW Australia from at least 42,000 cal BP may be ascribed to highly adaptable foragers who comfortably switched between productive coastal, sub-coastal and often more resource-patchy interior habitats. The settling of a vast and relatively arid continent by people armed with advanced marine skills acquired from the increasingly poor terrestrial returns of the Wallacean islands (O'Connor and Veth 2000) sets the conditions for both early coastal resource exploitation and penetration deep into the continent. These early dates for use of dietary marine molluscs are the oldest for AMHs (anatomically modern humans) outside of Africa and are seen to be one of the signatures for the 'super-charged' dispersal of modern humans along the southern coastal route. They also raise the spectre of coastal colonisation models again (e.g., Bowdler 1977), as these were previously found wanting due to older dates being returned from interior sites (Veth et al. 2007). As coastal chronologies match those of the interior, and which now appear to contain some dietary molluscan assemblages (in addition to utilitarian and ornamental uses), the centrality of marine competencies in the dispersal of humans through southern and South East Asia must again be critically re-examined.

**Acknowledgements** The authors wish to thank Jaqueline Austermann, Harvard University, for sea-level models of the NW Shelf used to create Fig. 24.2, Mark Basgall, David Zeanah and David Glover, California State University, Sacramento, for supplying Fig. 24.3, and Fiona Hook, Archae-aus Pty Ltd. for radiocarbon data used to create Fig. 24.5. The authors also thank Sean Ulm for his positive review of the manuscript.

## References

Basgall ME, Zeanah DW, Glover DW (2014) Technological organisation of artefact surface scatters on Barrow Island: Barrow Island Archaeology Project (2013–2015) Western Australia. In: Paper presented to the 79th Annual Meeting of the Society for American Archaeology Austin, TX

Bates MR, Bates RC, Briant RM (2007) Bridging the gap: a terrestrial view of shallow marine sequences and the importance of the transition zone. J Archaeol Sci 34:1537–1551

Benjamin J, Bonsall C, Pickard C, Fischer A (eds) (2011) Submerged prehistory. Oxbow, Oxford
Bird C, Hallam SJ (2006) A review of archaeology and rock art in the Dampier Archipelago. A report prepared for the National Trust of Australia (WA)
Bowdler S (1977) The coastal colonisation of Australia. In: Allen J, Golson J, Jones R (eds) Sunda and Sahul: prehistoric studies in Southeast Asia, Melanesia and Australia. Academic Press, London, pp 205–246
Bowdler S (1995) Offshore islands and maritime explorations in Australian prehistory. Antiquity 69:945–958
Bradshaw E (1995) Dates from the archaeological excavations on the Pilbara coastline and islands of the Dampier archipelago, Western Australia. Aust Archaeol 41:37–38
Brocx M, Semeniuk V (2011) The global geoheritage significance of the Kimberley coast, Western Australia. J R Soc West Aust 94(2):57–88
D'Alpoim Guedes J, Austermann J, Mitrovica J (2016) Lost foraging opportunities for East Asian hunter-gatherers due to rising sea level since the last glacial maximum. Geoarchaeology 31(4):255–266
Dawson H (2012) Archaeology, aquapelagos and island studies. Shima: Int J Res Res Island Cult 6(1):17–21
Dickinson WR (2003) Impact of mid-Holocene hydro-isostatic highstand in regional sea level on habitability of islands in Pacific Oceania. J Coast Res 19:489–502
Donaldson M (2011) Understanding the rocks: rock art and the geology of Murujuga (Burrup peninsula). Rock Art Res 28(1):35–43
Dortch C (2002) Preliminary underwater survey for rock engravings and other sea floor sites in the Dampier archipelago, Pilbara region, Western Australia. Aust Archaeol 54:37–42
Erlandson JM (2001) The archaeology of aquatic adaptations: paradigms for a new millennium. J Archaeol Res 9(4):287–350
Erlandson JM, Fitzpatrick SM (2006) Oceans, islands, and coasts: current perspectives on the role of the sea in human prehistory. J Island Coast Archaeol 1:5–32
Fa DA (2008) Effects of tidal amplitude on intertidal resource availability and dispersal pressure in prehistoric human coastal populations: the Mediterranean–Atlantic transition. Quat Sci Rev 27:2194–2209
Fleming K, Lambeck K (2004) Constraints on the Greenland ice sheet since the last glacial maximum from sea-level observations and glacial-rebound models. Quat Sci Rev 23:1053–1077
Green N (1998) Aboriginal affiliations with the sea in Western Australia. In: Gray F, Zann L (eds) Traditional knowledge of the marine environment in northern Australia. Proceedings of a workshop held in Townsville, Australia. Workshop Series No. 8, pp 19–29
Hayward P (2012) Aquapelagos and aquapelagic assemblages. Shima: Int J Res Res Island Cult 6(1):1–10
Jones HA (1973) Marine geology of the northwest Australian continental shelf. Bur Miner Resour Geol Geophys Bull 136:102
Kendall RA, Mitrovica JX, Milne GA (2005) On postglacial sea level: II. Numerical formulation and comparative results on spherically symmetric models. Int Geophys J 161:679–706
King P (1827) Narrative of a survey of the inter-tropical and western coasts of Australia, performed between the years 1818 and 1822 (2 volumes). John Murray, London. Facsimile reprint-Adelaide: Libraries Board of South Australia, 1969
Kojan CJ (1994) The geology and mineral resources of the proposed Dampier Archipelago National Park. Record 1994/2. Geological Survey of Western Australia, Dept. Minerals and Energy, p 43
Manne T, Veth P (2015) Late Pleistocene and early holocene exploitation of estuarine communities in northwestern Australia. Quat Int 385:112–123. doi:10.1016/j.quaint.2014.12.049
McDonald J (2011) Study of the outstanding universal values of the Dampier Archipelago Site, Western Australia. Report to the Australian Heritage Council. Available at: http://www.aasdenvironment.gov.au/heritage/ahc/national-assessments/dampier-archipelago/pubs/outstanding-universal-values.pdf. Accessed 12 Apr 2012
McDonald J (2015) I must go down to the seas again: or, what happens when the sea comes to you? Murujuga rock art as an environmental indicator for Australia's north-west. Quat Int 385:124–135. doi:10.1016/j.quaint.2014.10.056
McDonald J, Veth P (2009) Dampier archipelago petroglphys: archaeology, scientific values and national heritage listing. Archaeol Ocean 2009(Supplement):49–69
McGlade J, van der Leeuw SE (2013) Time, process and structured transformation in archaeology. Routledge, London
Mitrovica JX, Milne GA (2003) On postglacial sea level: I general theory. Geophys J Int 154(2):253–267
Morse K (1988) Mandu Mandu creek rockshelter: pleistocene human coastal occupation of north west cape, Western Australia. Archaeol Ocean 23:81–88
Morse K (1993) New radiocarbon dates from north west cape, Western Australia: a preliminary report. In: Smith MA, Spriggs M, Fankhauser B (eds) Sahul in review: Pleistocene archaeology in Australia, new Guinea and island Melanesia, Occasional papers in prehistory, vol 24. Department of Prehistory, Research School of Pacific Studies, ANU, Canberra, pp 155–163
Mulvaney K (2013) Iconic imagery: pleistocene rock-art development across northern Australia. Quat Int 285:99–111
Mulvaney K (2015) Murujuga Marni: rock art of the macropod hunters and mollusc harvesters. University of Western Australia Press, Crawley

Nunn PD (2009) Vanished islands and hidden continents of the Pacific. University of Hawai'i Press, Honolulu
O'Connor S (1999) 30,000 years of aboriginal occupation: Kimberly, North-West Australia. Canberra: Department of Archaeology and Natural History and Centre for Archaeological Research, Australian National University. Terra Australis 14
O'Connor, S, Veth P (2000) The world's first mariners: savanna dwellers in an island continent. In: O'Connor S, Veth P (eds) East of Wallace's Line: studies of past and present maritime societies in the Indo-Pacific region. Modern Quaternary Research in South East Asia 16:V–VI. A.A. Balkema, Rotterdam. pp 99–137
Peltier WR (2004) Global glacial isostasy and the surface of the ice-age earth: the ICE-5G (VM2) model and GRACE. Annu Rev Earth Planet Sci 32:111–149
Przywolnik K (2002) Patterns of occupation in Cape Range Peninsula (WA) over the last 36,000 Years. Unpublished PhD thesis, Department of Archaeology, University of Western Australia, Nedlands
Przywolnik K (2005) Long-term transitions in hunter-gatherers of coastal northwestern Australia. In: Veth P, Smith M, Hiscock P (eds) Desert peoples: archaeological perspectives. Blackwell, Oxford, pp 177–205
Semeniuk V (1980) Quaternary stratigraphy of the tidal flats, king sound, Western Australia. J R Soc West Aust 63:65–78
Semeniuk V, Wurm PAS (1987) The mangroves of the Dampier archipelago. J R Soc West Aust 69(2):29–87
Semeniuk V, Chalmer PN, Le Provost I (1982) The marine environments of the Dampier archipelago. J R Soc West Aust 65(3):97–114
Smyth D (1997) Saltwater country aboriginal and Torres Strait islander interest in ocean policy development and implementation. Socio-cultural considerations, Australia's ocean policy issues paper 6. Environment Australia, Canberra
Smyth D (2007) Sea countries of the North-West. Literature review on Indigenous connection to and uses of the North West Marine Region. Report for Australian Government Department of Environment and Water Resources. Available at http://www.environment.gov.au/resource/sea-countries-north-west. Accessed 18 Sept 2014
Squire P, Joannes-Boyau R, Scheffers AM, Nothdurft LD, Hua Q, Collins LB, Scheffers SR, Zhao J-X (2013) A marine reservoir correction for the Houtman-Abrolhos archipelago, East Indian Ocean, Western Australia. Radiocarbon 55(1):103–114
Stein JK (1993) Scale in archaeology, geosciences and geoarchaeology. Geol Soc Am Spec Pap 283:1–10
Veth P (1993) The aboriginal occupation of the Montebello Islands, Northwest Australia. Aust Aborig Stud 2:39–50
Veth P, O'Connor S (2013) The past 50,000 years: an archaeological view. In: Bashford A, Macintyre S (eds) The Cambridge history of Australia, vol 1. Cambridge University Press, Melbourne, pp 17–42
Veth P, Aplin K, Wallis L, Manne T, Pulsford T, Chappell A (2007) Late quaternary foragers on arid coastlines: archaeology of the Montebello Islands, Northwest Australia, British archaeological reports international series S1668. Archaeopress, Oxford
Veth P, Ditchfield K, Hook F (2014) Maritime deserts of the Australian northwest. Aust Archaeol 79:156–166
Veth P, Ward I, Manne T (2016) Coastal feasts: a pleistocene antiquity for resource abundance in the maritime deserts of north West Australia? J Island Coast Archaeol. doi:10.1080/15564894.2015.1132799
Vinnicombe P (1987) Salvage archaeology of the Burrup peninsula. Aust Archaeol 25:53–79
Ward I, Larcombe P (2003) A process-orientated approach to archaeological site formation: application to semi-arid northern Australia. J Archaeol Sci 30:1223–1236
Ward I, Larcombe P, Mulvaney K, Fandry CA (2013) The potential for discovery of new submerged archaeological sites near the Dampier archipelago, Western Australia. Quat Int 308–309:216–229
Ward I, Larcombe P, Veth P (2015) A new model for coastal resource productivity and sea-level change: the role of physical sedimentary processes in assessing the archaeological potential of submerged landscapes from the northwest Australian continental shelf. Geoarchaeology 30:19–31
Ward I, Veth P, Manne T (2016) To the islands born: unique records of past landscapes and human habitation from the islands of northwestern Australia. In: Harff J, Bailey G, Lüth F (eds) Geology and archaeology: submerged landscapes of the continental shelf. Geological Society, London, Special Publication, 411, pp 251–263. First published online September 19, 2014, updated October 12, 2015. http://dx.doi.org/10.1144/SP411.4
Zeanah DW, Basgall ME, Seah I (2014) Lithic landscape of Barrow Island: a view from the open scatters: Barrow Island Archaeology Project (2013–2015) Western Australia. Paper presented to the 79th Annual Meeting of the Society for American Archaeology Austin, TX

# Part V
# Outreach and Management

# Chapter 25
# Education and Engagement: Developing Understanding and Appreciation of Submerged Prehistoric Landscapes

**Julie Satchell**

**Abstract** The position of submerged prehistoric landscapes beneath the water and often beneath the seabed makes them difficult to access. Scientific investigation and research are providing data on prehistoric peoples and their associated drowned landscapes which is important for illuminating aspects of the past. Development of academic study and management approaches to the resource are vital for professional growth and appropriate protection. Alongside this is a large public fascination with past landscape change which provides an important route for engaging people with this otherwise hidden aspect of their past. Prior to the SPLASHCOS project the approach to education and outreach relating to submerged prehistoric landscapes across Europe had been patchy and inconsistent. This paper explores examples from the Maritime Archaeology Trust (MAT) of how education and outreach can promote this understudied area of the historic environment. It then details initiatives of the SPLASHCOS project that provided learning opportunities, field schools, engagement with industry and wider management aspects. Experience from this work is drawn upon to consider future challenges for expanding understanding and appreciation of submerged prehistoric landscapes.

## 25.1 Introduction

The understanding and appreciation of submerged prehistoric landscapes (SPLs) is currently at a very early stage, whether within the professional community of archaeologists and marine scientists, or more widely amongst marine managers, heritage professionals, policy makers, politicians and the general public. The fact that SPLs are not only underwater but also often further buried under seabed sediments presents a number of obstacles to ease of access and understanding. The 'remoteness' of SPLs, both in terms of the period of human history they refer to and their physically inaccessible location, provides particular challenges for engaging a wider audience.

With a limited number of preserved and stratified sites so far located and relatively few organisations undertaking any physically intrusive investigation of sites and landscape deposits, there are limited examples of well-studied and conserved material available for access, research and wider dissemination through archives, museums and other media. Moreover, the research itself requires specialist equipment for survey, archaeologists with professional diving qualifications, and complex logistics. The low number of experienced practitioners currently active in the field highlights the relative immaturity of the discipline.

However, the location and investigation of submerged archaeological sites and landscapes is ongoing and expanding, as is clear from other chapters in this volume, and these are revealing the huge

J. Satchell (✉)
Maritime Archaeology Trust, National Oceanography Centre, Empress Dock, Southampton SO14 3ZH, UK
e-mail: julie.satchell@maritimearchaeologytrust.org



G.N. Bailey et al. (eds.), *Under the Sea: Archaeology and Palaeolandscapes of the Continental Shelf*, Coastal Research Library 20, DOI 10.1007/978-3-319-53160-1_25

potential of the underwater environment to add to the understanding of early human history in a number of key areas, including sites with unusual conditions of preservation and from very early periods. This as yet largely unexplored frontier of archaeological research will certainly provide significant discoveries in coming years and it is only through active efforts to promote increased education, engagement and incorporation within the wider public consciousness that this information will become available for all, and thus reinforce wider public and governmental support for ongoing research and conservation.

The aim of this chapter is to examine the key issues associated with communicating the significance of SPLs to a wider public and the different types of audience that need to be addressed, to describe the activities of the Maritime Archaeology Trust as an example of developing education and outreach programmes, and to summarise the activities developed as part of the SPLASHCOS Action, which included a Working Group specifically dedicated to communication, outreach, training and collaboration with industry.

### *25.1.1 Challenges and Benefits*

The need to understand SPL deposits, and the need to communicate that understanding to a wider audience, is acute due to a range of humanly-induced threats through development and commercial use of seabed deposits and to natural threats from erosion, hydrodynamics and geomorphological processes. Without quality data on these deposits it is not possible to target investigations and manage the resource effectively.

In addition to the access issues of the underwater environment, there are other key obstacles to wider understanding. Concepts of large-scale landscape and environmental change can be difficult for the non-specialist to grasp, given the scale of the physical processes involved and the extent to which they have dramatically reshaped landscapes which now bear no resemblance to modern coastlines and international borders. An additional difficulty in conveying the nature and significance of the research to a wider audience is that the discipline is still developing and trying to understand where and how these deposits are preserved and how to locate them. Another barrier to engagement and understanding, although this applies equally on land and is not confined to underwater material, is that the most commonly available artefacts for study are stone tools. These artefacts often need additional interpretive materials demonstrating the specific application of different type of tools to allow their full significance to be appreciated.

On the positive side, underwater archaeological material and deposits are now becoming available for study in increasing numbers even if they are localised in nature and difficult to access. The preservation potential in the underwater environment can result in recovery of organic artefacts such as wooden structures, boats, fish traps, wooden implements, basketry, cordage and other organic materials, which would not normally survive on land-based archaeological sites, and which convey a more lively and detailed picture of past day-to-day life. Also, the rapidly changing evidence base does demonstrate 'science in action'; as often with archaeological research, the actual detective work involved in the process of discovery and reconstruction is a source of perennial fascination to the non-specialist, and this is especially the case with underwater research, providing many opportunities for engaging the interest of non-specialist audiences.

## 25.1.2 Key Messages and Audiences

Inevitably there is a need to reach a range of different audiences when attempting to increase understanding. This task is especially important to pursue alongside the actual research itself, because increased understanding amongst a wider academic and scientific community as well as the wider public will help to ensure that the underwater cultural heritage will be valued, cared for and properly managed. For the SPLASHCOS project, it was important to identify the most important messages to convey to various audiences to maximise the international impact across Europe and beyond.

For the scientific academic audience there is a need to develop and promote this growing area of study, and especially the need for inter-disciplinary skills and approaches to solve research problems and develop future strategies. There is huge potential for the re-use of data and resources that are held in public and private archives to help develop research. This can help demonstrate the relevance of the study of SPLs to other areas of science, for example in contributing to an understanding of climate change or changes to the maritime environment.

Moreover, re-use of surveys and investigations undertaken by marine industries can help engage the commercial and industrial audience. Collaborations between companies and research organisations can help unlock the potential within marine datasets for understanding the extent and preservation of SPLs. This has been ably illustrated through examples such as the North Sea Palaeolandscape Project, which used 3D seismic industry data to map the early Mesolithic landscapes (Gaffney et al. 2009; Gaffney et al., Chap. 20), and the development of Rotterdam's Maasvlakte 2 port extension which included extensive assessment and mitigation work on Mesolithic archaeology and associated landscapes (Moree and Sier 2015; Momber and Peeters, Chap. 21). In the longer term, this can reduce risks related to unexpected discoveries or impacts on prehistoric archaeology resulting from industrial and commercial projects in the marine zone. While there is EU legislation and guidance documents that dictate minimum best practice in relation to consideration of prehistoric remains during the course of industrial work on the seabed, fulfilling and surpassing these minima can provide positive public relations opportunities for the companies involved.

Reaching government agencies and policy makers is also vital to help develop management, protection and funding. Some key elements for achieving this include highlighting how SPLs contain unparalleled information on early human populations and environmental change that is not preserved on land and is currently not well studied or integrated within research and management programs. Developing understanding of past climate change, particularly rising sea levels and their effects on prehistoric territories and human populations, has direct relevance for issues faced by modern coastlines and communities. From a management point of view, increased investment in research can develop understanding of SPLs—their location, extent and significance—and this in its turn will feed into marine spatial planning and associated licensing and protection. This can be particularly relevant for commitments under the UNESCO Convention on the Protection of the Underwater Cultural Heritage (2001), the Valletta Convention (1992) and the Council of Europe's European Landscape Convention (2007).

Messages to the wider public need to focus on a landscape that is now submerged and that can preserve fascinating evidence about daily life. A key aim is to provide a basic understanding that underwater investigations are providing a new dimension to knowledge of early human populations impossible to gain from studying only terrestrial sites and landscapes. SPLs can also be a valuable tool to engage children with a wide range of cross-curricular subjects, promoting engagement from an early age.

To deliver these messages effectively, I draw on experience from the UK, in particular the work of the Maritime Archaeology Trust, where education and outreach has been a major priority in actively raising the profile of submerged prehistoric landscapes, and then examine some of the initiatives developed within the SPLASHCOS Action.

## 25.2 MAT Experience in the UK

Within the UK, the Maritime Archaeology Trust, or MAT (formerly the Hampshire & Wight Trust for Maritime Archaeology) has extensive experience with research-led investigation of SPLs. This includes work on the nearby underwater site of Bouldnor Cliff (Momber et al. 2011; Momber and Peeters, Chap. 21) involving excavation, data gathering, analysis and interpretation, as well as dissemination and promotion of the significance of SPLs to audiences ranging from school children to marine managers and planners. The work of the Trust extends to all aspects of the maritime heritage including shipwrecks and coastal and underwater installations from more recent periods, but has a particularly strong focus on prehistoric underwater material.

As an independent charitable Trust, the MAT operates to deliver its broad aim of 'promoting interest, research and knowledge of maritime archaeology'. Specific objectives related to education and outreach include:

- Promoting public awareness, enjoyment, education and participation in maritime archaeological heritage
- Ensuring that maritime archaeology plays an important role in coastal planning, management and policies.

These objectives are achieved through formal and informal initiatives for different sectors and audiences. The lessons learnt from these education and dissemination activities provide examples at a regional scale that could be expanded nationally and internationally to promote this understudied area of the historic environment.

### *25.2.1 Direct Hands-on 'Training in the Field': Professionals and Volunteers*

One of the most important ways of gaining an understanding of SPLs is to directly experience them. The MAT provides opportunities to participate in research projects for professional colleagues and volunteers from the scientific and research community as well as the general public. In addition to being open to those qualified to dive, the Trust offers a range of related tasks for non-divers. This inclusive approach aims to develop public archaeology in its broadest sense through creating access to materials and experience of methodologies and techniques.

The Solent has been recognised as being of particularly high potential for submerged prehistoric landscapes (Fulford et al. 1997; Momber 2000; Momber et al. 2011). One of the most significant areas is off the north-west coast of the Isle of Wight at Bouldnor Cliff, where stratified Mesolithic occupation remains have been located in 12 m depth of water (Fig. 25.1). The site provides a significant opportunity to study well preserved prehistoric material including a range of organic artefacts and palaeoenvironmental evidence. Not only is the site contributing unusual knowledge of the Mesolithic within the UK (Smith et al. 2015), it also provides a range of opportunities for training.

This expanding area of archaeological research and investigation is demanding the development of innovative approaches to underwater recording, sampling and recovery (Fig. 25.2). The project has also demonstrated the need for training of archaeologists, either those used to terrestrial prehistoric sites, or those more familiar with shipwreck remains. The Bouldnor Cliff project hosted two Short Term Scientific Mission (STSM) placements as part of the SPLASHCOS project, providing the opportunity for early stage researchers to experience all aspects of underwater investigation of SPLs (http://splashcos.org/training/2010_Southampton).

There is also an ongoing requirement for the processing of a large number of samples of seabed deposits from the Bouldnor Cliff site. Due to the difficulties of conducting underwater excavation in

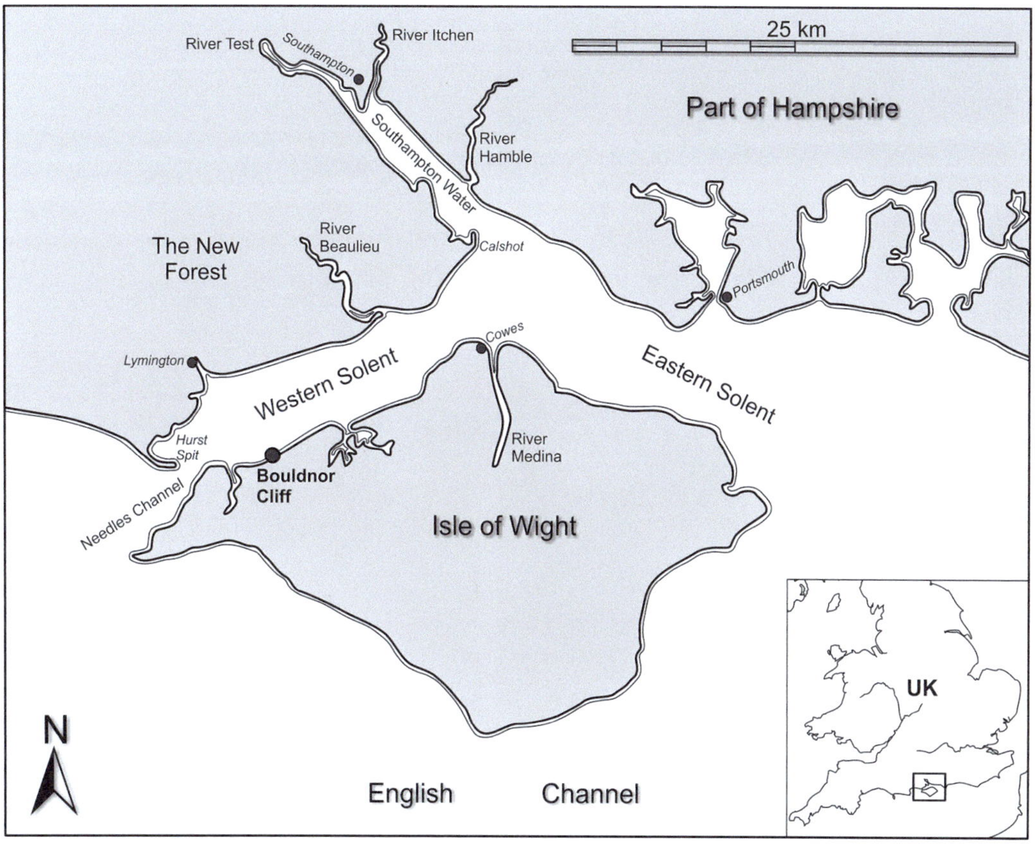


**Fig. 25.1** The location of the Bouldnor Cliff submerged Mesolithic site and landscape, off the North West coast of the Isle of Wight, UK (Maritime Archaeology Trust)

**Fig. 25.2** Diver records evidence of Mesolithic occupation at Bouldnor Cliff (Photo: Maritime Archaeology Trust)

situ because of the harsh tidal environment, the excavation strategy required the removal of blocks of sediment and their transportation back to dry land for detailed processing and the development of a sieving programme to maximise the recovery of data (Momber et al. 2011, pp. 24–33). Much of the data is in the form of palaeoenvironmental samples (e.g., bulk samples, cores, monoliths and timber

**Fig. 25.3** Volunteers of all ages have been involved with the recording and processing of material recovered from the submerged prehistoric landscape deposits from Bouldnor Cliff (Photos: Maritime Archaeology Trust)

samples), often with unusual conditions of preservation. Processing of such samples may be familiar to those used to working and volunteering on terrestrial sites, but they provide an excellent learning opportunity for non-specialists as well as for those more familiar with other types of maritime work, such as photography, survey or diving.

The examination, recording and recovery of material from these samples involved the training of non-diving volunteers and students who were keen to gain experience (Fig. 25.3). Although recognising the need for supervision and guidance while carrying out processing of samples, those involved have relished 'hands-on' work with prehistoric material.

The educational potential of this project was recognised from the outset, and has achieved positive results simply through the significant numbers who have participated. Those involved have gained practical experience in a range of activities:

- Underwater recognition of prehistoric landscape deposits, their stratigraphy and the potential of their associated palaeoenvironmental evidence
- Use of recording techniques such as photography on submerged prehistoric remains
- Adapting and developing survey and sampling approaches to a new environment
- Recognition of archaeological material including worked and burnt flints, worked wood, and food remains such as hazelnuts

The development of interest in SPLs is not confined to the archaeological field. There is huge public interest in this subject, which combines the study of early human populations with 'drowned' landscapes. This is further fuelled by challenges posed by modern climate change and sea level rise. It has been possible to link evidence from Bouldnor Cliff with these themes in interpretations and exhibitions and through this to attract wide interest in this area of marine archaeology.

### 25.2.2 *Formal and Informal Education and Learning: Children and the General Public*

Embracing the potential to inspire children and young adults with SPLs through formal and informal learning opportunities has resulted in the development of a range of education products. As archaeology is seldom taught in UK schools as a separate discipline, we have found it necessary to raise the profile of the subject through a number of innovative schemes in order to enhance its accessibility and appeal. We have introduced maritime archaeology and SPLs to junior-age children in order to encourage them to explore the subject further both outside a formal teaching environment and into higher education. At a basic level, simply providing children with an appreciation of this under-studied aspect of their cultural heritage imparts knowledge that is likely to stay with them throughout life.

Archaeology and the study of SPLs is a highly cross-curricular subject and is relevant to history, geography and sciences in addition to contributing to literacy and enjoyment of language. Teaching resources that we have provided for schools have included education packs made available on loan. The packs contain folders and notes for teachers that provide background information in addition to a range of activities linked to learning objectives, and a range of artefacts, core samples showing stratigraphy, and CDs containing interactive components for teaching. One example of an activity focusing on SPLs is 'Reconstruction of ancient landscapes and people' that encourages pupils to study artefact and environmental evidence through which they learn about prehistoric lifestyle, technology and culture.

Learning is not confined to the formal setting of schools, and there are a wide range of ways that information can be provided to children and adults. The writing and production of well-illustrated children's story books has been a route for reaching both children and parents. 'Derek the Dredger and the Underwater Archaeologists' (HWTMA 2008) explore aspects of archaeology and the marine aggregates industry including how traces from prehistoric landscapes such as remains of bones and flints can be discovered in sorting through dredged-up material.

Based on the experience of attending a range of events, shows and conferences to engage different audiences, the MAT developed the 'Maritime Discovery Bus', a travelling resource which show-cases maritime archaeology including SPLs (Fig. 25.4). The bus has undertaken tours within the UK and across Europe, most recently within the EU Funded 'Common Cultural Connections' project with partners in France and Spain (www.commonculturalconnections.org). Using onboard artefacts, information boards, videos and a screen-linked microscope, visitors learn about the process of archaeological investigation in addition to the types of evidence recovered.

The approaches taken by the MAT have demonstrated the advantages of taking a holistic approach to the investigation and dissemination of SPLs. In addition to practical activities for outreach, the MAT also works through representation on coastal and marine management groups and forums to help ensure that the wider appreciation of the resource is translated into effective management approaches and policies for protection. This experience was drawn on when working within the SPLASHCOS project to promote communication and wider engagement with marine industry.

**Fig. 25.4** The Maritime Archaeology Trust's Discovery Bus presenting information on submerged prehistory at a public event located at a coastal park, Plymouth, UK

## 25.3 SPLASHCOS: Training, Outreach and Collaboration

The aim of the SPLASHCOS project in pushing forward the development of the study, appreciation and management of SPLs included a number of objectives for delivering training opportunities, increasing collaboration with marine industries and broader outreach to stakeholders and the wider European public.

### *25.3.1 Training*

Targeted training and direct field experience was provided for early stage researchers. There were two key mechanisms for this – Short Term Scientific Missions (STSMs) for individuals or small groups, and Training Schools for larger groups of individuals (see Uldum et al., Chap. 5; Galili et al., Chap. 6). Eleven STSMs were delivered during the project. These allowed individual researchers to gain valuable experience at a host organisation within another EU state. The missions differed in duration from 5 days up to 1 month, the key aim of which was to foster collaboration and to learn new techniques (Fig. 25.5).

The training schools provided intensive specialist training on particular subjects for groups of researchers. SPLASHCOS delivered six training schools, which focused on a range of areas related to SPLs: palaeo sea-level modelling (Estonia), underwater excavation (Israel, Galili et al., Chap. 6), underwater geoacoustic modelling (Spain), underwater recording and conservation (Malta) and remote sensing of submerged landscapes (Rhodes).

During the project, 65 early stage researchers were able to benefit from these training opportunities.

**Fig. 25.5** An Early Stage Researcher from Portugal (Leandro Infantini) working in the laboratory on prehistoric sediment samples during a SPLASHCOS Short Term Scientific Mission (Photo: Maritime Archaeology Trust)

### *25.3.2 Engagement with Industry*

There is an urgent need for greater collaboration between SPL research and marine industry. For industries impacting the seabed there are requirements to ensure heritage is investigated prior to development, and this is embedded within European Directives. However, direct contact between companies and archaeologists is not always required as part of the regulatory process. The framework through which heritage legislation and management are delivered differs between countries with regulation and relations with industry being more developed in some states than others. The need for promotion of collaboration with industry was highlighted by Flemming's (2004) volume 'Submerged Prehistoric Archaeology of the North Sea: Research Priorities and Collaboration with Industry'. Over the following decade, growing awareness of SPLs, an increasing volume of research and new evidence from the seabed has focused attention on the need for appropriate evaluation and monitoring through development work. This situation has been examined by Salter et al. (2014), who summarise approaches to research and management of SPLs, emphasising the importance of industry collaboration and the need for greater international cooperation and coordination.

The existing survey data that is held by companies, developers and governmental organisations is of great value when developing understanding of SPLs. Making such data available can bring benefits to scientific and wider public understanding of past human occupation and environments. Facilitating this access requires increased cooperation and communication between industry and the heritage sector, which will be of mutual benefit. During the SPLASHCOS project a number of internationally significant projects were initiated or came to fruition which demonstrate the potential of such collaborations and are building a body of knowledge about best practice (see also Homlund et al., Chap. 4; Glorstad et al., Chap. 19; Gaffney et al., Chap. 20; Sturt et al., Chap. 28).

The largescale project to develop Maasvlakte 2, an expansion of the Port of Rotterdam, embedded archaeological assessment and mitigation throughout the project involving close cooperation from the beginning of the planning process through to completion between the Port authority, the government heritage agency, archaeologists and geoscientists. Evidence from SPLs within the development area included underwater Mesolithic sites, coring and excavation techniques adapted to the working conditions, and recovery of flint tools and palaeoenvironmental indicators (Moree and Sier 2015; Momber and Peeters, Chap. 21). The work included initiatives to communicate the results to the public, generating wide interest (https://www.maasvlakte2.com/en/index/show/id/665/archaeology-palaeontology).

In the UK, the Aggregate Levy Sustainability Fund enabled research and consultancy organisations to undertake projects to better understand the impacts of the aggregate extraction industry on the marine historic environment (Sturt et al., Chap. 28, http://archaeologydataservice.ac.uk/archives/view/alsf/). Notable SPL examples include the University of Birmingham's West Coast Palaeolandscapes Project, which used industry seismic data to explore Late Palaeolithic and Mesolithic deposits (Fitch and Gaffney 2009) applying approaches originally developed as part of the North Sea Palaeolandscape Project (Gaffney et al. 2007, 2009, see also Gaffney et al., Chap. 20). Further work by Wessex Archaeology in their Seabed Prehistory project developed methodologies for assessing the presence or absence of prehistoric archaeology within marine aggregate deposits (Russell and Tizzard 2011). The British Marine Aggregate Producers Association Protocol for Reporting Finds of Archaeological Interest has established a framework for direct contact between industries working in the marine environment and heritage specialists (BMAPA 2005), which has resulted in reports of flint tools, bones and submerged landscape evidence. This approach is also being applied to the fishing industry through a pilot project (https://fipad.org/) and has been very recently adapted into the Marine Antiquities Scheme through which divers, fishermen, boat operators and coastal visitors can report marine discoveries (https://marinefinds.org.uk/).

SPLASHCOS initiatives targeting relations with industry include a publication 'Marine Industry and Submerged Prehistoric Archaeology: Sharing Data, Developing Understanding and Delivering Best Practice', which was distributed through network members in all project countries and is publically available on the SPLASHCOS website (http://splashcos.org/sites/splashcos.org/files/SPLASHCOS_Marine_Industry_Guide_FINAL.pdf), and a conference with Industry entitled 'Offshore Industry and Archaeology: a Creative Relationship' held in Denmark in March 2013 and organised by Thijs Maarleveld (http://splashcos.org/events/splashcos-esbjerg-meeting). The 2-day event addressed key themes of existing data, the need for standardisation in fulfilling environmental regulations, and the involvement of small and medium sized enterprises (SMEs). Speakers included representatives from major industries in addition to those responsible for gathering and interpreting data from the heritage sector. The event provided opportunities for SMEs which specialise in survey, offshore archaeology services, data evaluation and consultancy to present their experience to representatives of offshore industries and the heritage management sector.

### *25.3.3 Wider Promotion*

The need to enhance understanding and appreciation of SPLs by scientists working in related disciplines and the public was achieved through a range of routes. For events and conferences, we created posters for display. For wider promotional opportunities, we designed and produced a leaflet (Fig. 25.6), and distributed 10,000 of these during the project. As with any current project, internet presence was important, and the project website at www.splashcos.org attracts several hundred visitors per month. A Facebook page was also established with posts regularly reaching over 500 people. Statistics available for the Facebook page indicate that the largest audience reached was the 25–34 age group, providing an important route to reach younger researchers.

**Fig. 25.6** The SPLASHCOS publicity leaflet. Over 10,000 were distributed during the project (Photo: Maritime Archaeology Trust)

## 25.4 Looking to the Future

As understanding of SPLs is a developing area of research, it is important to capitalise on any opportunities for raising awareness within the scientific community as well as more broadly with the general public. It is important for those involved in data gathering and analysis to think about how new results can be disseminated as widely as possible, making sure this goes beyond the research community. Examples from a number of European countries are showing how information can be used for maximum impact and to reach the widest possible audience.

As the SPL resource is physically difficult to access, this makes archaeologists' responsibilities towards promoting public archaeology even more pressing. While definitions of public archaeology, or community archaeology, are difficult to pin down to specifics, there is broad acceptance of the need to make results available for all to examine, interpret and reinterpret (Merriman 2004; Holtorf 2007).

The results of enhanced wider public understanding of the SPL resource and its importance in the story of human history can have wider effects. Increased public awareness helps raise the profile of the marine cultural heritage and this in turn influences the development of management plans focussed more generally on the coastal and marine environment at local and regional level. When local and regional approaches are translated into national and international frameworks, then more robust protection is possible.

While raising the baseline profile of SPLs within the population has been an important goal, it has also been vital to increase communication with marine industry, in order to increase understanding of the submerged heritage, the need for protection, and a positive dialogue that ensures appropriate treatment of the resource and collaborative partnership without compromising the commercial demands of underwater work.

Within the SPLASHCOS framework it has also been possible to take forward specialist training of early stage researchers and help to increase the numbers involved. Building a body of professionals who will be able to expand current levels of research, and some of whom will find employment in government agencies and archaeological companies responsible for implementation of environmental impact assessments and legislation, is an important legacy for the future.

Enjoyment and understanding of SPLs should be fostered at every opportunity, and all those involved in heritage need to promote this on a national and international scale. Developments over the past decade have meant a growth in understanding and awareness. It is now essential to ensure there are available opportunities to facilitate the future aspirations of those wishing to be directly involved in their heritage and to continue to build on this work.

## References

BMAPA & English Heritage (2005) Protocol for reporting finds of archaeological interest. British Marine Aggregate Producers Association and English Heritage, London. Downloadable from http://www.wessexarch.co.uk/projects/marine/bmapa/arch-interest.html

Fitch S, Gaffney V (2009) West coast palaeolandscape survey pilot. University of Birmingham. http://archaeologydataservice.ac.uk/archiveDS/archiveDownload?t=arch-1059-1/dissemination/pdf/Report/WCPP_V4.pdf

Flemming NC (ed) (2004) Submarine prehistoric archaeology of the North Sea. Research priorities and collaboration with industry, CBA Research Report, vol 141. Council for British Archaeology, York

Fulford M, Champion T, Long A (eds) (1997) England's coastal heritage. A survey for English Heritage and the RCHME, English Heritage Archaeological Report, vol 15. English Heritage, Swindon

Gaffney V, Thompson K, Fitch S (eds) (2007) Mapping Doggerland: the Mesolithic landscapes of the southern North Sea, BAR British Series, vol 31. Archaeopress, Oxford

Gaffney V, Fitch S, Smith D (2009) Europe's lost world, the rediscovery of Doggerland CBA Research Reports 160. Council for British Archaeology, York

Holtorf C (2007) Archaeology is a brand! The meaning of archaeology in contemporary popular culture. Archaeopress, Oxford

HWTMA (2008) Derek the dredger and the underwater archaeologists, Southampton. http://archaeologydataservice.ac.uk/archives/view/dredger_eh_2008/downloads.cfm

Merriman N (ed) (2004) Public archaeology. Routledge, London

Momber G (2000) Drowned and deserted: a submerged prehistoric landscape in the Solent, England. Int J Naut Archaeol 2(1):86–99

Momber G, Tomalin D, Scaife R, Satchell J, Gillespie J (eds) (2011) Mesolithic occupation at Bouldnor Cliff and the submerged prehistoric landscapes of the Solent, CBA Research Report, vol 164. Council for British Archaeology, York

Moree JM, Sier MM (eds) (2015) Interdisciplinary archaeological research programme Maasvlakte 2, Rotterdam. Bureau Oudheidkundig Onderzoek Rotterdam, Rotterdam

Russell J, Tizzard L (2011) Seabed prehistory: site evaluation techniques (Area 240) Synthesis. OASIS/ADS, Wessex Archaeology 70754:04. doi:10.5284/1008287

Salter E, Murphy P, Peeters H (2014) Researching, conserving and managing submerged prehistory: national approaches and international collaboration. In: Evans AM, Flatman JC, Flemming NC (eds) Prehistoric archaeology on the continental shelf: a global review. Springer, New York, pp 151–172

Smith O, Momber G, Bates R, Garwood P, Fitch S, Pallen M, Gaffney V, Allaby RG (2015) Sedimentary DNA from a submerged site provides evidence of wheat in the British Isles 8000 years before present. Science 347(6225):998–1001. doi:10.1126/science.1261278

# Chapter 26
# Arch-Manche: Using Archaeological, Palaeoenvironmental, Historic and Artistic Resources in Coastal Management

**Lauren Tidbury, Julie Satchell, and Garry Momber**

**Abstract** The coast of the English Channel (La Manche) and the Southern North Sea is a dynamic environment. Coastal erosion, increased storm frequency, flooding and instability are all providing challenges for managing risks associated with these threats. Understanding the long-term development of the coast is vital in order to understand how the present situation has arisen. Archaeological and palaeoenvironmental data from submerged and prehistoric landscapes along the coastlines and sub-tidal fringes of the European Continental Shelf represent previously under-used coastal indicators that can be applied as tools to inform long-term patterns of coastal change. The preservation of organic material and traces of flora and fauna within submerged and intertidal sites provide detailed evidence of the past environment including plants, animals and insects, the types of soils and crucially whether it was dry, damp or wet, saline or brackish. Recording changes to these environments demonstrates the impact of rising or falling sea levels and relationships with coastal adaptation. This chapter presents the results of the Arch-Manche project, which used archaeological, palaeoenvironmental, historical and artistic resources to advance understanding of the scale and rate of long-term coastal change.

## 26.1 Introduction

Coastal managers face an ongoing battle to moderate impacts from the sea in the face of a changing climate and pressures from human uses of the coastal zone. The challenges that lie ahead are forecast to increase while pressure on limited resources is also likely to increase. This paper explores the value of under-used coastal indicators that can be applied as tools to inform on long-term patterns of coastal change. This work forms part of the 'Archaeology, Art and Coastal Heritage—Tools to Support Coastal Management and Climate Change Planning across the Channel Regional Sea" (Arch-Manche) project. The project aims to advance understanding of the scale and rate of long-term coastal change using sources of evidence including archaeology, palaeoenvironmental data, works of art, historic maps, charts and photographs (see also Karle and Goldhammer, Chap. 15 for a similar approach).

Traditionally, coastal engineers, planners and decision-makers have rarely studied long-term changes to the coast. However, the roots of coastal instability often relate to progressive geomorphological evolution that dates back thousands of years. Evidence from coastal, submerged and intertidal sites can be used to model the position of past coastlines, understand the rate of past sea level change

L. Tidbury (✉)
Museum of London Archaeology, Mortimer Wheeler House, 46 Eagle Wharf Road, London N1 7ED, UK
e-mail: LTidbury@mola.org.uk

J. Satchell • G. Momber
Maritime Archaeology Trust, National Oceanography Centre, Empress Dock, Southampton SO14 3ZH, UK
e-mail: julie.satchell@maritimearchaeologytrust.org; garry.momber@maritimearchaeologytrust.org



G.N. Bailey et al. (eds.), *Under the Sea: Archaeology and Palaeolandscapes of the Continental Shelf*, Coastal Research Library 20, DOI 10.1007/978-3-319-53160-1_26

and map coastal erosion. Early archaeological evidence can demonstrate how people were impacted by coastal change in the past and how populations reacted to some large-scale landscape and climatic changes (Van de Noort 2013). Coastal zones have often been favoured for human occupation, and marine resources in particular are thought to have played a key role in the settlement patterns of hunter-gatherers from an early period (Bell and Warren 2013, p. 31). Much of this evidence, now submerged in the Channel and southern North Sea, has been investigated and analysed during the project in order to reconstruct past landscapes and to model changes over time.

Various palaeogeographical models exist for the English Channel and North Sea region (including Shennan et al. 2000; Brooks et al. 2011; Sturt et al. 2013) and this work has stressed the importance of these once inhabited landscapes, now submerged, in terms of what can be learnt about the rate of change in the Holocene and the impact on human populations (Sturt et al. 2013 p. 3964). The Arch-Manche project focusses on a number of case study areas in the Channel and the southern North Sea and aims to highlight the importance of such data in terms of coastal management and how an understanding of past change can help with planning for the future. The project looked at archaeological and palaeoenvironmental data ranging from the Palaeolithic to WWII, located in the marine, intertidal and coastal zones. However, it was clear that submerged prehistoric land surfaces have the highest potential to inform on long-term coastal change. Many of those investigated as part of the project contain long stratified and dated sequences which can demonstrate a record of the changing environment, with some spanning a series of geomorphological events.

The case studies included a variety of coastal frontages faced with different challenges in terms of management, physical conditions and available data resources. Project partners in the UK, France, Belgium and the Netherlands employed a variety of research and fieldwork techniques involving in-depth, inter-disciplinary investigations (Momber et al. 2014). The work aimed to determine sea level in relation to the coast at particular times, identify resources showing specific measurable change, and model coastal change through time. This chapter demonstrates the methodology used and presents some of the results from case studies on the English south coast at Langstone Harbour and the western Solent (Fig. 26.1).

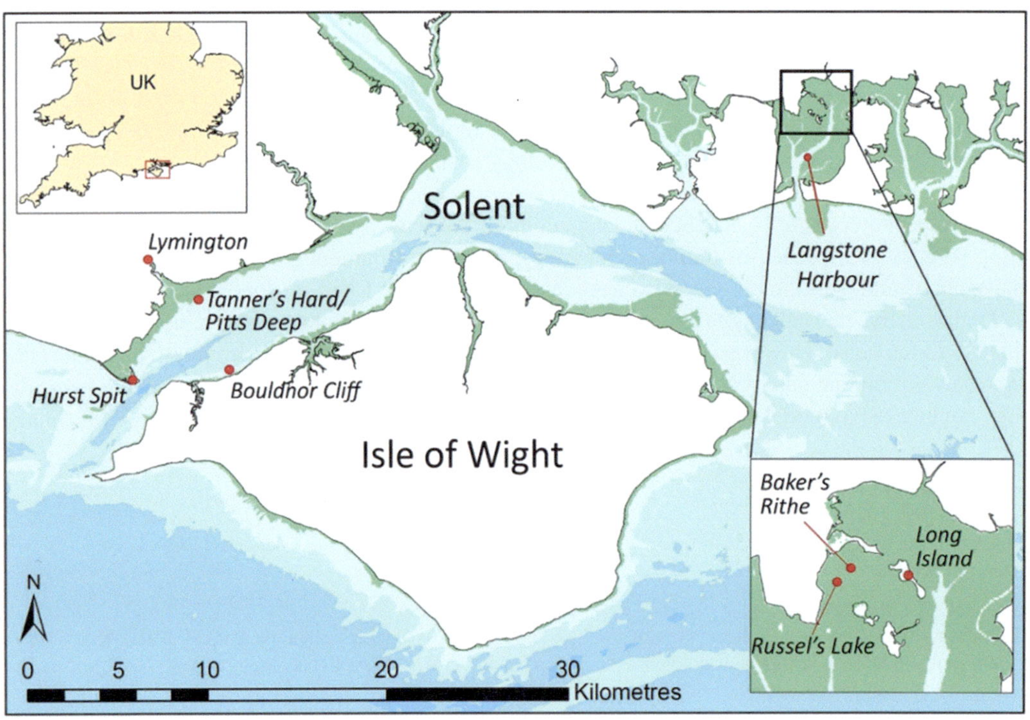


**Fig. 26.1** Map of the Solent showing key sites mentioned in the text from the western Solent and Langstone Harbour. The *green shading* indicates the extent of the intertidal zone exposed between high and low tide

## 26.2 Methodology

A large amount of data was assessed as part of the project, and the process of extracting information from archaeological and palaeoenvironmental sources has involved a phased approach to work. This included initial desk-based assessment, ranking of sites, features and samples, and collection of primary field data using a range of field investigations.

### *26.2.1 Desk-Based Assessment*

An initial desk-based survey identified areas of the coastline which have archaeological or palaeoenvironmental information that can help tell the story of past change, including monuments, fish traps, shipwrecks and submerged landscapes. A ranking system was developed in order to provide a relative value on the potential of each site to provide scientific information beneficial to practical decision-making in the long-term management and protection of the coastline. Particular importance was attached to information concerning the past behaviour of the coastline and to chronological information concerning the nature, scale and pace of sea level rise and coastal change.

Ranking methodologies were also developed for historical resources in order to assess their reliability. Such resources can provide important information for studying coastal development, for example, maps can be georeferenced and digitised to reconstruct former shorelines. However, the quality and detail can vary dramatically, and it was therefore necessary to evaluate these resources in terms of accuracy and reliability prior to using them for coastal research, using methods based on work carried out by Jongepier et al. (2016) and McInnes (2008).

In areas of coastal instability, archaeological and palaeoenvironmental sites have a particular role to play in establishing proven histories of localised change in coastal geomorphology and shoreline positions. These are particularly valuable in sectors subject to landslide movement and other coastal changes. Thus the ranking criteria sought to identify those sites that might best offer evidence for measuring the magnitude and rate of coastal change. Data were ranked according to the following criteria:

- Does the site contain evidence of changes in sea level?
- Does the site provide evidence of environmental change?
- Does the site contain archaeological material from different cultures/time-periods, showing evidence of temporal continuity?

The ranking did not place particular archaeological value on sites but was used to highlight the potential of a site to inform coastal managers of past changes to the coastline. The data was integrated into the project database and GIS to facilitate analysis and visualisation of areas of high potential. Although the process of ranking was subjective, it has revealed areas and environments where questions concerning the links between past and present coastal behaviour can be positively pursued.

This study has demonstrated that certain types of site and deposit can gain consistently positive scores for their potential to inform on coastal change. Some of these sites represent single or short-lived episodes. These might include a shipwreck or a prehistoric camp site. Sites of this kind can occur at a particular height, location or time that is pertinent to the understanding of shoreline-change. Other sites can offer a broader range or sequence of chronological and environmental information. They include biostratigraphical evidence such as pollen records in peat deposits, diatoms in stratified marine sediments and plant macro-fossils in river valley alluvium.

**Fig. 26.2** *Left*: Auger survey off long island. *Right*: Marine seismic survey, Langstone Harbour

### *26.2.2 Fieldwork*

The project employed a variety of fieldwork techniques to carry out detailed research on significant sites and areas of the coastline. This involved in-depth, inter-disciplinary research including fieldwork, scientific dating and analysis. The methods varied depending on the specific environment and the type of information being targeted, and included:

- Diving archaeological investigation
- Intertidal survey
- Archaeological excavation
- Geophysical and geotechnical sources

Here we discuss two case studies. The prehistoric landscape of Langstone Harbour was investigated using intertidal and marine seismic surveys (Fig. 26.2). Here the focus was on the islands present in the north of the harbour, which are known to be the last vestiges of the prehistoric landscape prior to inundation, as well as the main channels in the harbour. Six of the 14 case-study sites were subject to detailed field investigation, and full details can be found in the project technical report (Satchell and Tidbury 2014).

In the western Solent, the fieldwork included survey and sampling of submerged landscapes. Evidence from sites and deposits in this area provide high resolution data on the development of the Solent as a river, including coastal and climate change and human responses to this, which inform on current change and vulnerabilities.

## 26.3 Results

### *26.3.1 Langstone Harbour*

Langstone Harbour is located between Portsmouth Harbour to the west and Chichester Harbour to the east (Fig. 26.1). The site was selected as a case study area due to its rich and diverse archaeological record and its history of archaeological investigation, in particular a large multi-disciplinary project carried out over a 5-year-period in the 1990s (Allen and Gardiner 2000). The project revealed intermittent human activity commencing in the Mesolithic and adapting, at various times, to a rising sea-level and changes in areas of mudflats and saltmarsh.

Langstone Harbour today includes an area of around 1900 ha, of which 1700 ha are exposed at low tide, demonstrating the shallow nature of the harbour. Peat deposits in the Broom Channel at a depth of between −10.5 and −12 m OD (Ordnance Datum, equivalent to Mean Sea Level) were discovered within cores related to a housing development (Mottershed 1976). But other than these, the main evidence for the development of the harbour comes from work associated with the Langstone Harbour Archaeology project undertaken in the 1990s. Of particular relevance are two radiocarbon dates obtained from submerged tree remains embedded within peat deposits from the Baker's Rithe and Russel's Lake areas between the islands. From the former a peat deposit at −1 m OD provided a date of 2310–1950 cal BC (R-24993/2, 3735 ± 60 BP), and from the latter at −0.5 m OD a date of 3350–2910 cal BC (R-24993/1, 4431 ± 70 BP) (Allen and Gardiner 2000, pp 88–90), indicating the nature of changes that have occurred since the Neolithic and early Bronze Age period.

During auger work related to the excavation of the Langstone log boat in 2002, a substantial peat deposit was encountered at a depth of around −2 m OD off the north-west coast of Long Island. This peat remained undated until recently, but its substantial thickness of up to 2 m indicated that it was likely to be a different peat from those identified during the Langstone Harbour Project (Allen and Gardiner 2000; Scaife 2003). In 2012 this area was re-visited and the palaeochannel was tracked through hand augering along a transect south-west of the previous survey. Due to the difficulties of hand augering in the intertidal zone, the deepest core reached just −2.3 m OD. A sample from a wood horizon at −0.8 m OD located within the peat layer was sent for radiocarbon dating and was dated to 1660–1650 cal BC (Beta 344585).

In order to track the buried palaeochannels, a marine seismic survey was conducted in the harbour using a parametric echosounder. The survey was conducted at high water between the islands in the north of the harbour, focussing on the area of the auger survey and around the location of the submerged forests at Baker's Rithe and Russel's Lake. The survey was also conducted in the main channels of the harbour.

Between the islands the survey profiles revealed in high detail a network of channels and a number of features on the seabed which could relate to timber remains. These were outside of the known location of the submerged forests at Baker's Rithe and Russel's Lake, suggesting that the submerged forest could have been more widespread than originally presumed (Evangelinos et al. 2014, p 4).

Previous work in the harbour suggested that the area was dominated by a steep valley in the Mesolithic following the course of the present day Langstone Channel (Allen and Gardiner 2000 p 203). A network of profiles was recorded in the Channel including large palaeochannels (Fig. 26.3; Evangelinos et al. 2014, p 10).

A 4-D model of the topographical and environmental change in Langstone Harbour was constructed based on the results from the geoarchaeological data analysis including boreholes, sediment analysis, radiocarbon dating, the results of seismic survey and the data from the 1990s Langstone Harbour Project (Fig. 26.4). The model was produced using a CesiumWebGL cross-platform. Significant stages in the development of Langstone Harbour begin with a transition from a down-cut ravine with fresh water streams to a silted river valley, infilled by organic material, which became established during the Late Mesolithic/Early Neolithic. During the Bronze Age the area developed towards a more predominantly marine environment made up of salt marsh and tidal rivers. Fully marine conditions did not characterise the channels within the harbour until after 800 BC. There then followed periods of stasis and episodes of both accelerated erosion and deposition. The harbour was then altered significantly by anthropogenic processes in the post-Medieval period, particularly through land reclamation in the area of Farlington Marshes. These dynamic episodes show a relationship with concurrent changes in relative sea-level and shifts in local currents and the tidal system (Allen and Gardiner 2000, pp 186–198).

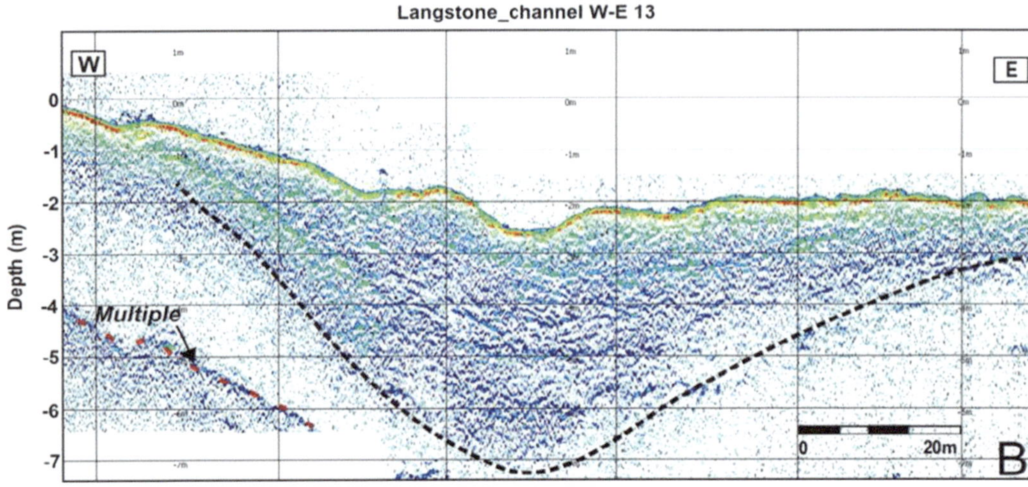


**Fig. 26.3** Seismic profile in Langstone Channel showing a large buried palaeochannel (From Evangelinos et al. 2014, p 10)

## *26.3.2 Western Solent*

The western Solent has been completely reconfigured over the last 8000 years by geomorphological changes and rising sea levels. The process has not finished and a stable equilibrium has yet to be reached. It demonstrates how the long-term evolution of a coastline can influence present and future patterns of change with important implications for ongoing management policies. Bouldnor Cliff and the wider landscape of the western Solent was chosen as a case study as it contains a long sequence of stratified prehistoric landscapes including Mesolithic occupation evidence (Momber et al. 2011).

Fieldwork was focused on the prehistoric and landscape features at Bouldnor Cliff, Tanners Hard and Hurst Spit (see Fig. 26.1) and included:

- Monitoring landscapes—many of the sites have been subject to previous survey and historic datum points were relocated to calibrate erosion over the past 10–15 years.
- Sampling landscapes—sediment samples were collected from Bouldnor Cliff and Hurst Spit to assess palaeoenvironmental changes and for radiocarbon dating.
- Excavation—an evaluation trench measuring 1 × 2 m was excavated at Bouldnor Cliff and lithic material and palaeoenvironmental samples were collected.
- Rescue artefacts—artefacts exposed through erosion were recovered from the seabed before they were lost.

Sediment archives are well preserved in this area, particularly at the site of Bouldnor, and the analysis of diatoms and formanifera together with analysis of bathymetry has revealed a sequence of events that saw final inundation by the sea around 6000 cal BC (Momber 2014, p 203). This was followed by the deposition of brackish estuarine sediments, which served to protect the palaeolandscape. Evidence suggests that the sea entered the system via the River Yar, and by 4500 cal BC rising sea levels eroded the barrier to the east of the basin, and breached the western barrier about 2000 years later. This formed the Solent, which changed from a sedimentary sink in the estuary to the new Solent channel cutting across the infill deposits and removing most of them (Fig. 26.5). Some of these deposits remain in sheltered areas to the north and south, including Bouldnor Cliff, although they are still subject to ongoing erosion (Momber et al. 2011).

The formation of the Solent dramatically remodelled the seabed by reshaping and transforming the submerged palaeolandscape. First, estuarine deposits covered and protected earlier land surfaces,

**Fig. 26.4** Screenshots from the 4D model of Langstone Harbour showing the development of the harbour from the Mesolithic to the present day based on archaeological and palaeoenvironmental evidence. The *coloured dots* show the location of archaeological sites and features, the larger *red dots* are the Baker's Rithe and Russel's Lake submerged forests. The view is taken from the north-east corner of the harbour looking south-west

and secondly sea level rise overtopped hills to the east and west allowing a new channel to be formed perpendicular to the original drainage pattern (Fig. 26.5). This masked the previous north–south flowing river. The results show that the centre of the Solent Channel has been eroded and is now deeper than would otherwise have been the case, demonstrating that the palaeolandscape was quite different from what might have been concluded by relying solely on the morphology of the present-day seabed (Momber 2014, p 203). Moreover, the accumulation of fine-grained fluvial and marine silts in an oxygen-free environment has preserved detailed, well stratified and well preserved remains, demonstrating the nature, rate and scale of change, and the fact of ongoing erosion of the palaeolandscape.

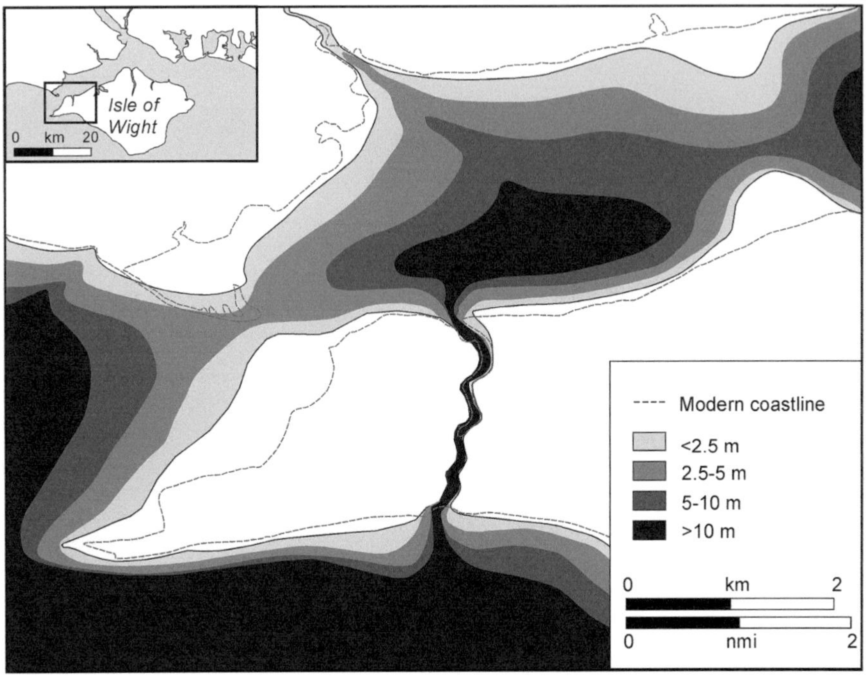


**Fig. 26.5** Formation of the western Solent, showing the effects of progressive sea-level rise and shoreline retreat. At 6000 cal BC, sea level was around 10 m below OD and the Isle of Wight was connected to the mainland. At 4500 cal BC, sea level reached 5 m below OD opening up a channel to the east. Around 1500–2000 cal BC, sea level rise crossed the 2.5 m contour and a second channel opened to the west. The breaching of this final land link to the Isle of Wight transformed a salt marsh into a channel that would eventually extend over 10 km long and up to 60 m deep. The strong tidal currents introduced when the channel was formed continue to erode the Solent shoreline today. The land shown in white indicates an estimate of the coastline when sea-level approached the present level. The difference with the modern coastline reflects widening of the Solent channel by coastal erosion (and locally, shoreline progradation). Image re-drawn by Geoff Bailey from an original by Garry Momber

## *26.3.3 Lymington, North-West Solent*

Alongside archaeological and palaeoenvironmental data, the project has also used historical resources including works of art, photographs, maps and charts to understand more recent change. This can be combined with the data from archaeological investigation to provide an understanding of coastal change both in the long and short term. A methodology was developed as part of the project to assess the reliability of historic maps and charts (Jongepier et al. 2016), and those which ranked more highly in terms of reliability were used to reconstruct past landscapes through digitisation and GIS-rectification.

Off Lymington in the north-west Solent, historic maps were used to assess the rate of saltmarsh and mudflat erosion (Fig. 26.6). The most reliable maps were georeferenced and the limit of the saltmarsh digitised demonstrating the rate of change from 1781 to the present day. Over a period of 153 years from the date of the first map (1781) to the second map (1934), the area witnessed around 500 m of regression of the saltmarsh. Over the next 57 years a further regression of 200–400 m took place, indicating an increased rate of change, and subsequently this accelerated more dramatically with saltmarsh erosion in just 22 years of up to 500 m in places. Clearly, the rate of erosion has dramatically increased in the last 100 years. The specific cause of saltmarsh regression is currently not well

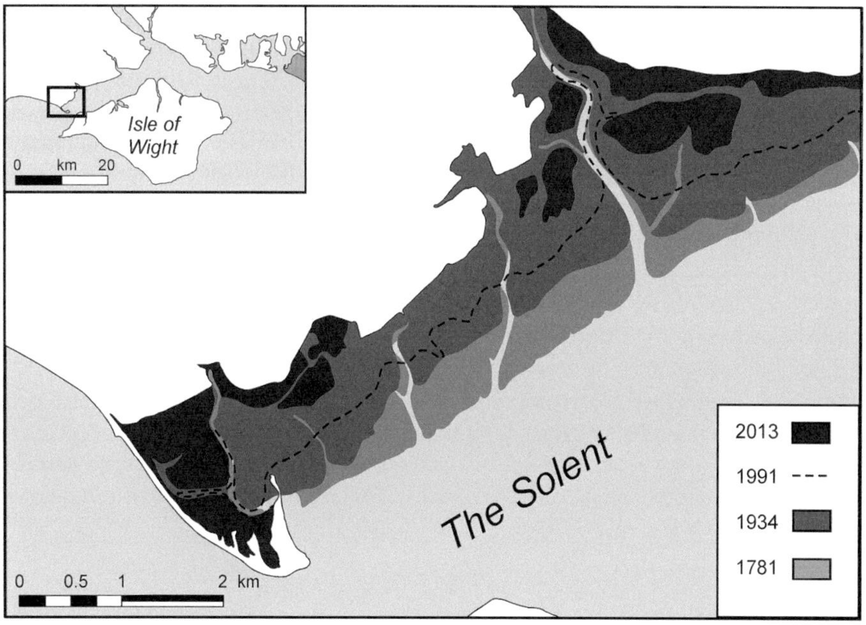


**Fig. 26.6** Saltmarsh and mudflat regression in the north-west Solent based on historical map analysis (Image re-drawn by Geoff Bailey from an original by the Maritime Archaeology Trust)

understood, but is thought to result from a combination of factors including wave action, lack of sediment supply, dieback of vegetation, tidal currents and sea level change (NFDC 2014). These maps cannot provide an answer to the cause of erosion, but they can provide high resolution detail on the rate and scale of change from the eighteenth century onwards.

## 26.4 Conclusions

All those involved in coastal management have a requirement for high quality data and a thorough understanding of the physical processes at work around the coastlines of the Channel and the southern North Sea. An appreciation of the impacts of coastal evolution and processes is fundamental in order to understand and manage coastal frontages in the most effective way. Long-term coastal monitoring is increasingly recognised as an invaluable data source to support coastal risk management, as well as providing information to assist, for example, the efficient design and construction of coastal defence measures.

Over 3000 sites across the English Channel and southern North Sea regions were assessed as part of the Arch-Manche project. Six areas were selected for detailed fieldwork and investigation, and the results from two of these have been outlined here. Information from these sites provides data on the scale and rate of coastal change in these areas. Sites from the Mesolithic period demonstrate evidence of rapid sea level rise as thousands of kilometres of inhabited land were lost to the sea. Since current climate change models predict future rise in sea level, inundated Mesolithic sites are of particular relevance to coastal management (Gaffney et al. 2009). Archaeological research on such sites can also contribute new sea-level index points, providing valuable information to improve existing climate change models.

Palaeoenvironmental data provide evidence of past landscapes from early prehistoric times through to the present. Analysis reveals evidence of the environment including plants, animals and insects, the types of soils, and whether it was dry, damp or wet, saline or brackish. Recording changes to these environments demonstrates the impact of rising or falling sea levels and relationships with coastal adaptations. Humans have used the coastal zone for thousands of years. The position of settlements shows the proximity to coastal areas, while specific features like trackways to cross marshy areas show adaptations to marine environments, such as those found off the north coast of the Isle of Wight dating to the Neolithic (Tomalin et al. 2012). Studying the archaeological record can demonstrate how humans adapted to change, and in more recent times, how they effected change.

The project has also looked at coastal heritage and historical resources to understand more recent change to the coastlines of the Channel and North Sea regions. Using a combination of data sources allows for maximum amounts of information to be extracted.

This project has highlighted the potential of these various sources of data to contribute to understanding the rate and scale of past change, their value in contributing to informed decisions about future management, and the scope for more detailed analysis along the same lines and in different areas to further refine interpretation and assessment of future risk associated with climate and sea-level change.

**Acknowledgements** This project was funded by the European Regional Development Fund through the Interreg IVA 2 Seas Programme. The authors would like to thank our partners, the CNRS, University of Ghent and Deltares and particularly Marie-Yvane Daire, Pau Olmos, Tine Missiaen, Iason Jongepier, Robin McInnes and Peter Vos for their contributions throughout the project. We would also like to thank Christin Heamagi and Brandon Mason of Maritime Archaeology Ltd. and Jason Sadler and Homme Zwaagstra of Geodata for their work on the 4-D model.

## References

Allen MJ, Gardiner J (2000) Our changing coast: a survey of the intertidal archaeology of Langstone harbour, Hampshire, CBA Research Report, vol 124. Council for British Archaeology, York

Bell M, Warren G (2013) The Mesolithic. In: Ransley J, Sturt F, Dix J, Adams J, Blue L (eds) People and the sea: a maritime archaeological research agenda for England, CBA Research Report, vol 171. Council for British Archaeology, York

Brooks AJ, Bradley SL, Edwards RJ, Goodwayn N (2011) The palaeogeography of Northwest Europe during the last 20,000 years. J Maps 7(1):573–587

Evangelinos D, Missiaen T, Tidbury L (2014) Acoustic mapping of Langstone Harbour. Unpublished Research Report

Gaffney F, Fitch S, Smith D (2009) Europe's lost world. The rediscovery of Doggerland, CBA Research Report 160. Council for British Archaeology, York

Jongepier I, Soens T, Temmerman S, Missiaen T (2016) Assessing the planimetric accuracy of historical maps (sixteenth to nineteenth centuries). New methods and potential for coastal landscape reconstruction. Cartograph J 53. http://www.maneyonline.com/doi/full/10.1179/1743277414Y.0000000095

McInnes RG (2008) Art as a tool in support of the understanding of coastal change in Hampshire and the Isle of Wight. The Crown Estate-Caird Fellowship 2008. Crown Copyright.

Momber G (2014) Submerged archaeology and cultural responses to climatic amelioration. In: Evans AM, Flatman JC, Flemming NC (eds) Prehistoric archaeology of the continental shelf: a global review. Springer, New York, pp 194–212

Momber G, Tomalin D, Scaife R, Satchell J, Gillespie J (eds) (2011) Mesolithic occupation at Bouldnor Cliff and the submerged prehistoric landscapes of the Solent, CBA Research Report, vol 164. Council for British Archaeology, London

Momber G, Tidbury L, Satchell J, Daire MY, Missiaen T, Vos P (2014) Coastal management: guide for using archaeology, palaeoenvironments, history and art. Maritime Archaeology Trust, Southampton

Mottershead DN (1976) The Quaternary history of the Portsmouth region. In: Mottershead, DH, Riley RC (eds) Portsmouth geographical essays, Vol 2. Department of Geography, Portsmouth Polytechnic, pp 1–21

NFDC (2014) Saltmarsh Erosion. New Forest District Council. http://www.newforest.gov.uk/index.cfm?articleid=7182

Satchell J, Tidbury L (eds) (2014) Arch-Manche: archaeology, art and coastal heritage—tools to support coastal management and climate change planning across the Channel-Regional Sea: Technical Report. Unpublished Research Report. http://archmanche.hwtma.org.uk/uploads/images/Documents/Arch_Manche_Technical_Report.pdf
Scaife R (2003) The Langstone Saxon log boat: a preliminary pollen analysis of the associated sediments. Unpublished Archive Report
Shennan I, Lambeck K, Flather F, Horton B, McArthur J, Innes J, Lloyd J, Rutherford M, Wingfield R (2000) Modelling western North Sea palaeogeographies and tidal changes during the Holocene. In: Shennan I, Andrews J (eds) Holocene land-ocean interaction and environmental change around the North Sea, Geological Society Special Publications, vol 166. Geological Society, London, pp 299–319
Sturt F, Garrow D, Bradley S (2013) New models of North West European Holocene palaeogeography and inundation. J Archaeol Sci 40(11):3963–3976
Tomalin DJ, Loader R, Scaife RG (2012) Coastal archaeology in a dynamic setting: a Solent case study, BAR British Series, vol 568. Archaeopress, Oxford
Van de Noort R (2013) Climate change archaeology: building resilience from research in the world's coastal wetlands. Oxford University Press, Oxford

# Chapter 27
# The SeArch Project: Towards an Assessment Methodology and Sustainable Management Policy for the Archaeological Heritage of the North Sea in Belgium

**Tine Missiaen, Marnix Pieters, Frank Maes, Pauline Kruiver, Philippe De Maeyer, and Jan Seys**

**Abstract** Large parts of the Belgian continental shelf (BCP) are affected by commercial activities. Close to the shore, major infrastructural works are also envisaged for the near future. All these activities constitute a serious threat for the underwater cultural heritage (UCH) but until recently solid regulation regarding UCH was (and largely still is) lacking in Belgium. The SeArch project tries to offer solutions to these challenges through the realisation of three objectives: (1) developing a reliable survey methodology based on remote sensing techniques that allows cost-effective evaluation of the archaeological potential of offshore, nearshore, and intertidal areas; (2) preparing correct implementation of the commitments imposed by international conventions and of comprehensive proposals for a transparent and sustainable management policy for UCH in Belgium; (3) offering guidance for the stakeholders from marine industry, government agencies, fisheries, and harbour authorities on how to implement the new methodology and management approach, and to increase the general awareness with regards to UCH. The 4-year SeArch project started in 2013 and involves partners from Ghent University, Flanders Heritage Agency, Deltares and Flanders Marine Institute. In this paper we discuss the main scientific challenges of the project and some first results.

T. Missiaen (✉)
Renard Centre of Marine Geology, Department of Geology, Ghent University, Krijgslaan 281 S8, 9000 Ghent, Belgium
e-mail: tine.missiaen@ugent.be

M. Pieters
Flanders Heritage Agency, Koning Albert II laan 19 bus 5, 1210 Brussels, Belgium
e-mail: marnix.pieters@rwo.vlaanderen.be

F. Maes
Maritime Institute, Department of Public International Law, Ghent University, Universiteitstraat6, 9000 Ghent, Belgium
e-mail: frank.maes@ugent.be

P. Kruiver
Department of Geology and Geophysics, Deltares, Princetonlaan 6, 3584CB Utrecht, The Netherlands
e-mail: Pauline.kruiver@deltares.nl

P. De Maeyer
CartoGIS, Department of Geography, Ghent University, Krijgslaan, 281 S8, 9000 Ghent, Belgium
e-mail: philippe.demaeyer@ugent.be

J. Seys
Flanders Marine Institute (VLIZ), Wandelaarkaai 7, 8400 Ostend, Belgium
e-mail: jan.seys@vliz.be


G.N. Bailey et al. (eds.), *Under the Sea: Archaeology and Palaeolandscapes of the Continental Shelf*, Coastal Research Library 20, DOI 10.1007/978-3-319-53160-1_27

## 27.1 Introduction

Compared to the neighboring countries, the general attitude in Belgium towards marine archaeology has long been marked by a genuine lack of scientific interest. The sparse archaeological research in the Belgian sector of the North Sea has up to now been very fragmented and focused almost exclusively on a few wreck sites (Missiaen et al. 2012; Zeebroek et al. 2010). Very little or no attention has been paid to submerged archaeological sites and remains, or to submarine landscapes and (pre)historical coastlines.

This lack of any structured investigation with regard to the UCH is largely due to the state-structure of Belgium, where heritage is a regional matter whereas the North Sea is largely a federal matter. As a result, until recently no management system has existed in Belgium for archaeological and other heritage in the marine area. This is a deplorable situation, especially regarding the ever increasing pressure on the North Sea due to infrastructural and commercial activities, which already occupy a very large part of the North Sea and nearshore area, and are in general carried out without any systematic and built-in consideration for the marine archaeological heritage (Douvere et al. 2007).

Several international regulations deal with marine archaeological heritage, such as the 2001 UNESCO Convention on the Protection of Underwater Cultural Heritage and the 1992 European Convention for the Protection of the Archaeological Heritage (so-called Malta or Valetta Convention). The latter has entered into force for Belgium in 2011 whereas the former was ratified by Belgium in 2013. At the onset of the SeArch project (January 2013), however, Belgium was still not adequately equipped to fulfil its international commitments to manage and protect the UCH present in the North Sea and nearshore (intertidal) zones.

## 27.2 The SeArch Project

The SeArch project (www.sea-arch.be) aims to provide the necessary tools that allow correct implementation of the international conventions. On the one hand this involves the development of a survey methodology which allows accurate, efficient and cost-effective evaluation of the archaeological potential of marine areas under development. On the other hand this involves clear proposals for a comprehensive, transparent and sustainable management policy for underwater cultural heritage in Belgium and further development and implementation of a legal framework. The project also aims to provide practical guidance on how to implement the new methodology and management approach. The 4-year SeArch project will finish in December 2016.

The results of the SeArch project will bring about important benefits to a wide range of stakeholders (Fig. 27.1), ranging from marine industries, harbor authorities and government agencies involved in marine heritage to non-governmental organization's, divers, tourism, the scientific community and the general public. For instance, an efficient survey methodology will allow serious cost reduction for marine industry during the preparatory phase (by better planning) and avoid costly damage and loss of valuable time during the operational phase of the works. Sustainable management of UCH within a well-defined legal framework (including licensing, monitoring and evaluation procedures) will allow government agencies and policy makers to work more efficiently. Furthermore, the project will provide adequate tools to prevent or minimise damage to the archaeological heritage not only for offshore works but also in the land-sea contact zone which is actually the focus of large infrastructural works. Finally it will result in better knowledge regarding the marine (pre)historic environment and the development of (pre)historic coastlines, which is crucial for our understanding of the present-day changes related to sea level rise and climate change.

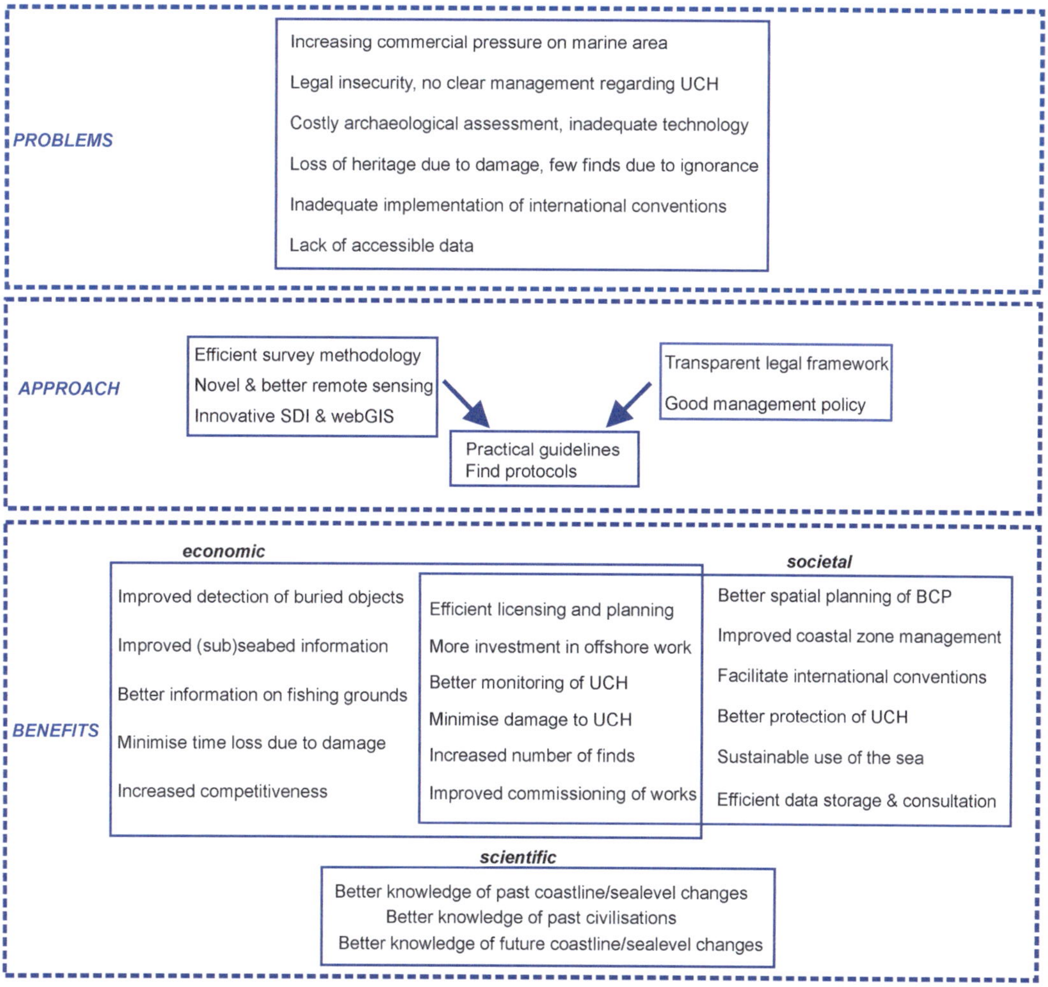


**Fig. 27.1** Overview of the current problems related to UCH in Belgium, the approach of the SeArch project to tackle these problems, and the expected benefits for the different stakeholders (*BCP* Belgian continental shelf, *SDI* spatial data infrastructure)

## 27.3 Survey Methodology

To date no efficient marine survey methodology exists that is specifically aimed at archaeological assessment studies. Standard geophysical and remote sensing techniques are used on an *ad hoc* basis and often they are not well adapted for archaeological investigations, since they were in the first place aimed at other purposes (Missiaen et al., Chap. 2). Moreover, the available techniques are frequently ineffective in large parts of the near-shore zone due to the presence of biogenic gas in the sediments, and their application in intertidal areas poses major technological challenges due to the shallow water depth, strong currents and wave action. A cost-efficient survey methodology that is based on effective geophysical and remote sensing techniques suited for both deeper and very shallow water is therefore urgently needed.

As a first step to achieve this goal a thorough review was made of all state-of-the-art remote sensing techniques that can be applied for detailed imaging of the seafloor and sub-seafloor environment in a 3D geological/palaeo-environmental model. Conventional techniques (e.g. reflection seismics) as well as techniques that are not commonly employed at sea (e.g. shear-waves, non-acoustic techniques, Lidar) were evaluated with archaeological and geological indicators in mind (De Clercq et al. 2013; Kruiver et al. 2013; Stal et al. 2013; Zurita Hurtado et al. 2013).

The most promising remote sensing techniques are to be subsequently tested on designated test sites (Fig. 27.2). The latter cover a wide variation in environment, archaeological features, economic importance and ground-truth data. The main offshore test site is a large, buried palaeovalley off the coast of Ostend (so-called Ostend Valley) and potential hot-spot for prehistoric life, as indicated by the palaeontological finds in the area (Vermeersch et al. 2015). The palaeovalley, which is covered by a large sand bank, is also a known sand extraction area (Mathys 2009). The sand bank poses an important challenge because sound waves are quickly absorbed in the heterogeneous sandy sediments. The

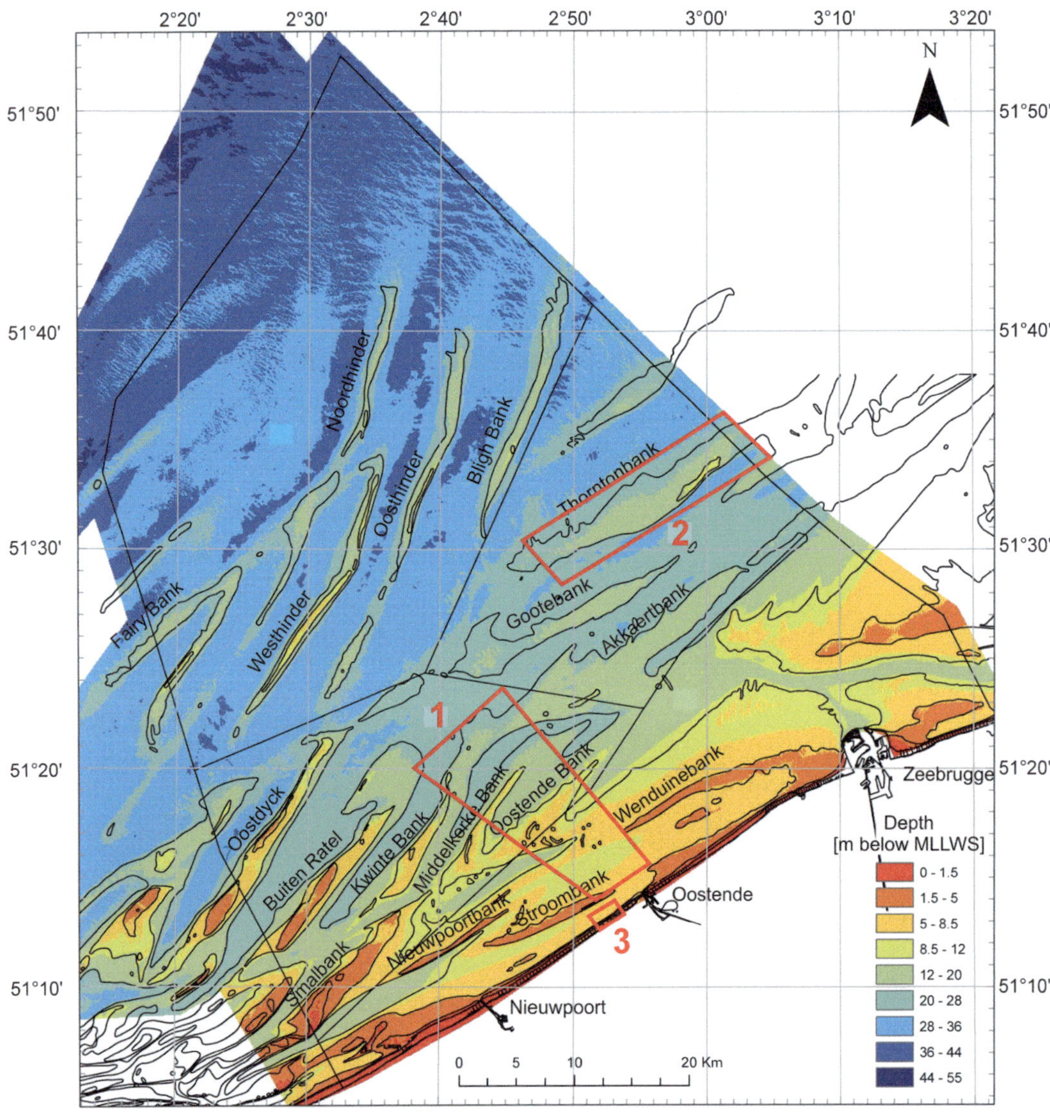


**Fig. 27.2** Overview of the different test sites in the SeArch project: 1 Ostend Valley, 2 Thornton Bank, 3 Raversijde Background map is the current seafloor topography in m below MLLWS (mean lower low water at spring tide) (© Flemish Hydrography)

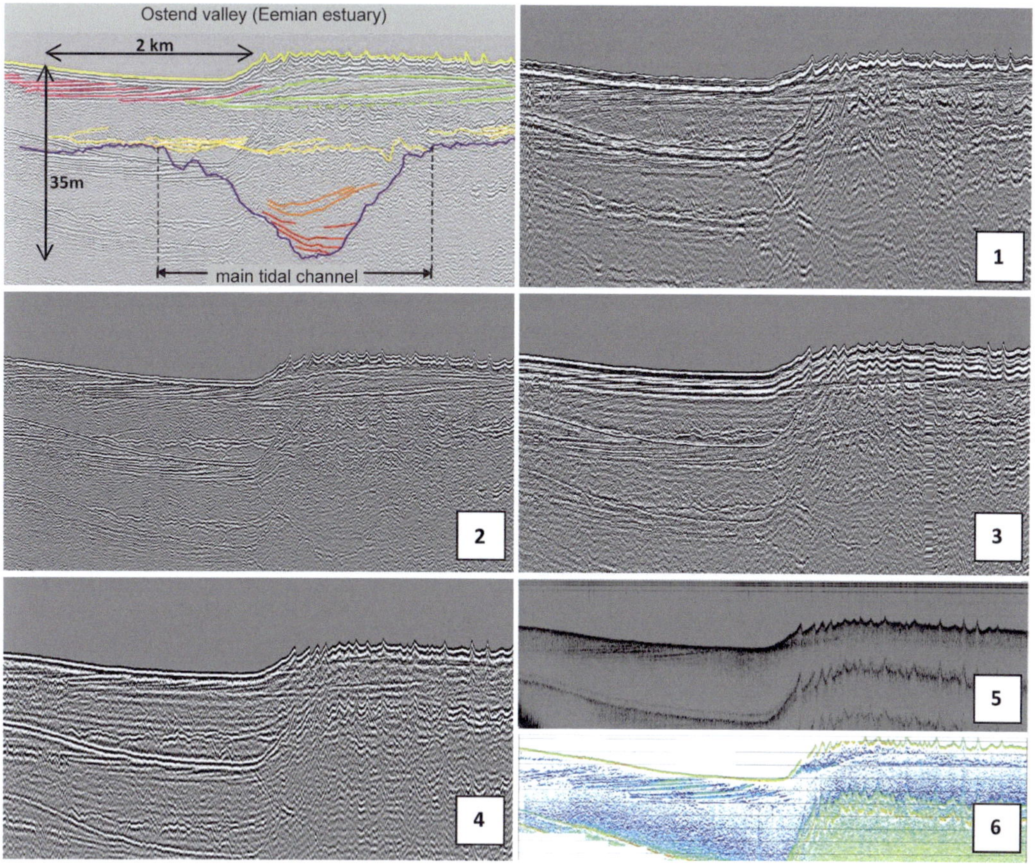


**Fig. 27.3** Comparison of seismic profiles obtained with different acoustic sources over the Ostend Valley. *Top left*: geological interpretation. Single channel sections obtained with *1* SIG sparker, *2* Centipede sparker, *3* Seistec boomer, *4* 500 J boomer, *5* X-Star chirp, *6* Parametric echosounder. Same dimensions for each section (see *top* image) (after Zurita Hurtado et al. 2014)

main intertidal test site is Raversijde beach, which is known for the buried remains of a drowned Medieval village, Roman and Medieval peat and salt extraction, and historic dyke structures (Pieters et al. 2006, 2010).

During the surveys, major attention is paid to smart acquisition design and inventive data processing to enhance cost efficiency (Incoul et al. 2014). The results so far indicate that for offshore areas marked by sandbanks the multi-electrode sparker seems to allow the best trade-off between resolution and penetration (Fig. 27.3). Even when the target depth is only a few tens of metres, and especially in relatively shallow water, multichannel acquisition furthermore offers considerable advantages over single channel data. In addition to 2D imaging, also true 3D subsurface imaging is tested using a recently developed multi-transducer parametric echosounder system (Innomar Technologie 2015). The results of a recent try-out at Raversijde were astonishing: for the first time buried peat and salt excavation patterns could be observed in the highest detail (cm to dm resolution in three dimensions, Fig. 27.4). Moreover, the observed features match perfectly with old aerial photographs (Missiaen 2015). These results set a new standard for detailed three-dimensional sub-bottom imaging, and open up significant perspectives for archaeological prospection in shallow water.

An additional focus of the survey methodology concerns the presence of biogenic gas in marine sediments, which is known to obscure conventional (reflection) seismic data (Missiaen et al. 2002).

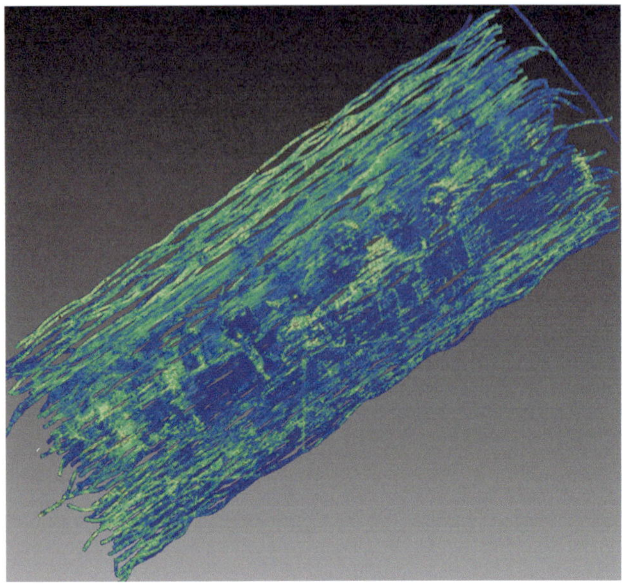

**Fig. 27.4** Horizontal depth slice from the 3D volume obtained at Raversijde beach with the multi-transducer parametric echosounder SES-2000 Quattro (slice dimensions 180 × 80 m, depth below seafloor ~1 m). The characteristic pattern of rectangular excavation pits, long (diagonal) trenches and circular depressions stands out clearly

Within the Search project, various experiments involving shear waves, surface waves and refracted waves are carried out at sea and in the intertidal zone in order to obtain reliable subsurface information in gas-rich areas. This research is technically challenging and highly novel and the results are expected to present an important step forward in the exploration of the sub-seafloor, especially in nearshore areas which are often prone to shallow gas.

Based on these different survey results a generic methodology will be put forward, preferably with a flexible structure based on a step-by-step approach. It is not the objective of the SeArch project to come up with a list of 'good' or 'bad' techniques since the potential of each technique will depend to a high degree on the environmental setting and the archaeological feature. The generic methodology, which is intended to be a guide for all those involved in commercial activities at sea as well as the international scientific community, will be complemented with a specialized Spatial Data Infrastructure (SDI) and web-GIS to allow efficient integration and visualisation of archaeological and environmental data in a user-friendly way.

## 27.4 Sustainable Management

The second goal of the SeArch project is to develop proposals for a sustainable and efficient management regime for the Belgian underwater cultural heritage. This management regime must fit into a legal framework, and includes a marine spatial planning process and a set of specific procedures and rules to deal with different types of heritage in an internationally acceptable way. The management regime for underwater cultural heritage will at the same time provide legal security for industry to invest in activities at sea.

In order to achieve this goal, a thorough evaluation of legislation, directly or indirectly, that affects the legal status of underwater cultural heritage has been carried out. It includes international law, EU legislation, as well as national legislation (Maes and Derudder 2014; Derudder and Maes 2014). Attention was also paid to bridge the gap between the legal and management approach between heritage on land and heritage at sea (Khakzad 2015).

With regard to national legislation several important objectives have (at least partly) already been achieved. In August 2013 Belgium, as the 45th member state, ratified the 2001 UNESCO Convention for the protection of UCH. Merely a year later, in June 2014, Belgium implemented in its national law most of the commitments imposed by that convention. This, in combination with a new law on marine spatial planning which takes heritage concerns into consideration, means that Belgium is now in a position to deal more efficiently with UCH in its waters. At this moment eight historic wreck sites (among which the HMS Wakeful, a British destroyer sunk in 1940 during the evacuation of Dunkirk, and the West-Hinder, a late nineteenth century lightship sunk in 1913) have received heritage status (Demerre et al. 2014). This is a major achievement and an important step forward for Belgium; moreover since 2014 underwater heritage has become an official stakeholder in all North Sea matters. With this, Belgium is finally joining the UK and the Netherlands, where official UCH management has been established for a number of years.

With regard to the international level, the most important provisions for UCH protection at this moment are found at international level rather than at the level of the Council of Europe. The 2001 UNESCO Convention is by far the most valuable convention to create an international framework for UCH protection, as well as a basis for national UCH legislation. However, so far only Belgium and France as North Sea States have ratified the Convention. In contrast to this, all the States bordering the North Sea have ratified the United Nations Convention on the Law of the Sea (UNCLOS) 1982, obliging them all at least to protect UCH and to cooperate for this purpose. A downside to both Conventions dealing specifically with UCH is the incorporation of constructive ambiguities as a result of compromises made to achieve a consensus. As a result the provisions in UNESCO 2001 and UNCLOS 1982 are sometimes general and vague, possibly leading to different interpretations (Maes and Derudder 2014).

Based on the above mentioned studies and the outcome of a large-scale consultation of all possible stakeholders, definitions and criteria for tailoring the management regime are currently being set up. It is the aim to come up with a coherent and inclusive management regime for underwater cultural heritage in the Belgian part of the North Sea, thereby looking both at 'unknown' zones (i.e. without information on heritage) and at already identified sites. The final objective is to deal with underwater cultural heritage in an internationally acceptable way and to provide legal security for industry to conduct relevant socio-economic activities at sea.

Within the SeArch project, an approach is also developed to include the protection of underwater cultural heritage in the process of marine spatial planning. In 2003 Belgium was one of the first countries in Europe to develop a marine spatial plan (MSP) for the North Sea, the so called 'Master plan North Sea'. In 2014 an updated plan was developed, which is committed even more to following the planning process (comparable to what is happening in the regions regarding spatial planning on land) (Maes and Seys 2014). Whereas underwater heritage was not in focus in the first marine spatial plan for the North Sea of 2003, in the new version of 2014 underwater cultural heritage is well integrated in the annexes to the new law.

Sustainable management of underwater cultural heritage highly depends on optimal access to data. At this moment, correct and efficient storage of submerged archaeological and environmental data is still lacking in Belgium. The SeArch project therefore aims to provide the first steps towards a comprehensive, reliable and accessible national data base. Different aspects and requirements related to the data type, resolution, accessibility, legal status, usefulness, accuracy, data format, and above all the compatibility with national data and international data base systems are thoroughly analysed. Major attention is also paid to the associated legal issues in order to reach a clear understanding on the terms and conditions regarding access and ownership of the data.

## 27.5 Guidance and Outreach

A last important goal of the SeArch project is to provide optimal guidance for commercial industries and government agencies involved in marine work. To this end a 'best practice' manual including advice and procedures for all stages of development and operation is being developed. The guidance manual will stress the importance of pro-active concern for underwater cultural heritage and provide specific advice and procedures ('best practice') on how to implement UCH in every stage of planning, development and operation of marine works (including licensing and monitoring).

In addition a set of protocols for the identification, reporting and handling of archaeological finds in the Belgian part of the North Sea is being prepared. The protocols are tailored to the specific needs of the end-users (marine industry, harbour authorities, fisheries and recreational divers). They describe the actions needed after discovery of an archaeological artefact, and include clear guidelines on how to identify finds of archaeological interest, as well as rules for safe handling and storage. A preliminary draft of the protocols is currently under discussion with the various stakeholders.

The guidance manual and protocols represent a significant step in ensuring that the marine historic environment in Belgium is considered in the marine development process, which will help to prevent or minimize damage to the archaeology during activities at sea. With this, Belgium joins the UK and the Netherlands, where guidance rules and reporting protocols already exist and where they have resulted in a spectacular increase in number and frequency of reports and finds, ranging from aircraft remains and cannon balls to whale and mammoth bone fragments (e.g. Wenban-Smith 2002; BMAPA and English Heritage 2003; Wessex Archaeology 2007).

In order to support practical use of the manual and protocols, it is crucial to raise awareness and appreciation. Therefore various measures are being included such as practical handouts, targeted workshops, visits to vessels and survey companies, as well as demonstration sessions. The need for such awareness is clearly demonstrated in the UK, where an extensive awareness programme has resulted in a pro-active and positive approach on the part of the marine industry, perfectly illustrated by a publication from the aggregate dredging industry stating that 'a potential threat can be turned into an opportunity through an open and constructive partnership approach' (Russell and Firth 2007). In the Netherlands, however, no awareness-raising was done and as a result much less attention is given there to UCH in offshore and infrastructure works.

## 27.6 Conclusions

Until recently Belgium was not adequately equipped to fulfill its commitments to manage and protect the underwater cultural heritage. Timely action was needed in view of the ever increasing industrial activities at sea such as aggregate extraction, wind farms, dredging, cable & pipeline projects, intensive fishing, and nearshore infrastructural works (in particular for coastal protection). It was not only the scientific community who were aware of this urgent need, but it was also expressed by the representatives of marine industries in Belgium, who are in need of clarity in legislation. In close collaboration with stakeholders from the marine industry and governmental and societal organisations involved in marine issues, the SeArch project (www.sea-arch.be) is preparing the way for efficient management of submerged cultural heritage in Belgium. At this moment a number of important objectives have already been realised with regard to survey methodology, management and legislation. Further work will focus on how to translate these results into a 'best practice' guidance manual and protocols for finds that are specifically targeted at the different marine industries.

**Acknowledgements** The SeArch project (www.sea-arch.be) is funded by the Flanders Agency for Innovation by Science and Technology (IWT).

## References

BMAPA and English Heritage (2003) Marine aggregate dredging and the historic environment: guidance note. British Marine Aggregate Producers Association and English Heritage, London

De Clercq M, Vos P, Missiaen T, Pieters M (2013) Archeologische en geo-archeologische indicatoren. Unpublished Internal Report SeArch project, Ghent University

Demerre I, Pieters M, Van Haelst S (2014) Gekelderd Erfgoed. West-Hinder HMS Wakeful: Allereerste Beschermde Scheepswrakken Ruimte 22:65–66

Derudder T, Maes F (2014) Assessment of the actual legal situation in Belgium. Unpublished internal report SeArch project, Ghent University

Douvere F, Maes F, Vanhulle A, Schrijvers J (2007) The role of marine spatial planning in sea use management. The Belgian case. Mar Policy 31(2):182–191

Incoul A, De Maeyer P, De Wit B, De Wulf A, Missiaen T, Nuttens T, Stal C, Van den Berghe M (2014) Preserving the archaeological heritage of the North Sea using innovative data acquisition techniques and webGIS. In: Mees J, Seys J (eds) Book of abstracts – VLIZ Young Scientists' Day, 7 March 2015, Brugge. Vlaams Instituut voor de Zee – Flanders Marine Institute (VLIZ) Special Publication 67, Oostende, p 69

Innomar Technologie (2015) SES-2000 multi-transducer sub-bottom profiler system for 3D investigation of buried structures. http://www.innomar.com/ses2000quattro.php

Khakzad S (2015) Integrated approach in management of coastal cultural heritage. Unpublished PhD thesis, University of Leuven

Kruiver P, Zurita Hurtado O, Diaferia G, Missiaen T (2013) Non-conventional survey techniques for marine archaeological investigations. Unpublished Internal Report SeArch project, Ghent University

Maes F, Derudder, T (2014) Assessment of international and European law related to or affecting underwater cultural heritage. Unpublished internal report SeArch project, Ghent University

Maes F, Seys J (2014) Ruimtelijke Planning op zee: waar België goed in is. Grote Rede Nieuws Over Onze Kust Zee 39:15–21

Mathys M (2009) The Quaternary geological evolution of the Belgian continental shelf, southern North Sea, geology and soil science. PhD thesis, Ghent University

Missiaen, T (2015) Efficiency of marine acoustic techniques for the archaeological prospection of intertidal areas: results from a test study on the Belgian coast (Ostend-Raversijde). IEE/OES Acoustics in Underwater Geosciences Symposium, 29–31 July 2015, Rio de Janeiro (Brazil), Conference Proceedings

Missiaen T, Murphy S, Loncke L, Henriet J-P (2002) Very high resolution seismic mapping of shallow gas in the Belgian coastal zone. Cont Shelf Res 22(16):2291–2301

Missiaen T, Demerre I, Verrijken V (2012) Integrated assessment of the buried wreck site of The Dutch East Indiaman 't Vliegent Hart. Relicta – Archeologie Monumenten- Landschapsonderzoek Vlaanderen 9:191–208

Pieters M, Demerre I, Zeebroek I (2006) Dijk uit de Romeinse tijd aangesneden onder het middeleeuwse vissersdorp Walraversijde. In: Bosman A, De Clercq W, Hoevenberg J (eds) Romeinendag. Jaarlijks Belgisch congres voor Romeinse archeologie, pp 93–97

Pieters M, Demerre I, Lenaerts T, Zeebroek I, De Bie M, De Clercq W, Dickinson B, Monsieur P (2010) De Noordzee: een waardevol archief onder water Meer dan 100 jaar onderzoek van strandvondsten en vondsten uit zee in België: een overzicht. Relicta – Archeologie Monumenten- Landschapsonderzoek Vlaanderen 6:177–218

Russell M, Firth A (2007) Working alongside the marine historic environment – an aggregate dredging industry perspective. CEDA Dredging Days 2007. Central Dredging Association, Rotterdam

Stal C, Nuttens T, De Maeyer P, De Wulf A (2013) Digital surface modelling of the intertidal zone (Lidar). Unpublished Internal Report SeArch project, Ghent University

Vermeersch J, Demerre I, Pieters M, Van Haelst S, De Hauwere, N (2015) Paleontologische resten afkomstig uit het Belgische deel van de Noordzee of aangetroffen op de aanpalende stranden. Unpublished Internal Report SeArch project, Onroerend Erfgoed

Wenban-Smith F (2002) Palaeolithic and Mesolithic Archaeology on the sea-bed: marine aggregate dredging and the historic environment. Wessex Archaeology Report

Wessex Archaeology (2007) Historic environment guidance for the offshore renewable industry sector. Commissioned by COWRIE Ltd (project ARCH-11-05). http://www.thecrownestate.co.uk/energy-and-infrastructure/downloads/cowrie/

Zeebroek I, Pieters M, Andrés-Lacueva C, Caluwé D, David J, Deforce K, Haneca K, Lamuela-Raventós R, Lenaerts T, Medina Remón A, Mees F, Missiaen T, Muylaert L, Op de Beeck E, Streel M, Van den Haute P, van Hees M, Wauters E (2010) Een 18de-eeuwse wraksite op de Buiten Ratel-Zandbank (Belgische territoriale wateren)(I): multidisciplinair onderzoek van het vondstenmateriaal. Relicta – Archeologie Monumenten – Landschapsonderzoek Vlaanderen 6:237–327

Zurita Hurtado O, Missiaen T, Kruiver, P (2013) Acoustic survey techniques for marine archaeological investigations. Unpublished Internal Report SeArch project, Ghent University

Zurita Hurtado O, Missiaen T, De Clercq M, Versteeg W, Kruiver P, de Kleine M (2014) Comparative seismic source study for buried palaeolandscape investigations in the southern North Sea. EAGE First Applied Shallow Marine Geophysics Conference, 14–18 Sept 2014, Athens (Greece), Conference Abstract 06

# Chapter 28
# The History of Industry-Linked Research in English Waters: Lessons for the Future

**Fraser Sturt, Justin Dix, and Michael J. Grant**

**Abstract** This chapter charts the shifting relationship between industry and archaeology offshore. We argue that, just as on land, collaboration with developers has changed the scope and scale of our investigations, transforming our understanding of the submerged continental shelf. However, through considering the development of the subject we argue that there is a need to actively avoid complacency and to continually develop new approaches which better reflect our shifting interests and capabilities as archaeologists. In this light, the challenges of working offshore are shown to be one of its great strengths, in that it forces us to consider what we want to know, and how best to find it out.

## 28.1 Introduction

Offshore industry and a developing understanding of the distribution and potential of submerged landscapes have been inextricably linked since the birth of archaeology as an academic subject. Nowhere is this more apparent than for the waters that surround England. Here both the history of research and current practice have been shaped by the nature of industrial activity. While this relationship can be seen as longstanding and mutually beneficial, it is also complex and worthy of careful consideration.

In this paper we consider the history of archaeology's relationship in England with offshore industry, its impact on the discipline and how we might proceed in the future. We argue that, just as on land, collaboration with developers has changed the scope and scale of our investigations, transforming our understanding of the submerged continental shelf. However, to paraphrase Levi-Strauss (1964, p 89), we also contend that one of the most significant outcomes from work investigating submerged landscapes in English waters is recognition of the fact that they are 'good to think' with.

F. Sturt (✉)
Centre for Maritime Archaeology, University of Southampton, University Road, Southampton SO17 1BJ, UK
e-mail: F.Sturt@soton.ac.uk

J. Dix
Ocean and Earth Science, National Oceanography Centre Southampton, University of Southampton, European Way, Southampton SO14 3ZH, UK
e-mail: j.k.dix@soton.ac.uk

M.J. Grant
Coastal and Offshore Archaeological Research Services (COARS), National Oceanography Centre Southampton, University of Southampton, European Way, Southampton SO14 3ZH, UK
e-mail: M.J.Grant@soton.ac.uk



G.N. Bailey et al. (eds.), *Under the Sea: Archaeology and Palaeolandscapes of the Continental Shelf*, Coastal Research Library 20, DOI 10.1007/978-3-319-53160-1_28

The act of investigating these now partially obscured places helps us to confront the changing nature of the world, and the people that lived within it. Importantly, this forces us to consider what our most important research questions are, how we might best answer them, and what industry's relationship to this might be. Thus, as we argue below, while the goals of archaeology and offshore industry have the capacity to be strongly aligned, this alignment should not be taken for granted and needs to be actively maintained.

## 28.2 History and Development of Thought: The Early Twentieth Century

The recognition and investigation of submerged landscapes in English waters has long been directly connected to industrial activity. Reid (1913, p 1) in his ground-breaking book on submerged forests extended his understanding of their distribution beyond the inter-tidal zone through discussion with fishermen. This practice in turn built on earlier nineteenth century antiquarian relationships with North Sea trawler men, where the faunal remains frequently dragged up by their nets were sold to interested collectors (Bynoe 2014). Similarly, Crawford (1927), in the first paper in the first edition of the journal *Antiquity*, identified the presence of an extensive preserved prehistoric landscape below the high water mark in the Isles of Scilly. While the majority of his paper derives from first-hand observations of anthropogenic features and lithic artefacts, he also noted (1927, p 12) the significance of knowledge gained by fishermen. In each of these early twentieth century examples there is a recognition of the need to engage with the people who spend time working at sea, and to value the knowledge they garner.

One of the best early examples of the impact that can be generated from engaging with offshore industry, and the significance it can have for our understanding of British Prehistory, is seen in the recovery of a single find: the Leman and Ower Harpoon. In 1932 the trawler Colinda dredged up a well preserved Maglemose harpoon from the Leman and Ower banks (Fig. 28.1), 25 miles off the Norfolk Coast. Its recovery received a brief paragraph in the journal *Man* (Burkitt 1932, p 118), but was to have a far more significant impact on the research trajectory of J.G.D. Clark, who in turn shaped the nature of hunter-gatherer archaeology in the United Kingdom.

As Gaffney et al. (2009, p 19) observe, Clark quickly recognised the significance of this find, coming at a time when he was focused on writing about the Mesolithic cultures of northern Europe (Clark 1936). The Leman and Ower harpoon allowed him to populate the forest environments of the now submerged North Sea described by Reid (1913) with mobile groups of Mesolithic hunter-gatherers. Within his 1936 volume on the Mesolithic settlement of northern Europe, Clark mapped out the extent of the submerged landscapes beneath the North Sea, the finds that had been made and the implications they had for social connectivity. In Clark's work the North Sea moved from being a simple blue area on the map that divided prehistoric populations, to a space that had previously connected them in a meaningful manner. The catalyst for this change had been the product of industrial activity.

Significantly Clark did not work alone, and nor did he place pre-eminence on material culture as a line of evidence for archaeology. For Clark, the significance of the harpoon lay as much in what it indicated about the archaeological potential of the North Sea as it did in materialising Mesolithic connectivity. In the same year that the Leman and Ower harpoon was being recovered from the North Sea, the Fenland Research Committee was formed to work on the complex stratigraphy and associated remains found within the fenland basin of East Anglia (Smith 1997). Here the sedimentary archive told the story of Holocene transgression and regression, and of the fluctuating size, shape and dynamics of the broader North Sea basin. The formation of this group of scholars reflected the observation made by Reid (1913, p 2) that the investigation of prehistoric environments needed to be an interdisciplinary affair. It was as part of this team that Clark developed his working relationship with Harry Godwin, the father of Quaternary research in the UK.

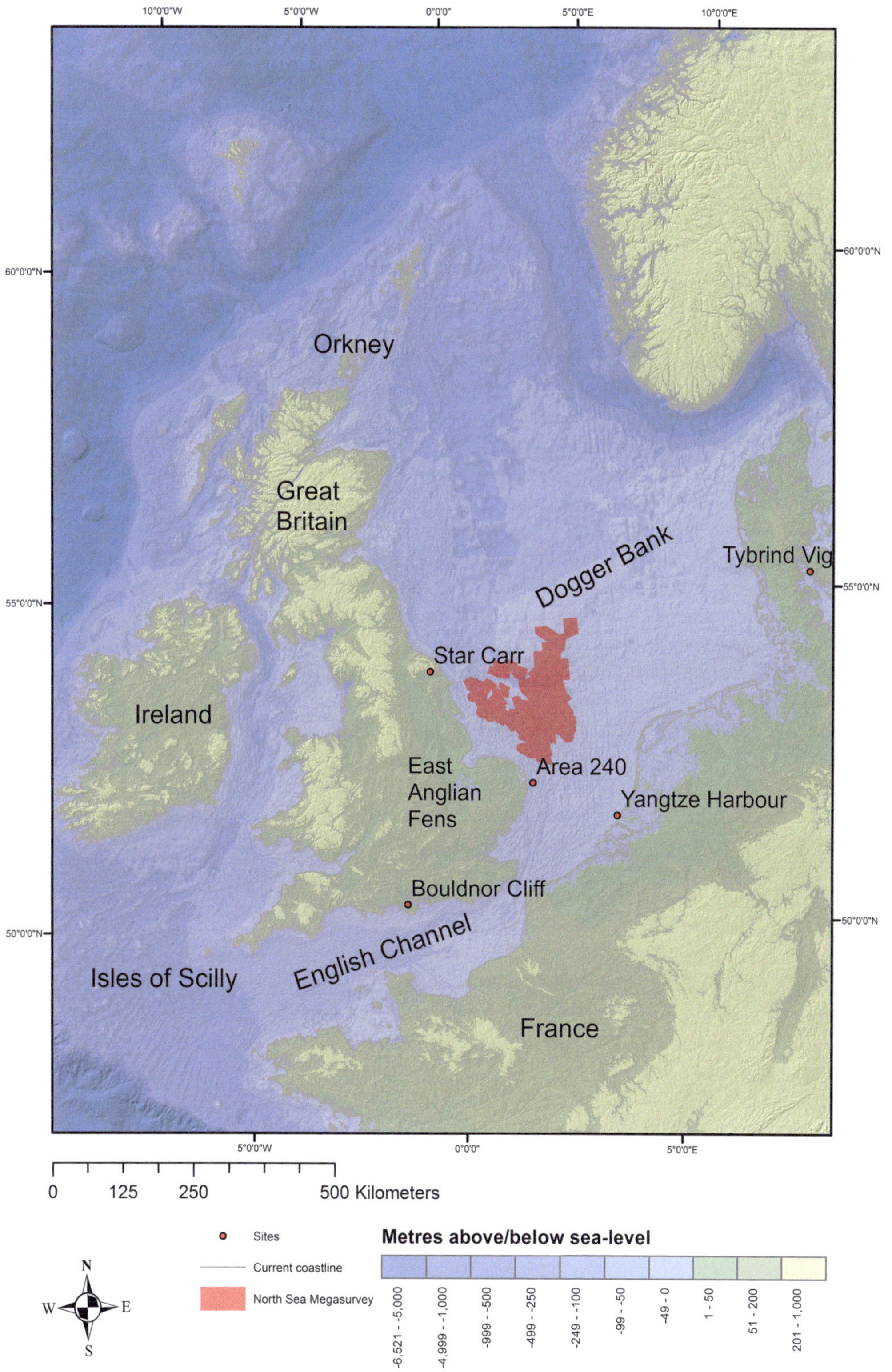


**Fig. 28.1** Map showing location of sites mentioned in the text and the area covered by the North Sea Megasurvey (used within the North Sea Palaeolandscapes Project) (Topographic and bathymetric data from GEBCO14 (www.gebco.net) data)

On the discovery of the harpoon, Clark and Godwin were not satisfied simply to plot its point of discovery on a map. Instead, they returned to the area from which it was recovered in order to sample the seafloor. At this location Harry and Margaret Godwin (Godwin and Godwin 1933) used a trawl net to recover organic material from the seabed. They subsequently carried out pollen analysis in order to act both as a form of relative dating, and as an environmental proxy for the context from which the harpoon had been recovered. For Clark, Godwin and Godwin, the peat layers that the Leman and Ower Harpoon were recovered from linked neatly to the sequences they were exploring on land in the Fenland basin. Through engaging with submerged prehistory Clark and the Godwins were removing a modern land/sea boundary to better understand the dynamics of the past.

The story of the Leman and Ower harpoon is more than just an historical aside. It charts the beginning of an improved understanding of submerged landscapes preserved in English waters and, perhaps more significantly, the value of data from this region for transforming our understanding of the past. However, it may also have contributed to the development of a responsive mode of investigation. Where Reid (1913) and Crawford (1927) had actively set out to research a specific topic, Clark and the Godwins can be seen to be responding to a spectacular chance find. In essence this may have helped to establish a modus operandi for the integration of archaeology and industrial activity for a number of years thereafter. Furthermore, it created a very high bar in terms of expectations and assumptions with regard to what constitutes a find of significance, a legacy still felt today.

Sadly, the history of research into submerged landscapes in English waters is not one of simple intensification of study, or indeed systematic engagement, from the time of Reid (1913) or Clark (1936) onwards. In contrast to developments on the continental mainland, engagement between archaeologists and offshore industry in the UK dwindled through the 1940s, 1950s and 1960s. As Coles (1998) and Sturt et al. (2013) document, there were few researchers in England who followed up on the significance of offshore finds, or thought deeply about what the shifting palaeogeography of North West Europe might have meant for the dynamics of prehistory. Thus work on finds from the Brown Bank (Louwe Koojimans 1971) and on the potential significance of changing continental connectivity (Jacobi 1976) failed to get the recognition it deserved. Nor was consideration given to how offshore industries might be impacting on this record. This stands in stark contrast to the pioneering developments that were taking place in the United States (Gagliano 1977), where the potential for submerged prehistory was being carefully thought through. Sadly, as Bailey (2014, p 292) documents, this excellent early research in the United States failed to have the wider impact it deserved. This suggests that as a community we need to be cautious of presuming that advances in one region directly translate to another.

Part of the reason for this may have been disengagement with offshore industry by academics (lack of communication of chance finds) and the absence of a regulatory framework. On dryland the post-war development boom in England led to recognition of the impact of construction on the archaeological record. Local archaeology societies quickly formed in the 1950s and 1960s and attempted to rescue what they could ahead of any development (Everill 2009). By 1971 this trend led to the formation of RESCUE: the British Archaeology Trust, a body determined to generate support and mitigate the impact of development. The success of RESCUE and the positive relationship which grew between archaeologists and developers over this period paved the way for new frameworks to be put in place. In 1990, Planning Policy Guidance 16 was introduced, with the principle that developers should be financially responsible for the mitigation of destruction to archaeology encountered during their works.

The success of RESCUE was remarkable; however it remained focused on the heritage found above the low-water mark. The absence of physical engagement with submerged prehistory (or indeed most underwater cultural heritage) in England over this period is in some ways understandable. While there had been spectacular discoveries of submerged prehistory by SCUBA divers in the calmer waters of the Baltic during the late 1950s and early 1960s (Andersen 2013), and a history of scuba-based research into wrecks and submerged cities in the Mediterranean, the murkier waters, complex tidal systems, sedimentary sequences and erosion patterns off England meant that the diving community became better acquainted with the shipwreck record. Similarly, the catalysts for change with regard to attitudes to heritage assets in English waters were all shipwrecks—the Association, Amsterdam and Mary-Rose—with no prehistoric finds generating the necessary public interest to drive shifts in policy. It wasn't until 1988 with the formation of the Joint Nautical Archaeology Policy Committee (JNAPC) that pressure was brought to bear and things began to slowly change.

The result of the above history is that the profile of submerged prehistory was not raised as highly as it might have been. Thus, while in 1979 Søren Andersen was beginning to carefully excavate Tybrind Vig, maritime archaeology in Britain was only just beginning to take shape, with Keith Muckelroy's (1978) foundational text eschewing the study of submerged landscapes altogether.

It is tempting to see such comparatively deep academic history as colourful background and little more. However, it helped shape our expectations for, and knowledge of, the waters around England. As such, despite Reid (1913), Crawford (1927) and Clark (1932, 1936) clearly demonstrating the significance of the offshore zone for understanding prehistory, it was not actively pursued as an area of research by the broader archaeological community in England nor was its significance disseminated to the public at large for nearly a century. It is for this reason that sporadic dredged finds of lithics and faunal material failed to gain broader recognition, resulting in only two high profile cases of direct ground truthing: the Mesolithic site of Bouldnor Cliff from the 1980s, but in earnest since 1999 (Momber et al. 2011) and Palaeolithic finds from Area 240 in 2008 (Bicket 2011; Tizzard et al. 2014).

## 28.3 The Late Twentieth and Early Twenty-First Century

While site-specific investigations did not develop in the same way as on the continent, archaeologists working in England did eventually re-engage with the broader topic of submerged prehistory. Coles' (1998) seminal 'speculative survey' of the submerged southern North Sea ('Doggerland') demonstrated to the wider archaeological community the potential of offshore contexts. Importantly this potential was not only seen to lie in the recovery of material, but in relating the broad-scale narratives of landscape change to those of social impact. This played to the strengths of British archaeology in landscape approaches (Johnson 2008) and allowed the broader academic community to realise the significance of what Coles was discussing.

The awareness this article generated was timely as it coincided with increased offshore development, acquisition of new data and followed on from a spectacular achievement by the JNAPC. In 1995, the JNAPC published an agreed voluntary code for seabed developers, establishing that archaeology be taken account of during the development process. In what was a comparatively rapid development of policy in 2002, Historic England built on this agreement and succeeded in resolving a peculiar regulatory gap, extending their remit out to the 12 nautical-mile limit, where previously it stopped at the water's edge (Roberts and Trow 2002). This saw a formalisation of the need for offshore developers to account for archaeology within their planning and mitigation works. As Bicket et al. (2014, p 218), Salter et al. (2014), and Firth (2015, p 1) note, this change in regulatory stance helped to accelerate offshore research, encouraged improving standards and firmly embedded the investigation of submerged landscapes in the planning process.

One key mechanism through which improvement of the knowledge base was achieved was through the Aggregates Levy Sustainability Fund (ALSF). The ALSF redistributed money derived from aggregate extraction licencing to research. A range of projects was funded between 2002 and 2011, which cast new light on the archaeological potential of the waters surrounding England. The majority of this work operated at the large regional scale (Wenban-Smith 2002; Westley et al. 2004; Gaffney et al. 2007; Dix and Sturt 2011; Bicket 2011; Bynoe et al. 2013), making best use of the data that the offshore industries supplied. This process also served to consolidate a nascent community of researchers, providing much needed experience and access to resources. Critically in this instance, resources refer both to the underwater archaeological record directly, and to the funds and equipment required to reach it.

At the same time the need to re-engage those working in offshore industries was recognised, with a series of guidance notes commissioned by the British Marine Aggregate Producers Association (BMAPA and EH 2003, 2005) and Collaborative Offshore Wind Research into the Environment (COWRIE 2007, 2011) in collaboration with English Heritage. These documents served to communicate to those working offshore the sorts of information required to make a substantial contribution to our understanding of the archaeological record. There was a clearly perceived need to ensure that increasing research offshore did not just equate to gathering more data, but harnessed careful thought as to what 'useful' data was. Significantly the BMAPA (2005) document re-instated the relationships that had enabled Reid (1913) and Clark (1936) to progress their research. It established a finds reporting protocol and, just as importantly, a means to disseminate that data.

This drive to evaluate the state of the art led to a series of significant publications that directly addressed the links between industry and archaeological research (Flemming 2004), region-specific priorities (Peeters et al. 2009), broader national strategies (Ransley et al. 2013; Sturt and Standen 2013) and wider-reaching documents working at the continental and global scale (Bailey 2011; Fischer et al. 2011; Benjamin et al. 2011; Bailey et al. 2012; Flemming et al. 2014). The challenge was perceived to lie (and still lies) in rapidly changing understandings of what might be achieved through investigating the submerged record. These publications all sought to map what was currently achievable with the available technology and methodology onto archaeological research priorities, as well as giving a speculative look at future possibilities.

The pioneering work of Gaffney et al. (2007, 2009) on the North Sea Palaeolandscapes Project, and the analysis by Gupta et al. (2004, 2007) of United Kingdom Hydrographic Office Data, act as another important turning point. Together these projects served to demonstrate the difficulties inherent in overly didactic agenda-setting. While local and regional reconstructions had previously been derived from industry-sourced data (Westley et al. 2004), the work of Gaffney et al. (2009) and Gupta et al. (2004, 2007) served not only to advance the methods adopted for data integration and visualisation, but also to capture the public's imagination. What Coles (1998) had done for the academic community, Gaffney et al. (2009) and Gupta et al. (2007) did for a broader audience. They thus served to focus attention on what could be achieved and why it was worth investing in such investigations. In this sense they stepped outside of what was considered possible at the time, advancing scholarship beyond what more regional and local agendas had considered achievable. As such, these projects helped to remind researchers to be ambitious and achieve more.

Significantly, both groups also demonstrated the disruptive potential of research, demonstrating the large advances that could happen when datasets were made accessible. The bathymetric maps rendered by Gupta et al. (2007) and the 3D seismic visualisation of Gaffney et al. (2009) were an order of magnitude greater in range and scope compared to what had been published previously, but, the data had been there (at least in part), untapped, in the background for a number of years. As such, one of the challenges when considering the future of the relationship between industry and archaeology is not to be constrained by the *status quo*, but to look at how innovative practices might transform the field of study.

## 28.4 Current Practice and Emerging Agendas

As the discussion above makes clear, in England there has been a deeper history of development-led archaeology onshore than offshore. Through that history, complicated relationships have formed between developers, archaeologists working in commercial units and archaeologists in Universities, non-governmental organisations, charities and research institutes (Satchell, Chap. 25). With regard to work onshore, Bradley (2006, p 2) describes 'unwelcome divisions between those who undertake field archaeology and those who remain disengaged'. He notes that University research 'often happens in a parallel universe and its results are disseminated in meetings that few outsiders attend (Bradley 2006, p 2), while on the other hand it can be hard for those working outside of commercial archaeology to engage with the scale, scope and results from the projects being undertaken. Thankfully, as commented on below, research into submerged prehistory appears, so far, to have largely bridged these divides.

If we examine the extent of commercial activity in English (and broader UK) waters in 2015 (Fig. 28.2), for the majority of the points, lines and polygons given on the map an environmental impact assessment has taken place. Subsequently, as part of mitigation works, additional analysis of seismic and core data has been commissioned. As Arnott and Baggaley (2012), Salter et al. (2014), and Firth (2015) clearly establish, archaeological considerations are now well integrated into this process, ensuring that the data are reviewed (and at times commissioned) from an archaeological perspective.

In contrast to the extent of commercial activity shown in Fig. 28.2, research into submerged prehistory undertaken with research funding by University-based archaeologists would be barely visible at this scale. The vast majority of work is being driven by commercial and environmental agendas (but see Gaffney et al., Chap. 20; Bailey et al., Chap. 23). As such, development-led and policy-mandated research represents the larger part of activity in this field. Fortunately, due to the principles established during the ALSF period of funding, and continued guidance from Historic England as the regulator, the research/development divide that Bradley (2006) was commenting on is not so readily apparent offshore. Put simply, those working in commercial units and on developer-funded projects have been producing high quality academic outputs, helping to shape the discipline and disseminate the knowledge gained (e.g., Russell and Tizzard 2011; Bicket 2011; Tizzard et al. 2014; Bicket and Tizzard 2015; Firth 2015).

Conversely, the authors of this paper have recognised that in order to advance an understanding of submerged prehistory they need to engage with developer-funded research from within a University context. It is only through doing so that it has been possible to work with both regional geophysical and local geotechnical data (Dix and Sturt 2011; Sturt et al. 2014; Griffiths et al. 2015).

This dialogue between groups working in different institutional settings has perhaps helped to slow down what Firth (2015, p 8) identifies as one of the biggest challenges to the discipline: that the process of developer-funded work will stagnate by becoming familiar and routine. Firth argues that this will see research reduced to a series of 'copy and paste' responses growing out of a sense of repetition. Essentially Firth is pointing out that what drove the advances seen to date was in part their novelty and the challenge of working within a new environment. As that challenge recedes so too does the need to question best practice and consider the direction in which the discipline is heading. In effect, industry standards become 'standard' and the pace of progress slows.

Flatman (2007) makes a related and equally interesting argument that we need to consider the ethics of our mode of engagement with submerged prehistory in England. Flatman (2007, p 144) argues that with the influx of data and new knowledge that occurred at the start of the ALSF we may have overlooked the issue of what the best/most effective means of investigating submerged prehistory might be. Instead he suggests that perhaps we too readily settled for adopting approaches used by other disciplines (ecologists and geologists) which were already integrated into commercial data collection. Flatman effectively makes the point that these strategies at least need to be questioned and debated more openly. Are vibrocores, seismic surveys and grab samples the most effective means of identifying submerged palaeolandscapes? Are sampling strategies robust enough? Will they generate the data we need to answer the questions posed in research agendas, academic books and journal articles?

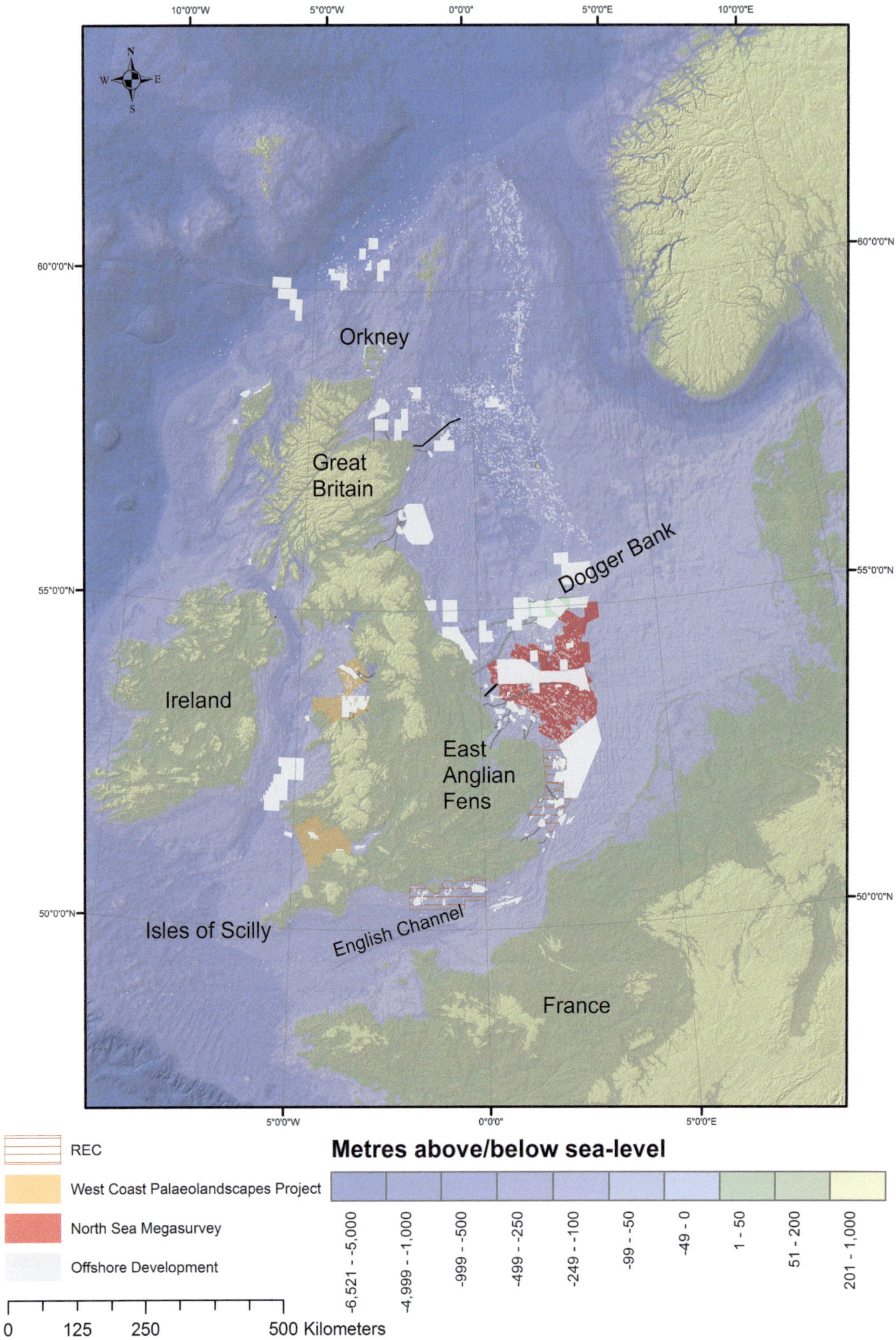


**Fig. 28.2** Maps showing the extent of offshore development activity from major projects around the UK in 2015. The extent of large regional environmental characterisation (REC) projects is also indicated (Topographic and bathymetric data from GEBCO14 (www.gebco.net), offshore development data from the Crown Estate and the Department of Energy and Climate Change (DECC) – contains public sector information licensed under the Open Government Licence v3.0)

As Firth (2015, pp 7–8) notes, there is no simple answer to these questions. In some instances established protocols are entirely appropriate and add substantially to our understanding of the past. However, both Firth (2015) and Flatman (2007) are right to raise concerns. Archaeology is constantly developing, with new research questions coming to the fore, alongside innovative techniques to answer them. As such, we should fear complacency and the emergence of a copy-and-paste mentality in developer-funded work. Each project needs to be assessed on its own terms and considered in light of the questions we want to (and can) answer.

A key part of this process lies in aspiring to be surprised by the archaeological record, and in acknowledging the necessity of trialling new approaches. This might be as simple as expanding areas of investigation. For example, the work of Conneller et al. (2012) at the terrestrial site of Star Carr altered our understanding of the Mesolithic at a site widely believed to be well understood. Through excavating a wider area and looking at the site in its broader context they managed radically to alter our thinking about Mesolithic communities. At the risk of grossly simplifying the work they undertook, by sampling a wider area they found more and new data. In a similar vein the recent work at Yangtze Harbour in Rotterdam (Moree and Sier 2015) demonstrates the gains in knowledge achievable by intense sampling and analysis of the seabed underpinned by funding from an industrial developer.

Moree and Sier's (2015) report details fine-grained geoarchaeological research and careful material cultural analysis of a submerged site. An area of high intensity prehistoric activity was identified through geophysical survey matched with numerous cores. This led to the development of an innovative approach to the recovery of material from the seabed in a controlled manner via grab sampling on an industrial scale with purpose-designed equipment made available by the industrial developer. This work should serve to have the same transformative impact that Gaffney et al. (2007) had with regard to expectations for regional survey. Here, reconstructions of the prehistoric terrestrial landscape were achieved over a large area of the sea floor, but the authors recognised that excavation by diver might not be appropriate. In essence, a middle ground between landscape reconstruction and traditional site-based excavation was identified. The result was a substantially improved understanding of the archaeology of the area, and, of how investigations of the submerged prehistoric landscape might progress. The data generated allowed consideration of landscape change through time in contrast to more traditional archaeological issues of material culture variability (see also Gaffney et al., Chap. 20).

However, as Firth (2015) identifies and Flatman (2007) has demanded, we cannot assume that the work of Moree and Sier (2015) will cause such a transformative realignment to occur within English waters. Firth (2015, p 1) notes that all archaeological work is underpinned by three principles:

1. Research contributes to human understanding based on physical evidence
2. Conservation seeks to safeguard the survival of material evidence in the future
3. Engagement is conducted for and on behalf of the public

It is this latter issue that has perhaps been too readily overlooked and may most threaten advances in submerged landscape research in English Waters. With the vast majority of fieldwork being development led, any request for more intensive research has to be supported by a broad mandate. Without a wider understanding of what is trying to be achieved, it will become all too easy to fall back on the newly established 'standard' as the best for all possible circumstances.

In recognition of this need to widen access to outputs from developer-funded work, we recently included examples from our work in a free massive open online course (MOOC). The Shipwrecks and Submerged Worlds MOOC has had 20,852 participants to date (February 2016, see https://www.futurelearn.com/courses/shipwrecks). Most significantly for this discussion, over 4500 specific comments were made on the week exploring submerged prehistory. Our aim in referring to this is not to offer a detailed content-analysis of these comments, but to document and reflect on a latent desire within the broader population to engage with this subject. If the material is made accessible, then there is an appetite to engage. This would appear to support Firth's (2015) observation, that if we want to see change in development-funded work we should not neglect the people on whose behalf it is being undertaken. Thus while it might be hard at present to envisage a project in English waters of the scale and scope reported by Moree and Sier (2015), this should not prevent us from preparing the way through opening that dialogue.

## 28.5 Conclusions

Research into the submerged landscapes of English waters has always had a connection to industrial activity. The nature of this relationship has varied considerably through time, from independent recovery of material dredged from the sea floor through to active partnerships collaborating in research. As Firth (2015) and Flatman (2007) have argued, the challenge that exists now is to recognise that this relationship is never fixed—nor should it be.

The investigation of submerged landscapes will always require a dialogue. This dialogue shapes what we aim to get from these spaces and what we are willing to expend to realise those goals. As Bailey (2004, p 9) notes, the challenge for submerged prehistory is not to re-create landscapes or find in-situ material underwater, it is to demonstrate the difference it can make to our understandings of prehistory. Thankfully this is the one thread that ties together each element within the history of research given above. When we have taken the time and effort to consciously investigate the submerged landscapes in English waters, the results have paid considerable dividends in terms of our broader understanding of prehistory, from Reid's (1913) consideration of submerged forests, through Clark's (1936) improved understanding of Mesolithic Europe, to transforming our accounts of palaeolandscape change and connectivity (Coles 1998; Gaffney et al. 2007). Submerged landscapes are clearly good to think with, posing challenges that promote new ideas and approaches.

Within this trend rests another subtle but equally important thread—the importance of communication. Writing a paper on submerged palaeolandscapes in English waters always had a faintly ridiculous feel. It may reflect current practice with regard to regulatory regimes and development-led research, but, modern national boundaries clearly don't relate adequately to the subject of study. Equally, knowledge of how to approach the study of submerged landscapes is not confined to a single nation's academic history. It is fitting that this paper sits within a volume that has come about because of the SPLASHCOS initiative, a European project to draw researchers together across many different national boundaries. In operating in that wider way, we radically increase our pool of knowledge and refresh our expectations. This is fundamental for combating the threat of complacency highlighted by Firth (2015).

Finally, submerged landscapes and the results from development-led work are not only good for archaeologists to think with. When the opportunity is given, a much wider range of people will actively choose to engage with this topic. This should not be seen as an interesting aside, or validation of established arguments as to public support for archaeological research. Rather, it should be taken as a challenge to heighten the quality and detail of debate in public circles, and as a mandate to push for greater open access to resources. It is only through maintaining these different dialogues that we will ensure the continued progress and relevance of research into submerged landscapes.

## References

Andersen SH (2013) Tybind Vig. Submerged Mesolithic settlements in Denmark, Jutland Archaeological Society Publications 77. Jutland Archaeological Society, Højbjerg

Arnott SHL, Baggaley PA (2012) The role of archaeology in offshore site investigation for renewable developments. Offshore site investigation and geotechnics: integrated technologies-present and future. Society of Underwater Technology, OnePetro. https://www.onepetro.org/conference-paper/SUT-OSIG-12-11

Bailey G (2004) The wider significance of submerged archaeological sites and their relevance to world prehistory. In: Flemming NC (ed) Submarine prehistoric archaeology of the North Sea. Research priorities and collaboration with industry, CBA Research Report, vol 141. Council for British Archaeology, York, pp 3–10
Bailey G (2011) Continental shelf archaeology: where next. In: Benjamin J, Bonsall C, Pickard C, Fischer A (eds) Submerged prehistory. Oxbow Books, Oxford, pp 311–331
Bailey GN (2014) New developments in submerged prehistoric archaeology: an overview. In: Evans, Flatman, Flemming (eds) Prehistoric archaeology on the continental shelf: a global review. Springer, New York, pp 291–300
Bailey GN, Sakellariou D, members of the SPLASHCOS network (2012) Submerged prehistoric archaeology & landscapes of the continental shelf. Antiquity 86(334). http://antiquity.ac.uk/projgall/sakellariou334/
Benjamin J, Bonsall C, Pickard C, Fischer A (eds) (2011) Submerged prehistory. Oxbow Books, Oxford
Bicket A (2011) Submerged prehistory: research in context. Wessex Archaeology
Bicket A, Tizzard L (2015) A review of the submerged prehistory and palaeolandscapes of the British Isles. Proc Geol Assoc 126:7–9 . doi:http://dx.doi.org/10.1016/j.pgeola.2015.08.009
Bicket A, Firth A, Tizzard L, Benjamin J (2014) Heritage management and submerged prehistory in the United Kingdom. In: Evans A, Flatman J, Flemming N (eds) Prehistoric archaeology on the continental shelf. Springer, New York, pp 213–232
BMAPA & English Heritage (2003) Marine aggregate dredging and the historic environment: guidance note. British Marine Aggregate Producers Association and English Heritage, London
BMAPA & English Heritage (2005) Protocol for reporting finds of archaeological interest. British Marine Aggregate Producers Association and English Heritage, London. Downloadable from http://www.wessexarch.co.uk/projects/marine/bmapa/arch-interest.htm
Bradley R (2006) Bridging the two cultures – commercial archaeology and the study of prehistoric Britain. Antiqu J 86:1–13
Burkitt MC (1932) A Maglemose harpoon dredged up recently from the North Sea. Man 32:118
Bynoe R (2014) The great fossil mine of the southern North Sea: exploring the potential of submerged Palaeolithic archaeology. Unpublished PhD thesis, University of Southampton
Bynoe R, Dix J, Sturt F (2013) The submerged Palaeolithic: an opportunity not a problem. In: Daire M, Dupont C, Baudry A, Billard C, Large J, Lespez L, Normand E, Scarre C, Bar S (eds) Ancient maritime communities and the relationship between people and environment along the European Atlantic Coasts, British Archaeological Reports International Series, vol 2570. Archaeopress, Oxford, pp 583–594
Clark JGD (1932) The Mesolithic age in Britain. Cambridge University Press, Cambridge
Clark JGD (1936) The Mesolithic settlement of Northern Europe: s study of the food-gathering peoples of Northern Europe during the early post-glacial period. Cambridge University Press, Cambridge
Coles BJ (1998) Doggerland: a speculative survey. Proc Prehist Soc 64:45–81
Conneller CC, Milner N, Taylor B, Taylor M (2012) Substantial settlement in the European Early Mesolithic: new research at Star Carr. Antiquity 86:1004–1020
COWRIE (2007) Historic environment guidance for the offshore renewable energy sector. Wessex Archaeology on behalf of COWRIE Ltd (ARCH-11-05), Salisbury
COWRIE (2011) Offshore geotechnical investigation and historic environment analysis: guidance for the renewable energy sector. COWRIE
Crawford OGS (1927) Lyonesse. Antiquity 1:5–15
Dix JK, Sturt F (2011) The relic palaeo-landscapes of the Thames Estuary. Southampton University (MALSF Open Report 09/ P126), 201, Southampton
Everill P (2009) The invisible diggers: a study of british commercial archaeology. Heritage Marketing & Publications Ltd, Oxford
Firth A (2015) Risks, resources and significance: navigating a sustainable course for marine development-led archaeology. Bull Aust Inst Marit Archaeol 39:1–8
Fischer A, Benjamin J, Pickard C, Bonsall C (2011) Epilogue. In: Benjamin J, Bonsall C, Pickard C, Fischer A (eds) Submerged prehistory. Oxbow Books, Oxford, pp 332–338
Flatman J (2007) The origins and ethics of maritime archaeology – part II. Publ Arch 6:77–97
Flemming NC (ed) (2004) Submarine prehistoric archaeology of the North Sea. Research priorities and collaboration with industry, CBA Research Report, vol 141. Council for British Archaeology, York
Flemming NC, Cagatay MN, Chiocci FL, Galandiou N, Lericolais G, Joens H, Missiaen T, Moore F, Rosentau A, Sakellariou D, Skar B, Stevenson A, Weerts H (2014) Land beneath the waves: submerged landscapes and sea level change. A joint geoscience-humanities strategy for European continental shelf prehistoric research. In: Chu NC, McDonough N (eds) Position paper 21 of the European Marine Board. European Marine Board, Ostend
Gaffney V, Fitch S, Smith D (2009) Europe's lost world, the rediscovery of Doggerland CBA Research Reports 160. Council for British Archaeology, York
Gaffney V, Thomson K, Fitch S, Finch S (2007) Mapping Doggerland: the Mesolithic landscapes of the Southern North Sea. Archaeopress, Oxford

Gagliano S (1977) Cultural resources: evaluation of the northern Gulf of Mexico continental shelf. National Park Service, Washington, DC
Godwin H, Godwin ME (1933) British Maglemose harpoon sites. Antiquity 7:36–48
Griffiths S, Sturt F, Dix JK, Gearey B, Grant MJ (2015) Chronology and palaeoenvironmental reconstruction in the sub-tidal zone: a case study from Hinkley Point. J Archaeol Sci 54:237–253
Gupta S, Collier J, Palmer-Felgate A, Dickinson J, Bushe K, Humber S (2004) Submerged palaeo-Arun River: reconstruction of prehistoric landscapes and evaluation of archaeological resource potential: integrated projects 1 and 2 (Project 3543 MAIN). English Heritage, London. Downloadable from http://ads.ahds.ac.uk/catalogue/adsdata/arch-838-1/dissemination/pdf/complete_report/Final_EH_Report.pdf
Gupta S, Collier J, Palmer-Felgate A, Potter G (2007) Catastrophic flooding origin of shelf valley systems in the English Channel. Nature 448:342–345
Jacobi R (1976) Britain inside and outside Mesolithic Europe. Proc Prehist Soc 42:67–84
Johnson M (2008) Ideas of Landscape. Blackwell, Oxford
Levi-Strauss C (1964) Totemism. Merlin Press, London
Louwe Koojimans LP (1971) Mesolithic bone and antler implements from the North Sea and from the Netherlands Berichten van de Rijksdienst voor Oudheidkundig Bodemonderzoek, 20–21 (1971), pp 27–73
Momber G, Tomalin D, Scaife R, Satchell J, Gillespie J (eds) (2011) Mesolithic occupation at Bouldnor Cliff and the submerged prehistoric landscapes of the Solent, CBA Research Report, vol 164. Council for British Archaeology, York
Moree JM, Sier MM (eds) (2015) Interdisciplinary archaeological research programme Maasvlakte 2, Rotterdam Harbour site – Rotterdam Maasvlakte, the Netherlands Early Holocene landscape development. Bureau Oudheidkundig Onderzoek Rotterdam, Rotterdam
Muckelroy K (1978) Maritime archaeology. Cambridge University Press, Cambridge
Peeters H, Murphy PL, Flemming N (2009) North Sea prehistory research and management framework. Rijksdienst Voor Cultureel Erfgoed, Amersfoort
Ransley J, Sturt F, Dix J, Adams J, Blue L (eds) (2013) People and the sea: a maritime archaeological research agenda for England, CBA Research Report, vol 171. Council for British Archaeology, York
Reid C (1913) Submerged forests. Cambridge University Press, London
Roberts P, Trow S (2002) Taking to the water: English Heritage's initial policy for the management of maritime archaeology in England. English Heritage, London
Russell J, Tizzard L (2011) Seabed prehistory: site evaluation techniques (Area 240) Synthesis. OASIS/ADS Wessex Archaeol 70754:04. doi:10.5284/1008287
Salter E, Murphy P, Peeters H (2014) Research, conserving and managing submerged prehistory: national approaches and international collaboration. In: Evans A, Flatman J, Flemming N (eds) Prehistoric archaeology on the continental shelf: a global review. Springer, New York, pp 151–172
Smith PJ (1997) Grahame Clark's new archaeology: the Fenland Research Committee and Cambridge prehistory in the 1930s. Antiquity 71(271):11–30
Sturt F, Standen T (2013) The social context of submerged prehistoric landscapes. University of Southampton, Southampton
Sturt F, Garrow D, Bradley S (2013) New models of North West European Holocene palaeogeography and inundation. J Archaeol Sci 40(11):3963–3976
Sturt F, Dix JK, Grant MJ, Steadman S, Scaife RG, Edwards R, Griffiths S, Cameron N, Thompson CEL, Bray S, Jones J (2014) Life below the waves: palaeolandscapes preserved within the sub-tidal Bristol Channel. Archaeol Severn Estuary 22:41–66
Tizzard L, Bicket AR, Benjamin J, De Loecker D (2014) A Middle Palaeolithic site in the southern North Sea: investigating the archaeology and palaeogeography of Area 240. J Quat Sci 29:698–710
Wenban-Smith F (2002) Marine aggregate dredging and the historic environment: Palaeolithic and Mesolithic archaeology on the seabed. BMAPA and English Heritage, London
Westley K, Dix J, Quinn R (2004) A re-assessment of the archaeological potential of continental shelves. English Heritage, London

MIX
Papier aus verantwortungsvollen Quellen
Paper from responsible sources
FSC® C105338

If you have any concerns about our products,
you can contact us on
**ProductSafety@springernature.com**

In case Publisher is established outside the EU,
the EU authorized representative is:
**Springer Nature Customer Service Center GmbH**
**Europaplatz 3, 69115 Heidelberg, Germany**

Printed by Libri Plureos GmbH
in Hamburg, Germany